Auto Brakes Technology

by

Chris Johanson
ASE Certified Master Technician

Martin T. Stockel
Automotive Writer

Publisher
The Goodheart-Willcox Company, Inc.
Tinley Park, Illinois

Library of Congress Catalog Card Number 99-046634
International Standard Book Number 1-56637-704-8
1 2 3 4 5 6 7 8 9 10   00 03 02 01 00 99

Important Safety Notice

Proper service and repair is important to the safe, reliable operation of motor vehicles. Procedures
recommended and described in this book are effective methods of performing service operations. Some
require the use of tools specially designed for this purpose and should be used as recommended. Note
that this book also contains various safety procedures and cautions, which should be carefully followed
to minimize the risk of personal injury or the possibility that improper service methods may damage the
engine or render the vehicle unsafe. It is also important to understand that these notices and cautions
are not exhaustive. Those performing a given service procedure or using a particular tool must first
satisfy themselves that neither their safety or vehicle safety will be jeopardized by the service method
selected.

This book contains the most complete and accurate information that could be obtained from
various authoritative sources at the time of publication. Goodheart-Willcox cannot assume any
responsibility for any changes, errors, or omissions.

Library of Congress Cataloging in Publication Data

Johanson, Chris.
    Auto brakes technology / by Chris Johanson, Martin T. Stockel.
      p.   cm.
    Includes index.
    ISBN 1-56637-704-8
    1. Automobiles--Brakes. I. Stockel, Martin T. II. Title.

TL269 .J65 1999   99-046634
629.2'046--dc21

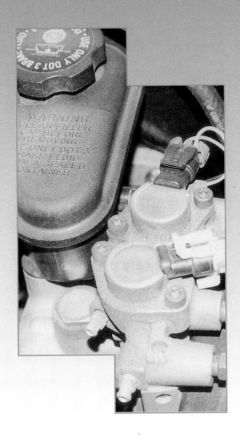

Introduction

If you are opening this copy of **Auto Brakes Technology,** you may be taking a course on automotive brake systems, or you may be a working technician seeking more information on brake service. **Auto Brakes Technology** is intended to give you a thorough understanding of the brake systems used on modern vehicles.

In the past, brake repair was a simple process, requiring few special tools. Most concepts could be understood simply by observing the brake system components. The brake systems on modern vehicles are more complex, making even slight defects more obvious. Diagnosis and service requires special tool and test equipment.

To diagnose and repair modern brakes, or any part of the vehicle, reading is a necessity. Repair information comes in all forms of printed material including this text, service manuals, and technical service bulletins. As a technician, you must know how to look up diagnostic procedures, test specifications, and other information. You must also know how to use a computer as more and more service data is now provided on compact disc and on the Internet. Understanding how computers work will also make understanding the on-board computer systems used on vehicles much easier.

While you will be learning to use modern test equipment and procedures, you will also learn test procedures used since the earliest days of the automobile. Your progress through this text will include learning about basic hydraulic and friction principles. You will also learn about brake electronics and computer control. Many concepts used in the brake system were discovered long before the automobile was developed. Other concepts, such as anti-lock brakes and traction controls, are relatively new. Both old and new concepts will be covered in **Auto Brakes Technology.**

Auto Brakes Technology has been carefully designed so that all parts and operating principles will be fully identified and explained before you begin work on the actual system. At no time will you be asked to service a part of the brake system before you are shown its theory of operation. Each chapter of the text begins with Objectives that provide focus for the chapter. Technical terms are printed in ***bold italic*** and are defined when first used. Figure references are printed in **bold**. Warnings and cautions are provided in cases where a danger to life, limb, or property may exist. Notes are provided to illustrate procedures that can make routine tasks easier. Each chapter contains a Summary, "Know These Terms" section, and Review Questions.

One reason you may be holding this text is ASE certification. **Auto Brakes Technology** covers all areas that you will be expected to know to pass the ASE Brakes test (A5). Each chapter has a section of ASE-type questions to help you prepare for this test. Experienced technicians can also use this text to review before taking the Brakes recertification test. Chapter 24 is devoted to information about ASE certification and passing the certification test.

A **Workbook for Auto Brakes Technology** is also available. It contains additional questions and jobs that test your hands-on ability. Each chapter in the workbook corresponds to the chapters in this text.

Chris Johanson
Martin T. Stockel

Table of Contents

Chapter 1

Introduction to Brake Systems

After studying this chapter, you will be able to:

- ❏ Identify the purpose of automotive brakes.
- ❏ Compare holding and stopping brakes.
- ❏ Describe the development of modern brake systems.
- ❏ Describe basic brake system hydraulic components.
- ❏ Describe basic brake system friction components.
- ❏ Describe power brake components and state their purpose.
- ❏ Describe parking brake components and state their purpose.
- ❏ State the purpose and major types of wheel bearings.
- ❏ Define the purpose of anti-lock brakes and traction control systems.
- ❏ Describe basic brake system operation.

Important Terms

Brake system

Holding brake

Stopping brake

Hydraulic system

Friction

Friction materials

Disc brakes

Drum brakes

Parking brake

Anti-lock brake
systems (ABS)

Traction control
systems (TCS)

This chapter is an introduction to brake systems, including the purpose of the brake system and its major components. Also included in this chapter is a brief overview of brake system development, from the earliest designs to modern anti-lock brake systems. Basic brake operation will be briefly discussed. This chapter is also an introduction to future chapters where brake system components and operation will be discussed in greater detail.

Purpose of the Brake System

Obviously, the purpose of the vehicle **brake system** is to stop the vehicle, **Figure 1-1**. However, hitting a wall will also stop the vehicle, therefore we must modify the first sentence to read that the brakes should stop the vehicle without damaging it or causing damage to anything else.

In addition, the brake system should stop the vehicle smoothly and consistently, so the driver will know what to expect when he or she applies the brake pedal. The braking force must be even from side to side, so the vehicle comes to a controlled stop. The components of the brake system must also allow the vehicle to be stopped over and over, thousands of times, before service is required. How the modern brake system performs these jobs is covered in this chapter.

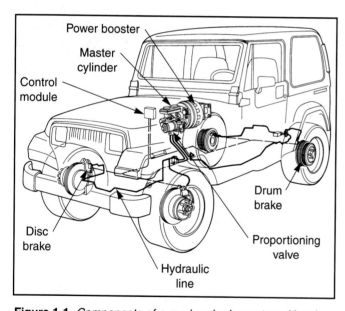

Figure 1-1. *Components of a modern brake system. (Jeep)*

Brakes for Holding and Stopping

Brakes can be divided into two general classes: the holding brake and the stopping brake. A **holding brake** is a simple mechanical device which keeps the vehicle from moving when it is already stationary. For instance, if you place a stick between the spokes of a bicycle wheel, the wheel cannot turn. The stick is a simple holding brake. A common example of an automotive holding brake is the parking gear or pawl used in automatic transmissions.

However, if you have ever accidentally placed an automatic transmission in *Park* before the vehicle was fully stopped, you know that this makes a very poor device for stopping a moving vehicle.

The **stopping brake** is designed to smoothly slow and eventually stop a moving vehicle. It relies on a complex set of parts that interact to convert the vehicle's movement to heat, which is released to the surrounding air. When two parts move against each other, the resistance to movement between them is called **friction.** Stopping brakes use the principle of friction to convert movement to heat. Since even the smoothest materials have microscopic imperfections, two parts will encounter interference when they slide against each other.

In all automotive powertrains, friction is minimized by machining parts to be as smooth as possible, using bearings, and by forcing lubricating oil between moving surfaces. However, in the brake system, friction is vital if the vehicle is to be stopped. The heat produced by friction is removed to the surrounding air. Removing heat allows the brakes to function for long periods of time. How this is done will be discussed in this and other chapters.

Brake System Power

To move a vehicle from 0-60 mph requires a complete drive train. Bringing the vehicle from 60-0 mph is the job of the brake system. A 3000 lb. vehicle traveling at 60 mph has stored roughly 250,000 pounds of kinetic energy, or momentum. To overcome this energy and stop the vehicle as quickly as possible, the brake system must convert all this momentum to heat within a few seconds.

The average vehicle needs about 1/4 mile (1320 feet) to accelerate from 0-60 mph. However, in an emergency, it must be stopped from 60 mph in 200 feet or less. Therefore, the brake system must be able to absorb energy at a much faster rate than the engine and drive train can produce. It is not unusual for a vehicle with a 100 horsepower engine to be equipped with a brake system capable of absorbing 600-700 horsepower.

Not only must the brake system absorb this energy and convert it to heat, it must disperse the heat before the brakes fade. **Brake fade** occurs when the friction elements become so hot they can no longer create friction. When this happens, the brake system requires greater pedal pressure to stop the vehicle. Eventually, no amount of pedal pressure will cause the brakes to operate. The brakes must be allowed to cool off before they will work normally. To prevent fading, the brake friction surfaces are heavy cast iron to absorb heat, and contains fins and openings designed to pass this heat to the surrounding atmosphere.

Evolution of the Brake System

The concept of braking may seem fairly obvious. However, the modern brake system is the result of many years of advancement in design, based on changing types of transportation, improvements in materials, and government involvement. It will be helpful to review some of the highlights in the evolution of the modern brake system.

The First Brakes

As soon as man began using horses, camels, and oxen to pull carts and chariots, the need for brakes became evident. Several crude braking systems were developed thousands of years ago. The first brake consisted of a stick placed in a hole in the turning wheel. The other end of the stick then contacted the body of the cart, **Figure 1-2.** This brake was more of a mechanical holding brake than a true stopping brake. However, it worked well enough at low speeds, and made a simple, but effective parking brake.

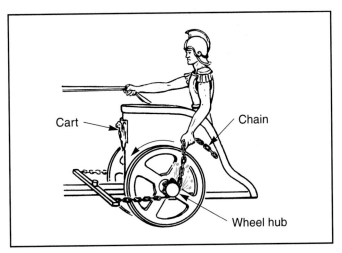

Figure 1-3. *A friction-type brake using a snubbing chain. (John Bean)*

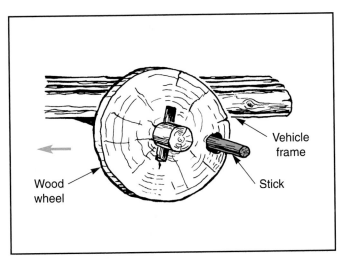

Figure 1-2. *An early brake method consisting of a stick pushed through a hole in the wheel. When the stick contacted the frame, it would jam, stopping the wheel rotation. (FMC)*

Later, two types of friction brakes were developed. The first was the snubbing rope, **Figure 1-3.** A length of rope, or sometimes chain, was attached to the vehicle and wrapped around the rotating axle or wheel hub. When the vehicle was required to stop, the driver would pull on the rope, causing it to firmly grip the hub. The braking force was multiplied as wheel rotation tried to wind the rope tightly to the axle.

Another type of friction brake was developed centuries later for use on horse drawn wagons. This brake consisted of a block of wood which was forced into contact with a metal band on the rim of the wagon wheel. Friction between the wood and rim slowed the wheel and the wagon. The wood block was applied by a lever operated by the driver. A variation of this brake was used to stop trains, **Figure 1-4.** None of these designs were efficient or smooth, but the top speed of animal-drawn vehicles was limited, so braking was not as critical. When the automobile came into existence, brakes had to be improved.

Figure 1-4. *Friction brake used on a steam engine's wheel.*

Automobile Brakes Evolve

Early automobile brakes were very crude compared to those on modern vehicles. The first manufacturers were more concerned with the problems involved in getting the car to move. Stopping the vehicle was a secondary consideration. The very first automobiles used the rim brake described earlier. This was soon discovered to be inadequate even at the low speeds of early cars, and became completely useless when rubber tires were introduced. It was soon replaced with slightly more efficient designs.

Many early cars had brakes only on the rear wheels. Often, the brakes themselves consisted of flexible bands which were installed around the outside of the axle hubs. Sometimes the vehicle used only one band, wrapped around a drum attached to a hub on the rear of the transmission output shaft. The bands were tightened against the hubs to create friction. Brake friction materials were simple: fabric, brass, leather, or even camel hair. The brake was sometimes operated by a hand lever instead of a brake pedal. Linkage from the hand brake to the wheels was usually a series of levers and cables. An early brake system is shown in **Figure 1-5.**

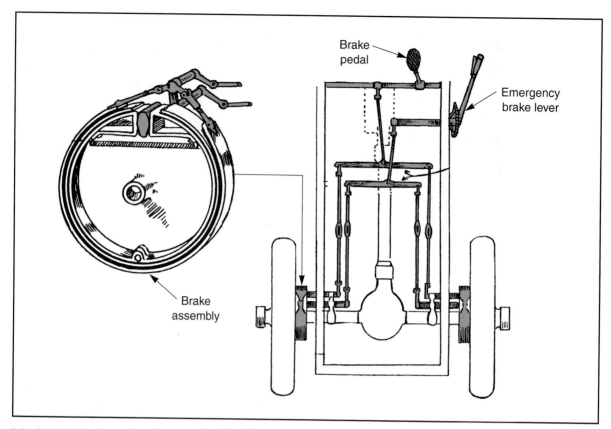

Figure 1-5. *An early brake system incorporating bands and manual operating linkage. (Hermans)*

In the 1920s, more efficient shoe and drum brakes, installed on all four wheels and operated by a foot pedal, became common. Heat-resistant friction materials made of asbestos reduced the problem of fade and increased lining life. At the same time hydraulic brakes began to appear, gradually replacing the mechanical brake system over the next 15 years. Hydraulic fluid was gradually improved to increase its resistance to boiling and corrosion. Power brakes were introduced during the late 1940s, and gradually became common on all vehicles.

Disc brakes were first used in the 1950s, but were not widely used until the 1970s. Safety devices such as dual master cylinders and failure warning lights were first installed on vehicles in 1967, as mandated by Federal law. Asbestos brake linings have gradually been replaced with other, less hazardous compounds. In response to increased safety concerns, anti-lock brakes have become standard or optional equipment on almost every new car or truck built during the last 10-15 years.

Modern Automotive Brake Parts

Most modern automobiles and trucks use the same basic brake system as the first automobiles, a combination of a hydraulic system and friction materials. The *hydraulic system* uses brake fluid to transmit force and movement by sealing the fluid in a closed set of tubes and actuators. *Friction materials* are heat resistant linings

and braking surfaces that perform the actual act of stopping the vehicle. Also included in the modern brake system are power assist units, a parking brake, and electrical and electronic components.

Every modern vehicle has front disc brakes and either disc or drum rear brakes. Efficient, safe, and long-lasting friction materials are used. The master cylinder is divided into isolated front and rear sections. Each section will operate, even if the other loses all hydraulic pressure. The hydraulic brake system incorporates valves which control brake pressures to maximize stopping power. Other hydraulic components warn of hydraulic system failure. The following sections identify the most common brake system components.

Pedal and Master Cylinder Assembly

The *brake pedal* is the most obvious part of the brake system. The brake pedal, along with its associated linkage is hung, or suspended, from a framework under the dashboard. Therefore, the modern brake pedal is known as a *suspended pedal assembly,* **Figure 1-6.** The pedal is connected to the *master cylinder* through linkage. Pressing on the pedal moves a pushrod into the master cylinder.

The master cylinder consists of three major parts: the *cylinder, pistons,* and the *reservoir.* The cylinder and pistons convert the movement of the brake pedal into hydraulic pressure. Movement of the pushrod moves the pistons in the cylinder, pressing on the fluid and creating

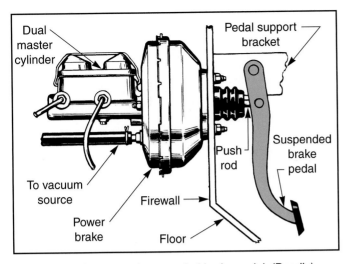

Figure 1-6. *One type of suspended brake pedal. (Bendix)*

hydraulic pressure. Modern master cylinders always have two separate pistons, and are usually called **dual master cylinders.** The dual master cylinder is actually two master cylinders in one housing. This design divides the brake hydraulic system into two separate systems.

The system can be divided into front and rear sections, or diagonally, with one front brake and one rear brake on opposite sides of the vehicle. Hydraulic failure in one system does not affect the other system, and the vehicle can still be stopped, although with more pedal effort and travel. A **pressure differential switch** built into the system warns the driver by illuminating a dashboard light.

The master cylinder reservoir contains additional fluid, in case it is needed by the other hydraulic system components. The fluid reservoir has a filler cap for adding fluid when necessary. Some reservoirs have a low fluid warning switch which illuminates a dashboard light if the fluid level becomes too low. A typical dual master cylinder is shown in **Figure 1-7.**

Power Booster

On all but the lightest vehicles, a **power booster,** sometimes simply called a *power brake,* is installed between the master cylinder and the firewall. The purpose of the power booster is to reduce the amount of pressure the driver must apply to the brake pedal. The power booster does this by adding force to the master cylinder pushrod when the brakes are applied. This additional force is provided by intake manifold vacuum or hydraulic pressure.

Most power boosters are large vacuum diaphragms, powered by the vacuum developed in the engine intake manifold. Some vehicles have a small electrically driven pump to provide additional vacuum when necessary.

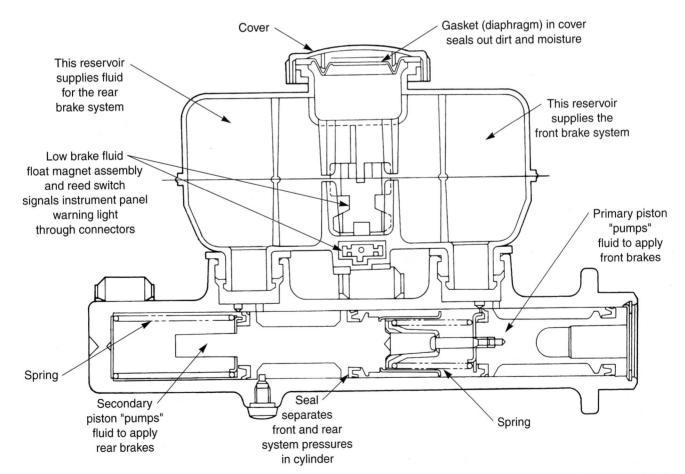

Figure 1-7. *A cross-sectional view of a dual master cylinder assembly. This unit incorporates a low brake fluid level warning float switch. (Ford Motor Co.)*

Some vehicles, especially those with diesel engines and a few vans or trucks equipped with heavy-duty towing packages use a hydraulic pressure operated booster. The hydraulic pressure is developed by the power steering pump or by a separate hydraulic pump operated by an electric motor. **Figure 1-8** illustrates a vacuum booster.

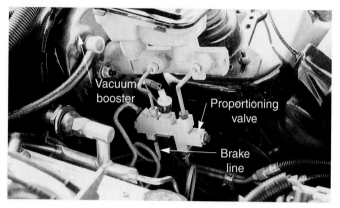

Figure 1-8. *Dual master cylinder, vacuum booster, proportioning valve, and lines as used on a modern vehicle.*

Lines and Hoses

Pressure developed by the master cylinder is delivered to the wheels by steel **hydraulic lines** and flexible **brake hoses.** The lines and hoses must transmit fluid at high pressures with no loss of pressure or leaks. Lines are used to connect parts which do not move in relation to each other, **Figure 1-8.** Steel is always used to minimize the chance of line rupture or leaking. The lines are clamped to the vehicle frame to minimize flexing. The clamps usually have rubber bushings to keep vibration from wearing holes in the line.

The brake hoses are the connection between the vehicle body and the wheels. Since the wheels move in relation to the body, these hoses must contain the high pressures developed in the hydraulic system while flexing as the vehicle moves over road irregularities. These hoses are constructed of several layers of fabric braiding and rubber, with high pressure ends to connect to the other system components. A typical brake line is shown in **Figure 1-9.**

Hydraulic Valves

To precisely control brake hydraulic system operation, several valves are installed in the brake lines or master cylinder. The hydraulic system may contain a **metering valve,** which delays application of the front brakes until the rear brakes have applied. In addition, a **proportioning valve** is used to limit the amount of pressure sent to the rear wheels. These valves are often combined into a single unit called a **combination valve.** A cutaway view of a combination valve is shown in **Figure 1-10.**

Wheel Hydraulic Units

The wheel hydraulic units receive the hydraulic pressure developed by the master cylinder and use it to operate the brake friction elements. While the master cylinder turns brake pedal movement into hydraulic pressure, the wheel hydraulic units turn hydraulic pressure into mechanical movement. Modern brake systems use two types of wheel hydraulic units: the caliper piston and the wheel cylinder. The **caliper, Figure 1-11,** consists of one or more movable pistons inside cylinders. The entire assembly is usually called simply a *caliper.* When pressure is applied to the cylinder, the piston moves outward, compressing the friction

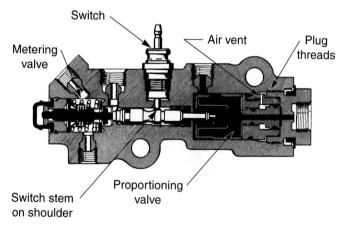

Figure 1-10. *A cross-sectional view illustrating a three-function combination valve. (Bendix)*

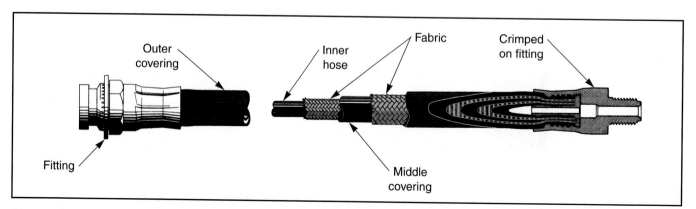

Figure 1-9. *Cutaway view of a two-ply rubber brake hose with crimped on fittings. (Wagner & Chrysler)*

Figure 1-11. *A disc brake caliper assembly. Hydraulic pressure is used to operate the caliper.*

elements against a moving rotor. Most modern calipers have one or two pistons, while the calipers on a few older systems and some high performance vehicles use four pistons, two on each side of the rotor.

Wheel cylinders, **Figure 1-12,** are pressure cylinders with pistons on each end. Pressure from the master cylinder enters the cylinder, pushing the pistons outward. Piston movement moves the brake shoes into contact with the rotating drum. Some older systems used two single piston cylinders.

Figure 1-12. *This illustrates a wheel cylinder installed on a drum brake assembly.*

Wheel Friction Units

Two major classes of wheel friction units are used on modern cars and trucks: *disc brakes* consisting of brake pads and a disc, usually called a rotor; and *drum brakes,* consisting of brake shoes and a drum. Some vehicles have disc brakes on all four wheels. More common is the use of disc brakes on the front axle with drums on the rear. Four-wheel drum brakes are found only on older cars and trucks.

Friction materials making up the pads and shoes were originally made of asbestos. However, asbestos was found to cause cancer, and is being replaced with Kevlar, polymers, resins, and other high temperature compounds. Rotors and drums are made of cast iron. They are relatively heavy to absorb the high heat created by the braking process. A few drums are made of aluminum with a cast iron liner.

Disc brakes are simple in design. The caliper and disc pads are stationary, mounted to the wheel spindle. The rotor is part of the rotating wheel assembly. Hydraulic pressure in the caliper assembly causes the brake pads to be pushed into contact with the rotor. The friction between the stationary pads and spinning rotor slows the vehicle. Heat is removed by air passing over the rotor surface. Many rotors have internal fins which force air through the inside of the rotor, removing more heat. **Figure 1-13** shows the relationship of the pads and rotor. Note the friction components are exposed to the atmosphere. The rotation of the rotor spins off water and dirt. A few rotors have holes drilled into the brake surface for additional cooling.

Drum brakes are sometimes referred to as internal expanding brakes, compared to the external contracting drum brake discussed earlier. Internal expanding brakes are still commonly used on the rear axles of modern vehicles.

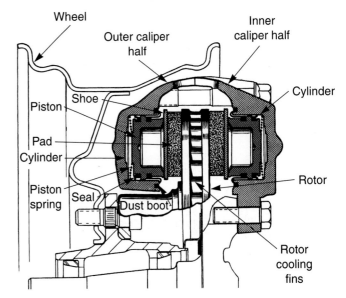

Figure 1-13. *Cross-sectional view of a disc brake assembly. Note the use of cooling fins in the brake rotor. (Bendix)*

The modern drum brake consists of two stationary shoes inside a rotating drum, **Figure 1-14.** Pressure from the master cylinder causes the wheel cylinder to push the shoes outward into contact with the drum. The friction between the stationary shoes and the rotating drum slows the vehicle. Heat is removed by the passage of air over the outer surface of the drum. The stationary parts of the drum brake assembly are mounted on a backing plate, which also protects the internal parts against water and dirt.

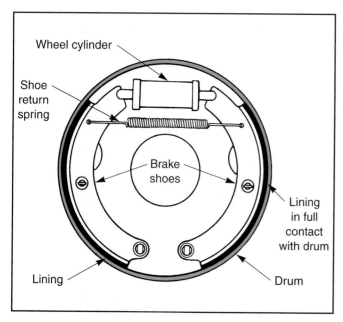

Figure 1-14. *The wheel cylinder forces the brake shoe lining into contact with the brake drum. (FMC)*

Wheel Bearings

The rotor or drum turns on the stationary spindle. Obviously this would produce unwanted friction and wear at the contact point of the rotating and stationary parts. To reduce friction, *antifriction bearings* are used. The three major types of antifriction bearings used on wheel assemblies are the *tapered roller, straight roller,* and *ball bearing.*

Tapered roller bearings are usually used on nondriving axles, typically the front wheels of a rear-wheel drive vehicle, and the rear wheels of a front-wheel drive vehicle. Tapered roller bearings are always adjustable, and can be regreased, **Figure 1-15.** Ball bearings are usually found on the front axles of front-wheel drive vehicles, **Figure 1-16.** Straight roller bearings are usually found on the rear axles of rear-wheel drive vehicles. See **Figure 1-17.** Ball and straight roller bearings are not adjustable, cannot be regreased, and must be replaced if loose or worn. On some vehicles, bearing placement may not follow the typical pattern discussed earlier.

Grease used in bearings must stand up to the extremely high temperatures developed by the brake

system. Usually, the grease designed for tapered roller bearings has different characteristics than grease used in ball and straight roller bearings. The bearings used in most rear-wheel drive vehicles are lubricated by the rear axle oil.

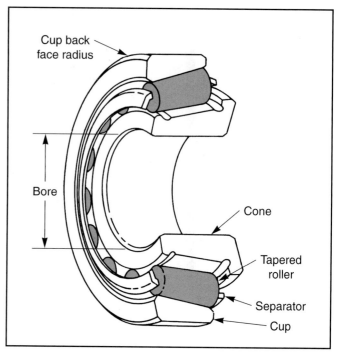

Figure 1-15. *A cutaway view of a typical tapered roller bearing. The bearing roller is cone shaped, much like a styrofoam cup. (Deere & Co.)*

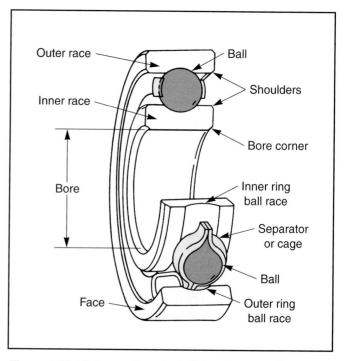

Figure 1-16. *Cutaway view of a ball bearing. Note the separator which keeps the individual balls equally spaced around the bearing. (Deere & Co.)*

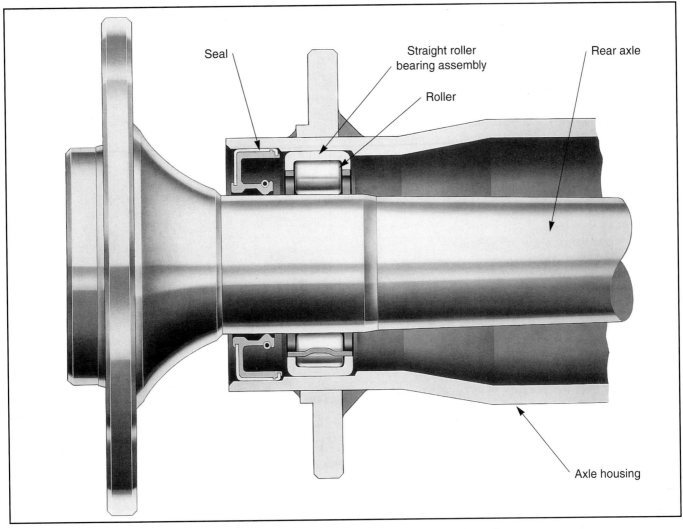

Figure 1-17. *A rear-wheel drive axle assembly which uses straight roller bearings. (Torrington)*

Parking Brake

To hold the vehicle stationary when it is not being driven, a **parking brake** is used. Almost all parking brakes are installed on the rear wheels. All vehicles with rear drum brakes, and a few with rear disc brakes, use the stopping brakes as parking brakes. Instead of operating the friction members with the hydraulic system, parking brakes use a separate mechanical linkage. The mechanical linkage consists of levers and cables, such as the design shown in **Figure 1-18.** The parking brake can be operated through a dashboard lever, foot pedal, or floor mounted lever, depending on the manufacturer.

If the rear wheel brakes are drum types, the mechanical linkage applies the brakes through linkage attached to the shoes. If the vehicle has rear disc brakes, a lever and screw assembly manually moves the caliper piston, pushing the brake pads into contact with the rotor.

On a few rear-wheel disc brake systems, the parking brake consists of a small set of brake shoes located inside of the rotor. The only purpose of these shoes is to act as a parking brake. The shoes are mechanically applied in the same manner as on rear drum brakes.

The parking brake can be used to stop the vehicle if the hydraulic brakes fail. For this reason the parking brake is sometimes called an **emergency brake.** However, the parking brake is not designed to be used regularly as a stopping brake. Under normal nonemergency conditions, the parking brake should only be used to hold a vehicle that is already stopped.

Anti-lock Brake and Traction Control Systems

Anti-lock brake systems, usually called **ABS,** are becoming common on late model vehicles. Anti-lock brakes were first used on large jets in the 1950s, and began to see use in cars and trucks in the 1980s. The ABS system is a set of electronic and hydraulic components which reduce skidding during hard braking. The ABS system controls hydraulic pressure, reducing pressure to any wheel in danger of locking up. When lockup is no longer detected, the system returns to normal braking operation. The ABS system can vary the system pressure many times per second. Early ABS

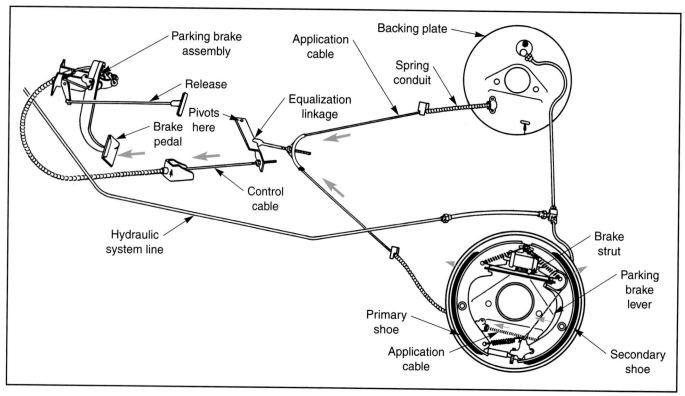

Figure 1-18. *Operation of a mechanical linkage parking brake assembly. As the parking pedal is depressed, the cables are tightened. They pull the parking brake levers which in turn, force the brake shoes and linings into contact with the drum, preventing it from turning. (Grey-Rock)*

systems controlled only the rear brakes. This system is still used on some pickup trucks, **Figure 1-19.** Most newer automobiles have ABS control on all four wheels, **Figure 1-20.**

On many newer vehicles, *traction control systems (TCS)* are used to reduce slippage on acceleration. In its

simplest form, traction control senses when a wheel is starting to slip under acceleration, and applies the brake on that wheel. Traction control uses many of the same parts as ABS, and the two systems are often combined into one unit. Many later traction control systems also reduce engine power when one or more drive wheels are

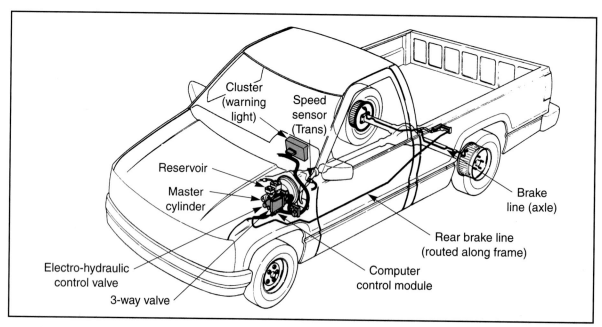

Figure 1-19. *A rear anti-lock brake system used on a pickup truck. A speed sensor located on the transmission is used in place of sensors at each rear wheel. (Chevrolet)*

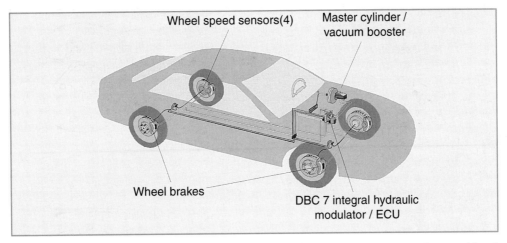

Figure 1-20. *Layout of a modern anti-lock brake system (ABS). This particular system is also capable of providing traction control. (General Motors)*

slipping. Engine power can be reduced by retarding ignition timing, reducing throttle opening, or selectively disabling engine cylinders.

Automotive Brake Operation

The following paragraphs give a brief description of how the system operates under normal braking conditions. This will familiarize you with the process of what happens when the brakes are applied. Knowing how a system operates is the first step in diagnosing and fixing a problem in the system. The operation of related equipment such as power boosters, parking brakes, and ABS systems will be discussed in later chapters.

Brakes Inactive

In **Figure 1-21,** the brake system is inactive. The pedal is not being depressed, and no force is applied to the master cylinder pushrod. No hydraulic pressure is created within the master cylinder, and therefore none is delivered to the lines and hoses. Without pressure, the caliper pistons and wheel cylinder cannot apply the friction materials. Spring and seal pressure keeps the brake pads and shoes in the released position. The brake system is not operating.

Brakes Applied

In **Figure 1-22,** the brake pedal is pushed, and the master cylinder pushrod is forced inward. This creates hydraulic pressure within the master cylinder. Pressure is

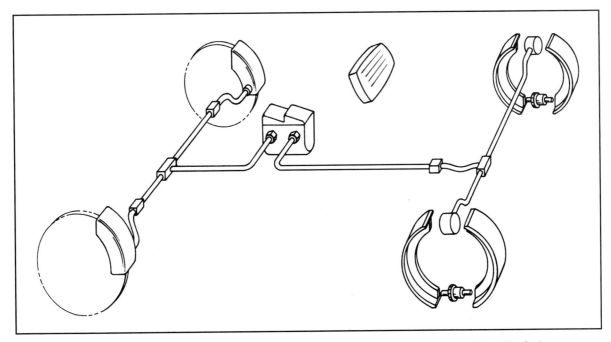

Figure 1-21. *A simple brake system schematic showing the brakes in the inactive mode. (Parker-Hannifin Co.)*

transmitted through the brake lines to the caliper pistons and wheel cylinders. Valves control front and rear brake pressures. Pressure causes the caliper piston to move, forcing the brake pads against the rotor, and forcing the wheel cylinder pistons outward, moving the brake shoes into contact with the brake drum. Contact of the stationary and rotating friction members causes the vehicle to be slowed. The heat resulting from friction is released into the surrounding air.

Brakes Released

In **Figure 1-23,** the brake pedal is released. When the pedal is released, hydraulic pressure drops off. With no pressure to oppose them, the caliper piston seals pull the piston back into the cylinder. This releases the brake pads and the rotor can turn freely. The wheel cylinder pistons are pushed back into the cylinder by the brake shoe return springs and the drum can turn with no

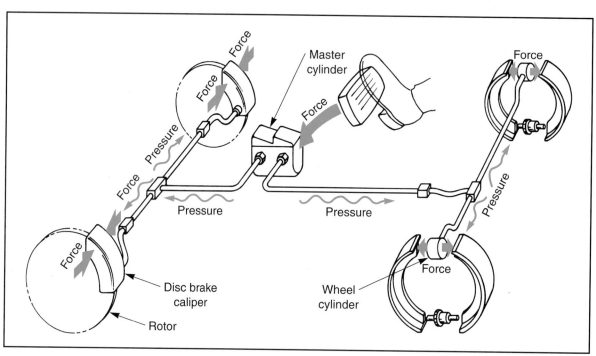

Figure 1-22. *Simple brake system shown in the applied mode. (Parker-Hannifin Co.)*

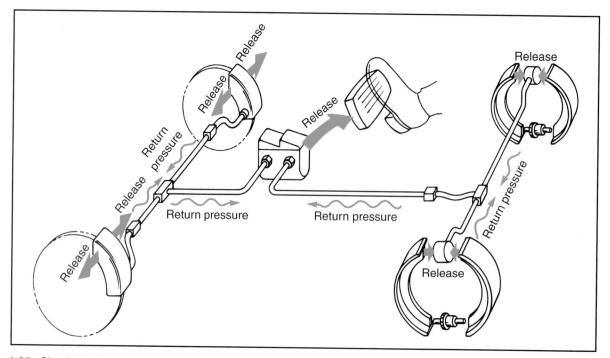

Figure 1-23. *Simple brake system in the released mode. (Parker-Hannifin Co.)*

resistance. As the caliper and wheel cylinder pistons are moved back into their cylinders, they displace fluid, which returns to the master cylinder. The fluid returning to the master cylinder helps an internal spring push the brake pedal back to its released position. Any extra fluid flows into the reservoir.

Summary

The brake system must stop the vehicle smoothly and consistently without causing damage, and must continue to do so over and over. Holding brakes keep a stationary vehicle from moving. Stopping brakes must bring a moving vehicle to a stop. The modern brake system must convert a great amount of momentum to heat through the use of friction. It must also pass this heat to the surrounding atmosphere so that the brake system can continue to function.

The first types of brakes were very inefficient, but worked adequately when all wheeled vehicles were animal powered. Early automobile brakes were very crude. However, they have been consistently improved to result in the modern brake system.

The modern brake system is a combination of hydraulic and friction units. Electrical and electronic parts are also used. Major components of the modern brake system include the brake pedal and associated linkage, and the master cylinder which converts pedal force into hydraulic pressure. On some vehicles, the pedal force is increased by the use of a power booster. Most master cylinders are part of a split system which divides the brake system into two independent units. From the master cylinder, pressure is sent through lines and hoses to the wheel hydraulic units. Various valves control the direction of pressure, and send warnings to the driver if a malfunction occurs.

The wheel hydraulic units consist of calipers and wheel cylinders. The hydraulic units cause the friction members to apply. Vehicle motion is changed to heat as the friction members reduce speed. This heat is transferred to the surrounding air.

Related components of the brake system include the parking brake. Most parking brakes make use of the existing rear wheel friction elements. A few vehicles have separate shoes used only as parking brakes. Wheel bearings are an integral part of the wheel brake assemblies. Various types of bearings are used. Some bearings can be lubricated and adjusted, while others must be replaced if they show any looseness.

Anti-lock brakes and traction controls are combinations of electronic and hydraulic devices which act to reduce skidding when braking, and increase traction when accelerating.

Review Questions—Chapter 1

Please do not write in this text. Write your answers on a separate sheet of paper.

1. A _____ brake uses friction to stop the vehicle, while a _____ brake is a simple mechanical device.

2. The parking brake on a vehicle with an automatic transmission is an example of a _____ type of brake.

3. To allow friction brakes to work for long periods of time, what must be removed from them as they operate?

4. Using a chain wrapped around a wheel hub on a cart is an example of a _____ brake.

5. Early brakes were installed only on the _____ wheels.

6. Improvements in hydraulic fluid reduced what two problems?

7. Hydraulic fluid can be used to transmit _____ and _____.

8. The brake pedal is always connected to the _____.

9. Power brakes are always operated by _____.

10. Brake hoses are used to connect brake parts that _____ in relation to each other.

11. What are the two main types of wheel hydraulic units?

12. What metal are rotors and drums usually made of?

13. Match the brake part with its description. Some answers may be used more than once.

    Rotor _____   (A) Stationary part of a
    Pads _____   disc brake assembly.
    Drum _____   (B) Rotating part of a
    Shoes _____   disc brake assembly.
    Backing plate _____   (C) Rotating part of a
                                        drum brake assembly.
                                   (D) Stationary part of a
                                        drum brake assembly.

14. To reduce friction, a _____ is always used between the spindle and rotor or drum.

15. Depending on the vehicle, three kinds of levers are used to operate the parking brake. Name them.

ASE Certification-Type Questions

1. A functioning brake system will quickly change vehicle movement into:
   - (A) skidding.
   - (B) friction.
   - (C) heat.
   - (D) fading.

2. All of the following are characteristics of early brake systems, EXCEPT:
   - (A) they were smooth.
   - (B) they were inefficient.
   - (C) they were good enough for animal-powered vehicles.
   - (D) they were replaced by later designs.

3. Every modern vehicle has front:
   - (A) drum brakes.
   - (B) disc brakes.
   - (C) band brakes.
   - (D) asbestos brakes.

4. Technician A says that modern brake pedals are called suspended pedal assemblies. Technician B says that the power booster is installed between the pedal and the pedal linkage. Who is right?
   - (A) A only.
   - (B) B only.
   - (C) Both A & B.
   - (D) Neither A nor B.

5. The master cylinder changes pedal force into:
   - (A) friction.
   - (B) vacuum.
   - (C) heat.
   - (D) hydraulic pressure.

6. All of the following statements about dual master cylinders are true, EXCEPT:
   - (A) dual master cylinders always have two separate pistons.
   - (B) the dual master cylinder is actually one master cylinder in two housings.
   - (C) the brake hydraulic system can be divided into front and rear or diagonal sections.
   - (D) a pressure differential switch controls a dashboard light.

7. Valves installed in the brake hydraulic system affect the pressure at the _____.
   - (A) front wheels
   - (B) rear wheels
   - (C) master cylinder
   - (D) Both A & B.

8. Technician A says that brake friction materials must stand up to high heat. Technician B says that the best modern brake friction materials include asbestos. Who is right?
   - (A) A only.
   - (B) B only.
   - (C) Both A & B.
   - (D) Neither A nor B.

9. Depending on the manufacturer, the parking brake is operated by a _____.
   - (A) dashboard lever
   - (B) foot pedal
   - (C) floor lever
   - (D) All of the above.

10. Which of the following units is directly controlled by the anti-lock brake system?
    - (A) Friction elements.
    - (B) Hydraulic system.
    - (C) Master cylinder reservoir.
    - (D) All of the above.

Chapter 2

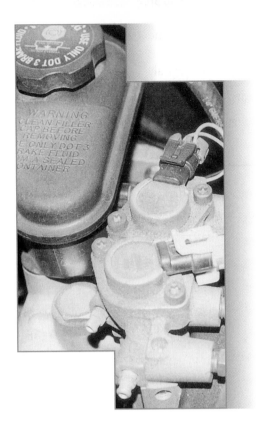

Shop Safety and Environmental Protection

After studying this chapter, you will be able to:
- ❑ Identify the major causes of accidents.
- ❑ Explain why accidents must be avoided.
- ❑ Identify brake dust hazards.
- ❑ Explain how to avoid exposure to brake dust.
- ❑ List ways to maintain a safe workplace.
- ❑ List safe work procedures.
- ❑ Identify types of environmental damage caused by improper auto shop practices.
- ❑ Identify ways to prevent environmental damage.

Important Terms

Safety glasses	Work gloves	Carcinogen	Material Safety Data Sheets (MSDS)
Face shield	Brake dust	High efficiency particle air (HEPA)	Environmental Protection Agency (EPA)
Safety shoes	Asbestos	Respirator	

There are many ways to do a job in an unsafe manner, and usually only one way to do it safely. This chapter covers reasons for and methods of doing things safely. The reason to make repairs in a safe manner is that it protects you, the vehicle, and the shop. Also covered in this chapter is the proper handling and disposal of waste products. Wastes must be handled properly to safeguard both you and the environment. This chapter examines various types of unsafe conditions and work practices, environmental violations, and how to correct or avoid them.

Causes of Accidents

No one usually tries to cause an accident. Accidents result in injuries that keep you from working or enjoying your time off. Some accidents can even kill. Even slight injuries are painful and annoying, and may impair your ability to work and play. Even if an accident causes no personal injury, it can result in property damage. Damage to vehicles or shop equipment can be expensive and time consuming to fix, and could even cost you your job.

The usual cause of accidents is the attitude that it is just too much trouble to do things correctly. Often, the end result of this attitude is an accident. Unfortunately even experienced technicians become rushed and careless. Therefore, falls, injuries to hands and feet, fires, explosions, electric shocks, and even poisonings occur in auto repair shops. Carelessness in the shop can also lead to long-term bodily harm from prolonged contact with harmful liquids, vapors, and toxic dust. Lung damage, skin disorders, and even cancer can result from contact with these substances. For these reasons, you must keep safety in mind at all times, especially when conditions tend to make it the last thing on your mind.

Personal Safety Equipment

To protect yourself from accidents and long-term harm, you should dress properly and have several types of personal safety equipment.

Proper Clothing and Protective Equipment

Always dress appropriately and safely. Do not wear clothes with long, loose sleeves, open jackets, or scarves. They can get caught in moving parts and pull you into the machine or engine. Do not wear a tie unless your duty position requires it. If you must wear a tie, tuck it inside your shirt. If you wear your hair long, keep it away from moving parts by tying it up or securing it under a hat. Remove any rings or other jewelry. Not only can jewelry get caught in moving parts, it can cause a short circuit, which can result in severe burns or start a fire if caught between a positive terminal and ground.

Eye Protection

The shop environment contains many things which could be hazardous to your eyes. Flying particles can injure immediately, and exposure to dust and chemicals can cause long-term damage. You should have two types of eye protection, **safety glasses** and a **face shield.** See **Figure 2-1.** Safety glasses should be worn at all times while you are in the shop. Face shields should be worn when you are performing grinding or cutting operations. Other types of protective eyewear are available for specialized operations, such as welding.

Figure 2-1. *Eye protection is a must when working with grinders, drills, air tools, etc. Remember the safety glasses can be replaced, but not your eyesight.*

Foot and Hand Protection

In any shop there is always the danger of objects falling on your foot or having your foot crushed between two heavy objects. Therefore, you should wear foot protection when working in the shop. **Safety shoes,** preferably with steel toe inserts, should be worn at all times. Most good quality safety shoes are constructed using materials that are oil and chemical resistant. Safety shoes have soles that are not only slip resistant and insulated, but also provide support and comfort.

Always have a pair of **work gloves** available to protect against hot objects, or parts with sharp or jagged edges. Gloves will also protect against skin rashes caused by exposure to chemicals.

The Danger of Brake Dust

One of the most common contributors to health problems for automotive technicians is long-term exposure to brake dust. **Brake dust** is nothing more than friction material, mixed with road dirt, that has been worn off the pads or shoes by the normal braking process. You can usually see brake dust simply by looking at the front wheels of a vehicle equipped with aluminum rims.

There is no way to totally eliminate the production of brake dust. Much of this dust enters the air when the brakes are applied. However, some dust collects on the components of the brake system, especially the interior of drum brakes, where less air passes. When any type of brake service is performed, this dust is disturbed, and enters the atmosphere. It has been estimated that *17 million brake dust particles* are released by nothing more than removing a brake drum!

Brake dust particles are extremely small; so small the body's natural filtering processes cannot stop them from entering the lungs. In the past, brake friction materials were made of **asbestos,** which is a **carcinogen** (a substance that can cause cancer). Asbestos, if inhaled over long periods, can cause lung cancer by irritating the lung tissues. Repeated exposure will cause irritation, weakening the body's defenses and making lung tissue more susceptible to the abnormal cell growth that can lead to cancer.

Asbestos exposure over long periods can also cause **asbestosis,** a disease of the lungs caused by the buildup of asbestos and other fibers in the lungs. The lung passages become clogged, and the body gradually becomes starved for oxygen. This condition is similar to *black lung,* a long-term exposure lung disease which affected coal miners for many years.

In addition, brake dust can cause eye and skin irritation. If brake dust is not removed from the skin and eyes promptly, eye damage, skin rashes, and over the long-term, possible skin cancer may result. There is also some evidence that ingesting brake dust for a prolonged period can cause digestive system disorders.

For these reasons, asbestos has been banned as a building material and insulator, and has been eliminated as a component of brakes and transmission clutches. However, many vehicles on the road are still equipped with brakes containing asbestos. Newer brake materials, while not as dangerous as asbestos, can still cause lung and other respiratory system damage if large quantities are inhaled, or if small quantities are inhaled over long periods.

Avoiding Brake Dust

It is impossible to totally avoid some contact with brake dust. However, you can take some simple steps to reduce exposure. The first step is to reduce the amount of brake dust entering the atmosphere where it can be easily inhaled.

Always clean wheel brake assemblies with liquid solvents, or with enclosed vacuum cleaners equipped with **high efficiency particle air (HEPA)** filters. When using brake lathes, make sure the filter assembly is installed and in good condition. Try to store used brake parts in closed containers, away from high traffic areas.

 Warning: Never blow brakes with compressed air, even after cleaning with liquid solvents.

Inevitably, some brake dust will enter the atmosphere. To keep this airborne dust out of your lungs,

always wear a **respirator, Figure 2-2,** whenever you are working on a brake system. Keep the filters in your respirator clean and change them regularly. After brake system repairs are complete, thoroughly wash your hands and face. If possible, change clothes after performing a brake service. At the end of the day, vacuum, and if possible, hose down the service area. Do not sweep the area, as this will send particles into the air.

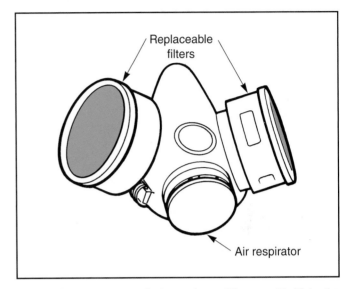

Figure 2-2. *One type of air respirator (filter mask). Note the replaceable filter elements. (Niehoff)*

Kinds of Accidents

The ways an accident can occur in the shop are many and varied. Most accidents are caused by one or a combination of two factors:
- Failure to maintain a safe workplace.
- Performing service procedures improperly.

Examples of failure to maintain a safe workplace are allowing tools and equipment to fall into disrepair, failing to dispose of old parts, containers, or other trash, and ignoring water or oil spills. Examples of improper service procedures are using the wrong tools or methods to perform repairs, using defective or otherwise inappropriate tools, not wearing protective equipment when necessary, and not paying close attention while performing the job.

The best way to prevent accidents is to maintain a neat workplace, use safe methods and common sense when making repairs, and wear protective equipment when needed. The following are some suggestions for reducing the possibility of accidents.

Maintaining a Safe Workplace

Return all tools and equipment to their proper storage places. This saves time in the long run, as well as reducing the chance of accidents, damage, and theft. Do not leave equipment out where others could trip, **Figure 2-3.**

Figure 2-3. *A neat, clean, and orderly shop work site. This is a sign of a top-notch technician and business.*

Keep workbenches clean. This reduces the chance of tools or parts falling from the bench onto the floor, where they could be lost or damaged. A falling tool or part could land on your foot. A clean workbench also reduces the possibility that critical parts will be lost in the clutter, or that a fire will start in oily debris.

Clean up spills immediately, before they get tracked around the shop. Many people are injured when they slip on floors coated with oil, antifreeze, or water. Gasoline spills can be extremely dangerous, since a flame or the smallest spark can ignite the vapors, causing a major explosion and fire.

Know what chemicals are stored on the shop premises. Chemicals include carburetor cleaners, hot tank solutions, parts cleaner, and even motor oil and antifreeze. Chemical manufacturers provide *Material Safety Data Sheets,* often called *MSDS,* for every chemical that they produce. These sheets list all the known dangers of the chemical, as well as first aid procedures for skin or respiratory system contact. There should be an MSDS for every chemical in the shop. You should read the MSDS for any unfamiliar chemical before working with it. **Figure 2-4** shows a typical MSDS.

Make sure that the shop is well lighted. Poor lighting makes it hard to see what you are doing, leading to accidental contact with moving parts or hot surfaces. Overhead lights should be bright and centrally located. Portable lights, or droplights, should be in operating condition and easy to use, **Figure 2-5.** Always use a "rough service" bulb in incandescent service lights. These bulbs are more rugged than normal lightbulbs, and will not shatter if they are dropped. Do not use a high wattage bulb in

a portable light. Lightbulbs get very hot and can melt the light socket or cause burns. Service lights that use cool fluorescent bulbs are now available from some tool manufacturers.

Do not overload electrical outlets or extension cords by operating several electrical devices from one outlet. An overloaded outlet is shown in **Figure 2-6.** Do not operate high current electrical devices through extension cords. Frequently inspect electrical cords and compressed air lines to ensure they are in good condition. Do not close vehicle doors on electric cords or air lines. Do not run electrical cords through water puddles, or use them outside when it is raining.

Ensure that all shop equipment, such as grinders, lathes, and drill presses, are equipped with safety guards, **Figure 2-7.** All shop equipment is equipped with guards by the manufacturer. These guards should never be removed, except for service operations such as changing the grinding wheels. When servicing any shop equipment, be sure it is turned off and unplugged. Read the equipment service literature before beginning any repairs.

Closely monitor tool and equipment condition and make repairs when necessary. This includes such varied things as replacing damaged leads on test equipment, checking and adding oil to hydraulic jacks, and regrinding the tips on screwdrivers and chisels.

Do not leave open containers of chemicals in the shop or outside. Most automotive chemicals will poison any animal (or person) who drinks it, and spills will create an extremely slippery floor.

Know where the shop fire extinguishers are located, and know how to operate them. Make sure you know what

ACME Chemical Company
Material Safety Data Sheet
Product Name: Acetylene   **Revised 3/3/99**

24-hour Emergency Phone: Chemtrec 1-800-424-9300   Outside United States 1-905-501-0802

Trade Name/Syn: Acetylene   **NFPA Ratings**
Chemical Name/Syn: Acetylene, Ethyne, Acetylen, Ethine   Health: 0
CAS Number: 74-86-2   Flammability: 4
Formula: C_2H_2 Reactivity: 0

Hazards Identification
Simple Asphyxiant. This product does not contain oxygen and may cause asphyxia if released in a confined area. Maintain oxygen levels above 19.5%. May cause anesthetic effect. Highly flammable under pressure. Spontaneous combustion in air at pressures above 15 psig. Acetylene liquid is shock sensitive.

Effects of Exposure-Toxicity-Route of Entry
Toxic by inhalation. May cause irritation of the eyes and skin. May cause an anesthetic effect. At high concentrations, excludes an adequate oxygen supply to the lungs. Inhalation of high vapor concentrations causes rapid breathing, diminished mental alertness, impaired muscle coordination, faulty judgment, depression of sensations, emotional instability, and fatigue. Continued exposure may cause nausea, vomiting, prostration, loss of consciousness, eventually leading to convulsions, coma, and death.

Hazardous Decomposition Product
Carbon, hydrogen, carbon monoxide may be produced from burning.

Hazardous Polymerization
Can occur if acetylene is exposed to 250°F (121°C) at high pressures or at low pressures in the presence of a catalyst. Polymerization can lead to heat release, possibly causing ignition and decomposition.

Stability
Unstable--shock sensitive in its liquid form. Do not expose cylinder to shock or heat, do not allow free gas to exceed 15 psig.

Fire and Explosion Hazard
Pure acetylene can explode by decomposition above 15 psig; therefore the UEL is 100% if the ignition source is of sufficient intensity. Spontaneously combustible in air at pressures above 15 psi (207 kPa). Requires very low ignition source. Does not readily dissipate, has density similar to air. Gas may travel to source of ignition and flashback, possibly with explosive force.

Conditions to Avoid
Contact with open flame and hot surfaces, physical shock. Contact with copper, mercury, silver, brasses containing >66% copper and brazing materials containing silver or copper.

Accidental Release Measures
Evacuate all personnel from affected areas. Use appropriate protective equipment. Shut off all ignition sources. Stop leak by closing valve. Keep cylinders cool.

Ventilation, Respiratory, and Protective Equipment
General room ventilation and local exhaust to prevent accumulation and to maintain oxygen levels above 19.5%. Mechanical ventilation should be designed in accordance with electrical codes. Positive pressure air line with full face mask or SCBA. Safety goggles or glasses, PVC or rubber gloves in laboratory; and as required for cutting or welding, safety shoes.

Figure 2-4. *A material safety data sheet (MSDS) from one manufacturer. Always read before using a chemical.*

Figure 2-5. *A "rough service" lightbulb. A clear rubberized coating adds extra strength. The filament also uses additional supports. Some of the bulbs have a brass base to help reduce corrosion in the socket.*

Figure 2-6. *An overloaded electrical outlet. This can lead to fires, blown fuses, and tripped circuit breakers.*

type of fire extinguisher is used on what type of fire. The chart in **Figure 2-8** shows the kinds of fire extinguishers and the fires they are designed to extinguish. Periodically check each fire extinguisher to ensure they are in working order and have them checked periodically by qualified personnel.

Performing Work Procedures Properly

Study work procedures before beginning any job that is unfamiliar. Do not assume the procedure you have used in the past will work with a different type of vehicle. Always work carefully. Speed is not nearly as important as

Figure 2-7. *A safety guard must always be used. Note that this guard is clear, providing you with a clear view of the work being done. (Ammco)*

doing the job right and avoiding injury. Avoid co-workers who will not work carefully, or who tend to engage in horseplay.

Use the right tool for the job. Using a screwdriver as a chisel or pry bar, or a wrench as a hammer, is asking for an accident or at least a broken tool. Never use a hand socket with an impact wrench. A hand socket can crack and shatter if used with an air tool. Do not use low quality tools or tools that are damaged. Use the right tool for the job.

Learn how to use new equipment before using it. This is especially true of impact wrenches, air chisels, and other air operated tools. It is also true for large electrical devices such as drill presses and brake lathes. These tools are very powerful, and can hurt you if they are used improperly. A good way to learn about new equipment is to start by reading the manufacturer's instructions.

When working on electrical systems, avoid creating a short circuit with a jumper wire or metal tool. Not only will this damage the vehicle components or wiring, it will develop enough heat to cause a severe burn or start a fire. Be careful when using a test light. Used improperly, they can cause as much or more damage than a non-fused jumper wire.

Lift safely. Make sure that you are strong enough to lift the object to be moved. Always lift with your legs, not your back. If an object is too heavy to lift by yourself, get help.

Figure 2-8. *Class A—Paper, wood, cloth. Class B—Liquids such as gasoline, oil, diesel fuel, etc. Class C—Electrical fires. (Amerex Corp.)*

Do not smoke in the shop. You may accidentally ignite an unnoticed gasoline leak. There are other less noticeable flammable substances around every vehicle. Batteries can produce explosive hydrogen gas as part of their normal chemical reaction. If a flame contacts escaping air conditioning refrigerant, it will produce poisonous ***phosgene gas.*** A discarded cigar or cigarette can also ignite any oily rags or paper debris that may be lying around.

When using any type of vehicle lifting equipment, be sure to place the lift pads at the vehicle frame or other points that the manufacturer specifies as lifting points. Do not attempt to raise a vehicle with an unsafe or undercapacity jack. Always support a raised vehicle with good quality jackstands. Do not use a car jack as they are designed for emergency use only. Never use boards or cement blocks to support a vehicle. When using a post or hydraulic lift, be sure to place the lifting pads under the frame or on a spot that can support the vehicle's weight, **Figure 2-9.**

Do not leave a running vehicle unattended. The vehicle may slip into gear, or overheat while you are away. Whenever you must work on a running vehicle, set the parking brake. Do not run any engine in a closed area without good ventilation, even for a short time. ***Carbon monoxide*** can build up quickly, cannot be seen or smelled, and is deadly. When working on or near a running engine, keep away from all moving parts. Never reach between moving engine parts for any reason. Seemingly harmless parts such as the drive belts and fan can seriously injure you.

When road testing a vehicle, be alert, and obey all traffic laws. Do not become so absorbed in the diagnosis process that you forget to watch the road. Be alert for the actions of other drivers. If you must listen, or observe a scan tool during a road test, get someone to drive the vehicle for you. Be aware of the type of tests and the effects they could have on the brake system. For example, some anti-lock brake systems are disabled whenever a scan tool is connected to the vehicle.

Accidents are not inevitable. It is up to you to notice and correct safety hazards and use safe work practices. This is the only way to prevent accidents.

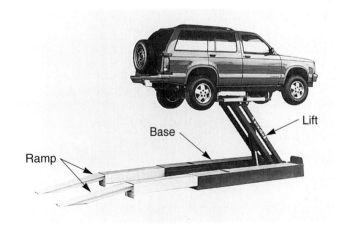

Figure 2-9. *Lifts can be very dangerous if they are not used properly. (Mohawk)*

Preventing Environmental Damage

Considerable environmental damage is caused by automotive repair shops. Shops are often guilty of carelessly disposing of solid and liquid wastes, and of causing damage to the atmosphere by improper repair procedures.

Typical solid wastes produced by automotive repair shops are scrap parts, tires, and cardboard boxes. Liquid wastes include antifreeze, brake fluid, cleaning solvents, motor oil, and transmission fluid.

Improper repair procedures include allowing refrigerant gases to escape into the atmosphere and repairing engines so they emit excessive amounts of pollutants.

Toxic materials lower the air and water quality, and can even affect food supplies. The effects of poisoned air, water, and soil may be noticed almost immediately, or may take decades to become apparent. The health and financial burdens of irresponsible waste disposal, if left unchecked, will grow ever larger. Even if we escape the consequences, future generations will not.

Preventing Solid Waste Contamination

The automotive service industry is a major emitter of solid and liquid wastes. The following are a few suggestions for dealing with solid wastes. In certain parts of the United States, these are rigidly enforced laws.

Do not allow solid wastes to build up around the shop. Recycle parts and scrap materials whenever possible. Do not throw away parts boxes, old tires, and salable scrap metals unless you are sure they cannot be recycled. While some solid wastes such as scrap metal, paper, and tires are not immediately damaging, they are unsightly and increase the burden on local landfills.

It makes good economic sense to recycle, since almost every rebuildable part has a return value, usually called a *core value*. Typical brake parts that have a core value include master cylinders, power boosters, calipers, drum brake shoes, drums, and rotors. Some anti-lock brake parts have a substantial core charge, sometimes more than the actual cost of the part, making their return a necessity.

The value of paper, scrap tires, and scrap metals such as aluminum, iron, and brass depends on the current market conditions, but it always has some value. Check with your local parts supplier to determine which parts can be sent back for rebuilding. Recyclers are often listed in the telephone directory, and can give you advice on what to do with recyclable materials. If solid wastes cannot be recycled, dispose of them responsibly, not by illegal dumping or burning.

Preventing Liquid Waste Contamination

Probably the most common way that automotive shops damage the environment is by pouring used motor oil, transmission fluid, antifreeze, brake fluid, or gear oil on the ground. This immediately contaminates the soil. In addition, these liquids sink further into the ground every time it rains, eventually contaminating the water table. This could be your local drinking water source.

Do not pour liquid wastes into drains. While brake fluid is water soluble, most municipal waste treatment plants cannot handle petroleum-based products. Most automotive fluids also contain poisonous additives and heavy metals absorbed from the vehicle during use. In most areas, such dumping is illegal.

In many areas, local waste management companies will accept used oil and antifreeze for recycling. Several 55-gallon drums should be kept in the shop to store liquid wastes. See **Figure 2-10.** The used oil and antifreeze is then reprocessed and reused. Some is burned by power plants to produce electricity, eliminating the used oil and reducing the dependence on imported oil. A recently developed process converts old motor oil into diesel fuel.

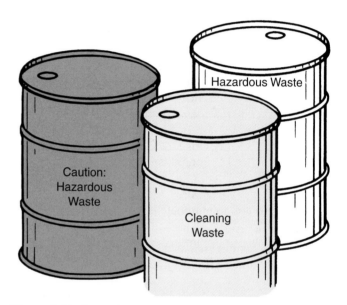

Figure 2-10. *Several 55-gallon drums used for hazardous shop waste. Keep the drums separated, and do not mix chemicals. (Dana Corp.)*

Preventing Damage to the Atmosphere

The major way in which automotive shops hurt the atmosphere is by making modifications which damage or defeat the purpose of the vehicle emission control systems. Adjusting carburetors for an excessive rich or lean air-fuel mixture, changing the manufacturer's timing settings, and disconnecting emission controls will all increase emissions.

Some seemingly harmless actions, such as installing a lower temperature cooling system thermostat or a nonstock air cleaner, can also cause a rise in engine and vehicle emissions. Not only is this illegal, they almost never increase power and mileage as much as hoped. While most brake work will have no effect on vehicle emissions, you should be aware that even seemingly minor work can potentially cause a major emissions problem.

Do not discharge air conditioner *refrigerant* into the atmosphere. Studies have shown that such as R-12 cause extensive damage to the ozone layer, leading to increased ultraviolet ray damage. Even so-called "safe" refrigerants, such as R-134a, still contribute to ozone layer loss. This is another area covered by Federal, and in some cases, state laws. Replacing refrigerants costs money. For instance, R-12 refrigerant costs many times more than it did just a few years ago. It makes good economic sense to recover and reuse refrigerants. If you do not have a machine which can recover and recycle refrigerant, send the customer to a shop which can do the job properly.

The Environmental Protection Agency

Modifying, disabling, removing, or otherwise tampering with engine emission controls is a crime, with severe penalties. Vehicle emissions laws are enforced by the **Environmental Protection Agency,** usually called the **EPA.** The EPA investigates suspected violations, and often conducts "sting" operations to catch violators. In addition, some states such as California have additional laws protecting the environment.

Additional information about waste disposal and vehicle emissions can be obtained from the Environmental Protection Agency. The EPA has ten regional offices and six field offices. For the address of the nearest EPA office, write or call:

Automotive/Emissions Division
United States Environmental Protection Agency
401 M Street S. W.
Washington, DC 20460

It is up to you to eliminate environmental damage by taking the proper steps to reduce the improper disposal of wastes. Always use common sense when working in the shop, and avoid people who do not.

Summary

The most common cause of accidents is carelessness. Many technicians get in a hurry and forget to do the job safely. An accident may result in personal injuries, long-term bodily harm, or injuries to equipment or property. No mature and competent automotive technician wants to be injured or cause property damage.

A common problem among brake technicians is exposure to brake dust. This dust can cause long-term damage to the lungs and respiratory system. The technician should take all possible precautions to avoid breathing in brake dust.

Many accidents are caused when technicians fail to correct dangerous conditions in the work area, such as oil spills or tripping hazards. Other accidents are caused when technicians try to take shortcuts instead of following proper repair procedures.

The best way to prevent accidents is to maintain a neat workplace, use proper methods of repair, and use protective equipment when needed. It is up to the technician to study the job beforehand, work safely and prevent accidents. Always use common sense when working on vehicles, and avoid people who do not.

Much environmental damage is caused by careless production and disposal of wastes. Wastes can take the form of liquids, solids, or gases. Anyone or any shop can be a cause of pollution. The two main ways in which an automotive shop can cause environmental damage are carelessly disposing of wastes, and repairing vehicles in such a way that they pollute the atmosphere.

Environmental rules should always be followed to prevent damage to the air, water, or soil. In many cases, proper disposal of wastes and proper vehicle repairs are required by federal and state law.

Review Questions—Chapter 2

Please do not write in this text. Write your answers on a separate sheet of paper.

1. Accidents are likely to happen when the technician becomes rushed and _____.

2. Brake dust contains a cancer causing material called _____.

3. Name some methods of reducing exposure to brake dust.

4. One of two factors, or a combination of the two causes most accidents. Name them.

5. Is a HEPA filter a wet or dry brake dust cleaner?

6. A can of brake cleaner is a type of _____ cleaner.

7. Why should the technician keep workbenches clean?

8. Give some examples of chemicals that can be found in an automotive shop.

9. A rough service lightbulb should be used in _____.

10. Match the tool or shop equipment with the potential safety hazard.

Electrical outlet _____	(A) Dull tip
Grinder _____	(B) Low oil
Test equipment _____	(C) Damaged leads
Chisel _____	(D) Overloaded
	(E) Guards missing

11. Describe what could happen if you wear loose clothing or do not secure long hair.

12. Speed is never as important as doing the job _____.

13. If the technician runs an engine in a closed area, _____ will build up quickly.

14. *True or False.* Pour liquid wastes into drains, not on the ground.

15. *True or False.* Installing a non-stock air cleaner can affect vehicle emissions.

ASE Certification-Type Questions

1. Technician A says that carcinogens are often present in brake dust. Technician B says that carcinogens are often present in clutches. Who is right?
   (A) A only.
   (B) B only.
   (C) Both A & B.
   (D) Neither A nor B.

2. Technician A says that the one way to prevent accidents is to avoid people who are careless. Technician B says that one way to prevent accidents is to keep an orderly workplace. Who is right?
   (A) A only.
   (B) B only.
   (C) Both A & B.
   (D) Neither A nor B.

3. A filter mask is a type of _____ protection.
   (A) eye
   (B) hand
   (C) respiratory
   (D) foot

4. All of the following statements about HEPA filters are true, EXCEPT:
   (A) a HEPA vacuum has one to two coarse filters ahead of the HEPA filter.
   (B) HEPA stands for High Efficiency Particulate Air.
   (C) the HEPA filter can trap all small particles except asbestos particles.
   (D) HEPA filters are the only recognized type of vacuum for brake cleaning.

5. A brake lathe is an example of which of the following?
   (A) A high current draw tool.
   (B) A tool which should be studied before using.
   (C) A tool which can injure you.
   (D) All of the above.

6. Material Safety Data Sheets are provided for all dangerous _____.
   (A) procedures
   (B) tools
   (C) chemicals
   (D) working conditions

7. If all tools and equipment are returned to their proper storage places, which of the following will occur?
   (A) They cannot be tripped over.
   (B) They will be easier to steal.
   (C) They will be hard to find the next time.
   (D) All of the above.

8. When should you use a 12-point socket with an impact wrench?
   (A) When an impact socket is not available.
   (B) When the bolt to be removed is not very tight.
   (C) When the air supply to the impact wrench is weak.
   (D) Never.

9. Which of the following is likely to happen if a standard lightbulb is used in place of a "rough service" bulb in a droplight?
   (A) Broken bulb.
   (B) Burns.
   (C) Both A & B.
   (D) None of the above.

10. Many brake parts can be:
    (A) recycled.
    (B) rebuilt.
    (C) glued back together.
    (D) Both A & B.

11. Spilled ethylene glycol antifreeze can cause all of the following, EXCEPT:
    (A) slipping.
    (B) poisoning.
    (C) fires.
    (D) environmental damage.

12. Carbon monoxide can cause all of the following, EXCEPT:
    (A) death from asphyxiation.
    (B) long-term respiratory system damage.
    (C) damage to paint and rubber materials.
    (D) environmental damage.

13. An automotive shop can cause environmental damage by _____.
    (A) carelessly disposing of wastes
    (B) repairing vehicles so that they pollute the atmosphere
    (C) discharging refrigerants into the atmosphere
    (D) All of the above.

14. Technician A says that R-134a can be discharged into the atmosphere, but not R-12. Technician B says both R-134a and R-12 can be safely discharged into the atmosphere. Who is right?
    (A) A only.
    (B) B only.
    (C) Both A & B.
    (D) Neither A nor B.

15. The _____ conducts sting operations to catch polluters.
    (A) EPA
    (B) ASE
    (C) DOT
    (D) DOE

Chapter 3

Brake Tools, Shop Equipment, and Service Information

After studying this chapter, you will be able to:

❑ Identify common specialty brake tools and explain their use.
❑ Identify brake system measuring tools and explain their use.
❑ Identify power tools and equipment used in brake service and explain their use.
❑ List general rules for the correct use and storage of tools.
❑ Identify tool safety rules where they apply.
❑ Select the proper tool for the job at hand.

Important Terms

Flare-nut wrenches
Bleeder wrenches
Tubing cutters
Snap ring pliers
Brake shoe adjuster
Spring removal tools
C-clamp
Brake cylinder hones
Filter mask
Feeler gauges

Micrometer
Shoe and drum gauge
Pedal effort gauge
Seal and dust boot installers
Pushrod height gauge
Breakout box
Scan tool
High Efficiency Particulate
  Air (HEPA) Vacuum
Brake cleaner

Wet cleaning station
Brake lathes
Brake rotor burnisher
Drum grinder
Brake shoe grinder
Hydraulic press
Pressure bleeder
Vacuum bleeder
Service information

This chapter will identify and explain the purposes of tools needed to perform brake diagnosis and service. As with any area of automotive service, the right tools are vital to doing the job right. This chapter will emphasize the special tools needed for brake service. Specific brake system measuring tools and test equipment will also be covered. Many common hand and air tools are also needed as part of brake service, and will be discussed when they are used.

The Importance of Quality Tools

The brake service technician must be aware of the large number of different tools needed to service modern brake systems. Having the correct tools for the job will make for quality repairs, faster service time, decreased comebacks, and overall customer satisfaction.

You should select the best tools possible. Quality tools will pay for themselves in many ways. Most quality hand tools come with a lifetime warranty from the manufacturer. While they may be more expensive, a quality tool will last longer than a less expensive tool. Usually, inexpensive tools do not have the same warranty as quality tools. If the less expensive tool breaks, you will have to buy a new one rather than simply turning it in for a replacement covered under a manufacturer's warranty.

Basic Brake Hand Tools

The following sections cover the basic tools that a technician must have to get started in brake service. The tools covered here will be discussed again in chapters where they apply.

Wrenches

Typical wrenches used in brake work include flare-nut wrenches and bleeder wrenches. **Flare-nut wrenches,** sometimes called tubing or *line wrenches,* are used to remove tubing fittings used on all brake line connections, **Figure 3-1.** They are also used to loosen fuel line and other soft fittings. **Bleeder wrenches** are used to open the bleeder screws found at various points throughout the brake system. Standard wrenches can be substituted for some bleeder wrenches.

Tubing Tools

Occasionally, you may be called on to replace or repair brake system tubing. To do this, tubing tools are needed. Typical tubing repair tools include **tubing cutters** and **reamers** to remove burrs from inside the cut tubing. For bending tubing, **bender springs** and

mechanical benders are needed. Brake tubing ends must always be flared using *double-lap* or *I.S.O.* (International Standards Organization) flares. See **Figure 3-2** for an illustration of typical tubing tools.

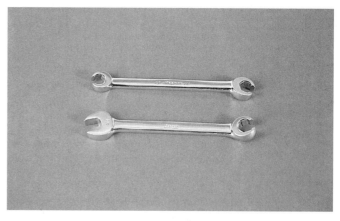

Figure 3-1. *Flare nut tubing wrenches. They usually come as a combination wrench or with an open-end.*

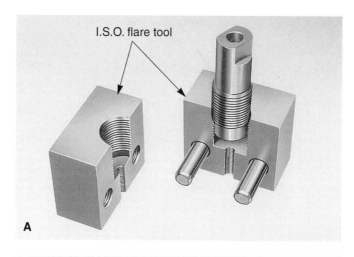

I.S.O. flare tool

A

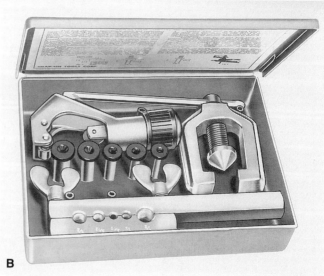

B

Figure 3-2. *Selection of tubing flare tools. A—I.S.O. flare tool. B—Single and double-lap flare tool set and storage case. (Kent-Moore & Snap-On)*

Brake Pliers

Brake service makes use of many designs and sizes of pliers. During many repair operations, various types and sizes of pliers are sometimes needed. In addition to the standard types, several kinds of specialty pliers may be needed. Some common specialty pliers used in brake repair are used to remove caliper piston and wheel cylinder retainers. **Figure 3-3** illustrates a wheel cylinder retainer remover. *Snap ring pliers* are also necessary to remove the snap rings used on some master cylinders and anti-lock brake hydraulic actuators.

Drum Brake Tools

Drum brakes must be periodically adjusted to compensate for wear. Although the self-adjusters will compensate for some wear, they are often not activated enough to maintain proper shoe-to-drum contact. The **brake shoe adjuster,** often called a *brake spoon,* is used to adjust the shoes. There are many types of adjuster spoons, shaped in different ways for use with different shoe and backing plate designs. They all perform the same basic job of turning the brake adjuster located between the brake shoes, **Figure 3-4.**

Spring removal tools are used to remove springs on drum brakes. The two major kinds of drum brake springs are shoe hold-down springs and shoe retracting springs. Each type requires a special tool, and different vehicles may require slightly different variations of the same basic tool. Other spring tools are used to retract parking brake tensioning springs, **Figure 3-4.**

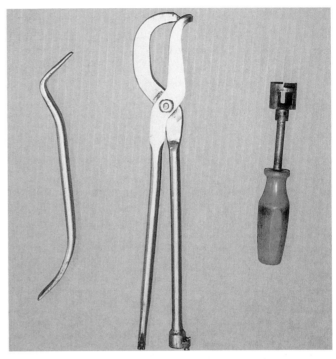

Figure 3-4. *Assortment of drum brake tools, brake spoon, brake pliers, and hold down spring tool.*

Allen and Torx® Bits

These tools are used to remove fasteners found on disc brake calipers. Instead of using standard hex head bolts, many calipers have attaching bolts with openings recessed into the bolt head. The two major types of

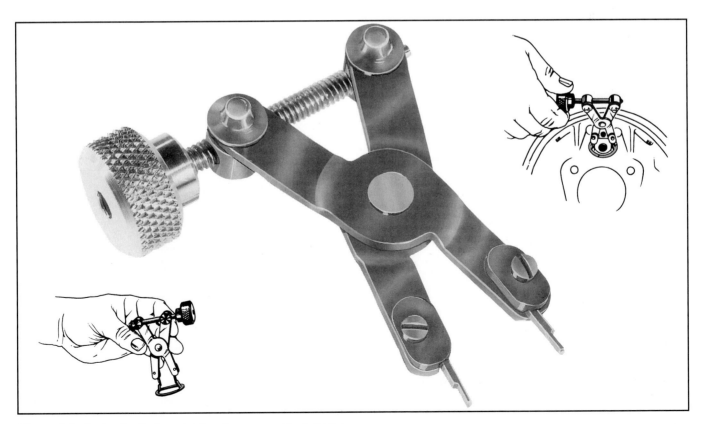

Figure 3-3. *A wheel cylinder retaining ring removal tool. (Lisle)*

recessed head bolts are the **Allen** and the **Torx®**. Special tool bits are needed to loosen and remove these screws. In addition, many other fasteners in the brake system have Allen or Torx® screws. As a brake technician, you will require an entire set of these bits. Typical Allen and Torx® bits are shown in **Figure 3-5.**

Some systems are equipped with a variation on the Torx® fastener. This fastener is referred to as a *tamper-proof Torx®*. Tamper-proof Torx® fasteners look like the standard Torx®, however, they have an additional raised point in the center. This prevents removal using a standard Torx® bit, **Figure 3-5.** These are used on some anti-lock brake systems.

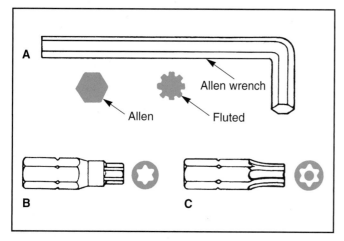

Figure 3-5. *A—Standard and fluted Allen wrench. B—Torx® bit. C—Tamper-proof Torx® bit. (Vaco Tools)*

C-clamps

A large **C-clamp,** such as the ones commonly found in hardware stores, is useful for pressing caliper pistons into the caliper. Other clamps can be used to press out rivets, and to remove and install pins at various locations in the brake system. A typical clamp is shown in **Figure 3-6.**

Figure 3-6. *A C-clamp can be used to press caliper pistons back into their bores.*

Brake Cylinder Hones

Brake cylinder hones are used to remove glaze and deposits from wheel cylinders, master cylinders, and disc brake caliper bores. The two general types of hones are the **rigid stone hone, Figure 3-7,** and the **flexible hone.**

The rigid stone hone has 2 or 3 spring-loaded fingers holding one stone each. The stones can be short, for use in caliper bores, or long for use in wheel and master cylinders. The total diameter of the hone is adjustable within a certain size range. Various hone sizes are available for different bore diameters. The flexible hone contains abrasive balls on the ends of flexible wires. Flexible hones are also available in different sizes.

> **Note: Not all cylinders can be honed. Some cylinders are coated or anodized and must be replaced instead of repaired. If any cylinder is heavily pitted or cannot be cleaned up by honing, it should be replaced.**

Both types of hones are operated by a portable drill, either electric or air operated. The hone is chucked into the drill, then inserted into the cylinder to be honed. The drill is started and hone rotation removes deposits and

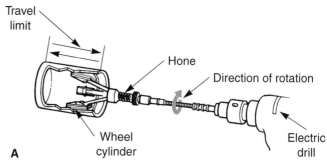

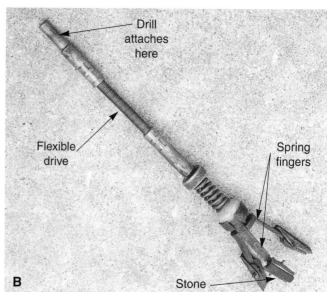

Figure 3-7. *One type of flexible brake cylinder hone. (Ammco & Niehoff)*

glazing. A small amount of brake fluid should be used as a lubricant. Do not exceed recommended hone rpm, if given. Do not allow the hone to leave the bore while it is still spinning as it can fly apart violently or be damaged.

Respiratory Protection

As discussed in Chapter 2, brake dust is a constant danger in the shop. You should have a **filter mask, Figure 3-8,** equipped with HEPA filters. Do not rely on painters' paper masks, since they will not protect against the entry of extremely small asbestos particles.

Caring for Filter Mask

Like any other tool, a filter mask must be cared for to be effective. Inspect the mask on a regular basis. Check for stretched or damaged head bands, torn or cracked face-piece, torn valve flaps and damaged filter housings. The mask should be disassembled and cleaned in a mixture of antibacterial soap and water. Replace the filters at every cleaning and anytime you notice increased resistance when you inhale.

Brake System Measuring Tools

Brake system repairs often call for making measurements. There are numerous specialty measuring tools. These include micrometers, shoe-drum gauges, dial indicators, and pressure checking devices.

Feeler Gauges

Gap gauges known as **feeler gauges** are thin strips of steel or brass, available in various thicknesses. These thicknesses are measured in thousandths of an inch, or fractions of a millimeter. Modern feeler gauges have both thousandths of an inch and millimeter markings. One thousandth of an inch, .001", is about .025 millimeters.

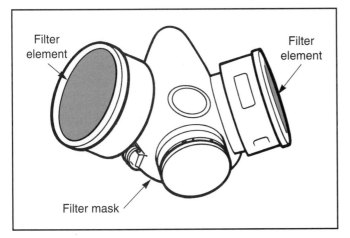

Figure 3-8. *Always wear an approved filter mask when working with parts containing asbestos. (TRW Inc.)*

Brass feeler gauges are used to set clearance on parts containing magnets, such as speed sensors. Brass feeler gauges resemble steel gauges, but can be identified by their distinct golden yellow color. Most brass gauge set sizes range from around .004" -.015" (.10-.38 mm). Feeler gauge strips are usually sold in sets to cover a range of uses. See **Figure 3-9.**

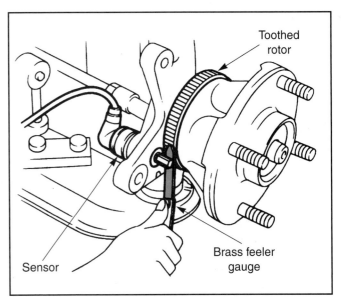

Figure 3-9. *Feeler gauges are needed for checking sensor clearance and other tolerances. Make sure you have both steel and brass gauges.*

Rotor and Drum Micrometers

The **micrometer** is a basic measuring tool for many automotive operations. Micrometers are used during brake service to measure rotor thickness and brake drum diameter. These measurements are extremely important, since one of the primary jobs of brake rotors and drums is to absorb heat. If the rotor or drum becomes too thin, it can distort, which may lead to overheating. Measuring with a micrometer will let you know whether the drum or rotor can be turned or must be replaced. **Figure 3-10** shows typical drum and rotor micrometers. Many shops are now equipped with electronic micrometers, **Figure 3-11.**

Using Micrometers

To use a micrometer correctly, you must first know how it is constructed. The following sections cover the design of both inch and metric micrometers.

Inch Micrometers

The barrel, or sleeve, of an inch micrometer is graduated in two ways. The numbers represent multiples of .100". The lines between the numbers represent .025". The thimble is also graduated in two ways. Each number represents a multiple of .005". The lines between the numbers represent .001". Refer to **Figure 3-12.**

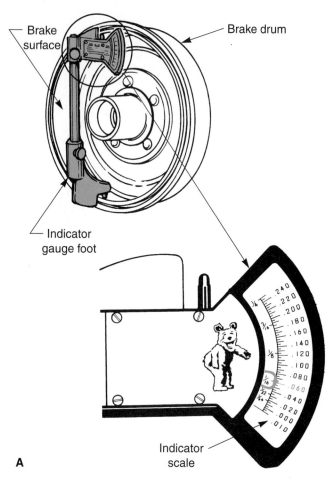

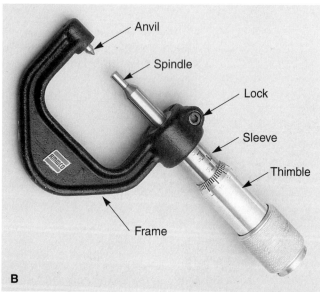

Figure 3-10. *A—A brake drum micrometer. B—Brake rotor micrometer. Handle these precision instruments with care. (Bear & Ammco)*

Metric Micrometers

The numbers on the barrel of a metric micrometer are read directly in millimeters. The lines between the numbers each represent one millimeter. Note the lines at the top and bottom of the centerline are placed at half

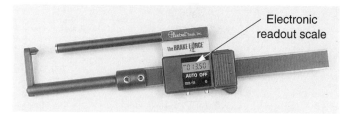

Figure 3-11. *A disc brake rotor electronic micrometer. This tool provides both inch and millimeter readouts, and will turn off automatically. (Central Tools)*

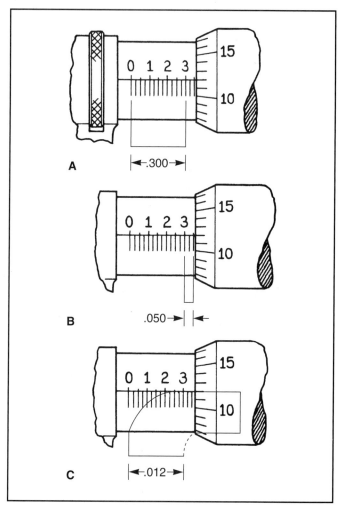

Figure 3-12. *Four steps in reading an inch micrometer. A—The first reading is .300". B—The second reading is .050". The third reading is .012". This totals .362, which is added to the size of the micrometer being used to get the final reading.*

intervals. This allows the technician to measure half millimeters. The numbers on the thimble are graduated in tenths of a millimeter. Each number is a multiple of .1 millimeter. Each line is equal to .1 millimeter. **Figure 3-13** illustrates a metric micrometer's scale.

Reading the Micrometer

Place the contact points against the part to be measured. Then turn the thimble until the anvils lightly contact

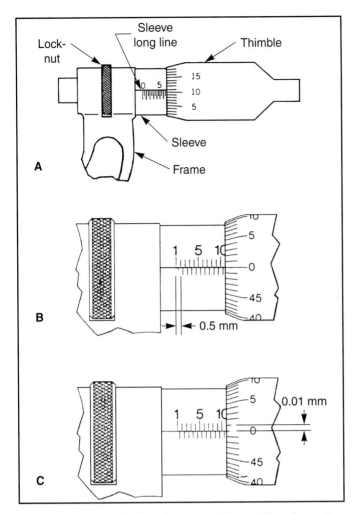

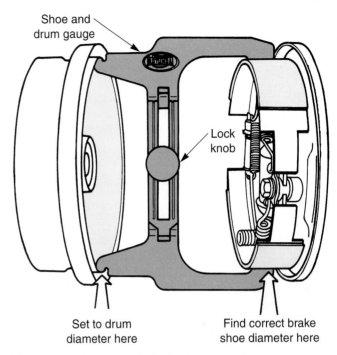

Figure 3-14. *One type of a brake drum and shoe gauge. (Ford)*

reduces the need to adjust the shoes after the drum is installed. To use the gauge, set the outer adjuster pointers to the maximum diameter of the drum. Then use the inner pointers to adjust the shoes.

Dial Indicator

Dial indicators are useful for checking for loose parts, and for out of round rotors. A dial indicator is shown in **Figure 3-15.** To check a rotor, mount the dial indicator on

Figure 3-13. *A—Scale of an outside metric micrometer. B—This micrometer requires two full revolutions of the thimble to equal 1 mm. C—It is read in a similar manner to an inch micrometer. (TRW, Deere & Co.)*

the part. Do not tighten the micrometer as this will cause a false reading and damage the micrometer. Once this is done, the micrometer can be read.

To read an inch micrometer, start by reading the numbers and lines on the barrel. Remember that each number is a multiple of .100″ and that each line represents .25″. If necessary, write down the reading, being sure to include the decimal point in the proper place. Then read the thimble numbers, remembering that each number is a multiple of .005″ and each line represents .001″.

To read a metric micrometer, read the numbers and lines on the barrel. Each number is a multiple of 1 millimeter. Also note that each line is one millimeter, and the spaces between upper and lower lines is .5 millimeter. Write down this reading. Next, read the thimble numbers. Each number represents a multiple of .05 millimeter, and each line equals .01 millimeter.

Shoe and Drum Gauge

The *shoe and drum gauge,* **Figure 3-14,** is used to adjust the shoes before the drum is reinstalled. This

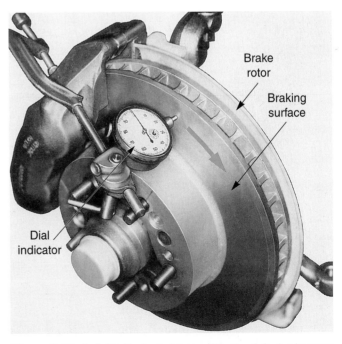

Figure 3-15. *A dial indicator being used to check for brake rotor warpage. (Chevrolet)*

the rotor surface and set it to zero. Then slowly turn the rotor. The indicator pointer will move as the rotor turns. If the pointer moves excessively, the rotor is warped.

Pedal Effort Gauge

The *pedal effort gauge,* **Figure 3-16,** is used to measure the amount of foot pressure needed to apply the brakes. The gauge is attached to the pedal, and the pedal is then pressed. Pedal pressure can be read directly on the gauge.

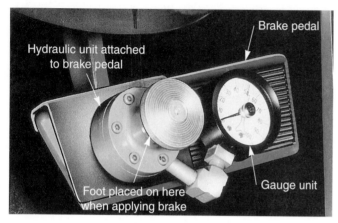

Figure 3-16. *A brake pedal effort pressure gauge. (Ford)*

Hydraulic Pressure Gauge

Hydraulic pressure gauges are often needed to measure the pressure produced by the master cylinder, and to check for internal system restrictions and leaks. Pressure gauges are also needed when working on anti-lock brake systems. Since brake system pressures can exceed 2500 psi (17 258 kPa) brake pressure gauges must be capable of reading a minimum of 4000 psi (27 580 kPa). **Figure 3-17** shows one type of hydraulic pressure gauge.

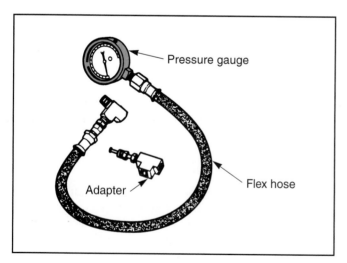

Figure 3-17. *Hydraulic brake pressure gauge and adapter. Because of high pressures involved, safety glasses must be worn. (Chrysler)*

Specialized Brake Tools

In addition to the tools above, many special brake tools make brake service easier. Some of the more common types are explained below.

Seal and Dust Boot Installers

Seal and dust boot installers, **Figure 3-18,** are useful when a caliper or wheel cylinder must be rebuilt, or when a wheel bearing grease seal must be installed. The installer will seat the new part without damaging it. Most installers are used by first placing the dust boot seal onto the installer. The installer is placed over the part, and the seal is then gently tapped into position. The tool is removed, leaving the seal properly installed. There are other special seal installers used when servicing power boosters and some anti-lock brake systems.

Wheel Bearing Service Tools

Wheel bearing service tools include packers, adjusters, and race removers. Some tools are needed to remove old grease from bearings and to pack grease into the cleaned or new bearings, **Figure 3-19.** While there are specialized bearing race installers, punches and brass drifts can be used to remove and install bearing races.

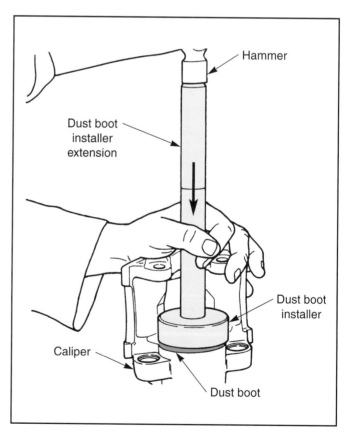

Figure 3-18. *Seating a caliper piston dust boot with the correct installation tool and hammer. (Plymouth)*

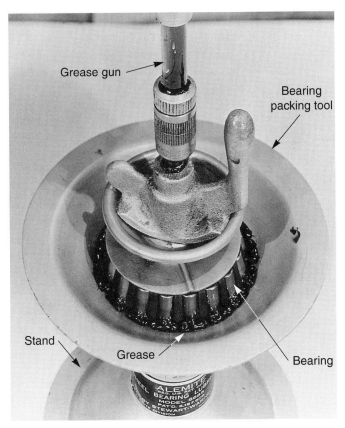

Figure 3-19. *Wheel bearings being lubricated with fresh grease while secured in one type of bearing lube fixture. (Stewart-Warner)*

Hydraulic System Tools

Special tools for use on the hydraulic system include *siphons* and *syringes*, **Figure 3-20.** These tools are useful for drawing out excess fluid from the master cylinder reservoir. In many cases, the master cylinder will become overfilled when the disc brake calipers are pushed back into their bores. The brake fluid should not be allowed to spill out the master cylinder, as it could cause damage to other engine parts or vehicle paint.

To make the job of bleeding brakes easier and neater, a clear jar and rubber hoses are used. One end of the hose is slipped over the open bleeder screw and the other end is placed in the jar, which is half full of fluid. When the brake pedal is pushed, air in the system is pumped through the hose into the jar, where it rises to the top. When the pedal is released, brake fluid is drawn into the system. **Figure 3-21** shows a hose and jar being used to bleed brakes.

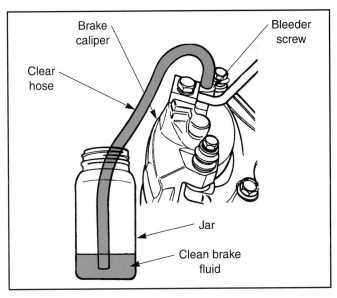

Figure 3-21. *Bleeding a brake caliper with a clear hose and jar to catch fluid. Always keep the jar end of the hose submerged in brake fluid. (Chrysler)*

Sometimes a brake system line or port must be left open while other brake service is performed. To keep water and dirt out of the system at these times, plugs are used to cap off the opening until the brake system can be reassembled.

Power Brake Service Tools

A typical tool needed to service vacuum power brakes is the *pushrod height gauge.* Other vacuum power brake tools include *spanners* which allow the two halves of the booster to be separated, and special seal removers and installers.

Special tools for use with power brakes are shown in **Figure 3-22.** These tools are needed to test, disassemble, and reassemble the power booster unit.

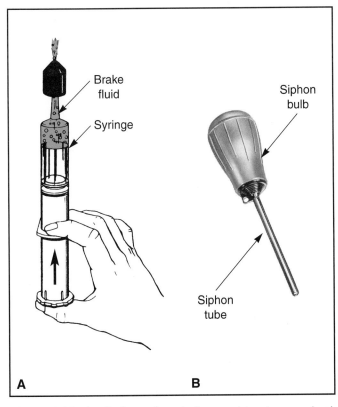

Figure 3-20. *A—Brake syringe being used to clean a check valve. B—A brake fluid siphon. (EIS & Snap-On)*

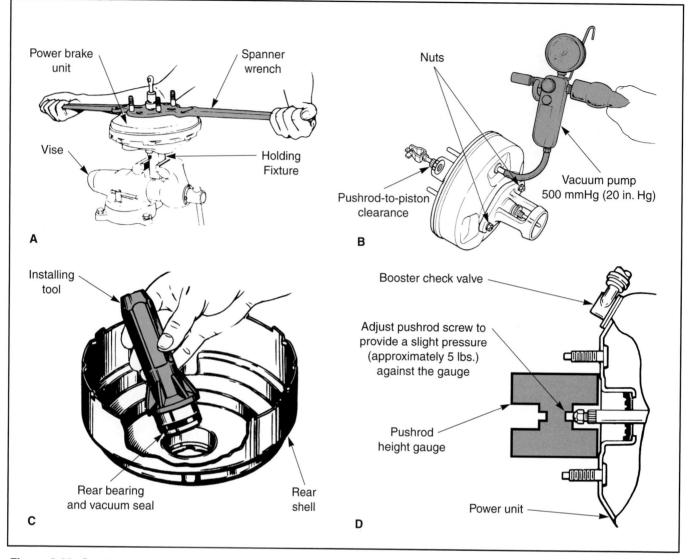

Figure 3-22. *Special power booster service tools. There are a large number of different tools. A—Spanner wrench to separate housings. B—Using a vacuum pump to check pushrod-to-piston clearance. C—Installing a bearing and vacuum seal. D—Pushrod height gauge being used to adjust pushrod. (Honda, General Motors, & Ford)*

Electrical and Electronic Testers

A few decades ago the only electrical device attached to the vehicle brake system was the brake light switch. Today, however, the brake technician needs a thorough knowledge of electrical and electronic theory, and a wide range of electrical/electronic testers. The following electrical testers are necessary to diagnose and service modern brake systems.

Jumper Wires

Jumper wires are short lengths of wire with alligator clips on each end. They are used to make temporary electrical connections during vehicle testing procedures. Jumper wires are often used to bypass electrical devices, such as relays and solenoids. Bypassing an electrical

device will enable you to determine whether it is defective. Jumper wires can also be used to access the computer memory on many vehicles by grounding the data link connector. They are useful for making an extended electrical connection, such as connecting an engine test lead to a test device inside the passenger compartment for a road test.

Although they can be purchased, jumper wires can be made easily from a length of wire and alligator clips or other metal terminals. Use copper wire that has a large enough gage to carry several amps of current, but small enough to be easily handled. 12 or 14 gage wire is a good size for most jumper wires. Cut about 2-3′ (60.96-91.44 cm) of wire from the roll, then strip about .5″ (12.7 mm) of insulation from each end of the wire. Crimp the alligator clips or terminals to the wire, making a secure joint at each end. For additional holding power and conductivity, the clips and wires can be lightly soldered together. Some examples are shown in **Figure 3-23.**

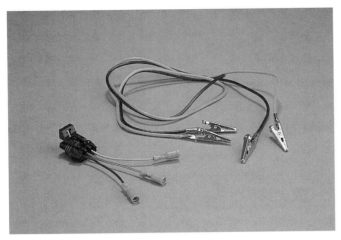

Figure 3-23. *Jumper wires can be used with meters to check sensor, actuator, and circuit properties.*

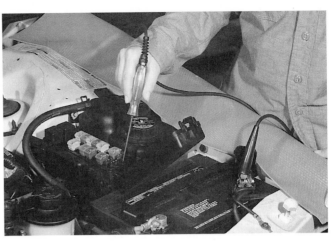

Figure 3-24. *12-volt non-powered test light. The sharp tipped probe fits easily into small areas. Never pierce wire insulation unless recommended by the manufacturer.*

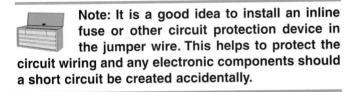

Note: It is a good idea to install an inline fuse or other circuit protection device in the jumper wire. This helps to protect the circuit wiring and any electronic components should a short circuit be created accidentally.

Another good source for jumper wires is the wiring harness in a wrecked vehicle. These can be found easily and inexpensively in most auto salvage yards. The advantage to having these is they come with the original terminals, which ensures a good connection to the device to be tested. They are handy for testing sensors and other parts that are difficult to access with a meter's test leads. Remove the terminal from the sensor or part and then cut the terminal from the harness. Strip and crimp a terminal on the wire(s) as described earlier. Additional wire can be added if the connector is used in a hard-to-reach location. These are handy if you only work on a few models, such as at a dealer or specialty shop.

Caution: When using jumper wires, make sure the current load does not exceed capacity. If the jumper wire insulation begins to discolor, blister, or smoke, remove the wire immediately using a shop towel or glove to prevent burns. Never grab a hot jumper wire with your bare hands. A severe burn could result. Never use a jumper wire to bypass a fuse or other circuit protection device.

Test Lights

Test lights are often used to check for the presence of voltage in the brake system or to determine whether a circuit is complete. The non-powered test light, **Figure 3-24,** can be used to probe electrical circuits to determine whether voltage is present. The powered test light, **Figure 3-25,** is used to determine whether a circuit is complete.

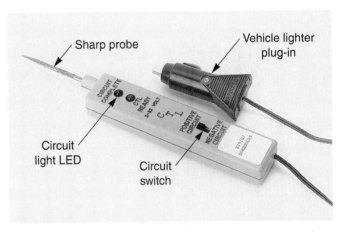

Figure 3-25. *Powered test light. When the circuit is complete, light will come on. (OTC Tools)*

Multimeters

Multimeters are combination meters capable of measuring ohms, volts, amps, and sometimes waveforms. A typical multimeter is shown in **Figure 3-26.** The multimeter should be used when exact electrical values must be determined. Typical uses for a multimeter include checking for high resistance in an electrical connection, locating small battery drains, checking the resistance of anti-lock brake wheel sensors, and determining the condition of various brake system electrical switches.

Ohmmeter

Ohmmeters are used to measure resistance in an electrical device or circuit. Resistance is the opposition to current flow that exists in any electrical circuit or device. This resistance is measured in ohms. Ohmmeters can only check an electrical circuit when no current is flowing.

An ohmmeter has two leads, which are connected in *parallel* to each side of the unit or circuit to be tested. Polarity (direction of current flow) is not important when

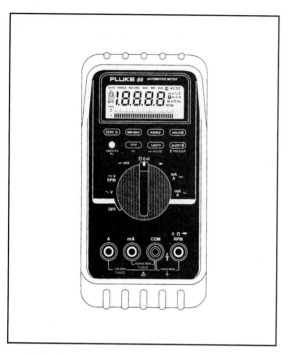

Figure 3-26. *A combination multimeter for volts, ohms, and amps. Handle with care. (Fluke)*

checking resistance, except in the case of diodes or some computer circuits. Consult the manufacturer's service manual before using an ohmmeter to check any computer control circuit.

Most ohmmeters have selector knobs for checking various ranges of resistance values. Analog ohmmeters have a special knob to adjust the needle to zero before checking resistance. The needle should always be adjusted to zero whenever the range is changed.

Voltmeter

Voltmeters are used to check voltage potential between two points in an energized circuit. The circuit must have a source of electricity available before voltage can be checked. On some voltmeters, different scales can be selected, depending on the voltage type and level to be measured.

Always observe proper polarity when attaching a voltmeter's leads. When a voltmeter is used on a modern vehicle with a negative ground, the negative lead should always be connected in *parallel* to the frame or other ground, and the positive lead to the positive part of the circuit.

Ammeter

An *ammeter* is used to check the current flow in a circuit. Ammeters can also be used to check the current draw of solenoids, motors, and other electrical devices, when they are suspected of drawing too much current. Most ammeters have two terminals for the positive lead. One is used to check low amperage, usually under 1 ampere. The other terminal is connected to the meter

through a high amperage fuse and can be used to check circuits with current loads up to 10 amperes. Unlike ohmmeters and voltmeters, the ammeter must be connected in *series* to measure current.

Heavy-duty ammeters are also used to check the amperage draw of starters or other motors, and when checking the charge level of a battery. Most ammeters used in automotive service are part of a multimeter.

Waveform Meters

The pulse and *waveform meter* is normally used to measure the control voltages applied to such devices as anti-lock brake solenoids. In most cases, the waveform frequency is controlled by the computer, and checking the waveform is a way of checking computer system operation. Abnormal waveforms or pulse rate changes in response to changes in sensor inputs indicate the input sensors, computer, and output devices are all working. Waveform meters are usually included as part of the more sophisticated digital multimeters, **Figure 3-27.**

Breakout Boxes

Early anti-lock brake systems did not have true self-diagnostics, which required technicians to use a breakout box to find the problem. A *breakout box* is simply a box

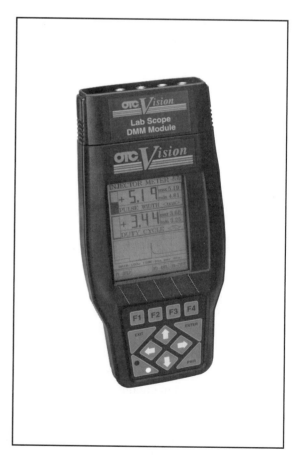

Figure 3-27. *Modern meters can read circuit waveforms and pulse signals. (OTC Tools)*

containing numbered test terminals and a harness designed to mate with a system's main wiring harness connector. Breakout boxes are designed to allow technicians to use a multimeter to check voltage and resistance values in a system's various circuits, **Figure 3-28.** The breakout box is connected to a system's harness, such as a computer wiring harness connector. The meter's terminals are then plugged into the various numbered terminals to check for proper voltage or resistance. A manual is used to outline the specific test procedures for each problem.

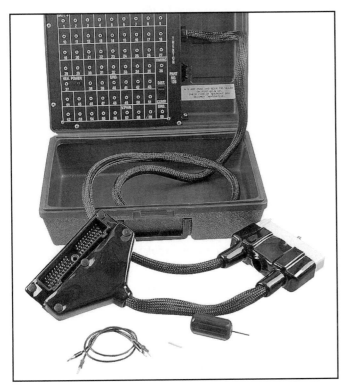

Figure 3-28. *Breakout boxes are used to diagnose early anti-lock brake systems. (OTC Tools)*

Scan Tools

The *scan tool* is used to check the operation of anti-lock brake and traction control systems. The scan tool is attached to a data link connector, usually the same connector used to retrieve engine and drivetrain information. The primary use of the scan tool in brake work is to retrieve trouble codes stored in the anti-lock brake and traction control computer. These trouble codes are stored when a malfunction occurs. Retrieving these codes before performing any tests makes the troubleshooting process much quicker and easier. Scan tools can also be used to operate parts of the anti-lock brake system, such as pumps and solenoids, to check system condition. A typical scan tool is shown in **Figure 3-29.**

The most obvious use of scan tools is to retrieve trouble codes. Many scan tools can be used to perform other diagnostic activities. For instance, scan tools can be used to access the wheel speed sensor input and output readings. These readings can be compared with factory

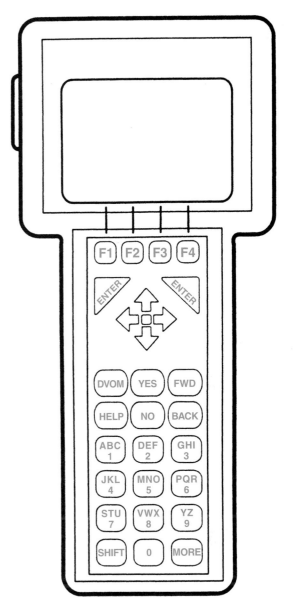

Figure 3-29. *One particular type of scan tool. Always use the proper tool and follow the troubleshooting process carefully. (Chrysler)*

specifications to determine whether or not the component is defective. The scan tool can also be used to override the control module and directly operate the hydraulic pump motor and solenoids in the hydraulic actuator. Forcing a component to operate allows you to determine the condition of the component and related wiring.

Types of Scan Tools

Some scan tools can be used with the computer control systems installed on only one manufacturer's vehicles, and cannot be used to diagnose the systems of other manufacturers. These are called *dedicated scan tools.* The earliest types of these tools could be used for trouble code retrieval only.

Multi-system and *generic scan tools* can be used to diagnose the systems of many different manufacturers. These scan tools make use of specialized cartridges and cables that are changed to match the vehicle being worked on. Some scan tools contain all internal information, and can be updated by an interface with a desktop or laptop computer or over a phone line to the manufacturer. Scan tools with this feature have a special terminal that can be connected to a phone line.

Most of the latest scan tools are **bi-directional.** A bi-directional scan tool is able to retrieve information from the vehicle computer, and can send data into the computer. This ability to transfer data two ways allows the scan tool to operate anti-lock brake and traction control components for diagnosis, and to reprogram the computer if updates are necessary.

Scan Tool Construction

Scan tools are small and light enough to be hand-held. All scan tools have a **keypad** with a set of function keys used to select various functions, **Figure 3-30.** Typical function keys include enter and exit keys, arrow keys, and number keys. Many scan tools have specialized keys to access specific diagnosis areas, or to interface with a remote computer or printer.

Most scan tools have a **liquid crystal display (LCD)** that can vary in definition, size, and shape. All the information generated by the scan tool, as well as menus and command functions, are displayed on this screen. Some scan tools have a light emitting diode (LED) or low-definition liquid crystal display. Newer scan tools have high-definition LCD displays that can show graphs and waveforms, just like an engine analyzer's oscilloscope.

The display screen is used to display a **menu** which shows all possible scan tool functions, often called **diagnostic routines.** The technician can pick one of these to begin using the scan tool. Some scan tools have indicator lights to tell the technician that the scan tool is operating, what mode the system is operating in, or to indicate an obvious system defect.

Various cables and adapters are provided with the scan tool to allow it to be attached to one or more data link connectors. Some scan tools also have battery clips or a cigarette lighter adapter to allow them to use the vehicle battery as a power source. Typical cables and adapters are shown in **Figure 3-31.** A few scan tools have an internal backup battery. Depending on the type of scan tool, it can have a variety of terminals to connect to other external devices, such as computerized analyzers.

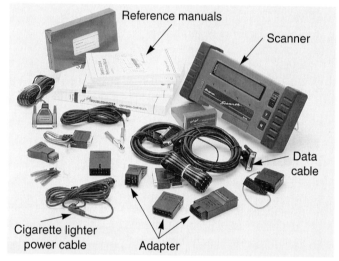

Figure 3-31. *Scan tools require many cables and adapters. (Snap-On Tools)*

Scan Tool Cartridges and Software

The scan tool, like a desktop computer, must have software in order to function. As manufacturers included greater self-diagnostic capability on more vehicle systems, technicians needed additional tools that could monitor and test all these systems. Tool manufacturers began offering scan tools that used small software cartridges often no larger than a deck of cards.

While these program cartridges allowed scan tools to monitor and control more system functions, there were some drawbacks. As new vehicle models came out or if a mid-year change was made to a vehicle system, the old program cartridges could not be updated and had to be replaced.

Scan tool manufacturers finally offered a solution to this problem with generic cartridges, sometimes referred to as **mass storage cartridges.** Mass storage cartridges allow the user to update the cartridge. This is done by connecting the scan tool to a computerized analyzer or personal computer (PC) and downloading an updated program from a CD-ROM disk to the cartridge through the scan tool.

The latest scan tools have self-contained programming on **personal computer memory card industry association (PCMCIA)** cards, that are about the size of a charge card. No external program cartridges are needed

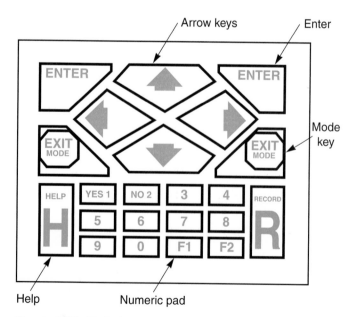

Figure 3-30. *Typical scan tool keypad. Some scan tools may use different keys and configurations. (OTC Tools)*

for the scan tool to operate. Most of these scan tools also have slots for one or more PCMCIA cards, **Figure 3-32.** This type of scan tool can be updated by downloading new information as needed from a personal computer or computerized analyzer.

Rules for Scan Tools Usage

The following are several hints for using scan tools properly and safely. A scan tool may look fairly harmless, but may cause damage or create a very dangerous situation if it is used or placed in the vehicle improperly.

❑ Make sure the cables are located away from the accelerator, brake pedal, and all moving parts.

❑ Do not shut off the ignition key unless prompted to do so by the scan tool. Any scanned information may be lost.

❑ Do not disconnect the scan tool cable from the data link connector during any programming or learn procedure. Damage to the vehicle's ECM may result.

❑ Do not remove the program cartridge or card from the scan tool while power is on.

❑ Do not expose the scan tool to excessive or direct sunlight for extended periods as it can damage the display screen.

❑ Do not hang the scan tool from the rear view mirror. This will place it in a location where the cables or the scan tool itself may become caught in the steering wheel. The tool's weight may pull the rear view mirror off of its mounting, possibly damaging the mirror or windshield.

❑ If the data link connector is located under the hood, use an extension cable that allows the scan tool to be located inside the passenger compartment.

❑ Be aware of any special operating conditions that exist while the scan tool is connected to the data link connector. The scan tool will usually warn you if certain system or vehicle functions will be disabled.

❑ Do not drive and monitor scan tool functions. You will not be able to devote full attention to either the scan tool display or the road. If it is necessary to monitor scan tool functions during a road test, have someone (not the vehicle's owner) drive the vehicle for you.

❑ Do not perform any actuator or scan tool tests while driving the vehicle. Doing so may disable the brake system.

❑ Do not perform any system learn functions during a road test.

❑ If you must leave the vehicle for more than a few minutes, disconnect the scan tool from battery power. Scan tools draw a great amount of power and can drain the battery if the engine is not running.

When you are finished using the scan tool, put it back in its case. Remove any external program cartridges, cables, and any adapters and place them in their respective compartments. Close the case and return it to the tool room or location where it is normally stored. This way, you or any other technician in the shop can easily find the scan tool the next time it is needed.

Brake System Cleaning Tools

For efficient brake work and personal safety, proper cleaning cannot be overemphasized. Not only is the finished job better if every related part is clean, the hazards of brake dust are reduced by safely removing as much of it as possible at the start of the repair process.

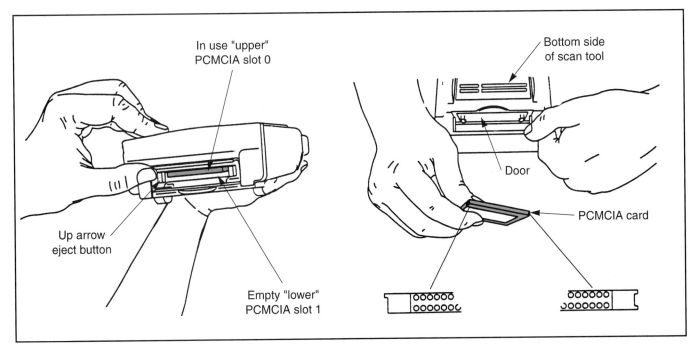

Figure 3-32. *Newer scan tools use PCMCIA cards, which slide into the scan tool and can be reprogrammed. (General Motors)*

Vacuum Cleaning Systems

Many shops use dry cleaning systems to remove brake dust from brake assemblies. These systems include standard **shop vacuum** and high efficiency particulate air (HEPA) vacuum systems. The traditional **shop vacuum, Figure 3-33,** operates in the same manner as a conventional household vacuum cleaner. Due to the extremely small size of asbestos particles, they can escape from the relatively inefficient filters used with the typical shop vacuum. Therefore, shop vacuums are no longer recommended for brake cleaning. However, most of these vacuum cleaners are wet and dry systems, and are useful for removing spills, or vacuuming up water after the shop is hosed down.

The **high efficiency particulate air (HEPA) vacuum** cleaner system, **Figure 3-34,** uses a filter called a HEPA filter. The HEPA filter can trap extremely small particles, including asbestos particles. To keep the HEPA filter element from quickly becoming clogged, the vacuum usually has one or two coarse filter elements which trap large particles before they reach the HEPA filter. The HEPA filter equipped vacuum is the only recognized method of vacuum cleaning brake parts. Before using a HEPA vacuum cleaner, check the filter condition and check all hoses for tightness.

Figure 3-34. *HEPA vacuum filtration system specially designed for cleaning brakes. (Nilfisk of America)*

Wet Cleaning Systems

The simplest type of wet cleaning system is the can of **brake cleaner.** The spray can is pointed at the brake assembly to be cleaned, and the brake dust is washed from the unit. For best results, spraying gently from top to bottom, wet down the assembly first, then use a stronger spray to wash the dust from the parts. Place a shop rag or waste cloth under the wheel and allow the solution to drip into it.

The **wet cleaning station, Figure 3-35,** uses water mixed with a solvent or detergent. Air pressure is used to spray the solution onto the wheel brake parts. After the assembly has been thoroughly wet down, the spray force can be increased to wash the dust from the wheel. The solution will collect in the pan where it can be disposed. This type of wet cleaner collects more dust than spray can cleaners.

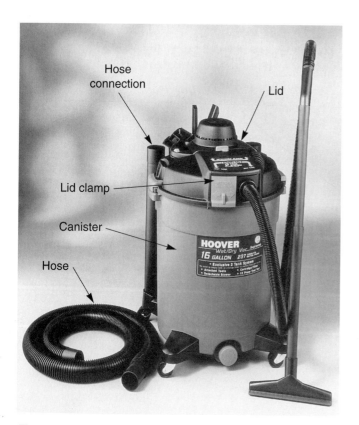

Figure 3-33. *A portable shop vacuum for wet or dry clean-up. (Hoover)*

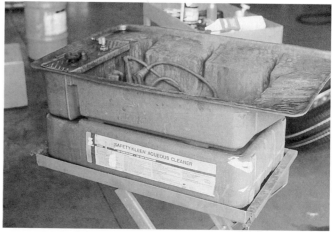

Figure 3-35. *A brake wet cleaning station. The cleaning solution is a mixture of water and organic solvent.*

Warning: The dust collected by any of these methods contains asbestos and should be treated as hazardous waste. Used shop rags, filters, and contaminated cleaning materials should be placed in a closed container and disposed of along with other hazardous waste.

Brake Service Equipment

The following equipment is needed to perform a complete repair to the brake system. These pieces of equipment represent a large investment, but must be available if brake service is to be performed correctly. These tools are considered shop equipment and are not normally purchased by the technician.

Brake Lathes

Brake lathes are used to remove the glazed and scored outer layer from rotors and drums. This removal process is usually called *turning.* There are two main divisions of disc and drum brake lathes, bench mounted and vehicle mounted.

Bench Mounted Brake Lathe

The most common type of brake lathe is the *bench mounted lathe.* The modern bench lathe is usually designed to turn both disc and drum brakes. Many modern bench lathes consist of two separate lathes so that both disc and drum brakes can be turned at the same time. The bench lathe can be permanently affixed to the shop floor, or can be equipped with wheels to allow positioning closer to the vehicle being repaired. A cutaway view of a bench mounted brake lathe is shown in **Figure 3-36.**

These lathes are equipped with cutting bits, attached to a boring bar. The bits are carefully positioned to remove small amounts of metal as the drum or rotor is turned. Drums are turned using a single bit, while rotors are turned using two bits which cut both sides of the rotor in one operation. A special cutter assembly is used to cut rotors.

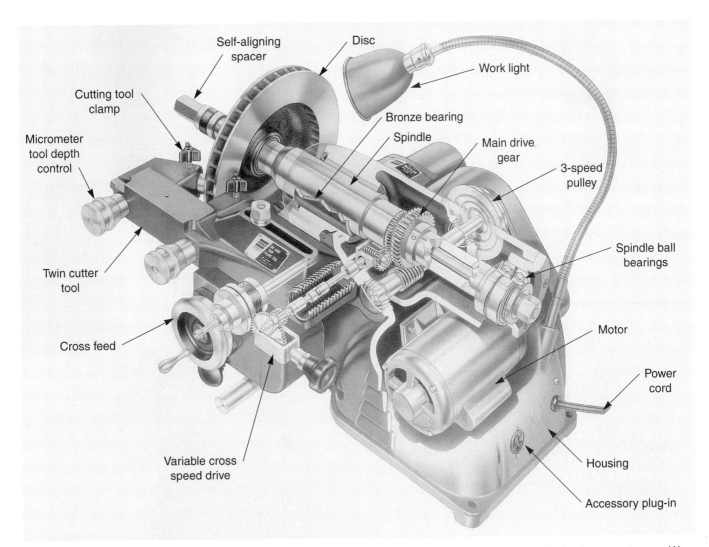

Figure 3-36. *Cutaway view of one particular type of a disc brake lathe. This unit cuts both sides of the brake rotor at once. Wear your safety glasses. (Ammco)*

Vehicle Mounted Brake Lathe

On some front wheel drive vehicles, the rotor assembly is pressed to the CV axle shaft, or for other reasons removing the rotor to turn it on a bench lathe would be very time-consuming. For these types of rotors, the *vehicle mounted brake lathe* is used. This lathe assembly bolts to the caliper mounting and the two bits are adjusted. The disc is turned by vehicle power if the rotor is installed on a drive axle, or by an electric motor drive if the rotor is installed on a non-driving axle. One type of vehicle mounted disc lathe is shown in **Figure 3-37.**

Figure 3-37. *An on-vehicle disc brake lathe being used to turn a rotor on the front axle of a four-wheel drive pickup. (Hunter)*

Brake Rotor Burnisher

The **brake rotor burnisher** is designed to burnish, or polish, the rotor braking surface after turning. Burnishing smooths the braking surface finish to manufacturers, specifications. The tool can be solidly attached to the lathe, or hand-held. Burnishing tool can be equipped with stones or abrasive discs. Some burnishing tools are operated by a hand drill. A typical brake burnisher is shown in **Figure 3-38.**

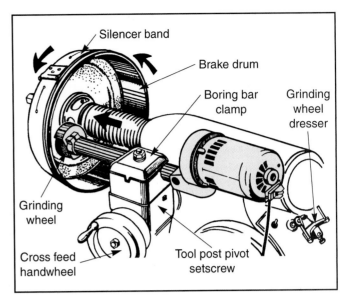

Figure 3-39. *One type of electrically driven brake drum grinder. For a proper finish, keep the stone dressed. Note the stone dresser. (Ammco)*

 Warning: If you must arc grind a set of brake shoes, be sure the vacuum motor and HEPA filters are operating and wear a respirator.

Hydraulic Press

It is almost impossible to do brake work without some sort of **hydraulic press.** Presses are used to replace broken studs, remove hubs from bearings, install brake shoe retaining pins, and many other jobs. The hydraulic press is a common piece of equipment found in most shops. **Figure 3-40** shows a common make of hydraulic press. To make the best use of a hydraulic press, it must be equipped with various adapters.

Pressure Bleeding Equipment

The **pressure bleeder** is designed to remove air and old fluid from the system. Pressure bleeders eliminate the need for a helper to pump the brakes, and the bleeding procedure takes less time. **Figure 3-41** depicts one type of pressure bleeder. To use the bleeder, fill it with new, clean brake fluid.

 Warning: Never mix glycol (DOT 3 and 4) and silicone (DOT 5) fluids.

Install the proper adapter to the master cylinder. After the tank is filled with fluid, use shop air to pressurize the tank, usually to about 15-20 psi (103-138 kPa). Then attach the bleeder using the proper adapters and open the fluid valves. The vehicle can then be bled normally.

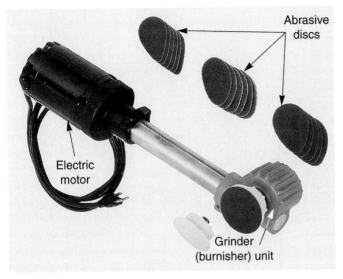

Figure 3-38. *One particular type of an electric brake rotor burnishing tool. Note the abrasive discs. (Ammco)*

Drum Grinder

A precision **drum grinder** is sometimes used to obtain a final finish on brake drums. The high-speed stone wheel is installed in place of the boring bar and bit. The wheel is turned by a high speed electric motor. The grinding wheel passes over the braking surface as the drum rotates, producing a very even finish. The grinder is often used to remove any hard spots, which are raised areas that cannot be removed with the regular cutting bits. A drum grinder is illustrated in **Figure 3-39.**

Brake Shoe Grinders

Some older shops may be equipped with a device known as a **brake shoe grinder** or *arc grinder*. Brake shoe grinders were used when individual shops relined brake shoes. These machines changed the arc, or radius, of a brake shoe to match the drum. When the shoe radius closely matched the drum radius, the brake shoes would wear in more quickly. However, almost all shops have stopped arc grinding brake shoes. Today, brake shoes are designed so the edges of the lining material are lower than the top, allowing the shoe to better wear to the drum. Also, the danger of asbestos particles makes shoe grinding a hazardous operation. If a brake shop has one of these grinders, it probably has not been used in many years.

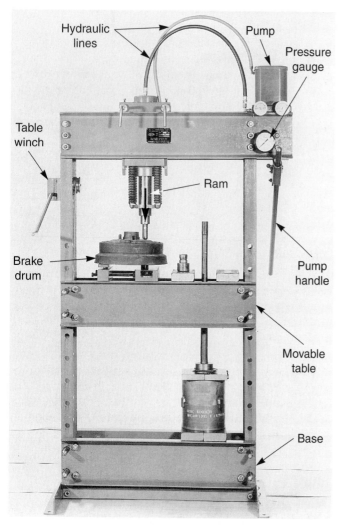

Figure 3-40. *A manually operated, floor mounted, hydraulic press setup to remove a broken lug bolt from a brake drum. (Walker)*

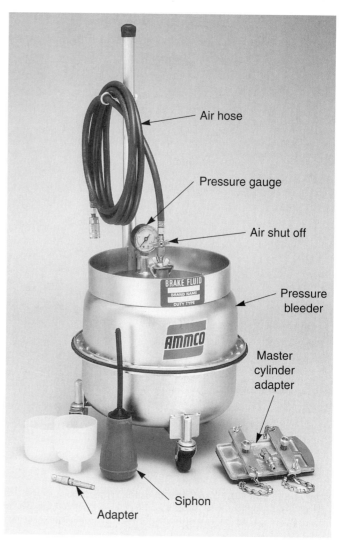

Figure 3-41. *An air powered, pressurized brake bleeder with several adapters and tools. (Ammco)*

Some pressure brake bleeders are electrically operated. An electric pump provides the bleeder pressure, eliminating the shop air connection.

Vacuum Bleeding Equipment

This hand operated *vacuum bleeder*, when pumped, creates a vacuum in the attached jar. It is connected to the brake lines via the bleeder screws and pumped to develop vacuum. Vacuum causes brake fluid to flow into the jar when the bleeder screw is opened. **Figure 3-42** depicts a vacuum bleeder.

Care of Tools

Just as important as obtaining the proper tools is maintaining them properly. Tools that are misused are much more likely to break, or to slip and cause injury. The following are a few rules for the proper use and care of your tools:
❑ Always use the proper tool for the job at hand.

❑ Store tools in an orderly manner.
❑ Clean tools before returning them to storage.
❑ Clean and lightly oil tools that are prone to rusting.
❑ Repair or replace any damaged tools.
❑ Place cutting tools such as lathe bits and carbide inserts in their protective cases when not in use.
❑ Do not drop or strike carbide tips and bits; they are brittle and can shatter easily.
❑ Store shop manuals in a storage cabinet. Do not leave them on benches where they can become dirty and damaged.

Service Information

Service information is often the most important tool in the shop. In the modern automotive environment, the importance of service manuals and other information resources cannot be overstressed. There are too many types of vehicles, all with complex, easily damaged systems to allow for any kind of guesswork.

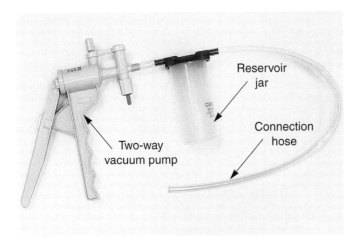

Figure 3-42. *A hand-held vacuum brake bleeder. Pumping the handle creates vacuum in the jar. When the hose is attached to the bleed screw, and screw is opened, brake fluid will be drawn into the reservoir jar. (Mac Tools)*

Although it may seem strange to classify manuals and other sources or service information under tools, they are often the technician's most valuable tool. Without these sources of information, as well as your own knowledge and experience, all other tools are useless. Service information provides the technical data needed to successfully perform diagnosis and parts replacement.

In the past, service information was generally ignored by technicians. Vehicles were simpler in design and completely new systems were rare enough they could be learned about in the course of performing other repairs, or often ignored completely. Today, more information is needed by the technician to perform even the most basic service. The major classifications of service information are service manuals, schematics, troubleshooting charts, technical service bulletins, manufacturers hotlines, and computers. The three major types of service manuals are the factory manual, the general manual, and the specialized manual.

Factory Manual

The *factory manual* is published by the vehicle manufacturer or a publishing house contracted by the manufacturer. It contains all necessary service information for that one vehicle. **Figure 3-43** shows some typical factory service manuals. Most modern factory service manuals now come in volume sets for one vehicle. The major drawback to the factory manual is its relatively high cost, compared to the limited range of vehicles it can be used with. While this type of manual is extremely detailed, it may not be the best choice if only one system, such as the brakes, is to be serviced.

General Manual

The *general manual,* **Figure 3-44,** contains the most commonly needed service information about many different makes of vehicles, such as brake, engine, and transmission specifications, fuse replacement data, and sensor locations. General manuals also contain procedures for preventive maintenance and minor repairs.

At one time, general service information for every vehicle could be covered in one manual. Today, due to the large number of different vehicles available, this is no longer possible. Modern general manuals are divided into automobile and light truck editions. Publishers further divide their general automotive manuals into US, European, and Asian models.

Figure 3-44. *General manuals are handy if you perform many different repairs on more than one line of vehicles.*

Figure 3-43. *Factory service manuals come in variety of sizes and usually in multiple volumes.*

The individual chapters of general manuals are grouped according to vehicle make, or several makes that are similar mechanically. Chapter subsections are devoted to particular areas of each make. General manuals also contain separate sections covering repair procedures that apply to all vehicles, such as engine overhaul, air conditioning service, and starter/alternator overhaul. The major disadvantage of these manuals is the necessity of eliminating most of the information on specialized vehicle equipment, sheet metal, and interior.

Specialized Manual

Specialized manuals cover one common system of many types of vehicles. These manuals are often used to cover such topics as computerized engine controls, electrical systems, or brakes. They combine some of the best features of the factory and general manuals. They are often a good choice for servicing one particular system on many different makes and models of vehicles. Some examples of these types of manuals are shown in **Figure 3-45**.

Schematics

Schematics are pictorial diagrams which show the path of energy through a system. This energy can take the form of electricity, vacuum, air pressure, or hydraulic pressure. Most schematics used for brake diagnosis show the flow of electricity or the inputs and outputs to a computer, **Figure 3-46**. Schematics do not show an exact replica of a system, but instead indicate the flow or process within the system. Some schematics show the exact flow of a form of energy while others show the general process of a particular system. Schematics are often included as part of a service manual, or may be supplied separately.

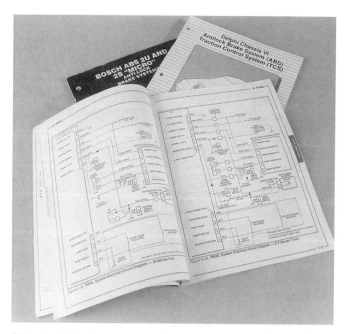

Figure 3-45. *Three specialized manuals covering anti-lock brake system service for a variety of vehicles.*

Tracing the flow through a schematic makes diagnosis easier by showing the exact path of electricity or other form of energy. Each line represents a single wire in the vehicle's wiring harness. The schematic lines are labeled with numbers or colors to correspond with a specific color, or color and color stripe combination on the actual wires. The path can be traced by carefully following the lines from component to component. Always carefully note the color designations of the wires and any stripes or bands to ensure you are following the correct wire.

Troubleshooting Charts

Troubleshooting charts are summaries, or checklist versions, of the troubleshooting information about a particular vehicle or system. Although the information is found in a longer form elsewhere in a service manual, the troubleshooting chart allows the technician to quickly reference the problem, the possible cause, and the solution. **Figure 3-47** shows a typical troubleshooting flowchart. Some troubleshooting charts are arranged with the problem on the left-hand side of the page, the possible cause in the middle, and the corrective action on the right-hand side.

Technical Service Bulletins

Frequently, manufacturers issue **technical service bulletins (TSB),** for newer vehicles to their dealership personnel. These bulletins contain repair information that is used to describe a new service procedure, correct an unusual or frequently occurring problem, or update information in a service manual. Many of the phone hotline and computerized assistance services receive these bulletins. They are a very good source of information to repair an unusual or frequently occurring problem. Subscriptions to these bulletins are also available through various services.

Manufacturer Hotlines

To further assist service personnel, various vehicle and aftermarket equipment manufacturers have established direct telephone links with service personnel. These telephone links are usually called **hotlines** or **technical assistance centers.** These hotlines connect with central information banks. Dialing one of these central information banks connects the technician to persons with access to troubleshooting and service information, or to computer memories called **databases.**

The various types of hotlines reflect their purpose. Vehicle manufacturer hotlines are usually installed in dealerships handling that manufacturer's vehicles, and are intended for use only by the dealership service personnel. Some aftermarket part and equipment manufacturers provide toll free numbers for technicians in need of diagnosis or service information. Aftermarket hotlines are available to any technician and are provided as part of the marketing and advertising strategy of the aftermarket manufacturer. These numbers are available from the local outlets handling that brand of parts or equipment.

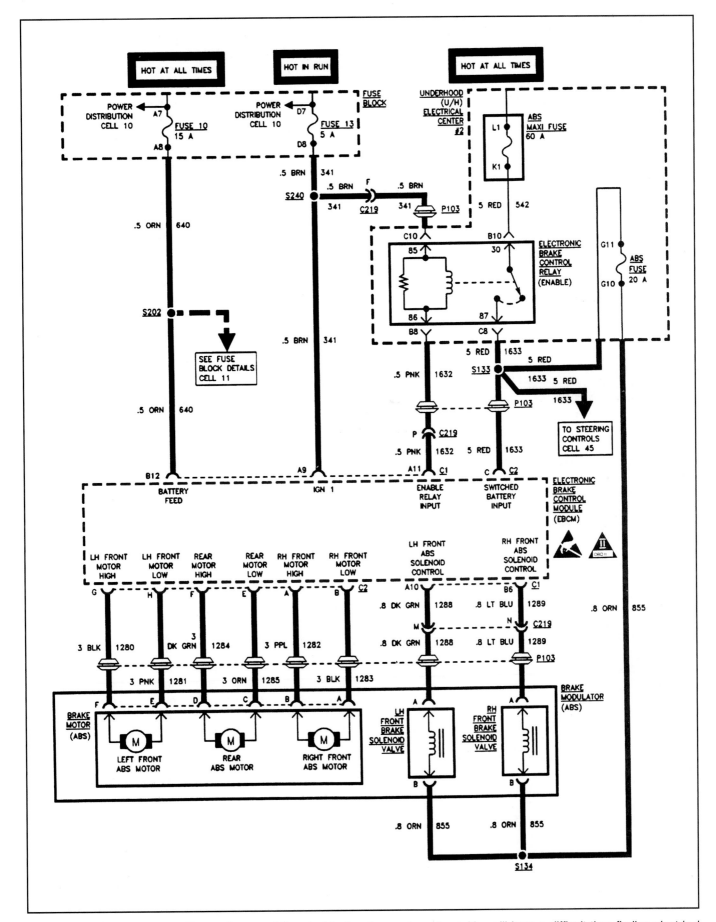

Figure 3-46. *Schematics are like road maps of the electrical and vacuum systems. You will have a difficult time finding electrical problems without the proper schematic. (General Motors)*

Symptom Table: Amber ABS Warning Indicator on Constantly, NO DTCs Stored				
Step	Action	Value(s)	Yes	No
1	Was the Diagnostic System Check performed?	—	Go to Step 2	Go to Diagnostic System Check
2	1. Turn the ignition switch to the OFF position. 2. Install a scan tool. 3. Turn the ignition switch to the RUN position. 4. Using the scan tool, select MISC TESTS. 5. Using the scan tool, select LAMP TEST. 6. Using the scan tool, turn the amber ABS warning indicator off. Did the amber ABS warning indicator turn off?	—	Go to Step 3	Go to Step 4
3	1. Using the scan tool, select MISC TESTS. 2. Using the scan tool, select LAMP TEST. 3. Using the scan tool, flash the amber ABS warning indicator. Did the amber ABS warning indicator flash?	—	Go to Step 7	Go to Step 4
4	Inspect all connectors and terminals for poor terminal contact and evidence of corrosion. Is there evidence of poor terminal contact or corrosion?	—	Go to Step 5	Go to Step 6
5	Replace all terminals that exhibit signs of poor terminal contact or corrosion. Is the repair complete?	—	Go to Diagnostic System Check	—
6	Repair the instrument panel cluster. Refer to SECTION 8A-8I. Is the repair complete?	—	Go to Diagnostic System Check	—
7	Malfunction not present at this time. Refer to Diagnostic Aids for more information.	—	—	—

Figure 3-47. *Typical troubleshooting chart found in most service manuals. (General Motors)*

Computerized Assistance

A few years after computer control came to the automobile, manufacturers began to put computer driven analyzers into the service departments of their dealerships. These were simply computers that were joined with a dedicated engine analyzer system designed to diagnose problems in that particular manufacturer's line of vehicles. Unfortunately, the first computer driven analyzers were not much better than traditional engine analyzers. Also, some technicians were very apprehensive to use the new analyzers, since many of them had little or no exposure or training in computer usage.

Improvements in software and computer technology have greatly improved the computerized analyzer. The newest computerized analyzers have user-friendly menus with touch screen capability or a standard computer mouse, which can be used to "point-and-click" on any particular menu selection. The newest analyzers can also

be used on more than one manufacturer's line of vehicles. These computer driven analyzers, **Figure 3-48,** are able to communicate with the various computers on most late-model vehicles, as well as perform all standard diagnostic procedures. The analyzer's information can be updated through a computer-to-computer modem connection over the telephone to a central computer at the manufacturers' service headquarters.

One feature of these analyzers is that technicians seeking service information can access the manufacturer's main computer and type in the VIN number of the vehicle in question. The computer then provides a printout of all service information, including any technical service bulletins or recall campaigns applicable to that particular vehicle. Aftermarket tool manufacturers have begun to computerize their engine analyzers, as well as their alignment equipment and exhaust gas analyzers.

Unfortunately, most small shop owners cannot afford these large, expensive computerized diagnostic equipment nor are they allowed to interface with a manufacturer's

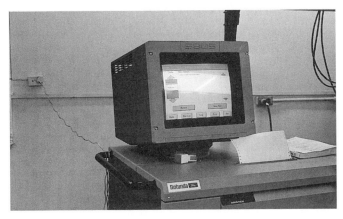

Figure 3-48. *Computerized analyzers are becoming more and more common even in the most specialized shop.*

central computers. However, some software and tool companies are beginning to offer kits that contain harnesses to access data link connectors, along with related hardware and software that can be installed in a personal computer. These kits can turn a normal desktop or laptop computer into a computer driven analyzer, with many of the same capabilities as the large, expensive shop analyzers. In addition, many shops now write their repair and purchase orders and keep their billing and accounting in order with computers.

Complete service, parts, and even labor time manuals are now available in digital format on a compact disc called a CD-ROM. A CD-ROM can be used with a desktop or even a laptop computer equipped with a CD-ROM drive, **Figure 3-49.** One CD-ROM can provide the same amount of information found in a complete series of printed manuals. Many CD-ROM manuals cover several model years of a particular manufacturer's line of vehicles. Some of the newest CD-ROM manuals show actual, step-by-step footage of certain repair operations. The one drawback to CD-ROM manuals is they are much more expensive than a printed manual.

On-line Diagnostic Assistance

By using a computer on-line service, small shops can access any one of several automotive central information banks over the information superhighway. These banks can offer diagnostic tips, technical service bulletins, and other service information similar to the telephone hotlines described earlier. Many of these services have user-friendly interfaces that make them easy for anyone to use. Most of these on-line assistance centers are operated by aftermarket companies, private organizations, and individuals.

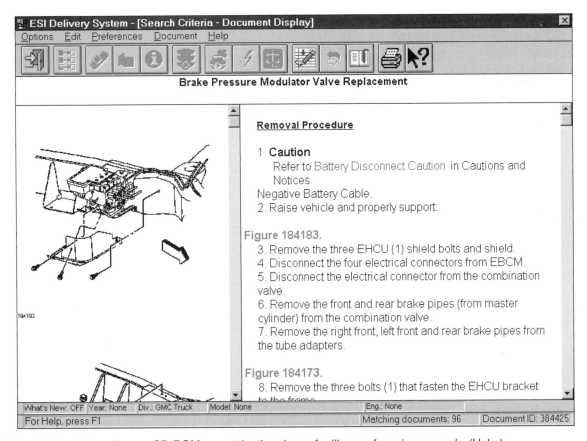

Figure 3-49. *Service information on CD-ROMs can take the place of a library of service manuals. (Helm)*

Summary

The professional brake technician should always choose quality tools and the correct tool for the job. Brake service requires many specialty tools. Basic brake hand tools include bleeder wrenches and flare-nut wrenches, tubing cutters, reamers, tubing benders, and flaring tools. In addition to standard types of pliers, special brake pliers that are often needed. Brake adjusters, spring removal tools, Allen and Torx® bits, and C-clamps are very useful.

To safely and efficiently service brakes, they must be thoroughly cleaned. The two types of cleaning systems are vacuum cleaners and wet cleaners. Precautions should always be taken to ensure that as little brake dust enters the atmosphere as possible. The best type of dry cleaning system uses a HEPA filter. Wet cleaning should be done gently at first to wet down the dust before washing it off.

Wheel cylinder, master cylinder, and caliper bores can sometimes be honed to smooth up any roughness. There are two types of hones, the rigid stone hone and the flexible abrasive type. Never hone coated or anodized cylinders unless specified by the manufacturer.

For precision brake work, numerous specialty measuring tools are needed. They include rotor and drum micrometers, brake shoe adjusting gauges, dial indicators, pedal pressure gauges, and hydraulic pressure gauges.

Brake specialty tools include hydraulic system service tools, seal and dust boot installers, wheel bearing service tools, and power brake tools. The modern brake technician needs several electrical and electronic testers, including test lights, multimeters, and scan tools. Tools should always be properly cared for and stored. Never use a tool the wrong way and always repair or replace tools that are damaged.

The two main types of brake system bleeders are pressure and vacuum operated. Brake lathes are made in two general types, bench or vehicle mounted. Brake drum grinders and burnishing tools are also used. New linings are sometimes ground on a brake shoe grinder. Be sure to prevent exposure to asbestos fiber dust during this procedure. A hydraulic press is often needed to perform some brake service.

Service manuals and other information resources are vital to servicing the modern brake system. The three main types of service manuals are the factory manual, the general manual, and the specialty manual.

Review Questions—Chapter 3

Please do not write in this text. Write your answers on a separate sheet of paper.

1. Quality tools will pay for _____.

2. To remove glaze and deposits from wheel cylinders, a brake _____ should be used.

3. A _____ is used to measure the thickness of rotors. If a rotor is too thin, it should be _____.

4. A dial indicator can be used to check for out-of-round _____.

5. A brake hydraulic pressure gauge must be able to read a minimum of _____ psi or _____ kPa.

6. To quickly test for voltage in a brake light system, use a _____.

7. *True* or *False?* Multimeters can check for voltage and hydraulic pressure.

8. *True* or *False?* Scan tools are used to test anti-lock brake systems.

9. *True* or *False?* Bench mounted brake lathes can only turn drums.

10. A rotor is turned using _____ cutting bits.

ASE Certification-Type Questions

1. Technician A says that a flare-nut wrench is used to open brake bleeder valves. Technician B says that flare-nut wrenches are sometimes called tubing wrenches. Who is right?
   (A) A only.
   (B) B only.
   (C) Both A & B.
   (D) Neither A nor B.

2. All of the following tools are used to service brake hydraulic system tubing, EXCEPT:
   (A) bleeder wrench.
   (B) tubing cutter.
   (C) reamer.
   (D) mechanical bender.

3. Brake adjusters are used to adjust _____ brakes.
   (A) drum
   (B) disc
   (C) parking
   (D) mechanical

4. Technician A says that Allen and Torx® are two types of recessed head bolts. Technician B says that Allen and Torx® bolts require special bits for removal. Who is right?
   (A) A only.
   (B) B only.
   (C) Both A & B.
   (D) Neither A nor B.

5. C-clamps can be used to perform all of the following, EXCEPT:
   (A) press out rivets.
   (B) pressing caliper pistons.
   (C) installing pins.
   (D) bleeding brakes.

6. Hones can be used to clean up all of the following, EXCEPT:
   (A) wheel cylinders.
   (B) caliper bores.
   (C) hose fittings.
   (D) master cylinders.

7. The only sure way to measure brake drum diameter is by using a _____.
   (A) shoe-drum gauge
   (B) micrometer
   (C) dial indicator
   (D) Both A & B.

8. A clear jar and rubber hoses are often used to make the job of _____ brakes easier and cleaner.
   (A) adjusting
   (B) removing
   (C) installing
   (D) bleeding

9. A pressure bleeder can be operated by air pressure or with a(n) _____.
   (A) hand pump.
   (B) foot pump.
   (C) electric pump.
   (D) All of the above.

10. Technician A says that a vacuum bleeder is hand operated. Technician B says that the master cylinder reservoir should be carefully monitored while vacuum bleeding. Who is right?
    (A) A only.
    (B) B only.
    (C) Both A & B.
    (D) Neither A nor B.

11. Technician A says that disc brake rotors can be turned without removing them from the vehicle. Technician B says that brake drums can be turned without removing them from the vehicle. Who is right?
    (A) A only.
    (B) B only.
    (C) Both A & B.
    (D) Neither A nor B.

12. A brake drum grinder is used to remove _____ after turning.
    (A) scoring
    (B) hard spots
    (C) oily spots
    (D) wheel studs

13. Technician A says that burnishing is performed to brake drums after turning. Technician B says that shoe grinding is performed to remove excess lining material. Who is right?
    (A) A only.
    (B) B only.
    (C) Both A & B.
    (D) Neither A nor B.

14. _____ are pictorial diagrams that show the path of energy through a system.
    (A) Schematics
    (B) Troubleshooting charts
    (C) Flowcharts
    (D) None of the above.

15. A specialized service manual covers which of the following?
    (A) Everything about one vehicle.
    (B) General information about many kinds of vehicles.
    (C) Specific information about one vehicle system.
    (D) Both A & B.

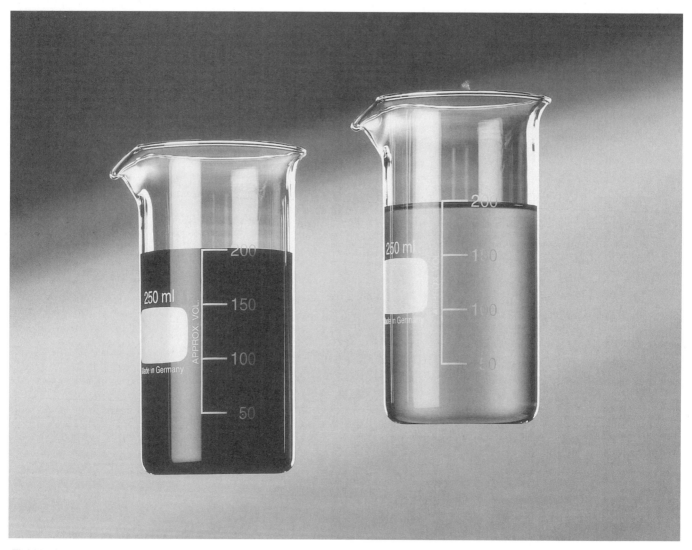

Fluid is the source of motion in the hydraulic system. The beaker on the left contains used brake fluid. Brake fluid turns dark due to heat. The beaker on the right contains new brake fluid. (Continental Teves)

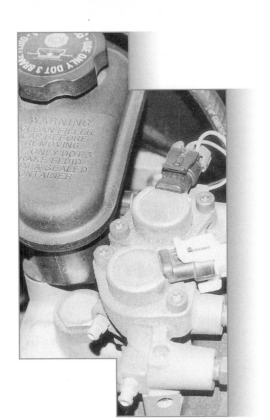

Chapter 4

Hydraulic System Fundamentals

After studying this chapter, you will be able to:

- ❏ Explain the use of hydraulic pressure in braking systems.
- ❏ Explain the incompressibility of liquids as the basis of hydraulic system operation.
- ❏ Explain how movement is transferred by way of liquids.
- ❏ Explain how cylinder size changes hydraulic force and distance traveled.
- ❏ Explain Pascal's law.
- ❏ Identify the components of a simple brake hydraulic system.
- ❏ Identify the qualities of brake fluid and fluid classifications.
- ❏ Explain the importance of using the proper brake fluid.

Important Terms

Hydraulics	Master cylinder	Lines	Department of Transportation (DOT)
Cylinders	Split brake system	Hoses	Society of Automotive Engineers (SAE)
Pistons	Wheel cylinders	Control valves	Hygroscopic
Pascal's law	Calipers	Hydraulic actuator	

You learned in Chapter 1 that the earliest brake systems were mechanical devices. You also learned that modern brake systems are operated by hydraulic pressure. Using hydraulics to apply the brakes is more efficient and convenient than mechanical linkages. The hydraulic system is designed to allow brake pedal pressure to be increased many times for greater braking power and to distribute braking force equally. This chapter will explain in detail the use of hydraulic pressure in braking systems.

The Basics of Hydraulic Systems

The study of liquids and how they work is called **hydraulics.** You will not require an extensive education in hydraulic theory, but you must know how hydraulic pressure is created and used. The information in this chapter is a brief overview of how hydraulic pressure is put to work in the modern brake system. If you understand what is presented here, you will be able to perform brake hydraulic system service.

Incompressibility of Liquids

Gases such as air can be **compressed,** that is, made to take up less space. Compressed air is used in the shop to power air tools and other equipment. However, liquids such as water, oil, and brake fluid cannot be compressed. The inability of liquids to compress is the basis of brake hydraulic system operation, as well as the operation of power steering and automatic transmissions.

Figure 4-1 compares the compressibility of gases and liquids. A weight is placed on the lids of closed containers of gas and liquid. When weight is applied to a container of gas, its volume is reduced. In other words, a gas under pressure takes up less space than it does when not under pressure. Placing the weight on the liquid increases its pressure, but does not reduce its volume.

Transfer of Movement

Since liquids are not compressible, they can be used to transfer movement. **Figure 4-2** shows a simple hydraulic system consisting of two **cylinders** and two **pistons,** connected by tubing called a **hydraulic line.** Both the right and left cylinders and pistons are the same size. The entire system is filled with liquid. If the piston on the left side is pushed, it pressurizes the liquid. This pressure on the liquid is transmitted through the hydraulic line to the other piston, causing it to move.

It is important to note that moving the first piston caused the second piston to move exactly the same amount. This is because the force was transmitted through the liquid unchanged. As long as the input and output piston areas are the same, hydraulic pressure is transmitted undiminished. The cylinder and piston assembly shown in **Figure 4-2** is the basis of all hydraulic brake system design and operation.

Changes in Force

There are times when it is desirable to change the amount of movement of the input and output pistons. In a simple hydraulic system, changing the cylinder size changes both the distance traveled and the amount of force applied. In **Figure 4-3,** note that the input cylinder has an area of one square inch and the output cylinder is 10 in^2. If we apply a force of 1 lb. to the input cylinder, this produces a pressure of 1 pound per square inch, or 1 psi in the system. This pressure travels to the output piston. At the output piston, the 1 psi acting on the 10 square inch piston produces an output pressure of 10 psi. Note that by varying the size of the input and output pistons, we have increased the input force 10 times.

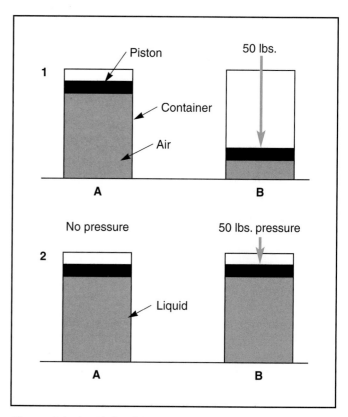

Figure 4-1. *1—Air (gas) is compressible. A—There is no pressure on the piston. B—Pressure has forced the piston down, compressing the air trapped in the container. 2—Liquids cannot be compressed. When pressure is applied to the piston in B, it does not compress the liquid.*

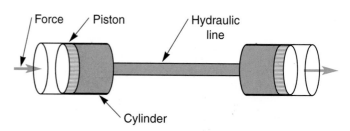

Figure 4-2. *Transfer of movement by using a liquid and hydraulic line (tubing).*

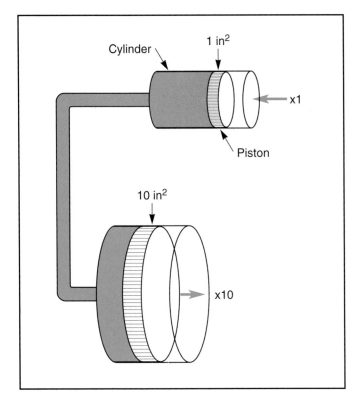

Figure 4-3. *Changing a cylinder's size will alter the piston travel and force.*

However, this increase in hydraulic force has a price. The input piston moved one inch while the output piston moved 1/10 of an inch. Therefore, the pressure was increased ten times, but the amount of movement was decreased to 1/10. Therefore, to increase force we must accept a decrease in movement. How this is principle is put to work in the modern brake system will be covered later.

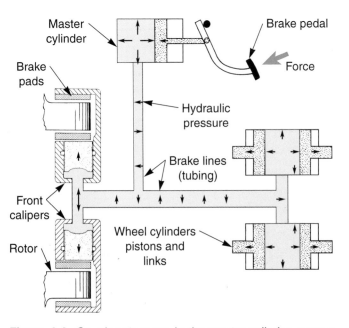

Figure 4-4. *Creating pressure in the master cylinder, causes pressure to build in the entire system. (Chevrolet)*

Pascal's Law

The simple brake system shown in **Figure 4-4** illustrates a basic law of hydraulics, called Pascal's law. *Pascal's law* states the pressure in a closed hydraulic system is the same everywhere in the system. This principle of hydraulics was discovered in 1653 by Blaise Pascal.

When the master cylinder at the top of **Figure 4-4** is pressurized by pushing on the piston, the pressure created is the same throughout the system. This principle allows you to check pressure at one point in the system with confidence that the pressure is the same elsewhere in the system. It also allows you to check variations in pressure and determine whether they are caused by design variations or system defects.

A Simple Brake Hydraulic System

The following illustrations show the design of a simple on-vehicle brake hydraulic system having front disc and rear drum brakes. Follow the illustrations and text to see how this simple system makes use of the principles of hydraulics.

Master Cylinder

In **Figure 4-5,** the *master cylinder* provides the pressure to operate the other hydraulic components. The master cylinder is in turn, operated by the foot pedal. Pushing on the pedal creates a force on the pistons, creating hydraulic pressure.

Note the master cylinder has two pistons. The purpose of this design is to split the brake hydraulic system into two parts. If one side of the system loses hydraulic fluid due to a leak, the other side will still function. In operation, pushing on the first piston causes the buildup of pressure in the first chamber. This, in turn, pushes on the second piston, causing it to pressurize the second chamber. In both systems, the pressure is sent to the wheel brake units. This design is used on all modern vehicles, and is called a *split brake system.* It is discussed in more detail in Chapters 5 and 6.

Wheel Cylinders and Calipers

The *wheel cylinders* and *calipers* are the output devices of the brake hydraulic system. In **Figure 4-5,** the disc brake caliper pistons are much larger than the wheel cylinder pistons. This allows the front brakes to be applied with more force. The reason for this is that inertia places more of the vehicle's weight on the front brakes during stopping, allowing the rear brakes to brake for control. These factors will be discussed in more detail in Chapter 11.

Also note the front caliper pistons are much larger than the master cylinder piston. This means the caliper pistons will multiply the brake pedal force many times. However, the brake pedal must be pushed a long distance

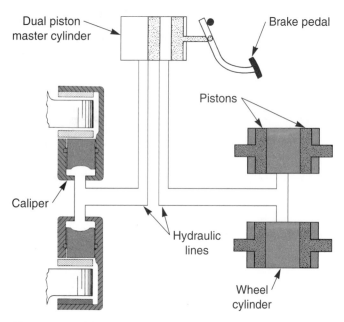

Figure 4-5. *The master cylinder generates the pressure needed to operate the calipers, wheel cylinders, valves, etc.*

to move the caliper piston a small amount. To keep the pedal travel from being excessive, the caliper assembly is designed so the pads and rotor are very close together when the brake is not applied. Therefore, less than .5" (12.7 mm) movement of the brake pedal will apply the brakes. Slightly misadjusted or worn brakes will greatly increase brake pedal travel.

Lines and Hoses

The various components of the brake hydraulic system are connected through *lines* and *hoses.* Lines are made of steel and hoses are made of braided rubber. Lines connect the stationary parts of the hydraulic system and hoses connect the parts which move in relation to each other.

Control Valves

The modern brake system contains several flow and pressure *control valves.* These valves are:
❑ Metering valve.
❑ Proportioning valve.
❑ Residual pressure valve.
❑ Pressure differential valve.

The job of these valves is to make the brake system more efficient, and to warn the driver when a failure occurs. They perform these jobs by using spring pressure to oppose hydraulic pressure. A brief explanation of these valves is given below. Note that two or more of these valves are sometimes installed into a single assembly called a *combination valve.*

Metering Valve

The *metering valve* is used to keep the front brakes from applying before the rear brakes. The front brake pads are not held in the retracted position by springs as

are the rear brake shoes. Therefore, if the same amount of pressure reached the front and rear brakes at the same time, the front brakes would do all the stopping and wear prematurely. Applying only the front brakes could also cause instability, or might throw the vehicle into a skid.

The metering valve assembly, **Figure 4-6,** consists of a small spring-loaded valve. The spring seats the valve against hydraulic pressure until a certain pressure is reached. At this pressure the valve is unseated and fluid can flow to the disc brake calipers. This delay in applying the front brakes gives the rear brakes time to overcome spring pressure and begin applying the rear brakes.

Proportioning Valve

If the brakes are applied hard during an emergency stop, much of the vehicle's weight is transferred to the front wheels. Hard braking can cause the wheels to lock up and skid. This is not only hard on the tires, it can cause dangerous instability, sometimes causing the vehicle to spin out of control.

To prevent this, a **proportioning valve** is installed in the rear brake line. A calibrated spring holds the proportioning valve open against brake pressure. **Figure 4-7** illustrates the proportioning valve. Under normal braking, brake fluid can flow to the rear brakes. Under a hard stop, however, brake pressure will exceed the calibrated tension of the spring, and increased pressure will close the valve against the opening. This limits brake pressure at the rear wheels to what is already in the rear hydraulic system. Preventing the development of additional pressure helps to prevent wheel lockup.

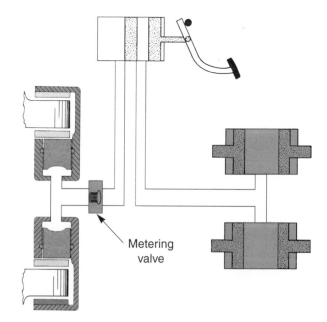

Figure 4-6. *A metering valve is added to the system to prevent the front brakes from applying before the rear brakes are actuated.*

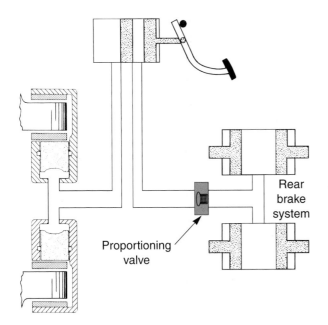

Figure 4-7. *The proportioning valve helps to limit brake pressure to the rear brake system, reducing the chance of wheel lockup.*

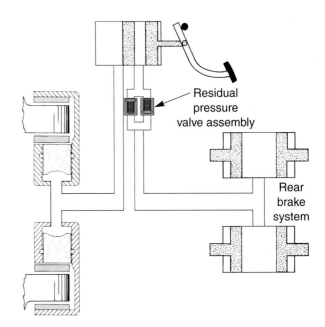

Figure 4-8. *The residual pressure valve helps to maintain a specific brake fluid pressure in the drum brake system.*

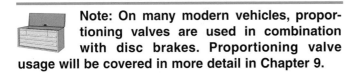

Note: On many modern vehicles, proportioning valves are used in combination with disc brakes. Proportioning valve usage will be covered in more detail in Chapter 9.

Residual Pressure Valve

If the vehicle has drum brakes, a small amount of pressure must be maintained in the system to keep the wheel cylinder lip seals from collapsing. The **residual pressure valve** is used on vehicles with drum brakes to maintain this pressure. **Figure 4-8** illustrates this valve design. Note that the actual valve assembly is a one piece unit.

In operation, the valve to the right is unseated and fluid flows to the rear wheels. When the brake pedal is released, the valve to the right closes, and fluid returns to the master cylinder through the valve to the left, which opens against spring pressure. When the fluid pressure becomes low enough (usually about 7-10 psi or 48.26-68.95 kPa), it no longer has enough force to keep the spring compressed. The spring then closes the valve, trapping a small amount of pressure in the rear brake system.

Pressure Differential Valve

The **pressure differential valve** is a warning device used in all split brake systems. The pressure differential valve is a double-sided valve. The valve is installed so that each line presses on one side of the valve. A typical pressure differential valve is shown in **Figure 4-9.**

When both sides of the hydraulic system are operating normally, the valve is centered and the switch has no way to complete the circuit. When one side of the system fails, pressing on the brake pedal will result in normal pressure on one side of the system, and lower than normal

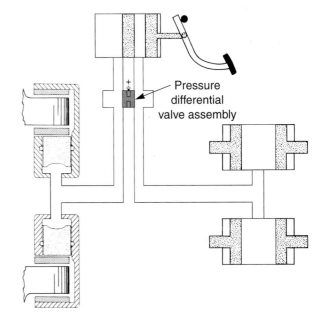

Figure 4-9. *Split brake system which incorporates a pressure differential valve unit.*

pressure on the other side. This causes the valve to move to the side with less pressure. When the valve moves, the switch is grounded. Electrical current flows through the switch and illuminates a dashboard light.

Basic Anti-Lock Brake Hydraulics

Detailed operation of the anti-lock brake system (ABS) will be covered in detail in Chapters 21 and 22. However, the anti-lock brake hydraulic system should be mentioned since it has many similarities and a few differences with the standard brake system. The major difference between the anti-lock and standard brake systems is the anti-lock brake hydraulic actuator.

The anti-lock brake *hydraulic actuator* contains control valves, similar to those discussed earlier. These valves are operated by electric *solenoids* or small motors. One type of anti-lock brake hydraulic actuator is shown in **Figure 4-10.** The valves can seal off parts of the brake hydraulic system, in a manner similar to the proportioning valve discussed earlier. The valves can also dump fluid back to the master cylinder, reducing system pressure. The valve solenoids are controlled by electrical signals from an on-board computer.

Some anti-lock brake systems contain a pump, driven by an electric motor to develop extra pressure. This pump is usually a *vane pump,* **Figure 4-11.** The vane pump consists of a set of movable vanes attached to a slotted drum, usually called a rotor. The rotor and vanes are installed inside an eccentric (egg-shaped) housing.

As the rotor turns, the vanes move in and out within the slots. Pressure behind the vanes keeps them in contact with the eccentric housing walls. As the rotor is turned by the electric motor, the vanes create a chamber which increases in size. This increase in size creates a suction at the intake port, drawing in fluid. As the vanes rotate through the eccentric housing, they reduce the size of the

chamber, creating pressure which is discharged through the pump outlet port. This pressure is then used in the anti-lock brake system.

Brake Fluid

Brake fluid is essential to the proper operation of the brake system. In the last 60 years, many types of brake fluid have been tried. Modern brake fluid is manufactured according to requirements of the **Department of Transportation,** usually called **DOT,** and the **Society of Automotive Engineers**, known as **SAE.** Other industry and government organizations also have standards for brake fluids.

Most brake fluids are derived from a mixture of non-petroleum fluids, such as polyglycols, glycoethers, and other additives which increase the fluid's reliability. Brake fluid is also used in manual transmission clutch or slave cylinders.

Brake Fluid Standards

Low quality brake fluid will work in a brake hydraulic system, as will water, alcohol, or many other fluids — for a short time. However, for maximum brake system life and the safety of the vehicle occupants, brake fluid must conform to high standards.

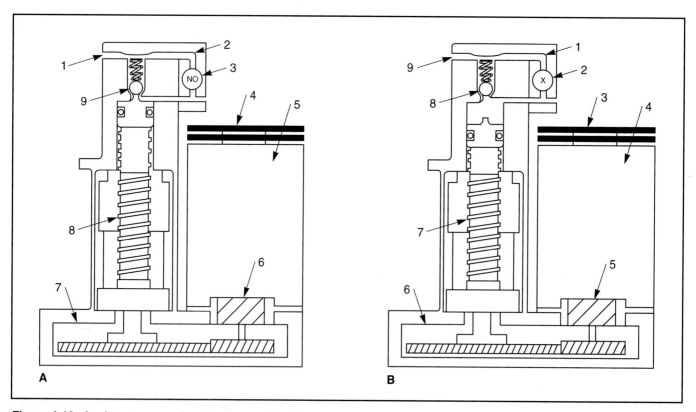

Figure 4-10. A—Actuator assembly during normal braking operation. 1—Applied master cylinder pressure. 2—Bypass brake fluid. 3—Normally open solenoid valve. 4—EMB braking action. 5—Dc motor pack. 6—ESB braking action. 7—Gear assembly. 8-Ball screw. 9—Check valve unseated. B—Actuator position during the anti-lock brake phase. 1—Trapped bypass brake fluid. 2—Solenoid valve activated. 3—EMB action released. 4—Dc motor pack. 5—ESB braking action released. 6—Gear assembly. 7—Ball screw. 8—Check valve seated. 9—Applied master cylinder pressure. (Delco Moraine)

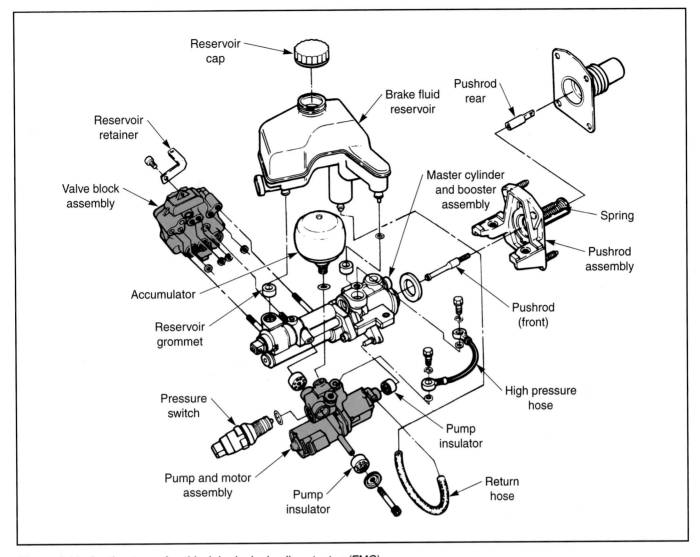

Figure 4-11. *Another type of anti-lock brake hydraulic actuator. (FMC)*

Caution: Never add petroleum-based fluids to the brake system. Any type of motor oil, transmission fluid, or other petroleum product will swell and destroy the rubber seals in the system.

Resistance to Boiling

The most important requirement of brake fluid is its resistance to **boiling.** If the brake fluid boils, it becomes a gas. As we learned earlier in this chapter, gases are compressible. Therefore, if the brake fluid boils, pressing on the brake pedal will simply compress gas instead of applying the brakes. Since frictional heat is transmitted to the hydraulic system during braking, fluid resistance to boiling is very important.

Water Absorption

Another important quality of brake fluid is its ability to absorb water. Brake fluids are intentionally designed to be **hygroscopic,** or able to absorb water. This would seem to be a bad quality, since water will lower the fluid's resistance to boiling. Should water get into the brake system, however, it would tend to collect at low spots in the system. Water collecting at one spot in the brake system could cause corrosion at that spot. Pure water would also freeze in cold weather, blocking off brake lines or sticking pistons. Designing brake fluid to absorb water helps to minimize these problems.

You must take precautions to limit the exposure of brake fluid to water. Brake fluid that has a boiling point of 446°F (230°C) when completely water free has a boiling point of 311°F (155°C) when it has absorbed the maximum amount of water it can hold, **Figure 4-12A.** Always keep containers tightly capped, and do not leave hydraulic system parts disconnected for long periods of time.

 Note: When servicing anti-lock and traction control systems, only use brake fluid from a newly opened container. This will minimize the chance of dirt and water entering the system.

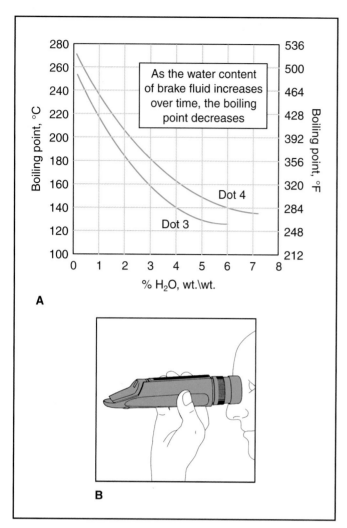

Figure 4-12. *A—Brake fluid boiling points decrease as they absorb more moisture. B—Tools are available for checking the moisture content of brake fluid. (Leica)*

There are several products on the market to check brake fluid for water contamination. One type is shown in **Figure 4-12B**. Another type uses chemically treated strips, which react in the presence of water or other fluids.

Other Brake Fluid Requirements

In addition, brake fluid must have other qualities. It must lubricate moving parts of the brake system, such as the pistons and seals. The fluid must help to prevent corrosion of the metal parts as much as possible. Good brake fluid should not damage rubber seals or any other part of the brake system, and must flow easily, even at very low temperatures.

Types of Brake Fluids

Brake fluid is classed by its DOT number. The two common types of modern brake fluid are glycol based and silicone based. *Glycol based* fluid is able to absorb water, and has a boiling point of over 400°F (200°C). Modern glycol based brake fluid is usually classified as DOT 3 or

DOT 4. Both types are clear fluids, with a slight amber tinge. DOT 5 brake fluid can be easily spotted since it is purple in color.

The difference between DOT 3 and DOT 4 is their viscosity and ability to resist heat. DOT 4 can absorb more heat than DOT 3, but has a slightly lower viscosity than DOT 3. *Silicone based* brake fluid is classed as DOT 5, and has a boiling point of over 500°F (260°C). Silicone brake fluid will not absorb any significant amount of water. For this reason, it is vital that no water or airborne moisture be allowed to enter a silicone fluid brake system or fluid containers.

> **Caution: Glycol and silicone brake fluids are not compatible, and should never be mixed. Do not use silicone brake fluid in a vehicle equipped with an ABS system.**

Federal and SAE Brake Fluid Standards

In addition to DOT classifications, there are other standards for brake fluid. These include Federal Motor Vehicle Safety Standard 116, or FMVSS 116 and Federal specification VV-B-680. The Society of Automotive Engineers (SAE) publishes specification J1703. Look for these specifications when purchasing brake fluid. Typical containers of brake fluid are shown in **Figure 4-13**.

Figure 4-13. *An assortment of brake fluid containers. Keep them tightly sealed. Moisture will lower the fluids recommended boiling point. (Wagner)*

Summary

The brake hydraulic system has evolved over many years and relies on the basic principles of liquids. The study of liquids and how they work is called hydraulics.

Liquids cannot be compressed. This property of liquids is used to cause liquids to transfer movement and pressure, as well as increase force. The operation of the brake hydraulic system is based on Pascal's law which states that pressure is transmitted unchanged through a closed hydraulic system. The basic components of a brake system are the master cylinder, wheel cylinders and calipers, connecting lines and hoses, and various control valves.

The anti-lock brake system is operated by the same hydraulic principles as the standard brake system. ABS hydraulic control valves are operated by electric solenoids controlled by an on-board computer. Pressure is often supplied by a motor driven hydraulic pump.

Some of the most important properties of brake fluid are its ability to resist boiling, its ability to absorb water, and its resistance to corrosion. Brake fluid must also lubricate moving parts and stay liquid at very low temperatures.

Brake fluid is classified and rated by the U.S. Department of Transportation (DOT) and the Society of Automotive Engineers (SAE). The two major classes of brake fluid are glycol based and silicone based. The two types should not be mixed. Brake fluid containers should be kept closed to reduce the amount of water absorption.

Review Questions—Chapter 4

Please do not write in this text. Write your answers on a separate sheet of paper.

1. A basic principle of hydraulics is that _____ cannot be compressed. Air and other _____ can be compressed.

2. Liquids can be used to _____ motion and pressure.

3. Tubing that connects two hydraulic units is usually called a hydraulic _____.

4. Changing the sizes of the input and output cylinders can change the amount of _____ applied.

5. State Pascal's Law.

6. What are the two brake hydraulic system output devices?

7. The metering valve keeps the _____ brakes from applying before the _____ brakes.

8. To keep the rear wheels from skidding during a panic stop, a _____ valve is installed in the rear brake line.

9. A residual pressure valve is used only on vehicles with _____ brakes.

10. ABS systems usually use a _____ pump to develop extra hydraulic pressure.

11. Standards for brake fluid are set by two organizations. Name them.

12. Brake fluid is designed to absorb _____.

13. It is very important that brake fluid resist _____ at high temperatures.

14. Brake fluids classified as DOT 3 or DOT 4 are _____ based. Brake fluids classed as DOT 5 are _____

based. This fluid can be easily spotted since it is colored _____.

15. Which of the above fluids should *not be* used in an ABS system?

ASE Certification-Type Questions

1. Which of the following cannot be compressed?
   (A) Air.
   (B) Water.
   (C) Brake fluid.
   (D) Both B & C.

2. Technician A says that changing the sizes of the input and output pistons can change the distance traveled. Technician B says that changing the sizes of the input and output pistons can change the amount of force from the output piston. Who is right?
   (A) A only.
   (B) B only.
   (C) Both A & B.
   (D) Neither A nor B.

3. According to Pascal's law, the pressure in a closed hydraulic system is the same _____.
   (A) only at the input piston
   (B) only at the output pistons
   (C) only at the input and output pistons
   (D) everywhere in the system

4. Technician A says that that the purpose of having two pistons in the master cylinder is to split the brake hydraulic system into two parts. Technician B says that the purpose of having two pistons in the master cylinder is to increase braking force. Who is right?
   (A) A only.
   (B) B only.
   (C) Both A & B.
   (D) Neither A nor B.

5. The wheel cylinders and calipers are the _____ devices of any brake hydraulic system.
   (A) input
   (B) output
   (C) fluid transfer
   (D) friction

6. Why are the front caliper pistons larger than the master cylinder piston?
   (A) To increase brake pedal travel.
   (B) To decrease brake pedal travel.
   (C) To increase brake pedal force.
   (D) To decrease brake pedal force.

7. All of the following statements about metering valves are true, EXCEPT:
   (A) the metering valve is used to keep the rear brakes from applying before the front brakes.
   (B) if the metering valve was not used the vehicle would be more prone to skid.
   (C) the metering valve prevents overuse of the front brake pads.
   (D) the metering valve keeps the front brakes from wearing out too soon.

8. During hard braking, the proportioning valve does which of the following?
   (A) Increases pressure to the rear brakes.
   (B) Prevents pressure increase in the rear brakes.
   (C) Increases pressure at the front brakes.
   (D) Bypasses fluid to the master cylinder.

9. Technician A says that the residual pressure valve is used on vehicles with disc brakes. Technician B says that the residual pressure valve keeps the piston cups from collapsing. Who is right?
   (A) A only.
   (B) B only.
   (C) Both A & B.
   (D) Neither A nor B.

10. What is the purpose of the pressure differential switch?
    (A) Restore equal pressure to each side of the hydraulic system.
    (B) Sense when unequal pressures are occurring.
    (C) Illuminate a dashboard light.
    (D) Both B & C.

11. The pump used in the typical ABS system is known as a _____ pump.
    (A) rotor
    (B) gear
    (C) vane
    (D) piston

12. ABS solenoid valves are operated by _____.
    (A) pump hydraulic pressure
    (B) signals from a computer
    (C) brake pedal pressure
    (D) feedback from the calipers and wheel cylinders

13. The most important requirement of brake fluid is its resistance to ____.
    (A) boiling
    (B) freezing
    (C) corrosion
    (D) thickening

14. Exposing brake fluid to water lowers its _____.
    (A) compressibility
    (B) boiling point
    (C) condensing point
    (D) Both B & C.

15. Silicone brake fluid carries what DOT number?
    (A) 3.
    (B) 4.
    (C) 5.
    (D) Varies with the fluid manufacturer.

Chapter 5

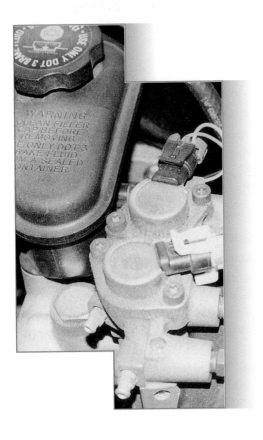

Master Cylinders, Calipers, and Wheel Cylinders

After studying this chapter, you will be able to:
- ❑ Describe the purpose and operation of brake pedal linkage.
- ❑ Explain the operation of dual master cylinders.
- ❑ Explain the operation of quick take-up master cylinders.
- ❑ Describe the operation of disc brake calipers.
- ❑ Identify the kinds of calipers.
- ❑ Describe the operation of drum brake wheel cylinders.
- ❑ Explain the purpose of bleeder screws.

Important Terms

Brake pedal linkage
Master cylinder
Dual master cylinders
Tandem master cylinders
Front/rear split hydraulic system
Diagonal split system
Pistons

Cups
Compensating port
Inlet port
Composite master cylinder
Quick take-up master cylinders
Single master cylinder
Caliper

Pressure seal
Dust boot
Single-piston floating caliper
Two-piston floating caliper
Four-piston fixed caliper
Wheel cylinder
Bleeder screws

This chapter is designed as a continuation of Chapter 4. In this chapter, the basic hydraulic system will be covered in detail, with emphasis on the main input and output devices—the master cylinder, calipers, and wheel cylinders. Completing this chapter will give you an understanding of the design, construction, and operation of the brake hydraulic system's major parts. Related hydraulic system parts will be covered later in this text.

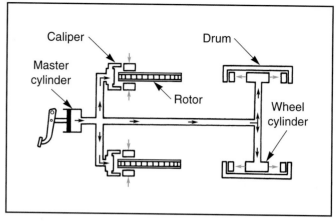

Figure 5-1. *A simple hydraulic brake system layout and operation. (TRW Inc.)*

Overall Hydraulic System

When the driver presses the brake pedal, the foot pressure is transferred through the pedal linkage to the pistons in the master cylinder. Moving the pistons generates hydraulic pressure. This pressure is then transmitted through the brake lines to the wheel cylinders and caliper assemblies.

The caliper pistons and wheel cylinders force the brake pads and shoes into contact with the brake rotor and drum. This process is shown in **Figure 5-1.** A detailed explanation of the master cylinder, calipers, and wheel cylinder is contained in this chapter. Other hydraulic components will be covered in later chapters.

Brake Pedal Linkage

Although it is not a hydraulic part, the brake pedal linkage is vital to the creation of hydraulic pressure. The **brake pedal linkage** increases foot pressure by simple mechanical leverage. As shown in **Figure 5-2,** the pedal is part of a simple lever arrangement. On a modern suspended pedal brake linkage, the connection to the master cylinder rod, usually called the **pushrod,** is closer to the pivot point than the pedal. This increases the pedal force by mechanical leverage. The force on the master cylinder is greater than the force placed on the brake pedal, but the distance traveled is less. If for instance, the ratio is one to four (1:4), pushing the pedal 1″ (25.4 mm) at a force of 10 psi (69 kPa), will result in a pressure on the master cylinder piston of 40 psi (276 kPa) with a movement of 1/4″ (6.4 mm).

The suspended pedal linkage pivots on bronze or plastic bushings. These bushings are usually prelubricated and do not require any maintenance. The brake pedal linkage on some vehicles may have a provision for adjustment. The adjustment mechanism usually consists of a threaded portion of the master cylinder pushrod, with a locknut to hold the adjustment.

The stoplight switch is usually installed on the brake pedal linkage. Other switches used with the vehicle cruise control, anti-lock brakes, or engine computer may be attached to the pedal linkage. The stoplight switch and anti-lock brakes switches will be discussed in later chapters.

Master Cylinder

The function of the **master cylinder** is to develop hydraulic pressure. Modern master cylinders have two separate hydraulic systems, using two **pistons** which operate in a single **cylinder bore.** If one of the hydraulic systems fail due to damage or other cause which results in fluid loss, the other system will continue to operate. This so-called split system provides extra safety for the driver and passengers. Master cylinders used with a split system are usually called **dual master cylinders** or **tandem master cylinders.**

Applying pressure to the master cylinder pushrod causes the rear piston to move forward in the cylinder. This exerts pressure on the brake fluid in the rear hydraulic system. The pushrod pressure acts on the front piston, causing it to move forward and create pressure in the front hydraulic system. The master cylinder rod movement creates the pressure in the rear hydraulic system. The rear hydraulic system creates the pressure in the front hydraulic system. **Figure 5-3** shows this process.

On many vehicles, the front piston operates the rear brakes and the rear piston operates the front brakes. The rear piston is sometimes called the primary piston, while the front piston is sometimes called the secondary piston. To avoid confusion with the piston cups (seals), we will refer to the pistons as the front and rear.

Notice that the rear piston has a solid front-facing projection which almost touches the front piston. If the front hydraulic system loses pressure, the projection will contact the front piston when the pedal is depressed. This allows the driver to stop the vehicle when the rear hydraulic system fails, although the pedal travel will be increased. If the front hydraulic system fails, the projection will keep the rear piston from moving excessively. A similar projection on the front piston keeps it from moving too far forward.

Because the wheel cylinders and master cylinder are connected by brake lines filled with fluid, any master cylinder piston movement is transmitted directly to the caliper pistons and wheel cylinders. Most master cylinder bores are the same size throughout their length. A few master cylinders have *stepped bores,* with the cylinder bore being two different sizes. A stepped bore is normally used in quick take-up master cylinders, discussed later in this chapter.

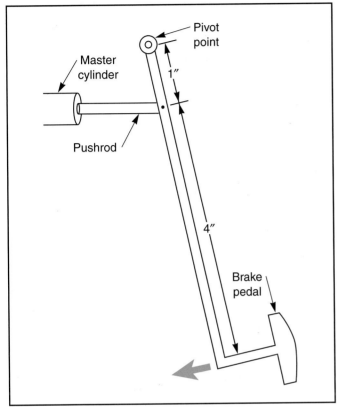

Figure 5-2. *Brake pedal mechanical linkage increases the foot pressure being applied by the driver.*

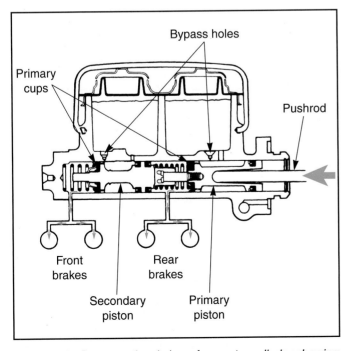

Figure 5-3. *Cross-sectional view of a master cylinder showing operation when the brakes are applied. (General Motors)*

Hydraulic Systems

On most older vehicles and a few modern brake systems, the rear piston operates the rear brakes, while the front piston operates the front brakes. This is known as a **front/rear split hydraulic system.** However, most new vehicles use a diagonal split system. In the **diagonal split system,** one of the hydraulic systems operates the left front and right rear brake system, while the other piston operates the right front and left rear system. The diagonal split system distributes the braking effort between the front and rear, resulting in better braking control.

Master Cylinder Components

The modern dual master cylinder has two separate pistons and chambers for pressure development. Older vehicles used a single piston master cylinder, discussed later in this chapter. Master cylinders are made of cast iron or aluminum. Most modern master cylinders are made of aluminum. Cast iron units are usually found in older vehicles and some current large cars and light trucks. Pressure development in either type is the same.

Most master cylinders are installed on the firewall or brake booster unit with studs passing through rear mounting holes in the master cylinder. The master cylinder is held by nuts threaded on the studs.

Master Cylinder Piston and Cups

The **pistons** used in the master cylinder are constructed of aluminum or high impact plastic. Installed on the pistons are rubber **cups,** sometimes called *lip seals,* **Figure 5-4.** The pistons also have springs which help to return the pistons and the brake pedal to their proper positions when the brakes are released. The cups and springs may be separate units, or may be assembled onto the piston forming a complete assembly.

Primary Cups

The inner (forward facing) end of each piston pushes against the **primary cup.** The primary cup prevents brake fluid from leaking past the piston, which could prevent the development of hydraulic pressure.

Secondary Cups

The rear of each piston has a **secondary cup.** The secondary cup on the front piston keeps pressure from leaking between the two chambers. The secondary cup on the rear piston keeps brake fluid from leaking out the back of the master cylinder.

When the pistons move forward, pressure forces the cups against the cylinder walls, **Figure 5-5,** and they seal in the developing pressure. When the brake is released, the pistons move toward the rear. This causes the pressure to drop and the cups flex inward, **Figure 5-6,** allowing fluid to enter from the rear of the cylinder. Fluid flow is increased by the use of bleeder holes in the piston, directly behind the cup.

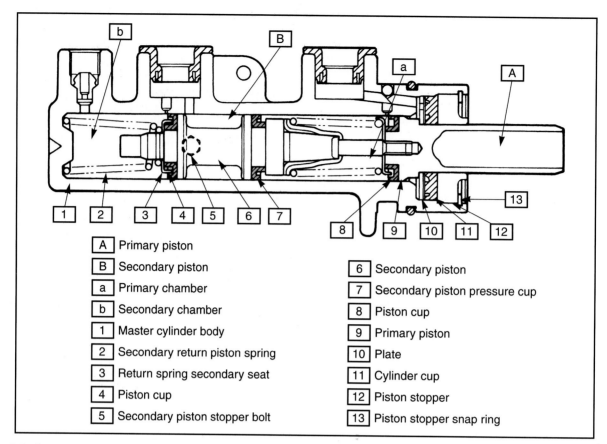

Figure 5-4. *Cross-sectional view of a master cylinder showing the piston cup seals. (General Motors)*

A Primary piston	
B Secondary piston	
a Primary chamber	6 Secondary piston
b Secondary chamber	7 Secondary piston pressure cup
1 Master cylinder body	8 Piston cup
2 Secondary return piston spring	9 Primary piston
3 Return spring secondary seat	10 Plate
4 Piston cup	11 Cylinder cup
5 Secondary piston stopper bolt	12 Piston stopper
	13 Piston stopper snap ring

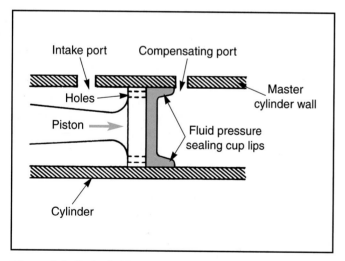

Figure 5-5. *Brake fluid pressure forces the seal lips against the cylinder wall. This causes brake fluid pressure to build. (FMC)*

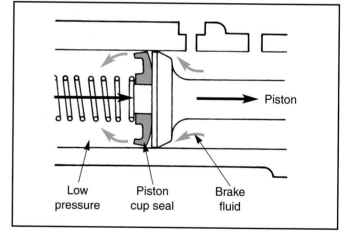

Figure 5-6. *As the piston returns to its resting position, brake fluid has forced the cup seal away from the cylinder wall, allowing fluid to travel by the seal. (TRW Inc.)*

Master Cylinder Ports

At certain times, brake fluid must pass freely between the cylinder and pistons and the reservoir. At other times, the reservoir and cylinder must be sealed from each other. This is accomplished by two ports drilled into the top of the cylinder, the compensating port, and the inlet port. **Figure 5-7** shows the relative position of these ports. Note that the cylinder contains four ports, one of each design for each pressure chamber.

Compensating Port

The **compensating port,** sometimes called the *bypass* or *vent port,* is located just ahead of the primary cup seal. It allows pressurized fluid in the hydraulic system to return to the reservoir when the brakes are released. When hydraulic pressure falls below atmospheric pressure, brake fluid can flow from the reservoir through the compensating port and into the piston cylinder. This keeps the system filled with fluid and ready to be pressurized again.

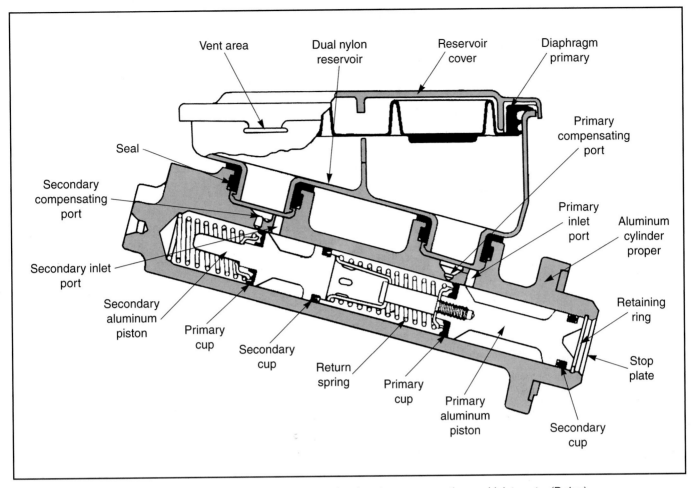

Figure 5-7. *A cross-sectional view of a dual master cylinder showing the compensating and inlet ports. (Delco)*

Inlet Port

The **inlet port,** sometimes called the *breather* or *intake port,* is located farther back towards the rear area of each piston, always behind the primary seal. This port is used to allow the entry of brake fluid into the rear section of the piston cylinder. When the fluid pressure drops during brake release and the lip seals flex inward, fluid is available to refill the pressure chamber.

Residual Pressure Valve

If the vehicle is equipped with rear drum brakes, some pressure must be maintained in the front system to keep the wheel cylinder cups from collapsing. To maintain pressure, a **residual pressure valve,** discussed in Chapter 4, is used. On some older master cylinders, the residual pressure valve was a piston return spring and check valve assembly. It was located at the front of the master cylinder, ahead of the front piston. See **Figure 5-8.**

The piston return spring holds the check valve shut after the brakes have been fully released. It is positioned between the piston primary cup and the check valve. On newer master cylinders, the residual pressure valve is installed under the tube seat of the outlet line, **Figure 5-9.** If the vehicle has four-wheel drum brakes, each outlet line will have a residual pressure valve.

The residual pressure valve is a two-way check valve that allows fluid to flow in and out of the master cylinder, but only as varying pressure conditions are met. When the brakes are applied, the piston forces fluid through the check valve into the brake lines. When the brakes are released, pressure created by the brake shoe return springs forces the fluid back through the check valve and into the piston cylinder.

The piston return spring will seat the check valve against its seat when the spring pressure is greater than the pressure of the returning fluid. This maintains a residual pressure of around 8-16 lbs (55.2-110.3 kPa) in the lines. This pressure is enough to keep the wheel cylinder cups expanded, but is not enough to overcome brake shoe return spring pressure and apply the brakes.

Master Cylinder Reservoir

As discussed earlier, each type of master cylinder uses a **reservoir** to hold extra brake fluid. As the brake pad and shoe linings become thinner, the apply pistons have to travel a longer distance to bring them into contact with the drums and rotors. To do this, the pistons must travel a longer distance, and require more fluid. This extra fluid is supplied from the reservoir.

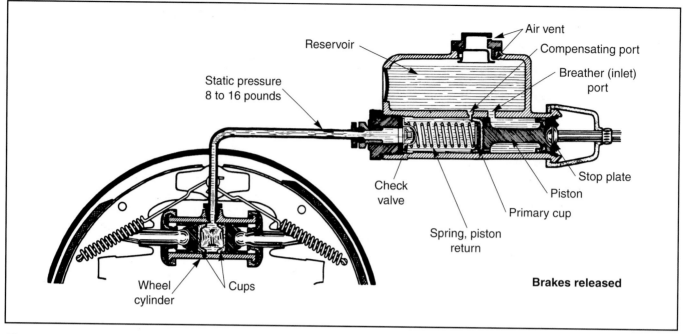

Figure 5-8. *Check valve location in an older style master cylinder. (General Motors)*

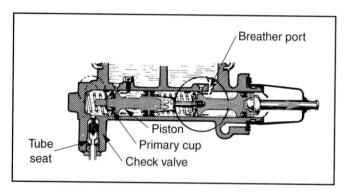

Figure 5-9. *Residual pressure check valve installed under the brake line tube seat. (General Motors)*

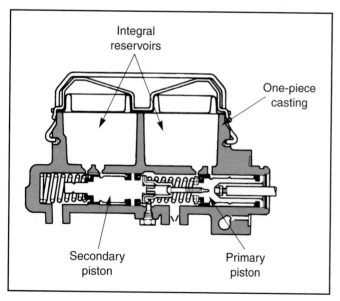

Figure 5-10. *A cross-sectional view of a one-piece cast master cylinder with an integral reservoir. (TRW Inc.)*

The extra fluid in the reservoir can compensate for minor leaks which may go undiscovered for a short period of time. The reservoir also acts as a chamber into which brake fluid can expand or contract with differences in temperature.

Many master cylinders have an *integral* (built-in) brake fluid reservoir, **Figure 5-10.** On some vehicles, a *detached*, or remotely located fluid reservoir is connected to the master cylinder by tubing. A detached reservoir system is shown in **Figure 5-11.** Modern master cylinders have plastic reservoirs which are installed directly on the master cylinder. A cylinder with this type of reservoir is sometimes called a **composite master cylinder.** O-rings seal the contact points between the reservoir and master cylinder. A composite master cylinder reservoir is shown in **Figure 5-12.**

The **solid baffle** reservoir has two separate compartments. One compartment is used to supply brake fluid to the front brakes and the other to the rear. A solid baffle reservoir is illustrated in **Figure 5-13.** A **split baffle** or **semi-baffle** reservoir is shown in **Figure 5-14.** This type has a semi-separate compartment that supplies brake fluid to both the front and rear brakes.

Inside the reservoir cover is a rubber diaphragm. When the cover is in place, the diaphragm seals the space between the reservoir body and cover. This diaphragm is flexible to allow the fluid level to drop without allowing air to enter the reservoir. Air must be kept out because it contains moisture. As the fluid level drops, air pressure pushes the diaphragm down. The reservoir cover is vented to allow air pressure to enter on top of the diaphragm. The reservoir cover may be held by clips or a wire bail, or it may be a plastic cap that tightens onto the reservoir cover like a jar lid. A few reservoir covers are held by a screw or bolt.

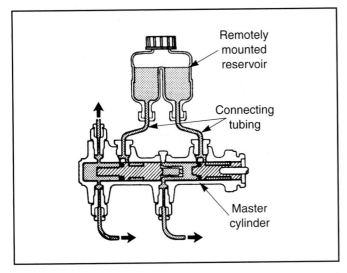

Figure 5-11. *A master cylinder which uses a remotely mounted brake fluid reservoir. (Delco)*

Master Cylinder Operation—Normal Braking

The following sections explain how the master cylinder operates. This information applies to a master cylinder used on a vehicle with front disc and rear drum brakes. The operation of all other designs, (four-wheel disc or four-wheel drum) is similar.

Brake Apply

In **Figure 5-15,** the pushrod is being forced forward by foot pressure on the brake pedal. The pushrod moves the rear piston forward into the cylinder. As the rear piston is forced forward, the primary cup seals the compensating port and pressure begins to build. This creates pressure in the rear brake system, which is transmitted to the front brakes. The pressure increase in the rear brake system pushes on the front piston. As the front piston moves

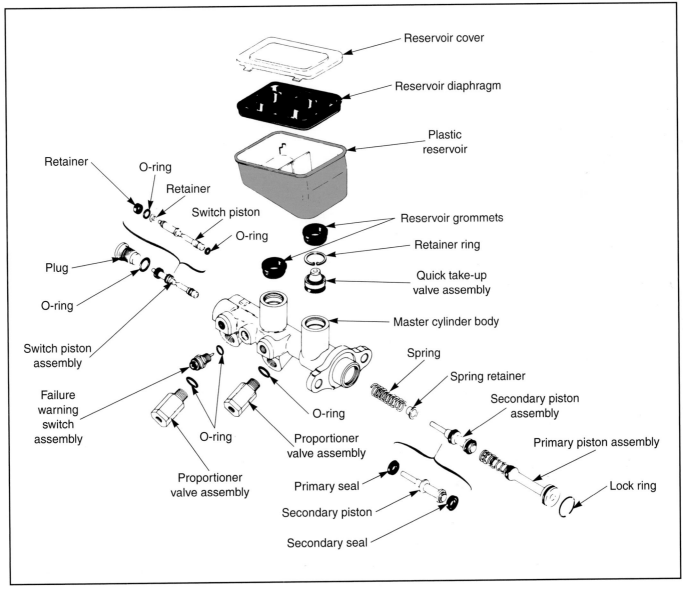

Figure 5-12. *Exploded view of a dual master cylinder incorporating a plastic (composite) fluid reservoir. (Buick)*

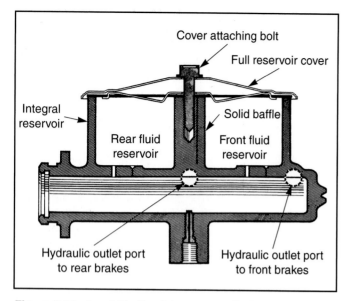

Figure 5-13. *A solid baffle style master cylinder. (Raybestos)*

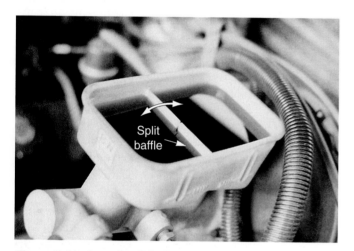

Figure 5-14. *A semi-split baffle master cylinder reservoir. Note that the front and rear reservoir fluid is free to travel from side-to-side over the top of the baffle.*

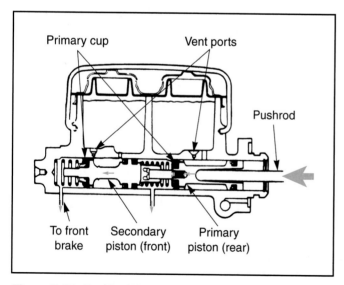

Figure 5-15. *Dual brake system master cylinder being applied by the pushrod. (Wagner)*

forward, the primary cup seals the compensating port and pressure builds up in the front brake system. When the hydraulic pressure increases to a level higher than the static pressure in the front system, the residual pressure valve opens, allowing pressure to be transmitted to the rear brakes.

Brake Release—Start

When the brake pedal is released, pressure on the pushrod is reduced. This allows the front and rear pistons to move back toward their normal released positions, assisted by the internal positioning springs. As the pistons move, brake fluid flows through the piston bleeder holes and the bent lips of the primary cups. This allows fluid to flow into the cylinder in front of the pistons. This action is shown in **Figure 5-16.**

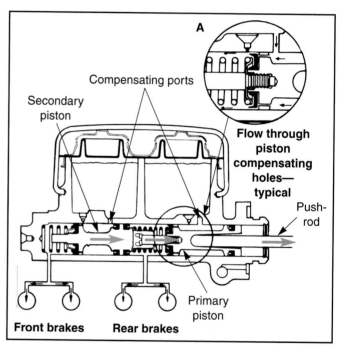

Figure 5-16. *Start of brake release in a dual master cylinder. Note the fluid flow arrows in view A. (General Motors)*

Brake Release—Finish

As the pressure continues to drop in the hydraulic system, the caliper seals and brake shoe return springs retract the pads and shoes and force the caliper and wheel cylinder pistons to the released position. As this occurs, brake fluid is displaced into the brake line and returns to the master cylinder.

As the fluid is forced back into the front piston area in the master cylinder, it pushes the residual pressure check valve inlet to the closed position. The returning hydraulic pressure check valve body is then moved from its seat and brake fluid travels into the master cylinder.

When the piston has returned all the way to its normal released position, the primary cup uncovers the compensating port and any excess brake fluid flows back into the reservoir. This action is shown in **Figure 5-17.** As the piston return spring overcomes fluid pressure, the residual pressure valve seats and static pressure remains in the hydraulic lines ahead of the residual pressure valve.

As the fluid is forced back into the rear piston area, it pushes the rear piston back toward its released position. When the piston has returned all the way to its normal released position, the primary cup uncovers the compensating port and excess brake fluid flows back into the reservoir. See **Figure 5-18.**

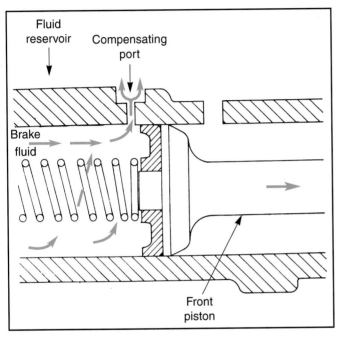

Figure 5-17. *As the piston returns to its seat, fluid travels back into the master cylinder reservoir. (Wagner)*

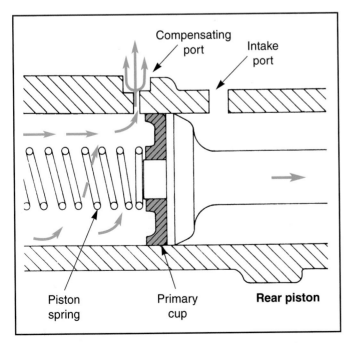

Figure 5-18. *Primary cup uncovers the compensating port, allowing excess brake fluid back into the reservoir. (Wagner)*

Master Cylinder Operation During Hydraulic System Failure

Master cylinder operation is affected when either the front or rear brake hydraulic system leaks. However, the design of the dual master cylinder prevents complete loss of braking. The operation of a master cylinder when either the front or rear brakes fail is covered in the following sections.

Front Hydraulic System Failure

When a seal failure or severe fluid loss occurs in the front hydraulic system of a master cylinder, the remaining brake fluid in the rear hydraulic system will still be able to stop the vehicle. When there is a failure in the front brake system, both pistons in the master cylinder will move forward when the brake pedal is pressed. However, because of the inability to build up hydraulic pressure, there is nothing to resist the front piston movement except the piston return spring at the front of the cylinder.

When the front piston reaches the end of the cylinder, it stops, and the rear piston can then create pressure in the rear hydraulic system. This creates enough pressure in two of the brakes to stop the vehicle, although braking distances will be longer. The brake pedal must move an excessive amount to move the front piston to the end of its travel and then pressurize the rear hydraulic system. Since the driver might not realize that excess pedal travel indicates a brake system problem, the pressure differential switch illuminates the dashboard brake warning light. Master cylinder operation during front hydraulic system failure is shown in **Figure 5-19.**

Rear Hydraulic System Failure

If a hydraulic failure occurs in the master cylinder rear pressure chamber, the rear piston is forced forward when the brake pedal is pressed, but does not create hydraulic pressure. Continued primary piston movement causes it to contact the

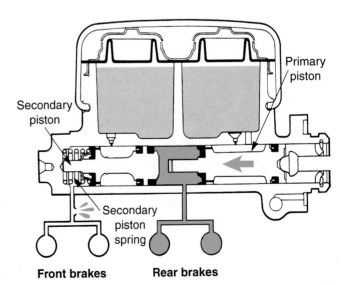

Figure 5-19. *Master cylinder operation during a front brake system failure. (Chevrolet)*

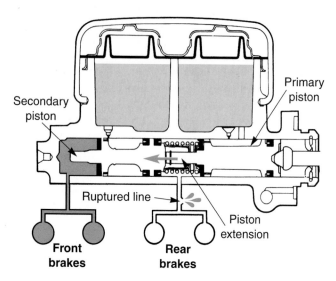

Figure 5-20. *Operation of the master cylinder as a rear brake system failure occurs. (Chevrolet)*

front piston and move it forward. The front piston then develops pressure in the front hydraulic system which operates the associated brakes. As with a front brake failure, pedal travel will be excessive, and the brake warning light will be illuminated. The operation of the master cylinder during rear hydraulic system failure is shown in **Figure 5-20.**

Quick Take-up Master Cylinder

Many modern master cylinders are equipped with additional valves to fully control the flow of hydraulic fluid. Some of these valves, such as the proportioning and metering valves, operate in the same manner as the valves discussed in Chapter 4, and are installed on the master cylinder for manufacturing convenience. Some master cylinders are designed with internal modifications designed to improve their operation. These master cylinders are usually called *quick take-up master cylinders.*

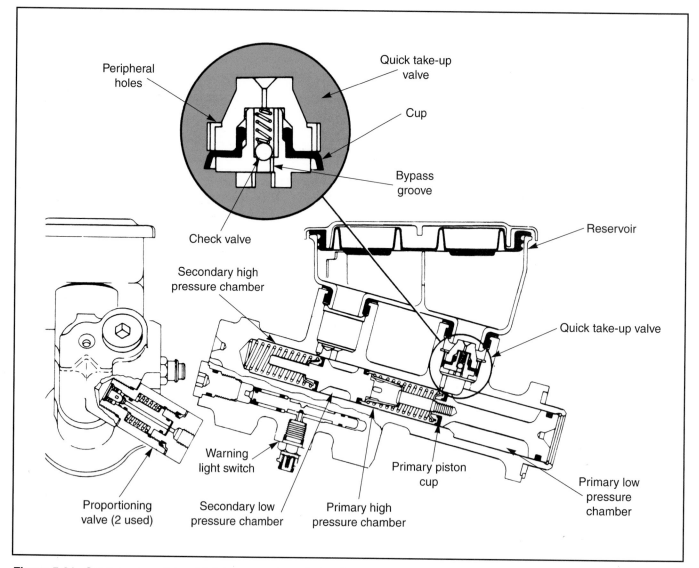

Figure 5-21. *Cutaway view of a quick take-up master cylinder. Study the construction of the cylinder and the quick take-up valve. Note the warning light switch location and the integral proportioning valves. (Chevrolet)*

Quick Take-up Design

The quick take-up master cylinder, sometimes called the *fast-fill master cylinder*, is installed on vehicles with no- or low-drag calipers. The *no-drag caliper* is designed to ensure the disc brake pads do not contact the rotor when the brakes are released. To accomplish this, the caliper piston seals retract more than piston seals in standard calipers. However, this extra clearance would cause a low brake pedal if the quick take-up valve was not used.

The quick take-up valve sends a large amount of brake fluid to the caliper pistons at low pressure during the initial brake application. The low pressure fluid quickly moves the piston outward, removing the extra clearance caused by seal retraction. A quick take-up master cylinder is shown in **Figure 5-21**. Refer to this illustration as you read the following paragraphs.

Stepped Bore

The main difference between the quick take-up cylinder and a standard master cylinder is the use of a **stepped bore.** In addition to the normal front and rear high pressure chambers, this master cylinder has a large third bore located at the rear of the cylinder. The job of this large bore and its accompanying piston is to create a low pressure, high flow chamber. The brake pedal pushrod contacts the piston contained in this large chamber. The large piston is physically connected to the rear high pressure piston.

Quick Take-up Valve

A quick take-up valve is connected to the low pressure chamber. The **quick take-up valve** contains a spring-loaded check valve that is normally closed. The pressure required to open the valve is controlled by spring tension. A narrow bypass groove runs along the side of the valve. This groove allows fluid to pass between the low pressure chamber and reservoir. The amount of fluid that can pass through this groove is too small to affect hydraulic pressure when the brakes are operated. The valve also has an external lip seal to allow quick refilling of the low pressure chamber when the brakes are released.

Quick Take-up Operation—Brake Apply

When the brakes are first applied, **Figure 5-22A**, the pushrod moves the large rear piston. Since this piston has a large area, slight piston movement causes high fluid flow, although pressure is low. Since most of the pushrod movement is used to displace the fluid in the large rear chamber, the front and rear high pressure pistons do not move much. Any slight pressure the high pressure piston develops is overwhelmed by the pressure and flow developed by the low pressure piston. Low pressure fluid from the rear chamber travels past the front and rear high pressure piston lip seals and into both high pressure chambers. From the high pressure chambers, it applies the wheel brakes. At this point, the entire brake system is pressurized, but only enough to bring the disc brake pads into light contact with the rotor.

Once the brake pads are in contact with the rotors, normal master cylinder operation begins and high pressure is created by the front and rear pistons. Any excessive pressure that builds up in the rear low pressure chamber overcomes spring tension and opens the quick take-up valve. Excess fluid then exits to the reservoir. Since the rear low pressure piston is directly connected to the rear high pressure piston, this loss of pressure has no effect on normal master cylinder operation. This is shown in **Figure 5-22B.**

Quick Take-up Operation—Brake Release

When the brake pedal is released, fluid returns from the front high pressure piston through the compensating and intake ports. Fluid returns from the rear high pressure

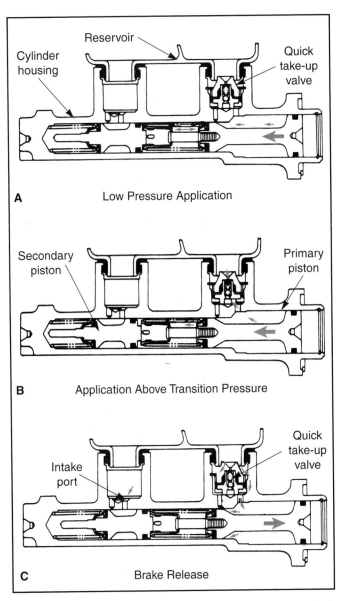

Figure 5-22. A—Initial brake application. B—During apply, it functions as a standard dual master cylinder. C—Brake release. Note the brake fluid flow through the quick take-up valve in views B and C. (Wagner)

piston through the front piston lip seal and through a vent hole located at the quick take-up valve. In most cases, the pressure in the releasing rear low pressure chamber will be less than reservoir pressure, and fluid will enter through the quick take-up valve lip seal. Refer to **Figure 5-22C.**

Master Cylinders Used with Anti-lock Brake Systems

Master cylinders used with most anti-lock brake systems (ABS) operate in the same manner as those used in non anti-lock systems. These master cylinders supply fluid to a hydraulic modulator, which distributes the fluid to the individual calipers and wheel cylinders. A few older ABS systems use master cylinders that are integral (one-piece) castings with other ABS components. Chapters 21 and 22 contain more information on ABS master cylinders.

Single Master Cylinder

You may occasionally encounter a **single master cylinder** on an older vehicle. The last new vehicles using single master cylinders were made in 1966. Starting with the 1967 model year, dual master cylinders were required by law.

A single master cylinder has only one piston. The hydraulic system of both the front and rear brakes are operated by pressure developed by this piston. Operating principles of single master cylinders is identical to one side of a dual master cylinder. Three single master cylinders are shown in **Figure 5-23.**

Disc Brake Calipers

All modern vehicles built in the last 25-30 years have disc brakes on the front axle, and many have discs on all four wheels. All disc brakes make use of a **caliper** containing at least one piston and cylinder. The basic function of the disc brake caliper cylinder and piston is to receive, contain, and convert hydraulic pressure from the master cylinder to mechanical force. The piston can then force the brake pads against the turning rotor.

Cylinder and Piston Construction

The caliper cylinder and piston assembly, **Figure 5-24,** is a basic hydraulic piston and cylinder assembly. Calipers are usually made of aluminum, but may be made from cast iron. Pistons are made of aluminum or phenolic (high strength) plastic. The caliper is drilled and threaded to accept a brake line connection and a bleeder screw.

Most caliper pistons have one **pressure seal** and a **dust boot, Figure 5-25.** The seal holds fluid under pressure, and the dust boot keeps dust and water from deteriorating the seal. Seals and boots are made of natural rubber. The brake pads may be attached to the caliper piston or held in place by other hardware.

Caliper Designs

In the past, many kinds of caliper designs were tried and discontinued for various reasons. On modern vehicles,

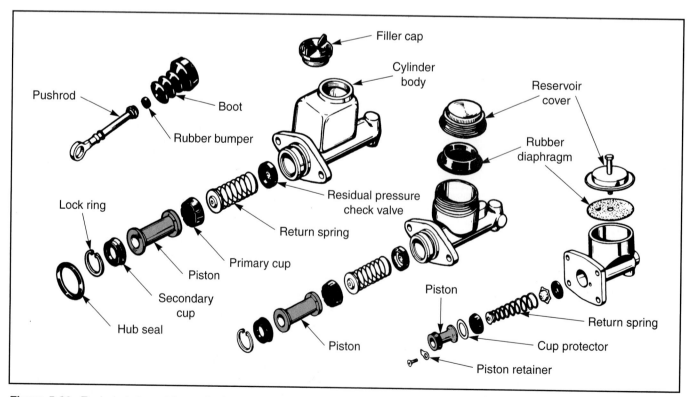

Figure 5-23. *Exploded view of three single piston master cylinders. (Bendix)*

Figure 5-24. *A single-piston disc brake caliper assembly.*

a few common caliper designs are used. The two major kinds of caliper mountings designs are the fixed caliper and the floating caliper. Modern calipers may be equipped with one, two, or four pistons. One and two piston calipers are floating types; four piston calipers are fixed designs. Other caliper design factors, as they apply to the friction system, will be covered in Chapter 12.

Single-Piston Floating Caliper

The *single-piston floating caliper* is a one piece casting with a single piston which can apply both brake pads. An example of this type of caliper is shown in **Figure 5-26.** This caliper is the most common design and is manufactured in many sizes depending on vehicle design. The piston and cylinder are mounted inboard (toward the middle of the vehicle). The caliper assembly can slide, or float, on

the fixed spindle mounting. This type of caliper is also known as a no- or low-drag caliper.

When the brakes are applied, hydraulic pressure pushes on the piston. The piston moves outward, pushing the inboard brake pad against the rotor. When the inboard pad has solidly contacted the rotor, instead of pushing the inboard pad farther outward, the piston pressure begins to push the entire caliper assembly inboard. As the caliper assembly slides inboard, the outboard pad comes into contact with the rotor. With both pads applied, friction begins to slow the rotor. When the piston moves forward, the piston seal is deformed outward, as shown in **Figure 5-27.**

When the brakes are released, hydraulic pressure is removed. Since the piston seal was deformed by piston movement, it tries to return to its original position. As the seal returns to its original position, it pulls the piston (and brake pad) away from the rotor.

Two-Piston Floating Caliper

A modern *two-piston floating caliper* uses either integral cylinders (part of the caliper) or removable cylinders. An integral cylinder caliper is shown in **Figure 5-28,** and a removable cylinder type is shown in **Figure 5-29.** Both pistons are located on the inboard side of the caliper assembly.

This caliper operates in the same manner as the single piston caliper. As the brakes are applied, both pistons move outward and push the inboard brake pad against the rotor. Once the inboard pad cannot move any farther outward, piston pressure pushes the caliper assembly inboard, causing the outboard brake pad to contact the rotor. When the brakes are released, the piston seal returns to its original position, pulling the piston away from the rotor.

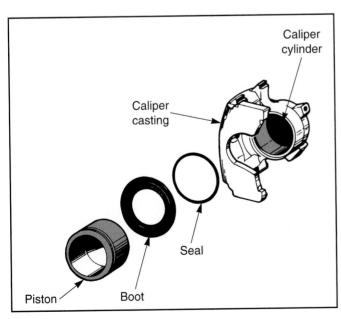

Figure 5-25. *A caliper assembly which contains a single piston. Only one fluid pressure seal and dust boot are used. (Plymouth)*

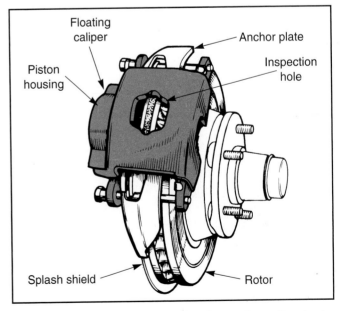

Figure 5-26. *A single-piston floating caliper disc brake assembly. Note the inspection hole which provides a clear view of the inboard brake pad. This is handy for quickly checking pad wear. (Bendix)*

Four-Piston Fixed Caliper

The *four-piston fixed caliper* was the first mass production disc brake design, and is still used on some vehicles. Two pistons are located on the inboard side of the rotor and two on the outboard side. All four pistons are the same size and shape. The two sides of the caliper are connected by internal fluid passages, or on a few vehicles, by external tubing. A four-piston fixed caliper assembly is illustrated in **Figure 5-30**.

When fluid enters the area behind the pistons, all four pistons move inward toward the rotor. This moves the brake pads into contact with the rotor. When the brakes are released, the piston seals return the pistons to the unapplied position.

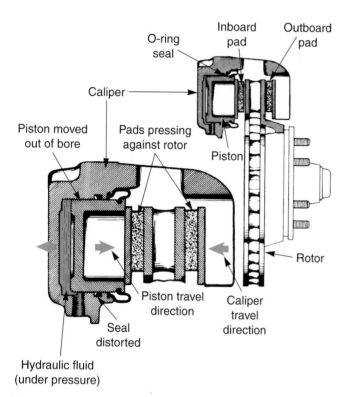

Figure 5-27. *Brake pressure being applied to stop the rotor. Note the distorted caliper piston seal. (Bendix)*

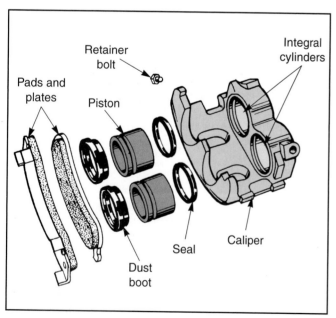

Figure 5-28. *Exploded view of a two-piston caliper assembly. The caliper and piston cylinders are cast as a one-piece unit. (Bendix)*

Wheel Cylinders

The function of the **wheel cylinder** is to change hydraulic pressure from the master cylinder into mechanical force that applies the brake shoes against the rotating drum. Many types of wheel cylinders were used in the past, but modern vehicles use one wheel cylinder design, the **single cylinder dual piston** or straight bore type. In this design, the cylinder bore is one size throughout its length, and has two pistons of equal size, one at each end. This type of wheel cylinder changes hydraulic pressure into movement in two directions.

Wheel Cylinder Construction and Operation

The wheel cylinder body is constructed of cast iron or aluminum. The pistons are usually made of aluminum or iron. Dust boots and piston seals are made of natural rubber. Both the front and rear pistons are the same size. Some wheel cylinders have separate pushrods which connect the pistons to the brake shoes, while on other brake systems, the shoes are designed to contact the pistons directly.

A hole is drilled and threaded near the center of the cylinder for a brake line connection. A screw hole is added to fit a bleeder screw for air removal. Typical wheel cylinder construction is illustrated in **Figure 5-31**.

As the brakes are applied, pressurized fluid enters the wheel cylinder from the brake line. The pressure forces each piston outward. They, in turn, force the brake shoes into contact with the brake drum.

When pressure on the brake pedal is released, the brake shoe return springs retract the shoes, and they no longer contact the drum. As the springs retract the shoes, the wheel cylinder pistons are forced back to their released position in the cylinder bore.

A small coil spring is used to keep the cups against the pistons and keep the internal parts in position. Some cylinders have metal cup expanders, **Figure 5-32**. These are used to force the cup lips into close contact with the cylinder wall.

Bleeder Screws

As you learned in Chapter 4, air is compressible. Air trapped in a brake system will compress instead of transmitting pressure to apply the brakes. Air can enter the system when a hydraulic line or component has been opened,

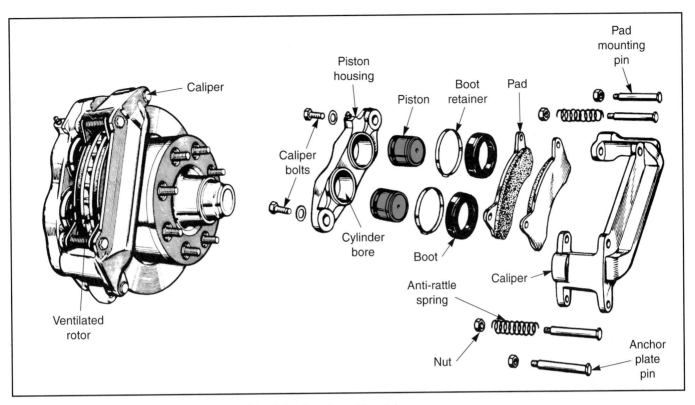

Figure 5-29. *A two-piece caliper with removable piston housings. (Bendix)*

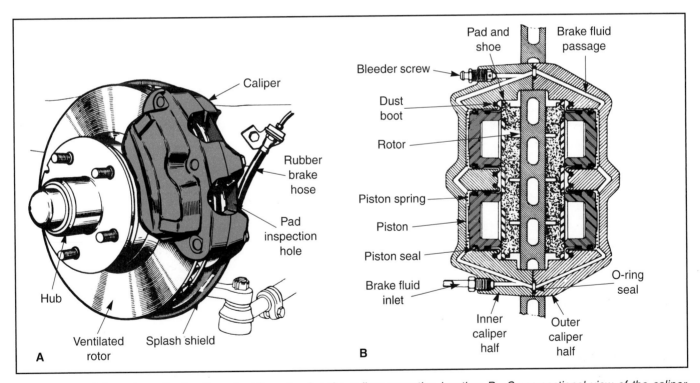

Figure 5-30. *A four-piston fixed caliper. A—View showing the caliper mounting location. B—Cross-sectional view of the caliper assembly. Note how the cylinders are connected with brake fluid passages. (Bendix)*

removed, or replaced, when the system has been drained and flushed, or if the level in the master cylinder becomes too low. Occasionally a defective seal will allow air to enter a wheel cylinder or caliper. For efficient brake operation, this air must be removed through the bleeder screws.

Bleeder screws are air removal devices installed at the highest point in a hydraulic device. See **Figure 5-33.** Calipers and wheel cylinders are always equipped with bleeder screws, and they are often installed on master cylinders and hydraulic valve assemblies.

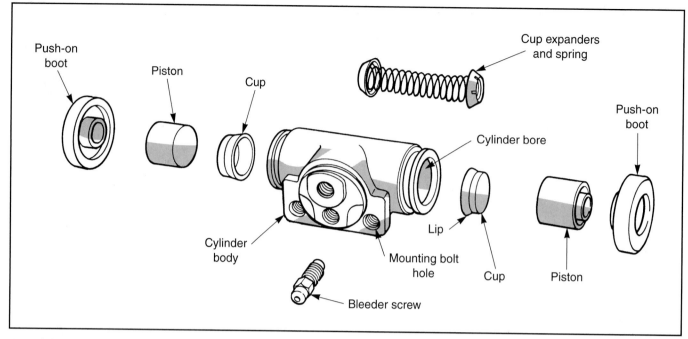

Figure 5-31. *Exploded view of a typical wheel cylinder. Note the cup expanders and spring which helps to keep the cups' sealing lip in contact with the cylinder bore wall. (Chrysler)*

To remove air from the hydraulic system, the bleeder screw is opened after pressure is developed in the hydraulic system by foot pressure on the brake pedal or by a pressure bleeder. Since the bleeder opening is at the highest point of the hydraulic system, the lighter air will be removed, or bled, from the system. While brake system pressure is maintained, the bleeder screw is closed to prevent the reentry of air. A detailed description of brake bleeding is covered in Chapter 6.

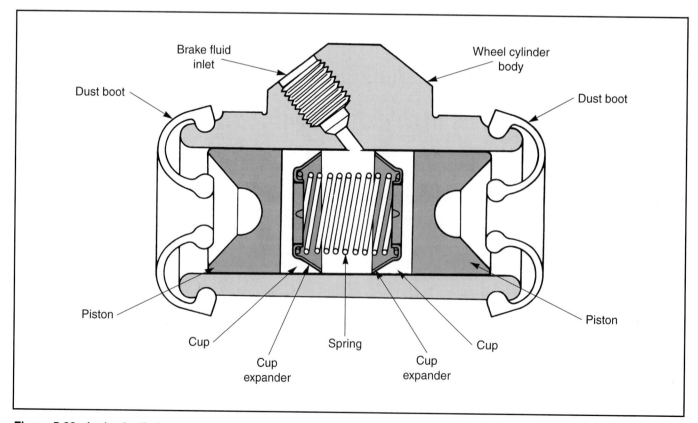

Figure 5-32. *A wheel cylinder that uses metal cup expanders and an expander spring. (Allied Automotive)*

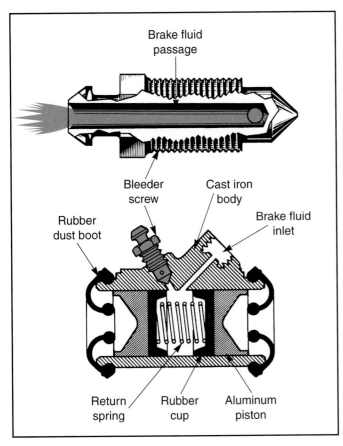

Figure 5-33. *A cross-sectional view of a wheel cylinder showing the brake fluid bleeder screw. (Wagner)*

The quick take-up master cylinder is used with low-drag disc brakes. It contains a large extra bore and a check valve. When the brakes are applied, the large bore and piston supply a large volume of fluid at low pressures, extending the brake pads until they contact the rotor. The master cylinder then operates like a conventional unit. The check valve releases excessive pressure from the low pressure chamber.

ABS master cylinders are identical in design to those used with non-ABS systems, but may be built as part of the ABS hydraulic unit. Single-piston master cylinders were used for many years. Single piston master cylinders have not been used since 1966.

The disc brake caliper cylinder and piston converts hydraulic pressure from the master cylinder to mechanical force. Disc brake calipers can be floating or fixed designs. The three most commonly used disc brake calipers are the single-piston, two-piston and four-piston.

Wheel cylinders are used to change hydraulic pressure from the master cylinder into mechanical force applying the brake shoes against the drum. Modern cylinders use a single size bore with pistons on both ends. Cylinders are made of aluminum or cast iron, with aluminum pistons. Rubber is used for dust boots and piston seals.

When air is trapped in a brake system, it can cause complete brake failure. Air can enter the system in many ways, and must be removed by bleeding. Bleeder screws are installed on all major hydraulic system parts.

Summary

The brake hydraulic system includes the master cylinder, wheel cylinders, and caliper cylinders. The hydraulic system is activated by the brake pedal through brake pedal linkage.

The function of the master cylinder is to generate hydraulic pressure. Modern master cylinders are dual or tandem types. The dual master cylinder is part of the modern split braking system. In the past, one of the problems with a hydraulic brake system was the fact that a loss of pressure usually resulted in a complete loss of braking ability. To improve safety, the dual cylinder, split system was incorporated. If one part of the hydraulic system fails, the other part can still stop the vehicle. The split system can be a front and rear type or operate one front and one rear brake on opposite sides of the vehicle.

The master cylinder body is made from cast iron or aluminum. Master cylinders may have a stamped steel reservoir, a plastic reservoir, or an integral unit with the cylinder housing and reservoir in one assembly. The master cylinder has two pistons for creating hydraulic pressure in the front and rear systems. Pressure on the rear piston causes pressure to be developed in the rear pressure chamber, and pushes on the front piston. This pressure is delivered to the wheel brakes. When the pressure is reduced, excess fluid is released through the compensating and vent ports.

Review Questions—Chapter 5

Please do not write in this text. Write your answers on a separate sheet of paper.

1. The brake pedal linkage is designed to increase foot pressure by the use of mechanical _____.

2. Modern master cylinders contain two _____ which operate in a single _____.

3. Since the hydraulic system is filled with brake fluid, any movement of the master cylinder _____ causes the caliper and wheel cylinder _____ to move.

4. Match the master cylinder part with its function:

   Primary cups _____   (A) Supplies extra fluid.
   Secondary cups _____   (B) Stops fluid from leak-
   Compensating _____   ing out of cylinder.
     port   (C) Allows pressurized
   Inlet port _____   fluid to escape to
   Residual _____   reservoir.
     pressure valve   (D) Stops fluid from leak-
   Reservoir _____   ing between pistons.
                                (E) Allows pressure
                                    chamber to refill.
                                (F) Keeps slight pressure
                                    on wheel cylinders.

5. During normal brake system operation, what moves the front piston in the master cylinder?

6. Describe the purpose of the split hydraulic system.

7. A quick take-up master cylinder has _____ low pressure chamber(s) and _____ high pressure chamber(s).

8. When the inner pad of a floating caliper is moved into contact with the rotor and cannot move any farther, the _____ moves.

9. Modern wheel cylinders have one _____ and two _____.

10. To remove air from the hydraulic system, _____ are installed at various points.

ASE Certification-Type Questions

1. All of the following statements about the brake pedal linkage are true, EXCEPT:
   (A) the pedal linkage is not a hydraulic part.
   (B) pedal linkage increases foot pressure by mechanical leverage.
   (C) the brake pedal linkage cannot be adjusted.
   (D) the brake pedal linkage contains the stoplight switch.

2. Modern master cylinders have _____.
   (A) one cylinder bore
   (B) two pistons
   (C) a reservoir
   (D) All of the above.

3. On a normally operating dual master cylinder, what moves the front piston when the brake pedal is depressed?
   (A) The pushrod.
   (B) The brake light switch.
   (C) Pressure from the rear piston.
   (D) The residual pressure valve.

4. If the primary cup fails, which of the following is most likely to happen?
   (A) External leaks.
   (B) No hydraulic pressure.
   (C) Brake lockup.
   (D) Both A & C.

5. Technician A says that the compensating port allows pressurized fluid to enter the master cylinder when the brakes are applied. Technician B says that the compensating port allows pressurized fluid in the hydraulic system to return to the reservoir when the brakes are released. Who is right?
   (A) A only.
   (B) B only.
   (C) Both A & B.
   (D) Neither A nor B.

6. Technician A says that the residual pressure valve is located at the front of the master cylinder, ahead of the front piston. Technician B says the residual pressure valve is installed under the tube seat of the master cylinder outlet. Who is right?
   (A) A only.
   (B) B only.
   (C) Both A & B.
   (D) Neither A nor B.

7. All of the following statements about master cylinder reservoirs are true, EXCEPT:
   (A) some reservoirs are built into the master cylinder assembly.
   (B) some reservoirs are located away from the master cylinder.
   (C) reservoirs are always made of cast iron.
   (D) a split baffle reservoir allows fluid to pass between the front and rear chambers.

8. When do the primary cups allow brake fluid to flow past them?
   (A) Start of brake apply.
   (B) Brakes held on.
   (C) Start of brake release.
   (D) End of brake release.

9. What might the driver notice if either chamber of a split braking system fails?
   (A) Increased stopping distances.
   (B) Increased pedal travel.
   (C) Brake warning light on.
   (D) All of the above.

10. The quick take-up master cylinder contains all of the following, EXCEPT:
    (A) stepped cylinder bore.
    (B) one low pressure piston.
    (C) one high pressure piston.
    (D) a quick take-up valve.

11. Technician A says that the purpose of the quick take-up master cylinder is to remove the air from the disc brake calipers. Technician B says that the purpose of the quick take-up master cylinder is to remove the clearance between the disc brake pads and rotors. Who is right?
    (A) A only.
    (B) B only.
    (C) Both A & B.
    (D) Neither A nor B.

12. All of the following statements about disc brake calipers are true, EXCEPT:

    (A) the floating caliper can slide on the fixed spindle mounting.

    (B) a fixed caliper cannot move at any time.

    (C) single piston calipers are always floating types.

    (D) two piston calipers have one piston on each side of the rotor.

13. What releases the piston when the hydraulic pressure is removed?

    (A) Return springs.

    (B) Piston seal flexing.

    (C) Dust boot flexing.

    (D) Both B & C.

14. All of the following statements about modern wheel cylinders are true, EXCEPT:

    (A) the cylinder bore is one size throughout the wheel cylinder length.

    (B) the cylinder bore has two unequal size pistons.

    (C) the wheel cylinder may be made of cast iron or aluminum.

    (D) the pistons are equipped with cup seals.

15. Bleeder screws are always installed at the _____ point of a hydraulic device.

    (A) highest

    (B) lowest

    (C) hottest

    (D) cleanest

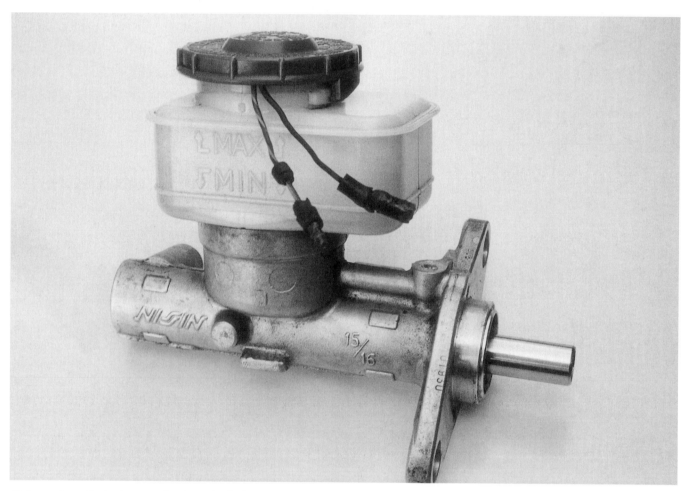

This master cylinder has an aluminum body. In most cases, the body cannot be honed and the master cylinder will require replacement.

Chapter 6

Master Cylinder, Caliper, and Wheel Cylinder Service

After studying this chapter, you will be able to:

❑ Check master cylinder condition.
❑ Remove, overhaul, and install master cylinders.
❑ Check disc brake caliper condition.
❑ Remove, overhaul, and install disc brake calipers.
❑ Check wheel cylinder condition.
❑ Remove, overhaul, and replace wheel cylinders.
❑ Bleed air from the hydraulic system.

Important Terms

Pedal travel

Overhaul kit

Vehicle identification
    number (VIN)

Honing

Anodized

Bench bleeding

Wheel cylinder kits

Brake bleeding

Manual bleeding

Pressure bleeding

Gravity bleeding

Flushing

In Chapter 5, you learned the basic operating principles of master cylinders, calipers, and wheel cylinders. In this chapter, you will learn how to service them. While the basic operation of all modern master cylinders, calipers, and wheel cylinders is similar, service and repair varies by manufacturer and year. This chapter will address these variations as they occur. By studying this chapter, you will learn how to service all current brake hydraulic units. This chapter also explains how to remove air from the hydraulic system by bleeding.

Master Cylinder Service

The master cylinder develops the hydraulic pressure needed to operate the other brake hydraulic system components. Without a properly operating master cylinder, the rest of the brake system is useless. Therefore, master cylinder service must be carefully performed. The following procedures explain how to remove, overhaul, and reassemble a master cylinder. These procedures are similar for both cast iron and aluminum master cylinders. Where differences exist, they will be noted.

 Note: Some master cylinders cannot be overhauled, and must be replaced as a complete unit. If you suspect a master cylinder cannot be successfully overhauled, replace it.

Checking Master Cylinder Condition

Most master cylinder problems fall into two categories: external and internal fluid leaks. External leaks can be spotted by the presence of brake fluid on the master cylinder housing or on the power booster face. In rare cases, master cylinders from vehicles without power brakes will leak through the firewall into the passenger compartment. This leak will be noticed as streaks of fluid under the brake pedal pushrod.

Seal failure at the rear of a master cylinder in a vehicle with vacuum power brakes may result in fluid disappearing from the reservoir with no visible leaks. This is caused by manifold vacuum drawing the fluid through the power assist unit and into the engine.

Internal leaks are caused by failure of the internal seals. This usually results in the pedal sinking to the floor when it is pressed, without the brakes applying. Internal leaks may occur at random intervals, with the brakes working normally for long periods. However, since there is no way of knowing when the seals will fail, any master cylinder suspected of an internal leak should be replaced or overhauled.

Occasionally, a master cylinder piston sticks in the bore, causing the brake pedal and brakes to remain in the applied position. This is usually caused when petroleum products such as transmission fluid or motor oil have been added to the master cylinder. This causes the seals to swell,

causing the piston to stick. In rare cases, the seals will swell to the point that the pedal cannot be depressed. When the master cylinder sticks, it must be overhauled or replaced.

Checking Pedal Travel

While not a hydraulic component, the brake pedal linkage is critical to the development of pressure in the master cylinder, and therefore, the operation of the rest of the hydraulic system. *Pedal travel,* sometimes called *free play,* is the amount of brake pedal movement before the hydraulic system is affected. The pedal travel can also affect the operation of the power brakes. This movement will be against very low resistance, compared to pedal feel when the brakes are actually being applied. Pedal travel on most vehicles is usually about 1/8″ (3.1 mm). Check pedal travel by pushing down on the brake pedal with your hand until effort increases. If the travel is not within specifications, it must be adjusted.

Master Cylinder Removal

To remove the master cylinder, begin by using a flare-nut wrench to loosen the hydraulic lines at the master cylinder body, **Figure 6-1.** Remove the hydraulic lines, position the lines so they are out of the way, and install plastic plugs in the outlet passages. If the reservoir is part of the master cylinder assembly, do not remove the reservoir cover. If the reservoir is remotely located, drain the reservoir and remove the supply hose at the master cylinder.

Some master cylinders may have one or more antilock brake system components mounted. Follow service manual guidelines when dealing with such systems. More information on ABS system service is located in Chapter 22. Remove any electrical connectors, then remove the

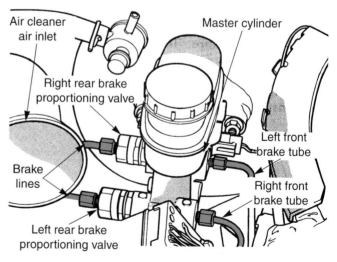

Figure 6-1. *Begin master cylinder removal by disconnecting the brake lines. Always use brake tubing wrenches. (Chrysler)*

mounting nuts and remove the master cylinder from the vehicle. See **Figure 6-2.** Carefully carry the master cylinder to a clean workbench.

 Caution: Brake fluid can damage the vehicle paint. Do not allow brake fluid to drip on the vehicle. Place rags or containers under the brake line fittings to catch any fluid. Clean off any spilled brake fluid immediately.

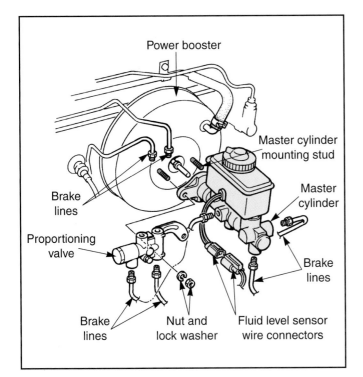

Figure 6-2. *Remove the electrical connection for the brake fluid level sensor, and then remove the master cylinder. (Mazda).*

Master Cylinder Disassembly

On the workbench, remove the reservoir cover and drain the brake fluid into a container, **Figure 6-3.** Dispose of the fluid properly.

 Caution: Do not reuse old brake fluid.

Install the master cylinder in a vise or other type of holding fixture. Be careful not to overtighten the vise and distort the master cylinder. If the master cylinder has a rubber boot covering the pushrod, remove it. If the pushrod is connected to the rear piston, remove it by pulling it out of the piston. Most pushrods of this type have a rubber O-ring which holds the rod to the piston. A new O-ring should be used when reinstalling the pushrod.

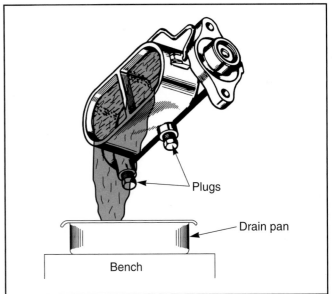

Figure 6-3. *Drain master cylinder fluid into a drain pan. Dispose of the old fluid properly. Do not reuse old brake fluid. (Bendix)*

Note: In some cases, leaving the pushrod attached will make removing the master cylinder primary piston easier.

If the master cylinder is the composite type, remove the reservoir by rocking and pulling upward on the reservoir. Do not use a screwdriver or other tool to pry on the reservoir. This will damage the plastic.

Remove the master cylinder pistons and seals next. Most master cylinders have a snap ring at the rear of the bore. This snap ring should be removed using snap ring pliers, **Figure 6-4.** A few older vehicles use a small metal retainer held by a screw, instead of a snap ring.

Before removing the snap ring, check for a bolt in the bottom or side of the cylinder, or inside of the reservoir. On some master cylinders, the front piston is held in place with a bolt located on the master cylinder body. This bolt should be removed *before* the snap ring is removed. If the bolt is removed after the snap ring, the pistons could fly out with great force.

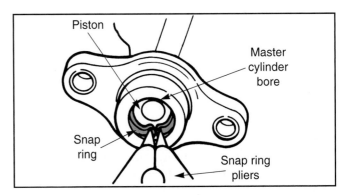

Figure 6-4. *Removing the snap ring from a master cylinder bore using snap ring pliers. (TRW Inc.)*

While removing the snap ring, keep pressure on the rear piston. Once the snap ring is removed, gradually release pressure on the rear piston and allow spring pressure to push the piston from the bore. Remove the other piston and related springs and seals, **Figure 6-5.** If the master cylinder has a bore mounted residual check valve, be sure to remove the seal at the bottom of the bore.

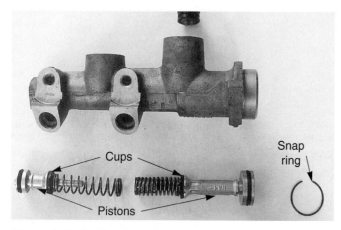

Figure 6-5. *Exploded view of a quick take-up master cylinder. Remove all parts during an overhaul.*

Note: Some rear piston assemblies are equipped with an adjusting screw. Do not remove this screw. It is factory adjusted and there is no procedure for readjustment.

If the master cylinder has proportioning valves or residual pressure valves installed in the outlet lines, remove them. Remove any electrical switches.

Note: Do not attempt to remove the quick take-up valve, when used. If the quick take-up valve is defective, replace the master cylinder.

Cleaning and Inspection

Clean all parts in a non-petroleum cleaner. Remove all seals from the pistons before cleaning to ensure the seal grooves are thoroughly cleaned. Blow-dry using compressed air. Do not use shop towels or rags to dry brake hydraulic parts, since they will leave lint. Lint can cause seal failure or clogging.

Caution: Do not use water to clean cast iron master cylinders. The cast iron will absorb moisture, then release it to the brake fluid once the master cylinder is reinstalled.

Once the master cylinder is thoroughly cleaned, check the cylinder bore for wear, scoring, and deposits, **Figure 6-6.** Note that slight wear is acceptable on some cast iron master cylinders. On aluminum master cylinders, shiny areas are acceptable. Discoloration caused by rubber seals will not affect operation and is not a defect.

Check the body for cracks, excessive corrosion, or signs of external leakage. Carefully inspect the pistons for wear, cracks, or corrosion. Check the internal springs for damage. Do not forget to check the diaphragm in the reservoir cover for leaks or other damage. Any damaged parts must be replaced. In many cases, it is quicker and cheaper to replace the entire master cylinder assembly.

Ordering Replacement Parts

After carefully inspecting the master cylinder, determine what parts will be needed. Many master cylinder parts are supplied in the form of an **overhaul kit** (a set of the most commonly needed seals, gaskets, and small metal parts). To order the proper kit or other parts, you may need to know the master cylinder manufacturer, since many vehicles use master cylinders from more than one manufacturer.

Note: Some master cylinders use a numerical or letter code that is stamped on the part or on a label attached to the master cylinder or power booster, Figure 6-7.

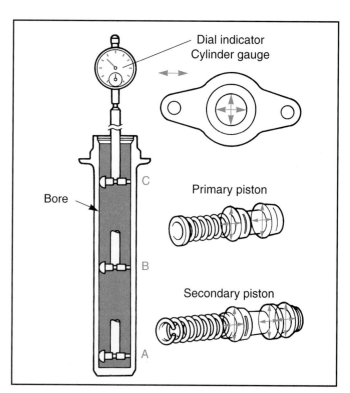

Figure 6-6. *Measuring a master cylinder bore with a dial indicator cylinder gauge. Pistons are measured with a micrometer. Follow vehicle and tool manufacturer's instructions. (Dodge)*

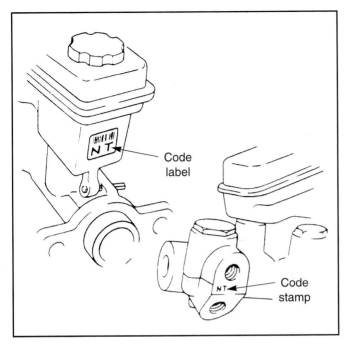

Figure 6-7. *Some manufacturers use code numbers or letters stamped, or on labels placed on their master cylinders and power boosters. Be sure to note this or any other code, if present, when ordering parts. (General Motors)*

In some cases, you may need the **vehicle identification number (VIN),** or the model number of the master cylinder. After the parts are delivered, carefully inspect them to ensure they are correct for the master cylinder.

Honing

Honing is the process of removing deposits from the master cylinder bore by spinning a set of abrasive stones inside of the bore. Before honing any master cylinder bore, check to be sure the manufacturer recommends honing. Many cast iron master cylinders are coated to prevent internal and external leaking. Honing will remove this coating which may lead to erratic operation or external leakage. Since aluminum is a relatively soft metal, all aluminum master cylinders have an **anodized** (electrically deposited) coating to reduce wear. Honing will destroy this coating, therefore, aluminum master cylinders should never be honed. If the master cylinder bore is scored and honing is not recommended, replace the master cylinder.

If a master cylinder can be honed, select the proper size hone and install it on an air or electric drill. See **Figure 6-8.** Determine whether there are manufacturer's specifications as to hone speed, if so, do not exceed them. Lightly wet the bore with brake fluid or honing lubricant specified by the hone manufacturer. Then carefully insert the hone and start the drill. Constantly move the hone up and down in the bore as it turns to create a crosshatch pattern and to reduce the possibility of excessive honing in one spot. Hone only enough to remove deposits and light scoring.

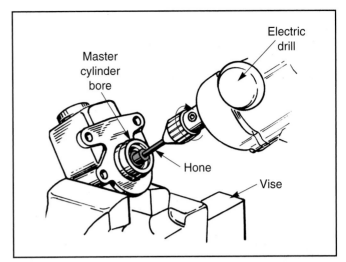

Figure 6-8. *An older style master cylinder being honed with a cylinder hone and electric drill. (FMC)*

 Caution: Allow the hone to stop spinning before removing it from the bore. A spinning hone may strike something, breaking the stones or causing damage or injury.

When honing is complete, clean the master cylinder. Clean thoroughly, since any remaining particles from the honing operation can cause seal failure and clogging.

After cleaning, re-inspect the cylinder bore. If the honing operation did not remove all deposits or scoring, replace the master cylinder. Some manufacturers recommend rechecking cylinder bore size after honing. If the bore is excessively oversized, the master cylinder should be replaced.

 Note: If a replacement master cylinder cannot be found, it is sometimes possible to bore the master cylinder oversize and install a sleeve to bring the cylinder to its original size. This operation should be performed by a specialty shop having the proper equipment.

Master Cylinder Reassembly

When reassembling a master cylinder or any hydraulic system part, use either brake fluid or an aerosol-based silicone spray. Some replacement parts manufacturers supply special seal lubricant, which comes with the overhaul kit. Carefully check that the replacement parts are correct, **Figure 6-9A.** Then lightly lubricate the replacement seals and install them over the pistons.

 Caution: Make sure the seals face in the proper direction(s). Consult the service manual during reassembly to be sure of correct seal positioning.

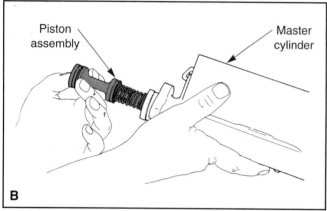

Figure 6-9. A—Assemble and lubricate the new piston assembly. B—Carefully install both piston assemblies into the master cylinder bore. Use piston to cylinder alignment marks if required. (TRW Inc.)

After the seals are installed on the pistons, the pistons can be installed in the cylinder bore. Refer to **Figure 6-9B.** Lightly lubricate the cylinder bore before beginning reassembly.

If the master cylinder uses a residual pressure valve, install the seal and valve assembly first. Then install the front piston and spring. On some master cylinders, the seals can be damaged by the outlet holes drilled in the cylinder. As you install the piston, carefully observe the seals as they pass the outlet holes. If necessary, use a small rod to slightly depress the seal as it passes the hole.

If the front piston is held in place by a bolt, press the piston down against spring tension, and install the bolt. Then install the rear piston assembly. Push the rear piston into the bore and install the snap ring.

If necessary, reinstall the reservoir using new O-rings. Then install any valves and electrical connections. Do not install the reservoir cover at this time.

Bench Bleeding the Master Cylinder

Before installing the master cylinder, it should be bench bled. **Bench bleeding** fills the master cylinder's internal passages with fluid, reducing the amount of time needed to bleed the brake system once the master cylinder is reinstalled.

To bench bleed a master cylinder, mount it securely in a vise or other holding fixture. Do not overtighten the vise or the master cylinder may be damaged. Connect one end of a short piece of tubing to each master cylinder outlet (or

bleeder screw if used) and place the other end in the reservoir. This is shown in **Figure 6-10.** Then fill the reservoir with brake fluid. Slowly push in and release the rear piston while observing the master cylinder reservoir. Bubbles should appear at the hoses, while the fluid in the reservoir will gradually go down. Add more fluid as needed. When bubbles stop appearing, the master cylinder has been bled.

Remove the hoses and place caps over the outlet ports. Reinstall the reservoir cover. Replace the pushrod if it is attached to the rear piston. Install the dust boot if used. The master cylinder is now ready to reinstall.

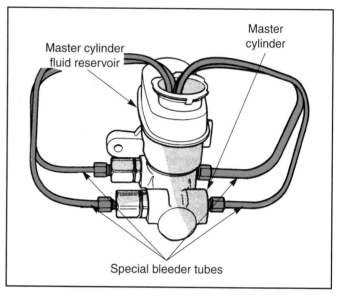

Figure 6-10. Bleeder tubes connected to the outlet ports. The tube ends must be completely submerged in the reservoir brake fluid. (Plymouth)

Master Cylinder Installation

To begin installation, place the master cylinder on the mounting studs and loosely install the mounting nuts. Then remove the outlet plugs, install, and tighten the outlet lines. Tighten the mounting nuts and replace any electrical connectors and low pressure hoses. Fill the master cylinder reservoir.

After the master cylinder has been installed, it should be bled according to the procedures given later in this chapter.

 Note: On some master cylinders, the pushrod length must be measured and if necessary, adjusted before reinstallation. This will be covered in more detail in Chapter 8.

Pedal Height Adjustment

On a few vehicles, the pedal height should be checked and adjusted. The procedure varies between manufacturers, but generally, the measurement is taken from a point on the floor. If the height can be adjusted, a threaded rod and locknut will be present on the pushrod. Loosen the locknut and make adjustments as needed, then tighten the locknut. Exact procedures and specifications are given in the service manual.

 Note: Once the pedal height is adjusted, check stoplight switch operation, and adjust if necessary.

Caliper Service

Caliper overhaul service is similar for every kind of caliper. Service procedures vary according to caliper material (cast iron or aluminum) piston material (steel or phenolic plastic), and whether the caliper has a provision for a parking brake. Variations, where they occur, will be noted in the text.

Checking Caliper Condition

Disc brake calipers are generally trouble free, but two problems occur with some frequency. A caliper can develop external brake fluid leaks, or the pistons can stick. An external fluid leak usually can be spotted by the presence of brake fluid on the caliper housing, rotor, or in the piston dust boot. Sometimes a caliper piston will stick in the caliper bore. The most common evidence of this is a wheel brake that either will not apply or will not release. Always overhaul or replace a leaking or sticking caliper.

Caliper Removal

To remove the caliper, first raise the vehicle in a safe manner and remove the tire and rim from the wheel hub. Then carefully check the temperature of the hub and rotor assembly. If the assembly is still hot, allow it to cool or use gloves to protect yourself from burns.

Loosen and remove the caliper hose. Cap the hose opening with a plastic cap or plug. If the vehicle has a pad wear sensor, disconnect the sensor electrical connector. If you are working on a rear wheel disc brake, it may be necessary to disconnect the parking brake lever from the apply linkage.

To ease caliper removal, it is necessary to back the piston off slightly. This can usually be done by prying. In some cases, a C-clamp may be needed to push the caliper piston back. If you were only replacing the brake pads, you would need to push the caliper piston all the way in. This will be covered in Chapter 13.

Remove the bolts holding the caliper to the spindle and remove the rotor. On some vehicles, the caliper is held in place by rubber or metal bushings. The screws holding the bushings should be removed and the bushings tapped out. On some vehicles, the pads will remain with the rotor, while on others, the pads are attached to the caliper

assembly, **Figure 6-11.** Once the caliper is removed, place it on a clean workbench.

If the caliper was sticking, check the condition of the rotor. If the rotor appears to have been overheated, check for cracks and blue spots. If they are present, replace the rotor. More information on this is located in Chapter 13.

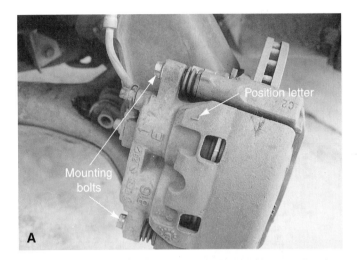

Figure 6-11. *A—Note the condition and direction of the caliper before removing it from the vehicle. This caliper has the letter "L" cast in its body, which denotes it is the left-hand caliper. B—Caliper with the pads removed.*

Caliper Disassembly

Before proceeding, loosen the bleeder screw to ensure it is not frozen in the caliper. If the bleeder screw is frozen, replace the caliper. To disassemble the caliper, remove the disc pads if necessary. Then remove the caliper piston or pistons with air pressure. On all types of calipers, the basic removal procedure consists of placing shop rags or blocks of wood and applying air through the inlet port, causing the caliper piston(s) to come out. This is shown in **Figure 6-12A.**

If the caliper uses more than one piston, precautions must be taken to ensure all pistons are partially removed before any one is completely removed. If any one piston comes out completely, it will be impossible to use air pressure to remove the other pistons. Therefore, when removing the pistons from a multiple-piston caliper, use pieces of wood or other devices to block the pistons and prevent them from coming out of their bores completely. This procedure is illustrated in **Figure 6-12B.** Make sure that all pistons are at least partially removed before proceeding.

Once the pistons are partially removed, they can be extracted by slight prying or by the use of a special caliper removal tool, **Figure 6-13.** After the pistons are removed, remove the dust boots and the piston seals. If the caliper is the fixed type, it can be disassembled into its two halves and the inner passage seals removed, **Figure 6-14.** These procedures basically complete the disassembly of the caliper.

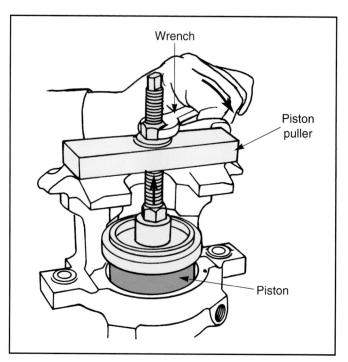

Figure 6-13. *Using a special caliper piston puller to remove the piston from its bore. (FMC)*

Rear Caliper with Parking Brake

On rear disc brakes equipped with a built-in parking brake feature, disassembly procedures are slightly different. In most cases, the parking brake components are located inside or behind the hydraulic piston. To begin removal of a typical adjuster mechanism, remove the nut holding the apply lever and remove the lever, seal, and washer from the rear of the caliper. Then remove the hydraulic piston retainer and rotate the parking brake actuator screw to remove the apply piston from the caliper, **Figure 6-15.** Other rear calipers may use a different parking brake mechanism, so follow service manual procedures.

Note: Some technicians will disconnect the parking brake lever while the caliper is still on the vehicle, leaving the lever attached to the parking brake cable. In many cases, this also allows you to retract the piston, easing caliper removal. If you do this, make sure you replace the lever seals before reconnecting the lever to the caliper.

On some parking brake systems, remove the caliper end retainer, then use a special tool to remove the piston, **Figure 6-16.** In other systems, the rear of the caliper can be disassembled and the piston pushed from the bore after the adjuster mechanism is removed. **Figure 6-17** shows this type of rear brake caliper. Since there are many variations in rear disc brake parking brake setups, consult the proper service manual for exact service information. Parking brake designs will be covered in more detail in Chapters 13, 18, and 19.

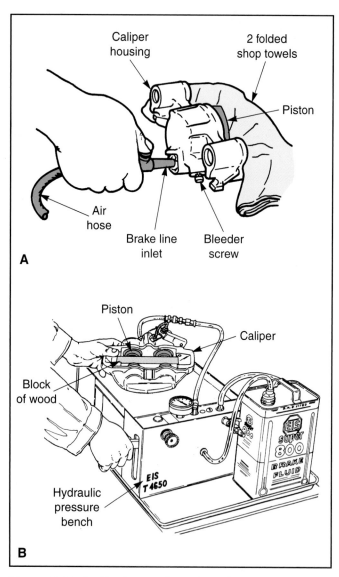

Figure 6-12. *A—Forcing a caliper piston from its bore with air pressure. Keep your fingers away from the piston(s). Use extreme caution when removing phenolic (plastic) pistons. They tend to stick, then fly out with force. B—Using a hydraulic pressure bench to force the caliper pistons out. Note the wooden block. This prevents metal-to-metal contact when the pistons come out. Wear safety glasses. (Ford and EIS)*

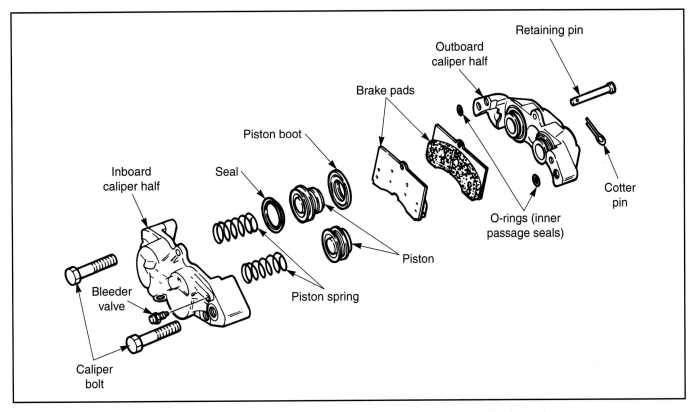

Figure 6-14. *A fixed caliper separated to allow for inner fluid passage O-ring seal removal. (Raybestos)*

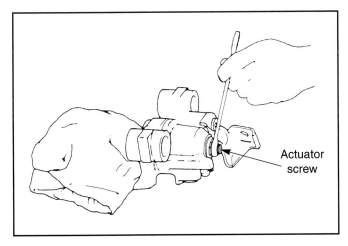

Figure 6-15. *Using a wrench to remove the piston from an actuator screw-type parking brake assembly. (Delco Moraine)*

Cleaning and Inspection

Clean all caliper parts in a non-petroleum cleaner. Remove all seals from the pistons before cleaning to ensure the seal grooves are thoroughly cleaned. Blow-dry using compressed air. Do not use shop towels or rags to dry brake hydraulic parts, since they will leave lint. Lint can cause seal failure, piston binding, or clogging.

Check the caliper bore(s) for cracks and pitting, deposits, and corrosion. Check the piston(s) for damage where they contact the disc pads. If the pistons are steel, check for cracks, deposits, scoring, and corrosion,

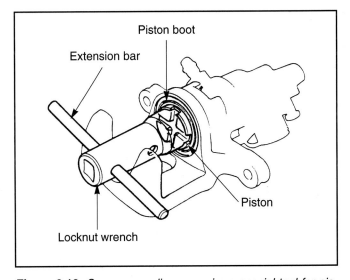

Figure 6-16. *Some rear calipers require a special tool for piston removal. This tool allows the technician to turn the piston out of the caliper bore. (Honda)*

Figure 6-18. If the pistons are made of phenolic plastic, check for cracks and other damage. Note that discoloration caused by the rubber seals is not a defect.

Ordering Replacement Parts

Most calipers require only seals and dust boots for a successful overhaul. If necessary, pistons can also be ordered. However, if an excessive amount of parts is

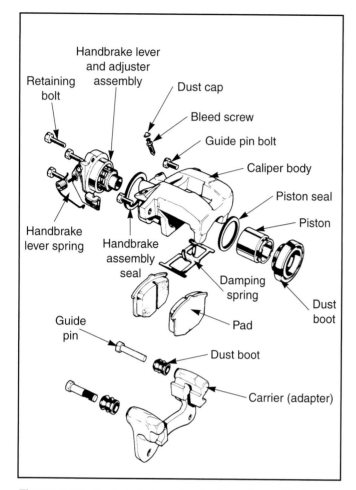

Figure 6-17. *Exploded view of a brake caliper with the rear adjuster assembly removed. (Girling)*

 Caution: Allow the hone to stop spinning before removing it from the caliper bore. A spinning hone may strike something, breaking the stones or causing damage or injury.

When honing is complete, thoroughly clean the caliper. Then re-inspect the cylinder bore. If honing did not remove all deposits or scoring, replace the caliper. Some manufacturers recommend rechecking caliper bore size after honing. If the bore is excessively oversized, replace the caliper.

Caliper Reassembly

To reassemble the caliper, make sure that you have the correct replacement parts. Use brake fluid or aerosol-based silicone spray to lightly lubricate the piston seal and install it on the piston or place the seal in its mounting groove in the cylinder bore, **Figure 6-20.** Make sure the seal is not twisted. If the caliper has a parking brake assembly, lubricate and install the parking brake mechanism in the bottom of the caliper bore, or on the piston as applicable. Install all new seals as provided in the rebuild kit.

Note: Some rear calipers have one or more additional retaining or locating rings, which fit around the piston. Follow service manual recommendations when installing this ring.

needed, it may be cheaper to order a new caliper. To order replacement parts, you may need to know who manufactured the caliper, or the vehicle model. Often, the VIN, production date, or options code is needed. Always check the new parts carefully to ensure they are correct.

Honing Calipers

Before honing a caliper bore, make sure the manufacturer permits honing. The bores of many cast iron calipers are coated, and many aluminum caliper bores are anodized. Honing will destroy these coatings. A badly scored caliper should be replaced. Do not rebuild a scored caliper.

If the caliper bore can be honed, select the proper size hone and install it on a power drill. Determine whether the manufacturer specifies a maximum hone speed, and do not exceed it. Lightly wet the bore with brake fluid or honing lubricant specified by the hone manufacturer. Then carefully insert the hone into the caliper and start the drill. See **Figure 6-19.** Constantly move the hone up and down in the bore as it turns to create a cross-hatch pattern and reduce the possibility of excessive honing in one spot. Hone only enough to remove deposits and light scoring.

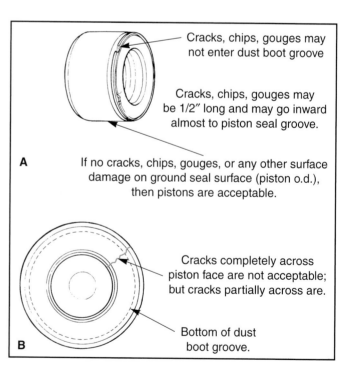

Figure 6-18. *A—Check the caliper piston bore and piston for signs of pitting, corrosion, cracking, etc. B—Inspecting phenolic (plastic) caliper piston for damage. Replace as necessary. (Toyota and Allied Automotive)*

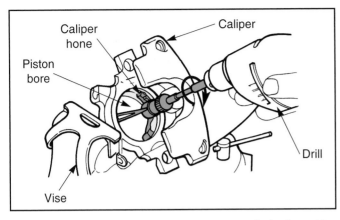

Figure 6-19. *Using a brake hone to clean a cylinder bore. Use clean brake fluid to lubricate the hone. Never insert or remove the hone while it is spinning. Wear safety glasses. (Niehoff)*

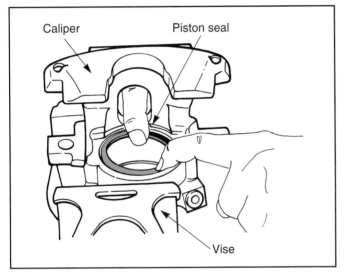

Figure 6-20. *Inserting a lubricated caliper piston seal into its groove. (Niehoff)*

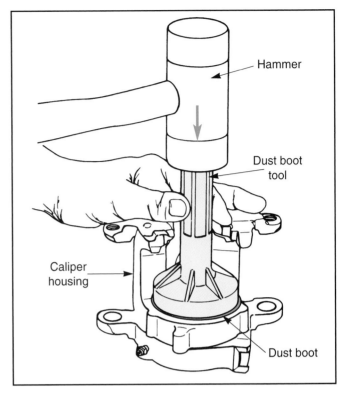

Figure 6-21. *Installing a dust boot with a special tool and hammer. Be sure the boot is completely seated in its groove. (Buick)*

Lubricate the caliper bore and install the piston by lightly pressing it into place. The piston should install easily with hand pressure or by using the end of a hammer handle. If the piston will not slide into the bore with a minimum of pressure, remove the piston and determine the cause. Do not hammer or use a C-clamp to install the piston. Once the piston is fully seated, install the dust boot using the proper installation tool, **Figure 6-21.** If the caliper has more than one piston, repeat the above steps for each remaining piston.

Caliper Installation

 Note: Do not switch calipers between the left and right sides of the vehicle. On many vehicles, reversing the calipers will place the bleeder screws in a position that makes complete bleeding impossible.

To reinstall the caliper, install the pads on the caliper or rotor as necessary. Then place the caliper over the rotor, **Figure 6-22,** and install the attaching bolts or bushings. Tighten the fasteners and ensure the rotor can turn freely with the caliper installed. Then install and tighten the brake hose. If a copper gasket is used at the connection, obtain a new gasket. If the vehicle has a pad wear sensor, reconnect electrical connector. On rear wheel disc brakes, reconnect the linkage and adjust the parking brake, **Figure 6-23.**

 Note: Some vehicles with rear disc brakes use a specialized parking brake adjustment procedure. See Chapters 13 and 19.

After all connections are made, bleed the brakes as explained later in this chapter. Then reinstall the tire and rim.

Wheel Cylinder Service

All wheel cylinders used on modern vehicles are single cylinder designs with two apply pistons. The only variation between modern wheel cylinders is bore size and whether it is made of cast iron or aluminum. Some manufacturers do not recommend overhauling the wheel cylinders unless they are leaking, as evidenced by brake fluid directly behind the dust covers or visible at the apply pins. However, many technicians prefer to overhaul the wheel cylinders whenever the brake shoes are replaced.

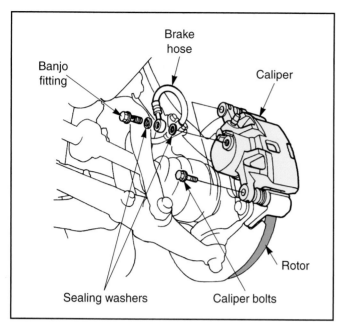

Figure 6-22. *Installing the caliper over the rotor. Insert and tighten the caliper bolts to specifications. Connect the brake hose, make sure you use new sealing washers. (Honda)*

Checking Wheel Cylinder Condition

Two problems are common on wheel cylinders: they can develop external leaks, or one of the pistons can stick. An external fluid leak can easily be spotted by the presence of brake fluid on the wheel cylinder housing and backing plate, **Figure 6-24.** Severe leaks will coat the entire inside of the brake drum. A wheel cylinder that is just starting to leak will leave brake fluid under the dust boot. Note that sometimes installing new shoes will cause the wheel cylinders to leak, since the new shoes push the pistons deeper into the cylinder, where the seals may be damaged by deposits or pitting.

> **Caution: Leaking axle seals on rear-wheel drive vehicles is often misdiagnosed as a wheel cylinder leak. Usually, an axle seal leak will coat the entire brake assembly, with additional fluid leaking out and coating the inside of the wheel. Make sure you correctly identify the source of any leak on rear-wheel drive vehicles.**

A sticking wheel cylinder piston is sometimes hard to spot, since the other piston will still move, and may at least partially apply the brake. The most obvious sign of a sticking piston is excessive wear on one brake shoe only. A leaking or sticking wheel should be overhauled or replaced.

Wheel Cylinder Removal

To remove the wheel cylinder, raise the vehicle and remove the wheel and tire. Then remove the brake drum and brake shoes. This will be covered in more detail in Chapter 15. Whether or not the wheel cylinder must be removed from the backing plate for disassembly depends on the backing plate design.

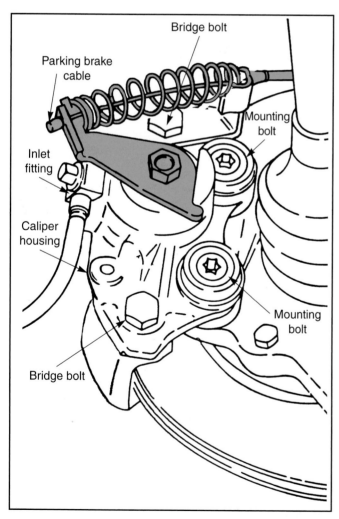

Figure 6-23. *Reinstall all parking brake components. Make sure the cable is secure in the actuating arm. (General Motors)*

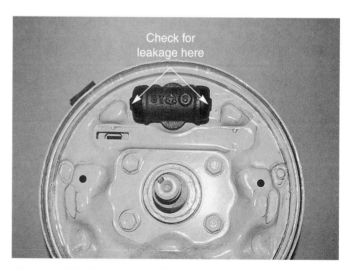

Figure 6-24. *Inspect the wheel cylinder at these points for leakage.*

Note: Be sure to follow all brake dust safety precautions during drum and shoe removal.

If the wheel cylinder is to be removed, begin by loosening and removing the brake hydraulic line. Place a pan under the backing plate to catch any dripping fluid. Then remove the bolts holding the wheel cylinder to the backing plate and remove the cylinder, **Figure 6-25.** On a few vehicles, the backing plate is held in place by a clip, and the clip can be removed to remove the wheel cylinder. At this point, the wheel cylinder is ready for disassembly or replacement.

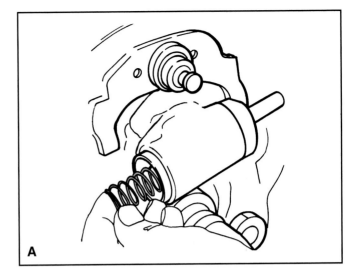

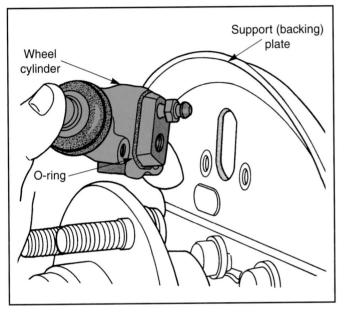

Figure 6-25. *Removing wheel cylinder from the backing plate. (Chrysler)*

Wheel Cylinder Disassembly

Wheel cylinders on most vehicles can be disassembled without removal. On some vehicles, however, the backing plate is designed to prevent the pistons from moving out of the cylinder bore, no matter how worn the brake shoes become. On these vehicles, the wheel cylinder must be removed for overhaul. Before starting the overhaul, make sure that the bleeder screw can be loosened. If the bleeder screw cannot be loosened or breaks off, replace the wheel cylinder.

Wheel Cylinder Disassembly on the Vehicle

If you decide to rebuild the wheel cylinder on the vehicle, place a pan under the backing plate assembly. Remove the brake line, then remove the dust boots from the ends of the wheel cylinder, **Figure 6-26.** Then push one piston through the cylinder while holding a rag over the other end of the cylinder to catch the internal components and brake fluid.

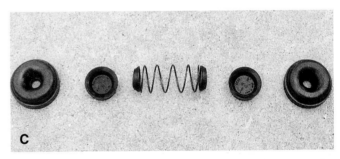

Figure 6-26. *A—Disassembling the wheel cylinder on the vehicle. B—Typical wheel cylinder pistons. These should be checked for scoring. C—Wheel cylinder boots, seals, and spring. These are normally part of a rebuild kit. (Honda)*

Wheel Cylinder Disassembly on the Bench

Remove and carry the wheel cylinder to a bench for disassembly. After the wheel cylinder is on the bench, remove the dust caps and push the internal parts from one side. As the parts exit from the other side, be ready with a rag to catch them and any brake fluid. **Figure 6-26** shows the internal parts of a typical wheel cylinder.

Cleaning and Inspection

Clean all wheel cylinder parts in a non-petroleum cleaner and blow dry with compressed air. Do not use shop towels or rags to dry parts. Rags and towels will leave lint deposits, leading to leaks and clogging. Remove the bleeder screw and make sure it is free of dirt and corrosion.

Check the cylinder bore for scoring, pitting, and deposits. Discoloration caused by the rubber seals is not a defect. If the internal spring and cup expanders are being reused, check for corrosion and damage. Check the wheel cylinder pistons for scoring and corrosion as shown in **Figure 6-27.**

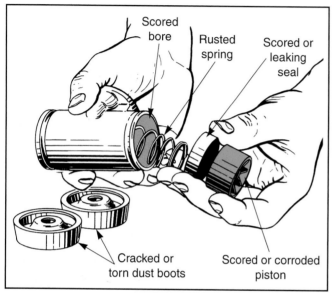

Figure 6-27. *Inspect wheel cylinder bore and pistons for corrosion, pitting, cracking, etc. Repair or replace as needed. (FMC)*

Ordering Replacement Parts

In most cases, **wheel cylinder kits** containing the most commonly replaced wheel cylinder parts are readily available. These kits usually contain the cups and center spring, plus cup expanders when used. However, you may need to obtain the vehicle model, VIN, axle or drum size, production date, or option code to obtain the correct kit. Carefully check the replacement parts against the old parts to ensure they are correct for the wheel cylinder that you are working on. Pay particular attention to the size of the replacement cups.

Honing Wheel Cylinders

Most cast iron wheel cylinder bores can be honed to remove deposits or light scoring. Aluminum wheel cylinders have an anodized coating and cannot be honed. Wheel cylinder bores with severe damage should be replaced. If the wheel cylinder can be honed, select the proper size hone. Install the hone on a power drill. Apply a light coating of brake fluid or honing lubricant to the wheel cylinder bore. Carefully insert the hone into the wheel cylinder. Do not start the drill until the hone is fully inserted. Once the drill is started, move the hone up and down in the wheel cylinder bore to create a crosshatch pattern and reduce the possibility of excessive honing in

one spot. Hone only enough to remove deposits and light scoring. **Figure 6-28** shows the honing process in a typical wheel cylinder.

 Caution: Allow the hone to stop spinning before removing it from the wheel cylinder. A spinning hone may strike something, breaking the stones or causing damage or injury.

When honing is complete, thoroughly clean the wheel cylinder and re-inspect the cylinder bore. If honing did not remove all deposits or scoring, replace the wheel cylinder. If recommended, recheck bore size after honing. If the bore is excessively oversized, replace the wheel cylinder.

Wheel Cylinder Reassembly

To reassemble the wheel cylinder, begin by lightly lubricating the bore with brake fluid or lubricant provided by the wheel cylinder kit manufacturer. Then reinstall one piston and cup assembly by pushing it in from one side of the cylinder, **Figure 6-29.** Then install the dust boot over that side.

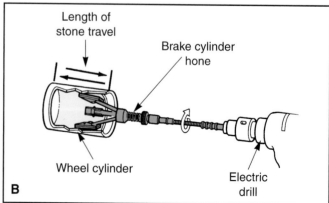

Figure 6-28. *A—Check the wheel cylinder for scoring. Light scoring can be removed by honing. B—Using a brake cylinder hone to remove light scoring and/or deposits. (Niehoff)*

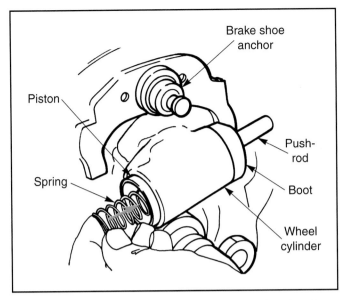

Figure 6-29. Reassembling a wheel cylinder that has been mounted to the backing plate. (TRW Inc.)

While maintaining pressure on the dust boot, install the spring and expanders, cup, and piston through the other side of the cylinder.

 Caution: Be sure the lips of both cups face inward (toward the spring). If the cups are installed backwards, brake fluid will leak out of the wheel cylinder.

Install the remaining dust boot. Remember that the assembly contains an internal spring, and care must be taken to ensure the spring does not push the other parts out of the cylinder. Clips or clamps for holding the cylinder in place are available. See **Figure 6-30.**

Wheel Cylinder Installation

If necessary, reinstall the wheel cylinder on the backing place by installing and tightening the mounting bolts, or by reinstalling the clip. Then reconnect and tighten the hydraulic line. If a copper gasket is used at the connection, obtain a new gasket. Use new wheel cylinder-to-backing plate gaskets or sealant as required.

Then reinstall the brake shoes and drum. Adjust them as outlined in Chapter 15, then bleed the brakes as explained later in this chapter. Then reinstall the tire and rim.

Bleeding the Brakes

After any service to the hydraulic brake system, *brake bleeding* is necessary. There are basically two ways to bleed the brake hydraulic system: manual bleeding and pressure bleeding. These methods are similar. Both should be performed carefully to ensure that all air is removed from the hydraulic system.

 Note: Some vehicles equipped with ABS systems require a specialized bleed procedure, which differs from the ones described here. For additional information, see Chapter 22.

Manual Bleeding

Manual bleeding requires no special equipment, but does require the presence of a helper. To manually bleed the brakes, fill the master cylinder reservoir with brake fluid and replace the cover. Then manually bleed the brakes by beginning at the wheel furthest from the master cylinder. Usually, the bleeding procedure will be right rear, left rear, right front, and left front. However, this may vary, and the manufacturer's service manual should be consulted. Also check the manufacturer's service information to determine whether the metering valve should be manually opened during the front brake bleeding operation.

Note: During any type of manual bleeding procedure, frequently recheck the master cylinder reservoir and add more fluid when necessary.

Station the helper in the passenger seat and direct him or her to pump (apply and release) the brake pedal at least ten times. If the system is full of air, the pedal will have little or no resistance. However, the pumping process does pressurize the system even if no fluid is present.

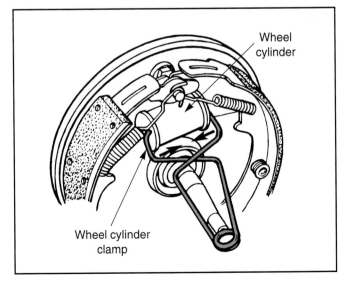

Figure 6-30. Using a wheel cylinder spring clamp to hold the wheel cylinder pistons in, until the springs and shoes can be reinstalled. Static line pressure tends to force the pistons out. (Allied Automotive)

Next, have the helper hold the pedal in the applied position (pushed toward the floor but not pumping). Then open the bleeder valve on the wheel farthest away from the master cylinder. This is usually the right rear wheel. Allow air to escape, then tighten the bleeder.

 Note: Instruct your helper to maintain foot pressure while the bleeder is open.

Have the helper pump and hold the brake pedal, then reopen the bleeder valve. Continue this procedure until only brake fluid flows from the bleeder when it is opened. When air exits the bleeder screw it will make a whistling or spitting noise and the fluid may appear aerated. When this noise is no longer heard, all air has been removed. Repeat this sequence at the other wheels.

Note: If no air or fluid escapes from the bleeder, it may be clogged with rust or road dirt. This is a common problem, especially if the brakes have not been serviced for several years. Remove the bleeder screw and check the passage for dirt buildup. If the passage is plugged, use a wire or small drill bit to remove the dirt, then blow the screw out with compressed air before reinstalling.

Bleeding Using a Jar and Hose

This procedure wastes less fluid, and reduces spillage. Begin by loosening the bleeder screw at the wheel farthest from the master cylinder. Attach a hose so that it fits snugly over the bleeder screw, and place the other end of the hose in a jar that is half full for new brake fluid. This is shown in **Figure 6-31.** Direct the helper to pump the brake pedal. As the pedal is pushed downward, air will be forced out of the bleeder, through the hose, and into the jar. The air will rise in the jar. When the pedal is released, brake fluid will be drawn into the hose, and enter the hydraulic system. Have the helper continue pumping until no air escapes from the hose. Then close the bleeder and repeat the process at the other wheels.

Pressure Bleeding

Pressure bleeding is similar to manual bleeding, with the pressure bleeder unit eliminating the need for a helper. To begin pressure bleeding, fill the bleeder unit with fresh brake fluid, as shown in **Figure 6-32.** Attach the proper bleeder adapter to the top of the master cylinder reservoir. Then attach the air supply (if needed) to the pressure bleeder, or start the electric pump. Regulate pressure to the master cylinder.

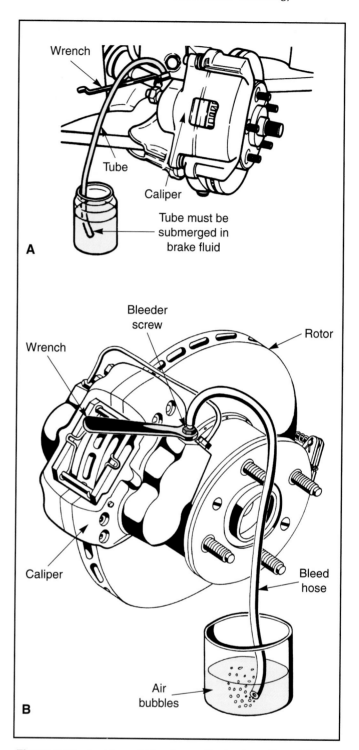

Figure 6-31. A—Manual bleeding a disc brake caliper. A bleeder tube is connected to the bleeder screw. Open and close the bleeder screw with a wrench only while there is pressure in the system. A soft hammering on the caliper cylinder bore area with a plastic tipped hammer will help to dislodge air bubbles clinging to the bore walls. B—Bleeding a rear wheel disc brake. Note the air bubbles coming out the end of the bleed hose. (Buick and Maserati)

 Caution: If the master cylinder has a plastic reservoir, do not pressurize the system to more than 10 psi (68.95 kPa).

After the system has been pressurized, go to the wheel farthest from the master cylinder and loosen the bleeder screw. Allow air and fluid to exit until no more air is heard exiting the bleeder screw. If no air or fluid escapes from the bleeder, it may be plugged. Refer to the cleaning steps under manual bleeding.

> **Note: If possible, place a pan or other container under the wheel to catch fluid.**

Repeat this process at all other wheels until all air is removed. Check the manufacturer's service information to determine whether the metering valve should be opened manually when bleeding the front brakes. If the bleeding process is prolonged, recheck the pressure bleeder brake fluid level and add more fluid if needed.

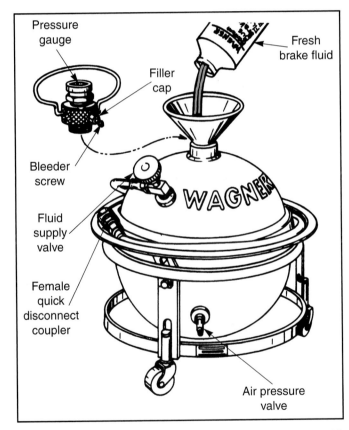

Figure 6-32. *Filling one particular type of pressure bleeder with new brake fluid. (Wagner)*

Bleeding the Master Cylinder

If the master cylinder was the only hydraulic system component removed, it is often possible to perform the bleeding process at the master cylinder. When performing manual bleeding, the master cylinder must be bled first so it can develop enough pressure to supply the rest of the system. To bleed the master cylinder, have a helper pump and hold the brake pedal, or use a pressure bleeder. Then open the lines at the master cylinder one at a time and allow air to exit.

> **Caution: Brake fluid can damage paint. Do not let fluid spray on the vehicle body.**

Gravity Bleeding

Some technicians will clamp off a brake hose leading to a caliper or wheel cylinders to be removed and then remove the clamp and allow the brake system to bleed itself. This process is referred to as *gravity bleeding.* The advantage of gravity bleeding is that it minimizes the amount of manual or pressure bleeding needed to purge the air out of the hydraulic system. It can only be used when a caliper or wheel cylinder is removed. Although some technicians believe all the air can be removed by gravity bleeding, it is not considered a substitute for either manual or pressure bleeding.

Brake Flushing

Many manufacturers recommend periodic flushing of the hydraulic system. *Flushing* is similar to brake bleeding and can be carried out by using the same methods. Manual or power flushing is possible, the only difference is that instead of removing air, the flushing procedure removes old fluid. Therefore, the technician should observe the fluid exiting from the bleeder screw. When the fluid is a clear, amber (purple if silicone fluid is used) liquid, the hydraulic system is considered to be adequately flushed.

Summary

Master cylinders, disc brake calipers, and wheel cylinders can be rebuilt or replaced. Many technicians replace these units instead of rebuilding them.

Repair procedures are similar for both cast iron and aluminum master cylinders. Some master cylinders cannot be overhauled. Common master cylinder problems are external and internal fluid leaks and piston sticking. Most cases of piston sticking are caused by the addition of petroleum products to the master cylinder. To check pedal travel, push down on the brake pedal with your hand until effort increases. Pedal travel can be adjusted

Master cylinder removal and disassembly is relatively simple, but should be done cautiously. After disassembly, the master cylinder should be carefully cleaned and inspected. If there is any possibility the master cylinder cannot be successfully rebuilt, it should be replaced. Before honing any master cylinder bore, check that the manufacturer recommends honing. Many master cylinders are coated and cannot be honed. If honing does not clear up all scoring, replace the master cylinder. Bench bleed the master cylinder before installing it.

Disc brake calipers are generally trouble free but can develop external leaks, or sticking pistons. In many cases, the brake pads are replaced without rebuilding the calipers. Calipers are removed from the vehicle for rebuilding. Removal procedures are relatively simple. If the rear caliper has the parking brake assembly built in, disassembly procedures are somewhat more complex. If disassembly of the caliper indicates that it cannot be rebuilt, it should be replaced as a unit.

Modern wheel cylinders are single cylinder designs with two apply pistons. Wheel cylinders can be made of cast iron or aluminum. All wheel cylinders can be rebuilt by following the same basic procedures. To rebuild the wheel cylinder, first remove the brake drum and brake shoes. Wheel cylinders of most vehicles can be rebuilt without removing it from the vehicle. On a few vehicles the wheel cylinder must be removed before it can be rebuilt.

After any service to the hydraulic brake system, the brakes must be bled. The two ways to bleed the brake hydraulic system are manual bleeding and pressure bleeding. Either method should be done carefully to ensure all air is removed. Manual bleeding requires the use of a helper. The most efficient, cleanest way to manually bleed is to use a hose and jar. Pressure bleeding requires a pressure bleeding unit. Pressure bleeding procedures are similar to those for manual bleeding. The same procedures can be used to flush the hydraulic system.

Review Questions—Chapter 6

Please do not write in this text. Write your answers on a separate sheet of paper.

1. Most master cylinder problems can be classed as types of internal and external _____.

2. If the brake pedal sometimes sinks to the floor, and works properly at other times, the master cylinder _____ are leaking.

3. Pedal travel is sometimes called _____.

4. Drain the reservoir before removing the master cylinder only if the reservoir is _____ mounted.

5. On some master cylinders, the front piston is held in the cylinder by a _____.

6. Solvents used to clean the master cylinder should contain no _____.

7. What should the technician do *first* when the rebuild kit for the master cylinder arrives?

8. Master cylinders made of _____ should never be honed.

9. List some precautions to take when installing new seals on the pistons and into the master cylinder.

10. To bench bleed the master cylinder, install it in a _____. Attach tubing to the outlet or _____ if used. As you operate the master cylinder pistons, bubbles will appear in the _____. The master cylinder is bled when the bubbles _____.

11. What are the two most common brake caliper problems?

12. To remove a caliper piston, _____ can be applied through the inlet port.

13. Hone a caliper bore only enough to remove _____ and _____.

14. If a wheel cylinder is beginning to leak, brake fluid will be found under the _____.

15. What are the advantages of bleeding brakes using a jar and hose?

ASE Certification-Type Questions

1. Technician A says that an internal fluid leak may cause the pedal to go to the floor unexpectedly. Technician B says that a power brake unit may draw brake fluid into the engine if a seal fails. Who is right?
   (A) A only.
   (B) B only.
   (C) Both A & B.
   (D) Neither A nor B.

2. If petroleum products are added to the master cylinder, which of the following could happen?
   (A) Pedal cannot be applied.
   (B) Pedal will not release.
   (C) Brakes will be locked.
   (D) All of the above.

3. All of the following statements about pedal travel are true, EXCEPT:
   (A) pedal travel is pedal movement before the hydraulic system is affected.
   (B) pedal travel is against very high resistance.
   (C) check pedal travel by pushing down on the brake pedal with your hand.
   (D) pedal travel can be adjusted.

4. Which of the following should *not* be removed from the master cylinder?
   (A) Quick take-up valve.
   (B) Front piston bolt.
   (C) Piston seals.
   (D) Residual pressure valve.

5. Do not hone a master cylinder if it is _____.
   (A) cast iron
   (B) aluminum
   (C) coated
   (D) Both B & C.

6. Bench bleeding reduces the amount of _____.
   (A) pedal travel
   (B) on-vehicle bleeding
   (C) pedal height
   (D) Both A & B.

7. On which of the following types of calipers should air pressure *not* be used to remove caliper pistons?
   (A) Single piston.
   (B) Dual piston.
   (C) Four piston.
   (D) None of the above.

8. Technician A says that shop towels should not be used to dry parts. Technician B says that air pressure should not be used to dry parts. Who is right?
   (A) A only.
   (B) B only.
   (C) Both A & B.
   (D) Neither A nor B.

9. If honing a caliper does not remove all deposits or scoring, what should the technician do?
   (A) Continue honing until the bore is clean.
   (B) Use a special grinder to remove scoring or deposits.
   (C) Carefully polish the bore with crocus cloth.
   (D) Replace the caliper.

10. After reinstalling front calipers, you find that you cannot successfully bleed the brakes. What should you do *next*?
    (A) Check that the caliper seals are not drawing in air.
    (B) Check that the caliper hoses are tight.
    (C) Check that the calipers have not been switched from side to side.
    (D) Check that the master cylinder is not defective.

11. All of the following statements about wheel cylinders are true, EXCEPT:
    (A) modern wheel cylinders can be made of cast iron or aluminum.
    (B) all modern wheel cylinders have a single cylinder with two pistons.
    (C) all modern wheel cylinders have the same bore size.
    (D) some manufacturers say not to rebuild the wheel cylinder unless it is leaking.

12. Technician A says that if brake fluid is visible at the apply pins, the wheel cylinder should be overhauled. Technician B says that the wheel cylinders should be overhauled every 20,000 miles (32 000 km), even if the brake shoes are good. Who is right?
    (A) A only.
    (B) B only.
    (C) Both A & B.
    (D) Neither A nor B.

13. If the cups of a wheel cylinder are installed backwards, what will happen?
    (A) Fluid will leak from the wheel cylinder.
    (B) The wheel cylinder will stick in the released position.
    (C) The wheel cylinder will stick in the applied position
    (D) All of the above.

14. Technician A says that a helper is needed to pressure bleed brake systems. Technician B says that using a jar and hose eliminates the need for a helper. Who is right?
    (A) A only.
    (B) B only.
    (C) Both A & B.
    (D) Neither A nor B.

15. If a master cylinder is not equipped with bleeder screws, loosen the _____ to remove air.
    (A) mounting bolts
    (B) hydraulic lines
    (C) reservoir cover
    (D) Both A & B.

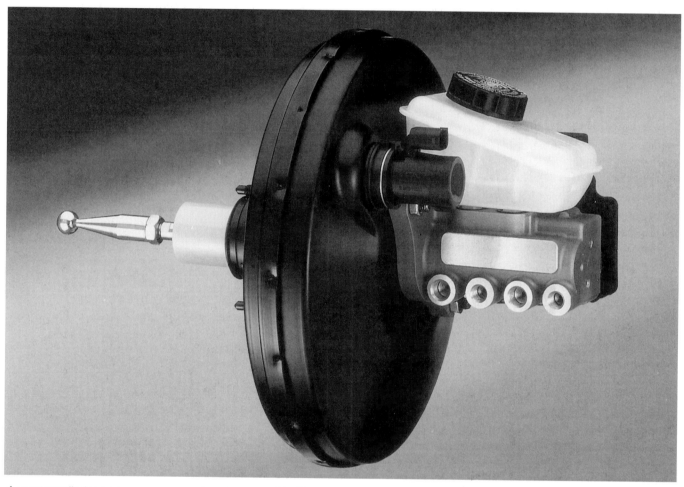

A master cylinder and vacuum power booster assembly. The pushrod on this booster is adjustable. (Continental Teves)

Chapter 7

Power Assist Units

After studying this chapter, you will be able to:

- ❑ Explain the purpose of brake power assist units.
- ❑ Identify vacuum and hydraulic power assist units.
- ❑ Explain how vacuum power assist units operate.
- ❑ Identify the components of vacuum power assist units.
- ❑ Identify the power sources and safety provisions of vacuum power assist units.
- ❑ Explain how hydraulic power assist units operate.
- ❑ Identify the components of hydraulic power assist units.
- ❑ Identify the power sources and safety provisions of hydraulic power assist units.

Important Terms

Power assist units

Vacuum boosters

Pressure differential

Atmospheric pressure

Diaphragm

Vacuum chamber

Single diaphragm vacuum
 booster

Tandem diaphragm vacuum
 booster

Control valve

Power piston

Check valve

Vacuum pump

Hydraulic booster

Hydro-boost

Lands

Grooves

Accumulator

Power steering pumps

Powermaster

This chapter identifies the purpose and operation of power assist units. The major types of power boosters, the major components of each, and how they receive power will be covered. Although their operating principles are similar, there are some variations in design and actual operation. By studying this chapter, you will learn the basic operating principles of power assist units before learning how to service them in Chapter 8.

The Need for Power Assist Units

The modern vehicle is designed for maximum driver control and ease of operation. In addition, larger cars and trucks develop considerable momentum at

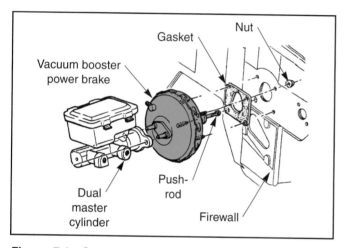

Figure 7-1. *One common type of vacuum operated power brake unit. (General Motors)*

highway speeds. Also, disc brakes, while safer and more efficient than drum brakes, require more pedal effort to apply the same braking force. For these reasons, *power assist units,* often called *power boosters,* or just *power brakes,* have become almost universal on all but the smallest cars.

Types of Power Assist Units

The two major types of power assist units are the vacuum booster, **Figure 7-1,** and the hydraulic booster, **Figure 7-2.** On all modern vehicles, the power assist unit is installed between the firewall (normally referred to as the *bulkhead*) and the master cylinder. Pressing on the brake pedal moves the pushrod through the power assist unit to apply the brake hydraulic system. Pushrod movement also causes the power assist unit to operate. In the following sections, vacuum and hydraulic booster operation will be discussed.

Vacuum Power Assist Units

Vacuum power assist units, or *vacuum boosters* were introduced over 50 years ago. Several variations have been used, but all modern units are basically the same design. They consist of a relatively large metal chamber assembly installed between the firewall (bulkhead) and the master cylinder. The booster unit stores enough vacuum for several stops. Most vacuum brake unit differences involve the control valves, maintaining brake feel, and the methods of ensuring an adequate supply of vacuum.

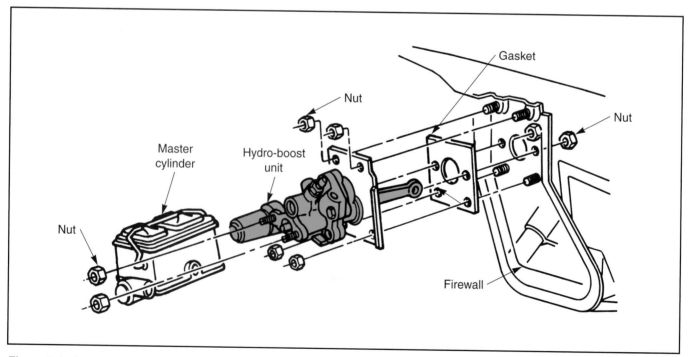

Figure 7-2. *A Hydro-boost power brake unit. (General Motors)*

Basic Vacuum Power Assist Operation

All vacuum power assist units operate by what is known as a **pressure differential.** A pressure differential is the difference in pressure between two sides of a sealed chamber. The key to understanding this process is knowing that at all times, we are surrounded by **atmospheric pressure.** An example of a pressure differential is blowing up a balloon, **Figure 7-3.** Blowing up a balloon causes a pressure differential between the inside of the balloon and the outside atmospheric pressure. The higher pressure inside the balloon causes it to expand.

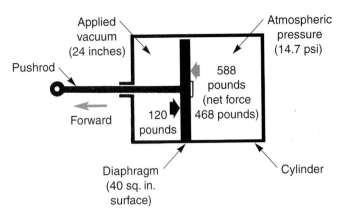

Figure 7-4. *A simple power brake showing atmospheric pressure forcing the diaphragm forward. Vacuum has created an area of lower air pressure on one side, so the atmospheric pressure net force on the opposite side will force the diaphragm forward, as it tries to fill in the low pressure area. (FMC)*

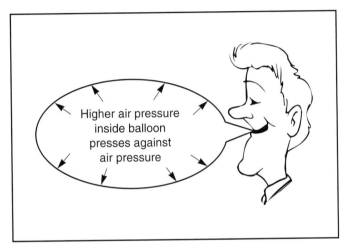

Figure 7-3. *An example of pressure differential by inflating a balloon with your mouth.*

The operation of the brake vacuum assist unit is similar to blowing up a balloon. Vacuum boosters contain a **diaphragm,** which is a flexible rubber sheet that divides two sides of a chamber. One side of the diaphragm has *negative pressure,* known as vacuum, and the other side has full *atmospheric pressure.* Atmospheric pressure will push or vacuum will pull the diaphragm toward the vacuum side. In either event, the final result is the same: the diaphragm is moved by the difference in pressure. The operation of a brake booster diaphragm by pressure differential is shown in **Figure 7-4.**

Vacuum Unit Construction

The vacuum assist unit consists of a two-piece metal housing which forms a **vacuum chamber** divided by a diaphragm. There is a return spring on the master cylinder side of the diaphragm. The flexible diaphragm assembly divides the housing into two sections, **Figure 7-5.**

Modern vacuum power brake boosters are known as **vacuum suspended systems.** In the vacuum suspended system, identical amounts of vacuum are present on both sides of the diaphragm when the brakes are not being used. When the brakes are applied, vacuum is bled from one side of the diaphragm, causing it to move toward the side which still has vacuum.

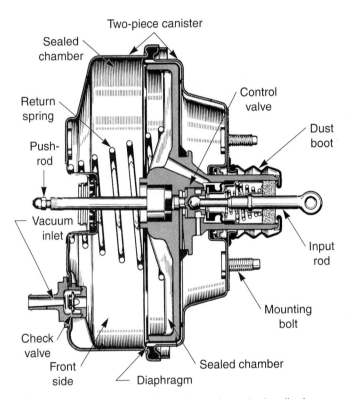

Figure 7-5. *A cross-section view of a single diaphragm, vacuum operated, power brake assembly. (Chevrolet)*

The brake pedal pushrod goes completely through the housing to push on the master cylinder pistons. The pushrod is solidly attached to a metal or plastic piston or collar installed at the center of the diaphragm. Outside air enters the vacuum unit through the passenger compartment where the brake pushrod enters the firewall. Air entering the booster passes through a filter and silencers, which fit around the control valve pushrod. The valve pushrod and filter are covered by a dust boot. A hose is connected between the booster unit and the intake manifold or vacuum pump, usually through a check valve.

Single Diaphragm Vacuum Boosters

The **single diaphragm vacuum booster** is the most common power booster unit installed. It is used in almost every type of vehicle equipped with power brakes.

The diaphragm in a single booster unit is attached to the diaphragm plate (or support plate) on its center, and to the front and rear housing on its outside lip. Since the pushrod is a solid rod which passes completely through the power assist unit, any power assist failure will not cause a loss of braking action. Also installed in the housing is the control valve assembly pushrod. A typical single vacuum booster unit is shown in **Figure 7-6A.**

Tandem Diaphragm Vacuum Booster

A variation of the vacuum power brake booster is the **tandem diaphragm vacuum booster,** sometimes called a *dual diaphragm* unit. Instead of having a single diaphragm, it consists of two diaphragms in a single housing, **Figure 7-6B.** The housing is somewhat longer and smaller in diameter than a single diaphragm housing. The control valve assembly is more complex, since it has to control vacuum and atmospheric airflow through two chambers.

As shown in **Figure 7-6B,** the two diaphragms and chamber assemblies are separated by a stationary housing divider. Vacuum enters the front chamber in the same manner as on a single diaphragm unit. Vacuum enters the rear chamber through the center control valve assembly. The valve is a two-piece unit that can bleed off the vacuum in both rear chambers at the same time.

When the brakes are applied, the valve is moved forward and allows air to enter the rear of both chambers. This causes both diaphragms to be pushed forward, assisting foot pressure to apply the brakes. Brake hold and release is done in the same manner as on a single diaphragm unit. Tandem boosters are installed on larger cars, light trucks, vans, and sport-utility vehicles equipped with a towing package.

Control Valve

The vacuum assist **control valve,** or **power piston,** is directly operated by the brake pushrod. When the brake pedal is depressed, pushrod movement is transferred through the power assist unit to operate the master cylinder and operates the control valve. The design of the control valve, **Figure 7-7,** allows it to perform two functions at once. The vacuum assist control valve can be thought of as two valves: an **air valve** which controls the flow of air into the diaphragm housing, and a **vacuum valve** that controls the build-up of vacuum in the diaphragm housing. Both the valves are connected to and operated by the pushrod.

When the brakes are applied, pushrod movement opens the rear chamber to atmospheric pressure while sealing the passage between the front and rear chambers. This allows vacuum to bleed off from the rear chamber while maintaining vacuum in the front chamber. When the brakes are released, the valve opens the passage between the front and rear chambers while sealing the

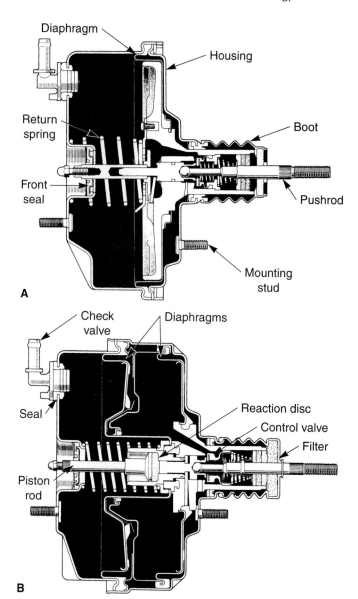

Figure 7-6. *Comparison of vacuum power boosters. A—Single diaphragm. B—Tandem diaphragm power brake unit. (Toyota)*

rear chamber from atmospheric pressure. This allows the vacuum to build up in the rear chamber, matching the vacuum in the front chamber.

Maintaining Pedal Feel

To give the driver a sense of feedback when depressing the brake pedal, two types of internal **reaction assemblies** are used: the lever or the disc. In the *lever reaction assembly,* the connection between the pushrod and the power brake valve is made through a reaction plate and levers. These components are located in the control valve assembly, **Figure 7-8.** When the brakes are released, the fixed ends of each lever come into light contact with the power piston, while the free ends are able to move. The free ends are under light spring tension.

When the driver first presses the brake pedal, the resistance to foot pressure is low. As the driver continues

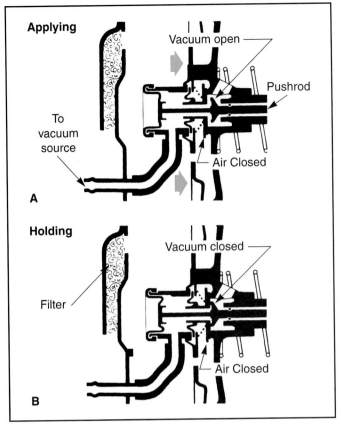

Figure 7-7. *A cutaway view of the vacuum valve. A—Apply position. B—Holding position. (Wagner)*

to press on the brakes, the lever free end springs will be deflected enough to allow contact between the movable ends of the levers and the control valve. The control valve begins to bleed out vacuum and allow atmospheric pressure to enter the rear of the diaphragm. Spring deflection causes the reaction force to the brake pedal to increase. This provides pedal feel similar to a non-power brake system.

The *disc reaction assembly* connects the pushrod and the power brake valve through a rubber disc. The rubber disc operates similarly to the lever design and produces pedal feel when the brakes are applied. **Figure 7-9** shows a disc reaction assembly.

Check Valve

To allow the power assist unit to continue working if vacuum is lost, a one-way **check valve** is installed in the line between the vacuum diaphragm unit and the intake manifold. Most check valves are installed on the vacuum booster housing, as shown in **Figure 7-10**.

A check valve is a device that allows airflow in only one direction, **Figure 7-11**. When the engine or vacuum pump is operating normally, air will flow out of the vacuum diaphragm unit. Air flowing out unseats the check valve, allowing air to leave the diaphragm, creating a vacuum. If the source of vacuum is lost, air will attempt to flow into the vacuum diaphragm. This seats the check valve, keeping the vacuum in the diaphragm.

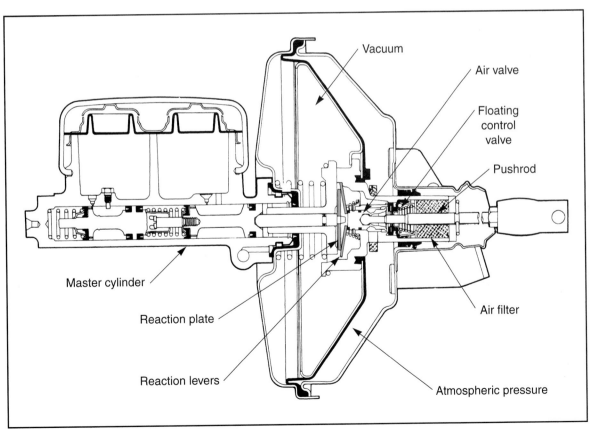

Figure 7-8. *A cross-sectional view of a vacuum power brake unit showing the reaction plate and levers. Unit is in the applying position. (General Motors)*

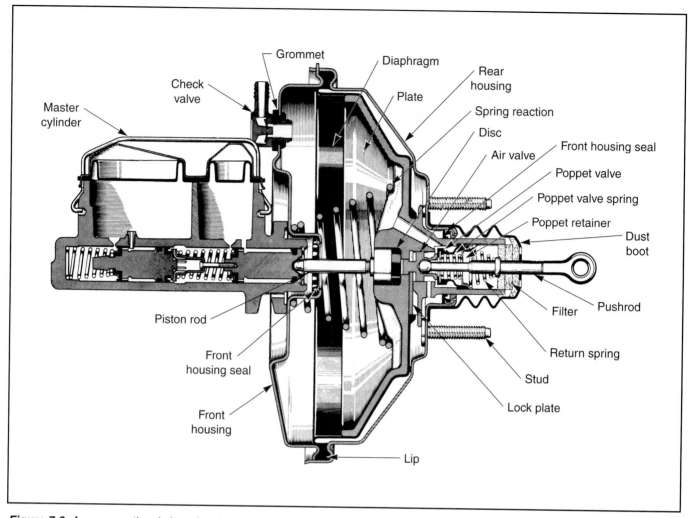

Figure 7-9. *A cross-sectional view showing a vacuum power brake assembly which uses a reaction disc.*

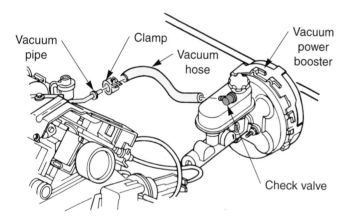

Figure 7-10. *A check valve located between the vacuum source and the vacuum power booster. (Buick)*

Vacuum Unit Operation

The basic operation of a vacuum suspended power assist unit is explained in the following paragraphs. While some design variations exist, this explanation applies to all modern vacuum assist units. Refer to **Figures 7-12** and **7-13** as you read the following paragraphs.

Brakes at Rest

When the brakes are not being used, both sides of the diaphragm are exposed to engine vacuum. The inlet to outside air is sealed off by the air valve, and vacuum enters from the inlet hose, removing air from the front vacuum chamber. The vacuum valve creates a passage between the front and rear chambers. This allows air to be removed from the rear chamber, causing the pressure to be the same on both sides of the diaphragm. With vacuum the same on both sides of the diaphragm, it remains stationary.

Brakes Applied

When the brakes are applied, the pushrod moves the control valve forward. This causes the components of the control valve to do two things:

❑ The air valve section opens the passage between the rear passage and the outside air.

❑ The vacuum valve section seals the passage between the front and rear chambers.

When the air valve section moves, atmospheric pressure is admitted to the rear chamber. As this happens, the closing of the vacuum valve section maintains vacuum in the front chamber. With the rear chamber at atmospheric

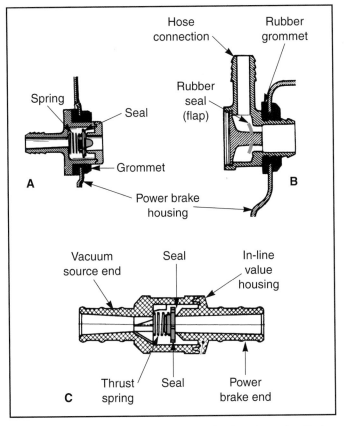

Figure 7-11. *A selection of vacuum check valves. A—Spring and seal type. B—Rubber seal or flap unit. C—Inline spring and seal valve. Valves A and B mount in the booster housing. (Delco & Pontiac)*

pressure, and the front chamber under vacuum, the diaphragm is moved toward the vacuum side. This action is shown in **Figure 7-13.** Since it is attached to the master cylinder pushrod, diaphragm movement adds additional pressure to the pushrod, lessening the amount of pressure needed from the driver to produce the desired braking effect.

Brakes Holding

When the driver holds the brakes in one position, the control valve must maintain an even pressure differential between the vacuum and atmospheric pressure sides of the diaphragm. This keeps the brake assist unit from overapplying the brakes. The air valve section of the control valve closes, sealing the passage between the rear chamber and atmospheric pressure. If additional outside air cannot enter the rear chamber, the pressure in the rear chamber will not be increased. At the same time, the vacuum valve section opens slightly to equalize the pressure between the two sides of the diaphragm. The pressure differential between the front and rear chambers stays the same, keeping the entire brake assist unit in a *balanced* position. The booster maintains consistent pressure, neither increasing or decreasing the pressure applied to the master cylinder.

Brakes Released

When the brake pedal is released, the control valve returns to its normal position. The air valve section closes the passage to atmospheric pressure, and air can be

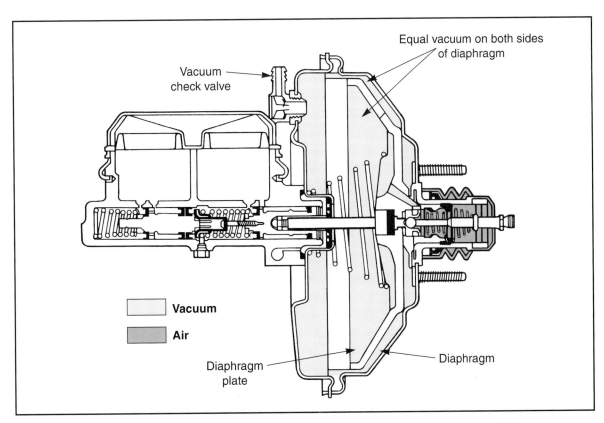

Figure 7-12. *A vacuum power booster at "rest." Vacuum is equal on both sides of the diaphragm. (Bendix)*

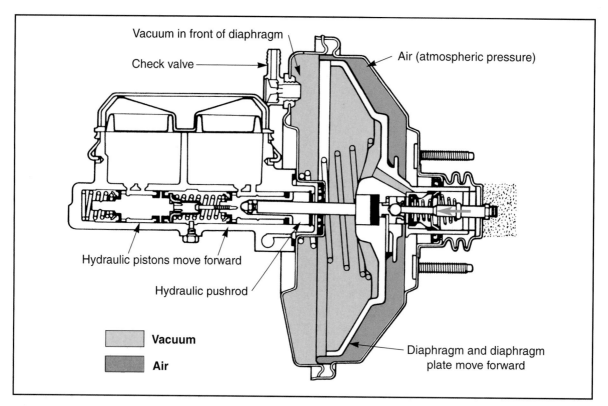

Figure 7-13. *A vacuum power booster shown in the brakes applied position. Compare to* **Figure 7-12.** *(Bendix)*

evacuated from the rear chamber. The same amount of vacuum is present in both chambers and the diaphragm returns to its normal balanced position.

Booster Operation During Loss of Vacuum

If the engine dies, or the booster or hose develops a leak, vacuum will be lost and the check valve will close. Most diaphragms are large enough to allow two or three stops before all vacuum is gone.

After all vacuum is gone, the vehicle can still be stopped. Remember, the pushrod passes through the diaphragm, forming a direct connection between the pedal and master cylinder. Pressing on the brake pedal will operate the master cylinder, although the pedal effort will be high.

 Note: If the vacuum booster system is losing vacuum, you will notice the engine idle will become erratic or possibly stall whenever you depress the brake pedal. This is due to the loss of engine vacuum.

Vacuum Sources

To operate the vacuum assist unit, a constant source of vacuum is needed. On most vehicles, vacuum is provided by the engine. On a few vehicles, the vacuum supply is provided by an electric pump. Sometimes the pump operates only when intake manifold vacuum is low. A few older vehicles have a vacuum pump operated by the engine, through a drive belt or gears.

Engine Vacuum

As the engine operates, the downward movement of the pistons in the cylinders creates a condition of less than atmospheric pressure, or a vacuum. As the piston moves down in the cylinder on the intake stroke, the intake valve is opened. The piston draws in air and fuel through the intake manifold and open intake valve. When decelerating, the throttle valve is closed, and the engine cylinders cannot draw in much air. The resulting piston movement creates a vacuum in the intake manifold. This process is shown in **Figure 7-14.** This vacuum is delivered by a hose to the vacuum unit through the check valve.

When the engine is under heavy load, manifold vacuum is low. However, during this period, the brakes are not normally used. The check valve maintains vacuum in the unit in case the brakes must be used before the engine vacuum returns to normal levels.

Vacuum Pumps

On many modern vehicles, the engine provides vacuum to operate the brake power assist, air conditioner controls, cruise control, transmission modulator, and other vacuum-operated devices. Some engines are unable to provide a consistent source of manifold vacuum to meet these demands. Other engines are overworked by trailer towing or operation at high altitudes.

To ensure the power assist unit has enough vacuum, many modern vehicles have an electric motor driven **vacuum pump.** A typical vacuum pump is shown in **Figure 7-15.** Most vacuum pumps are diaphragm types, although vane and gear pumps are sometimes used.

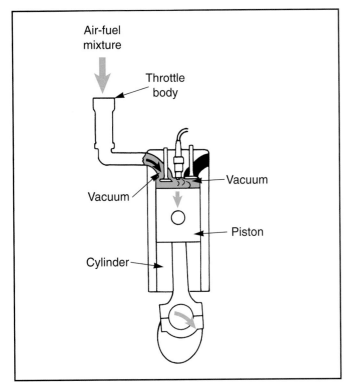

Figure 7-14. *Vacuum can be generated in the cylinder by the downward movement of the piston. (Deere & Co.)*

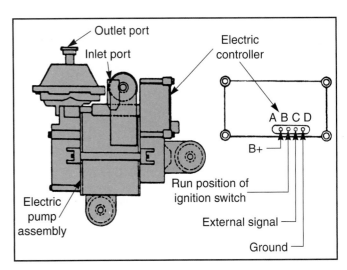

Figure 7-15. *Overall view of an electric vacuum pump and electric controller. (Wagner)*

The vacuum pump may be separate from the intake manifold, and supply vacuum to the brake booster only. However, most vacuum motors are used as backup systems, and are connected to the intake manifold by hoses. Some vehicles equipped with a vacuum pump have a vacuum switch mounted on the power assist unit housing. The switch will activate the dashboard brake warning light if vacuum becomes excessively low.

The vacuum pump motor is usually controlled by a relay operated by a low vacuum switch. If vacuum is not sufficient, the switch energizes the relay, which turns on the pump. When vacuum reaches a preset value, the switch and relay turns off the pump.

A few older vehicles have a belt- or gear-driven vacuum pump. Operation and service of these pumps is similar to the operation of electric motor drive diaphragm pumps. **Figure 7-16** shows a belt driven pump. **Figure 7-17** illustrates a gear driven pump.

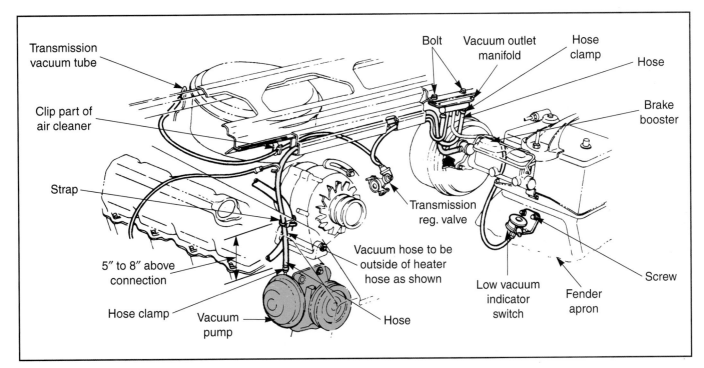

Figure 7-16. *Schematic of one particular belt driven, diaphragm type, vacuum pump and related components. Note the vacuum outlet manifold which directs the vacuum to other vacuum operated parts. (Ford)*

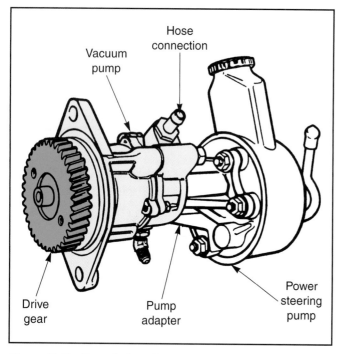

Figure 7-17. *Overall view of a gear driven vacuum pump. Note that the vacuum pump is connected to, and also drives the power steering pump. (Dodge)*

Hydraulic Power Assist Units

On some vehicles, vacuum brake assists are not practical. Many diesel engines are designed without a throttle valve, and engine speed is controlled by varying the amount of fuel injected. With no throttle valve, little or no vacuum is developed in the intake manifold. Also, the demands placed on some gasoline engines by hauling heavy loads can make the development of vacuum erratic. In these situations, manufacturers install a *hydraulic assist unit* or **hydraulic booster.**

There are two major types of hydraulic power assist units. The major difference between the two types is the source of hydraulic pressure. One type of hydraulic assist unit is powered by the vehicle's power steering pump. This system is usually called **Hydro-boost.** Other names for this unit are *Hydro-boost II, Hydra boost,* or *Bendix hydraulic booster.* This system has been largely discontinued in automobiles, but is still used extensively in vans and light trucks, especially those equipped with diesel engines or heavy-duty towing packages.

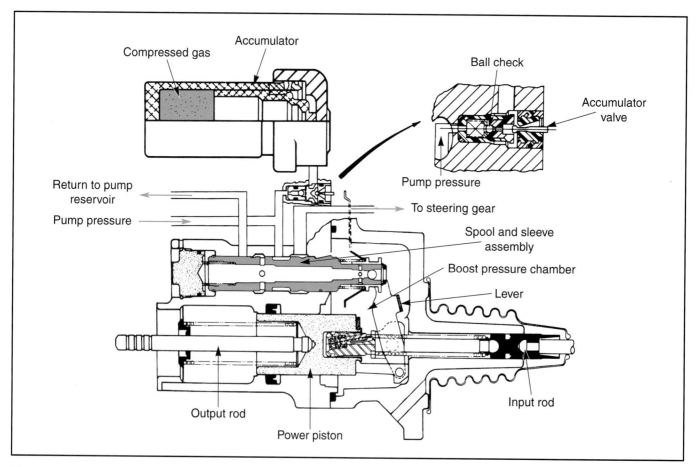

Figure 7-18. *A cross-sectional view of a Hydro-boost unit showing the spool valve and the various fluid passages. Note the accumulator on this unit uses compressed gas instead of a spring for pressure. The ball check is used to maintain pressure in the accumulator. (Ford)*

Hydro-boost Components

The following sections cover the components of the Hydro-boost system. While there may be variations between models and years, all Hydro-boost units have the following basic components.

Spool Valve and Power Piston

The Hydro-boost system uses a **spool valve** to control the flow of pressurized fluid to a **power chamber** containing a **power piston.** The spool valve is a hardened, highly polished valve which resembles a thread spool. The spool contains **lands** (high areas) and **grooves** (low areas), **Figure 7-18.**

The lands and grooves are designed to control fluid flow through the spool valve. The valve fits in a bore in the housing. The lands fit closely in the bore, sealing the grooved areas from each other. The spool valve controls fluid flow by moving to open flow passages between ports, as in **Figure 7-19.** Note that in **Figure 7-19A,** the lands allow flow to move from port A and B. When the valve is moved, **Figure 7-19B,** the lands, ports, and grooves line up so that fluid can flow from port A to port C.

In the Hydro-boost unit shown in **Figure 7-20,** the spool valve is attached to a lever. Note the lever is connected to the pushrod and pivots so that when the pushrod is pushed forward, the spool valve is pushed backwards. The power piston in **Figure 7-20** provides the actual power assist. The power piston is controlled by the spool valve. Springs in the valve and on the lever help to return the piston to the released position when the brake pedal is released.

Hydro-boost Accumulator

The Hydro-boost **accumulator** is a piston in a bore, attached to the Hydro-boost assembly, **Figure 7-21.** Power steering pump pressure pushes on a piston in the accumulator bore. The piston compresses an internal spring or a container of nitrogen gas. If the power steering fails, the spring or nitrogen gas pushes the piston as it tries to decompress. Piston movement supplies enough pressurized hydraulic fluid for two or three power assisted stops.

Power Steering Pump

Power steering pumps are usually rotor or vane pumps, belt driven from the engine. The same hydraulic fluid that operates the power steering is also used to provide power assist to the brake system. Typical power steering and brake hose routing for a Hydro-boost system is shown in **Figure 7-22.** A power steering pump used in a Hydro-boost system can produce pressures of over 2000 psi (13 800 kPa). Since the power steering system does not require this much pressure during normal driving, the pump can be used to operate the brake assist unit. The brake assist unit usually operates on about 150 psi (1020 kPa).

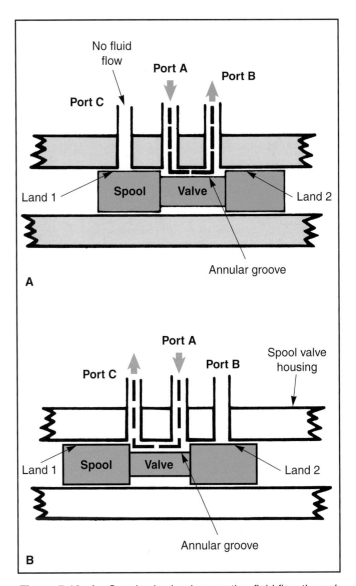

Figure 7-19. A—Spool valve land preventing fluid flow through port C. B—As the spool valve land allows fluid flow through port C, the annular groove is preventing fluid flow through port B. (Wagner)

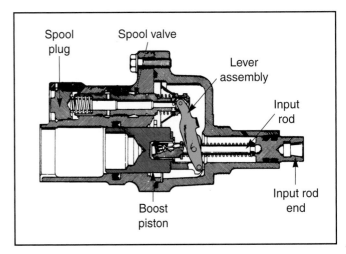

Figure 7-20. Spool valve connected to the lever assembly. The unit is shown in the released position. (Bendix)

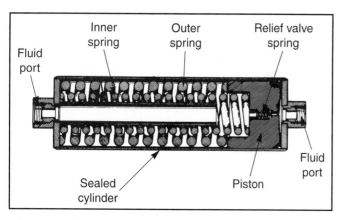

Figure 7-21. *A cross-sectional view of a spring pressure operated accumulator. (Bendix)*

Hydro-boost Operation

The following paragraphs describe the operation of the Hydro-boost system during the at rest, applied, and released positions.

Brakes at Rest

The operation of the Hydro-boost unit at the rest position is shown in **Figure 7-23.** Pressurized fluid enters the Hydro-boost assembly from the power steering pump. The fluid enters the internal passages around the spool valve. The spool valve is positioned so that fluid returns to the power steering system, and no pressure is applied to the power piston. As an extra precaution, the spool valve uncovers a port which vents any excess pressure in the power piston to the brake fluid reservoir. If you apply the brakes while the vehicle is in park, the brake pedal will tend to go to the floor.

Brakes Applied

When the brake pedal is applied, pushrod movement moves the lever, causing the spool valve to move. Movement of the spool valve closes the vent from the power piston to the reservoir. The spool valve then opens the pump port to allow pressurized fluid flow into the cavity behind the power piston, and the pressure causes the power piston to move. Piston movement assists driver foot pressure in applying force to the master cylinder. This process is shown in **Figure 7-24.**

Brake Holding

Often the driver presses on the brake pedal with steady pressure, such as when gradually decelerating or waiting at a stoplight. This is the holding position. In the holding position, the pushrod is stationary and the lever pivots slightly on the pushrod lever pin. This causes the spool valve to move rearward and close off the pressure inlet port. Since the vent port is already closed, the pressure in the power chamber cannot increase or decrease. With the power chamber sealed by the valve, constant pressure is maintained on the power piston as long as brake pedal pressure remains the same.

If the driver depresses the brake pedal further, the additional fluid pressure is vented back to the reservoir. This is done to prevent excess pressure from damaging the Hydro-boost unit.

Brakes Released

When the brake pedal is released, the spool valve returns to the released position, cutting off the supply of pressurized fluid and venting any remaining fluid pressure to the reservoir. Refer to **Figure 7-25.**

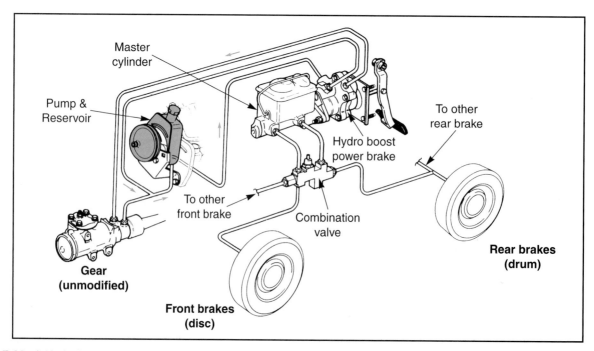

Figure 7-22. *A Hydro-boost system schematic. Power for this system is provided by the power steering pump fluid. (Chevrolet)*

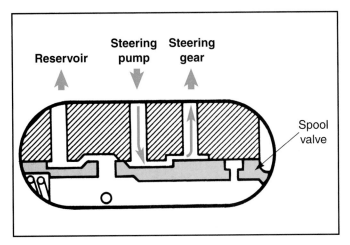

Figure 7-23. *Spool valve in the at rest position. Fluid is circulating from pump-to-gear, over and over. (Wagner)*

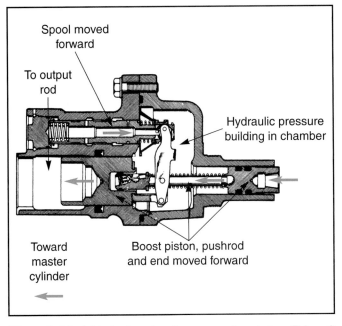

Figure 7-24. *A Hydro-boost unit cross-sectional view. This unit is depicted in the applied position. (Bendix)*

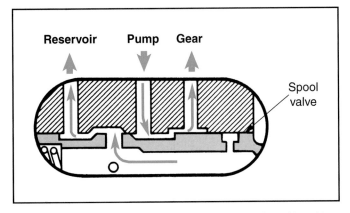

Figure 7-25. *Spool valve in the brakes released position. Note the remaining fluid pressure returning to the reservoir. (Wagner)*

Hydro-boost Operation During Loss of Hydraulic Pressure

Like the vacuum booster, a Hydro-boost system has enough pressurized fluid stored in its accumulator to provide 2-3 power assisted stops. However, unlike the vacuum booster, once the accumulator's pressure is gone, it is extremely difficult to apply the brakes. Since most Hydro-boost systems are used on trucks and vans, the vehicle's weight makes it extremely difficult for a driver to bring the vehicle to a stop without power assist.

> ⚠️ **Warning: It is highly recommended that any vehicle with a malfunctioning Hydro-boost system not be test driven. The lack of power assist could result in the vehicle being involved in an accident.**

Powermaster

The other type of hydraulic brake assist uses pressure supplied by an electrically driven hydraulic pump. Early designs of this type of hydraulic assist unit were called ***Powermaster***. This system was used on a few older vehicles. Most motor driven power brake assists, sometimes called *electrohydraulic assists,* operate in the same basic manner. The two basic designs are:

❑ Systems that provide additional braking force only. These have no connection to any other system. The most common system of this type is the Powermaster system.

❑ Systems that are connected to an anti-lock brake system (ABS), and use a common motor driven vane pump to provide both power assist and ABS operation.

The Powermaster system will be discussed in the following paragraphs. ABS operation will be discussed in more detail in Chapters 21 and 22.

Powermaster Components

All Powermaster systems contain the same basic parts. It is located in the same place as a conventional master cylinder and power booster. The Powermaster combines the master cylinder and power boosting functions into one electrically powered unit. Powermaster was installed on certain full-size General Motors vehicles during the mid-1980s.

Motor and Pump Assembly

The Powermaster power assist uses an electric motor and vane pump to supply brake fluid under pressure. The motor and pump assembly is mounted under the master cylinder, **Figure 7-26.** The brake fluid is supplied from the master cylinder reservoir. Operation of vane pumps was discussed in Chapter 4.

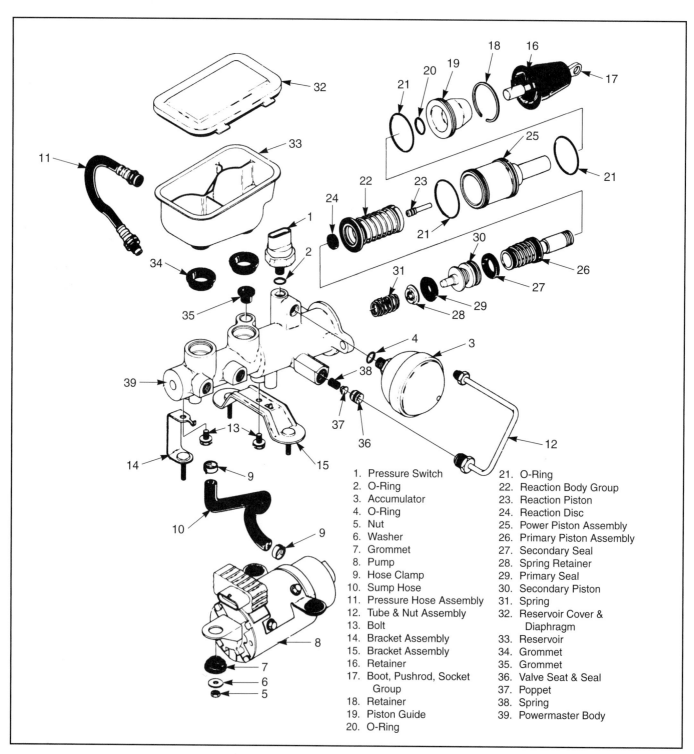

Powermaster Accumulator

The purpose of the Powermaster accumulator, **Figure 7-27,** operates the same as in the Hydro-boost system. The accumulator used on Powermaster systems is round (slightly larger than a baseball) and is usually painted black. The accumulator is pressurized by the pump, and supplies a reservoir of pressurized fluid should the pump fail. Most accumulators have sufficient pressure to allow two or three power assisted stops.

Pressure Switch

The *pressure switch,* **Figure 7-27,** ensures the pump is activated when the pressure falls below a set value. When accumulator pressure falls below a set value, the pressure switch closes, energizing the pump relay. The relay energizes the pump motor, which runs until pressure returns to the preset values. Once pressure returns to normal, the pressure switch de-energized the relay, which shuts off the pump motor.

1. Pressure Switch
2. O-Ring
3. Accumulator
4. O-Ring
5. Nut
6. Washer
7. Grommet
8. Pump
9. Hose Clamp
10. Sump Hose
11. Pressure Hose Assembly
12. Tube & Nut Assembly
13. Bolt
14. Bracket Assembly
15. Bracket Assembly
16. Retainer
17. Boot, Pushrod, Socket Group
18. Retainer
19. Piston Guide
20. O-Ring
21. O-Ring
22. Reaction Body Group
23. Reaction Piston
24. Reaction Disc
25. Power Piston Assembly
26. Primary Piston Assembly
27. Secondary Seal
28. Spring Retainer
29. Primary Seal
30. Secondary Piston
31. Spring
32. Reservoir Cover & Diaphragm
33. Reservoir
34. Grommet
35. Grommet
36. Valve Seat & Seal
37. Poppet
38. Spring
39. Powermaster Body

Figure 7-26. *An exploded view of a Powermaster assembly. Study the construction carefully. (General Motors)*

Power Piston and Valves

The power piston assembly and valves are built into the master cylinder. The valves control the direction of fluid to the power piston for hydraulic assist. Operation of the valves is controlled by the brake pedal pushrod. The power piston assembly is shown in **Figure 7-28.**

Powermaster Operation

The following paragraphs describe the operation of the Powermaster system during the at rest, applied, and released positions.

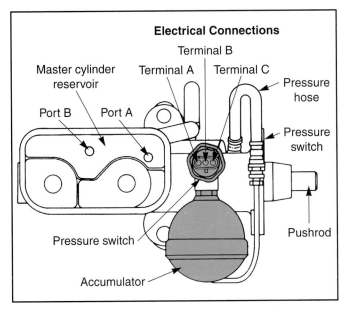

Figure 7-27. *Top view of a Powermaster showing the pressure switch and electrical connection terminals. (Raybestos)*

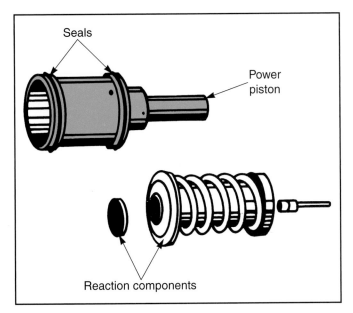

Figure 7-28. *A Powermaster power piston. The reaction components are positioned between the master cylinder and power piston to provide brake pedal feel. (Wagner)*

Brakes at Rest

When the brakes are not being applied, the Powermaster system inlet valve is held closed by a spring. With the valve closed, pressurized fluid cannot get to the power piston. Another spring holds the discharge valve open, allowing any fluid that leaks past the closed apply valve to be vented to the brake fluid reservoir.

Brakes Applied

When the brakes are applied, the brake pedal pushrod forces the discharge valve against spring pressure, closing it, **Figure 7-29.** Pushrod movement then unseats the apply valve, and the open valve allows pressurized fluid to enter the cavity behind the power piston. This pressure moves the power piston, helping to apply pressure to the master cylinder. The amount of apply piston opening is determined by the amount of force applied to the brake pedal.

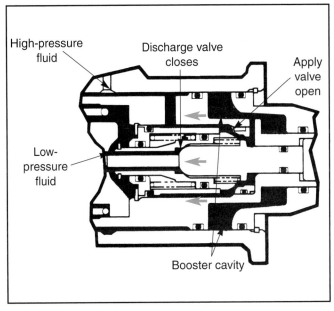

Figure 7-29. *Powermaster valves shown in the brakes applied position. In this position, brake fluid cannot travel back to the reservoir, and fluid pressure is building up. (Pontiac)*

Brakes Holding

In the holding position, the input pushrod is stationary. This holds the apply and discharge valves stationary. Since the valve body and power piston can move in relation to each other, fluid pressure causes the power piston to move a short distance in relation to the valve body. This movement causes the apply valve to close. The discharge valve remains closed. With both valves closed, the pressure on the power piston is constant.

If the driver leaves his or her foot on the pedal for a significant length of time, some pressure loss is possible. Once pressure drops below a certain point, the pump may activate to maintain pressure.

Brakes Released

When the brakes are released, pushrod force no longer holds the apply valve open. The apply valve spring then closes the apply valve preventing additional pressure from entering the booster cavity. The discharge valve spring opens the discharge valve, allowing booster cavity pressure to bleed off into the reservoir. This position is the same as the position in **Figure 7-30.**

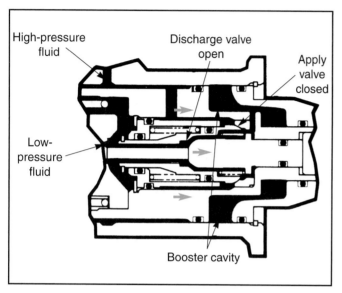

Figure 7-30. *Powermaster valves in the brakes released position. Compare to Figure 7-29. (Pontiac)*

Powermaster Operation When Pressure is Lost

Loss of assist in a Powermaster system is evidence that one of two conditions exist. Either there is a fluid leak in the brake system or the pump motor is not operating. As stated earlier, there is enough pressure in the Powermaster accumulator to provide two to three power assisted stops.

Once this pressure is gone, stopping a Powermaster equipped vehicle with no assist is similar to stopping a vehicle with no vacuum assist. If the loss of pressure is due to the lack of brake fluid, this will affect the hydraulic system, and a low pedal condition will also be noticed.

Power Brake Assist in ABS Systems

Anti-lock brake systems use either a vacuum booster or an electrical pump assist, depending on the type of system. ABS systems that use an electric pump and valves develop boost pressure similar to the Powermaster system. The difference is the pump and controls are operated through the ABS control module. For safety, the hydraulic actuator is usually operated by the pushrod, rather than by the control module. As on the Powermaster system, most ABS systems have an accumulator which provides reserve braking should the hydraulic pump fail. More information on ABS pump construction and operation is located in Chapter 21.

Summary

The two major types of power brake assists are vacuum and hydraulically operated. Vacuum units operate by the pressure differential between the vacuum developed on one side of the vacuum diaphragm and atmospheric pressure on the other side.

All modern vacuum boosters are vacuum suspended types, with vacuum present on both sides of a diaphragm. In addition to the diaphragm, the vacuum assist unit consists of a return spring, control valve assembly, pushrod, and associated seals. The control valve is composed of an air valve controlling the flow of atmospheric air into the diaphragm housing, and a valve controlling vacuum build-up in the housing. The valves are operated by the pushrod. To give the driver a sense of feedback when depressing the brake pedal, an internal reaction assembly is used. The two major types of reaction assemblies are the lever and disc.

To allow the power assist unit to continue working if the engine dies or the hose develops a leak, a one-way check valve is installed in the line between the vacuum diaphragm unit and the intake manifold. Most check valves are installed on the vacuum booster housing. In addition, the pushrod passes through the diaphragm, forming a direct connection between the brake pedal and master cylinder. A variation of the vacuum suspended power brake assist unit is the tandem or dual diaphragm unit. It has two internal diaphragms.

A consistent source of vacuum is needed to operate the vacuum assist unit. Vacuum is usually provided by the engine, but a pump is used on some vehicles. The pump is operated by an electric motor, or by the vehicle engine through a drive belt or gears.

There are two major types of hydraulic assist units, divided by the source of hydraulic pressure. The Hydroboost type of hydraulic assist unit is powered by the vehicle power steering pump. The other type is operated by pressure supplied by an electrically driven hydraulic pump. Both systems use an accumulator to provide for two or three stops should the source of hydraulic pressure fail. Motor drive systems also have a low pressure switch which energizes the pump when pressure falls below a set value. This system consists of a pressure switch and pump relay.

Review Questions—Chapter 7

Please do not write in this text. Write your answers on a separate sheet of paper.

1. Explain why power brakes are used on modern vehicles.

2. Vacuum power brakes operate on the principle of pressure _____.

3. The air pressure that surrounds us at all times is called _____ pressure.

4. Pressures are used to move the flexible _____ inside of the vacuum assist unit.

5. The vacuum assist control valve assembly consists of two valves. The air valve controls the flow of _____ pressure into the vacuum assist unit. The vacuum valve controls the buildup of _____ in the unit.

6. Reaction assemblies are used to give the driver a sense of _____ when braking.

7. Describe what the air valve and the vacuum valve in the vacuum assist control valve do when the brake pedal is depressed.

8. Most vacuum check valves are installed on the _____.

9. A tandem diaphragm vacuum power brake has two _____ in a single housing.

10. On most vehicles, vacuum is provided by the _____.

11. Hydraulic power brake assist units are used when vacuum is not available, such as on a vehicle with a _____ engine.

12. Fluid flow through the hydraulic power booster is controlled by a _____ valve.

13. If the power steering pump fails, what hydraulic device will provide for 2 or 3 power assisted stops?

14. An electrohydraulic assist gets its pressure from a pump driven by an _____.

15. A _____ allows the pump in the Powermaster system to monitor and maintain pressure.

ASE Certification-Type Questions

1. Technician A says that the most common type of power assist unit is the hydraulic unit. Technician B says that the power assist unit is always installed between the firewall and master cylinder. Who is right?
   (A) A only.
   (B) B only.
   (C) Both A & B.
   (D) Neither A nor B.

2. Atmospheric pressure is the pressure of the _____.
   (A) pressure differential
   (B) surrounding air
   (C) intake manifold
   (D) Both A & B.

3. A difference in pressure will move a diaphragm toward the side with _____ pressure.
   (A) more
   (B) less
   (C) stable
   (D) the same

4. A vacuum suspended power brake assist has identical amounts of vacuum present on both sides of the diaphragm when the brakes are _____.
   (A) being applied
   (B) being released
   (C) not being used
   (D) Both A & B.

5. The air valve is installed between the rear diaphragm and the _____.
   (A) front diaphragm
   (B) check valve
   (C) control valve
   (D) outside air

6. A vacuum check valve allows airflow in _____ direction(s).
   (A) one
   (B) two
   (C) three
   (D) None of the above.

7. Vacuum pumps can be operated by _____.
   (A) a gear drive
   (B) a belt drive
   (C) an electric motor
   (D) All of the above.

8. Technician A says that the Hydro-boost system is no longer used on automobiles. Technician B says that the Hydro-boost system has been largely discontinued in light trucks. Who is right?
   (A) A only.
   (B) B only.
   (C) Both A & B.
   (D) Neither A nor B.

9. All of the following statements about the Hydro-boost system are true, EXCEPT:
   (A) a spool valve controls the power piston.
   (B) the power piston is located in the spool valve bore.
   (C) the spool valve contains lands and grooves.
   (D) the spool valve moves to open passages between fluid ports.

10. The accumulator consists of a _____ under pressure.
    (A) spring
    (B) gas filled container
    (C) liquid filled container
    (D) Both A & B.

11. A power steering pump in good condition can produce a pressure of over _____.
    (A) 100 psi (680 kPa)
    (B) 1000 psi (6800 kPa)
    (C) 2000 psi (13 800 kPa)
    (D) 4000 psi (27 600 kPa)

12. Pressing on the brake pedal moves the spool valve by way of a _____.
    (A) screw jack
    (B) bellcrank
    (C) lever
    (D) inclined plane

13. Technician A says that the Powermaster system pressurizes power steering fluid to operate the power assist unit. Technician B says that the power piston of a Powermaster system is built into the master cylinder. Who is right?
    (A) A only.
    (B) B only.
    (C) Both A & B.
    (D) Neither A nor B.

14. On a Powermaster or ABS brake assist unit, the accumulator is charged by _____.
    (A) the brake assist hydraulic pump
    (B) brake pedal effort
    (C) a separate hydraulic pump
    (D) Both A & B.

15. The hydraulic pump is usually controlled by a _____.
    (A) relay
    (B) pressure switch
    (C) vacuum switch
    (D) Both A & B.

Chapter **8**

Power Assist Unit Service

After studying this chapter, you will be able to:
- ❑ Check and adjust brake pedal free travel.
- ❑ Diagnose problems in a vacuum power assist unit.
- ❑ Check vacuum power assist unit for leaks.
- ❑ Inspect vacuum hoses and check valves.
- ❑ Check electrical and mechanical vacuum pumps.
- ❑ Remove and replace a vacuum assist unit.
- ❑ Overhaul a vacuum assist unit.
- ❑ Diagnose problems in a Hydro-boost power assist unit.
- ❑ Overhaul a Hydro-boost power assist unit.
- ❑ Flush and bleed a Hydro-boost unit and related components.
- ❑ Diagnose problems in a Powermaster unit.
- ❑ Replace a Powermaster unit.

Important Terms

Pushrod length

Pushrod height

Height gauge

Overhauling fixture

Spanner wrench

Accumulator piston compressor

In the last chapter, you learned how vacuum and hydraulic power assist units operate. In this chapter, you will learn how to service power assist units. The following sections cover the testing and service of vacuum boosters, followed by service procedures for Hydro-boost and Powermaster units. Service procedures for auxiliary assist devices such as vacuum pumps and power steering pumps are also discussed.

Repair or Replace?

The major service question concerning power assist units, vacuum, hydraulic or electrohydraulic, is whether they should be repaired or replaced. Many repair shops replace the entire unit. Usually, this is the best option, since the cost of labor and parts to rebuild often exceeds the cost of a rebuilt unit.

However, in some cases, you may want to repair a power assist unit. Sometimes a new unit cannot be obtained or the cost may be excessive. In the case of a classic or rare vehicle, the owner may want to maintain the original factory equipment look. In these cases, it may be necessary to rebuild the unit. However, before disassembling the unit, make sure the necessary repair parts are available.

Vacuum Power Assist Service

The vacuum power assist is the most common power assist unit. The following sections explain how to test, adjust, and service single and dual vacuum power assist units. Note that several special tools are needed to disassemble the vacuum assist unit. *Do not* try to repair a vacuum assist unit without these tools.

Checking Vacuum Power Assist Condition

Before deciding the vacuum power assist unit is defective, check it for proper operation. Begin by pumping the brake pedal several times with the engine off. This will remove all vacuum from the assist unit. Then press on the pedal and start the engine. If the assist is working properly, you will feel the pedal move downward an additional distance. If you do not feel this additional movement, the vacuum booster is defective or is not receiving vacuum. If you hear an audible hiss accompanied by poor engine idle or stalling when you depress the brake pedal, the vacuum booster is leaking internally.

To test the vacuum hoses and check valve, shut off the engine and wait at least 10 minutes. Then press on the brake pedal. There should be power assist for at least one stop, with partial assist for the second stop. If the brake pedal is hard (no power assist), the check valve is defective. **Figure 8-1** shows several vacuum brake booster conditions and the problems that cause them.

 Note: Always check the condition of the brake hydraulic and friction elements before condemning the power assist. Never assume that a problem is in the power assist.

Common Vacuum Booster Problems

The most common symptom of a vacuum booster problem is lack of power assist. In many cases, this problem is not caused by the unit itself. Often, the vacuum hose has become disconnected or has developed a leak. A leaking hose is usually easy to spot when the engine is running. As the engine idles, leaking hoses will usually make a hissing sound, or a suction can be felt when the

Vacuum Power Assist Diagnosis		
Condition	Cause	Correction
Pedal does not go down when the engine is started.	Booster defective Booster not receiving vacuum	Replace or rebuild booster Check vacuum hose and check valve Check engine vacuum
Loud hiss heard when pedal depressed (engine also idles rough or stalls).	Booster defective	Replace or rebuild booster
No power assist after key off for 10 minutes.	Booster check valve defective.	Replace check valve and grommet
Brakes will not release.	Defective brake booster Defect in hydraulic system	Repair or replace booster Repair defect in hydraulic system

Figure 8-1. *Typical vacuum brake booster conditions and their causes.*

split section is touched. Be sure to check for leaks at the check valve and grommet, as this is a common source of leaks. See **Figure 8-2.** Often, engine idle and performance will also be affected, since a defective booster acts as a vacuum leak.

Sometimes the vacuum hose has collapsed, or has become internally plugged. This can be spotted by visual inspection or by removing the hose at the check valve and attaching a vacuum gauge to the hose end, **Figure 8-3.** If the vacuum gauge does not indicate manifold vacuum when the engine is started, the hose is plugged.

In a few cases, the engine may be in such poor condition that it cannot develop enough vacuum to operate the assist. Typical causes of low engine vacuum are retarded ignition or valve timing, extreme rich or lean fuel mixture, worn rings, burned valves, worn or high performance camshaft, or a clogged exhaust system. These problems must be corrected before the power assist will operate properly. If the assist unit receives vacuum from an auxiliary pump, check the pump's condition as well as the operation of pump relays and fuses. Also check for leaks at the pump output hose.

If sufficient vacuum is available to the booster, the problem is in the unit itself. Unit operation can be checked by using a pedal effort gauge. Possible causes of internal unit failure are a ruptured diaphragm, sticking or broken valves, or occasionally a clogged filter at the atmospheric pressure inlet valve.

If the brakes will not release, isolate the problem to the power assist unit. Ensure there is no binding in the pedal linkage. Then loosen the nuts holding the master cylinder to the power assist unit and pull the master cylinder away from the assist. If the brakes release, the trouble is in the assist unit. If the brakes do not release when the master cylinder is pulled away, the problem is in the brake hydraulic system.

Unusual noises from the vacuum power brake unit are rare. When the brake pedal is first pushed, a very low hissing sound may be heard. This is a normal noise, caused by the air rushing into the rear chamber. However, if the hissing sound is loud and continues when the brakes are in the holding position, the booster is defective, and repair is needed.

Checking and Adjusting Pushrod Length

Modern vacuum power assists are sensitive to **pushrod length.** Ordinarily, the original pushrod length will not change, and the unit can be reinstalled without checking. However, changing the master cylinder or power assist unit, or just normal wear can cause the pushrod length to change. Inadequate pushrod length can

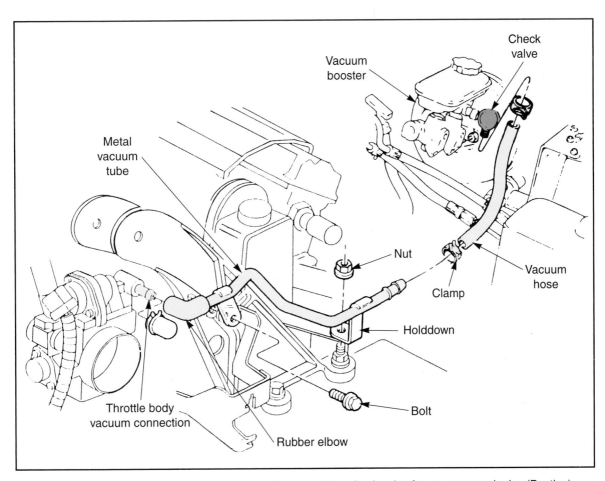

Figure 8-2. *Check the vacuum booster hose, hose connections, and the check valve for any vacuum leaks. (Pontiac)*

result in excessive pedal travel. An overly long pushrod can cause brake dragging.

Pushrod length can be checked before the master cylinder is reinstalled on the vacuum assist unit. Pushrod length is checked by checking the length of pushrod that sticks out from the vacuum booster, usually called **pushrod height.** For this test, you must have the proper **height gauge, Figure 8-4.** Place the gauge over the

pushrod and observe whether the gauge just touches the pushrod, as shown in **Figure 8-4.** Adjust the pushrod as necessary. The adjuster is usually placed at the end of the pushrod, **Figure 8-5.** In some cases, you may need a special tool or a vacuum gauge to perform this adjustment.

Caution: If you have to disconnect the pushrod from the brake pedal on any power assist unit, be sure to unplug the brake light switch or pull the fuse to the stoplamp circuit. Removing the power assist unit on most vehicles will cause the brake lights to stay on, possibly draining the battery.

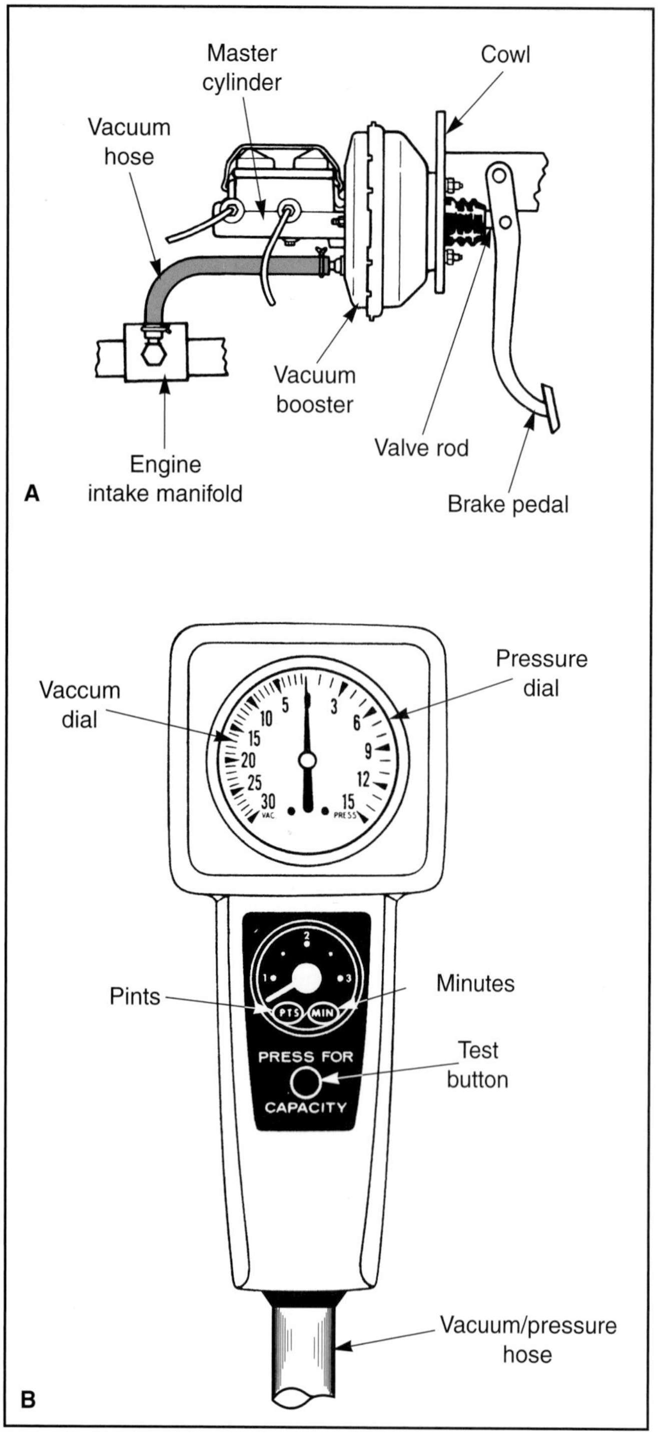

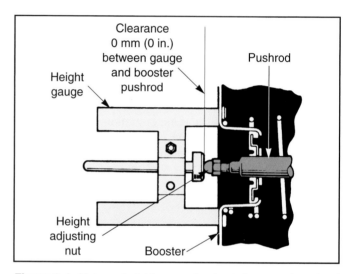

Figure 8-4. *Using a height gauge to check for correct pushrod length. (Chevrolet)*

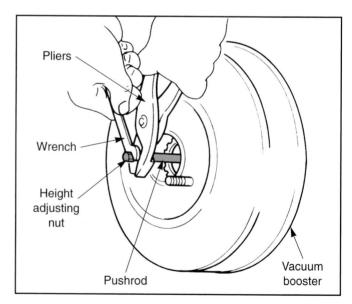

Figure 8-5. *Adjusting pushrod length. Hold pushrod from turning, while rotating the height adjustment nut with a wrench. (Chevrolet)*

Figure 8-3. *A—Power vacuum booster and vacuum hose to manifold. B—Vacuum gauge. Follow the vehicle and tool manufacturer's operating and testing instructions. (KAL Equipment)*

Testing the Check Valve

Before proceeding to replace or rebuild the vacuum booster, test the operation of the check valve. Remove the check valve from the booster housing. Then use lung pressure to blow into the intake manifold end of the valve. You should be unable to blow through the valve. Next, suck on the assist side of the valve. You should be unable to suck any air through the valve. If the valve fails either of these tests, replace it. Also check the check valve grommet for cracks and leaks. It should be replaced along with the check valve.

Vacuum Booster Removal

To remove the vacuum assist unit, remove the vacuum hose and remove the pushrod connection at the brake pedal linkage. If the unit has a low vacuum switch, disconnect it. Then remove the master cylinder.

> **Note: In some cases, it may be possible to move the master cylinder far enough to remove the vacuum booster without removing the hydraulic lines. However, if there is any danger of kinking the lines, remove the master cylinder completely from the vehicle.**

Most power assist units are held to the vehicle by nuts installed on threaded studs welded to the rear half of the booster. Remove the nuts and remove the power assist unit from the firewall. Refer to **Figure 8-6.** Be sure to unplug the brake light switch so as not to drain the battery while the booster is out of the vehicle.

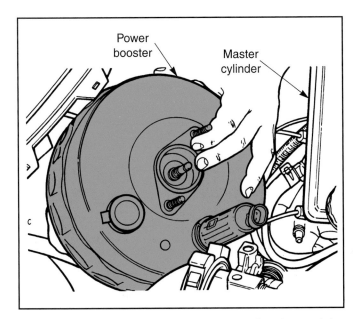

Figure 8-6. *Removing the power booster after the retaining fasteners have been backed off. (Chrysler)*

Vacuum Booster Overhaul

The following sections explain how to disassemble a typical vacuum booster and check the internal components. If you are replacing the vacuum booster, skip the next two sections.

Overhauling a Single Diaphragm Booster

Begin by placing the unit in an *overhauling fixture.* Then remove the dust boot, filter, and silencer from the rear of the housing. Remove the check valve grommet from the front housing and scribe matching marks between the front and rear housings to aid assembly. Loosen the booster tabs, then mount the booster assembly in the holding fixture, **Figure 8-7A.** In some cases, a special adapter tool is used to secure the booster, **Figure 8-7B.** Tighten the fixture screw to hold the rear housing in position, then rotate the housing counterclockwise to unlock the locking tabs.

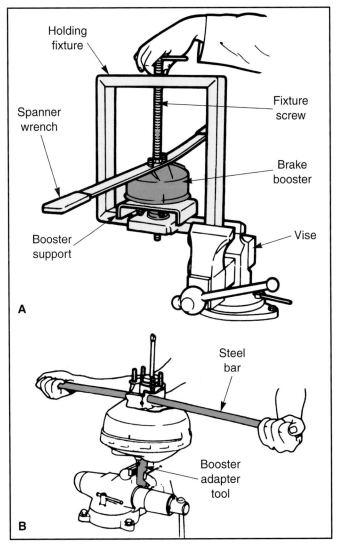

Figure 8-7. *A—The brake booster being secured in a special holding fixture. Do not try to separate a booster without this tool. They fit together tightly and are under spring pressure, which can cause them to fly apart if not secured. B—A special adapter and bar can also be used to separate the booster housing (Chevrolet)*

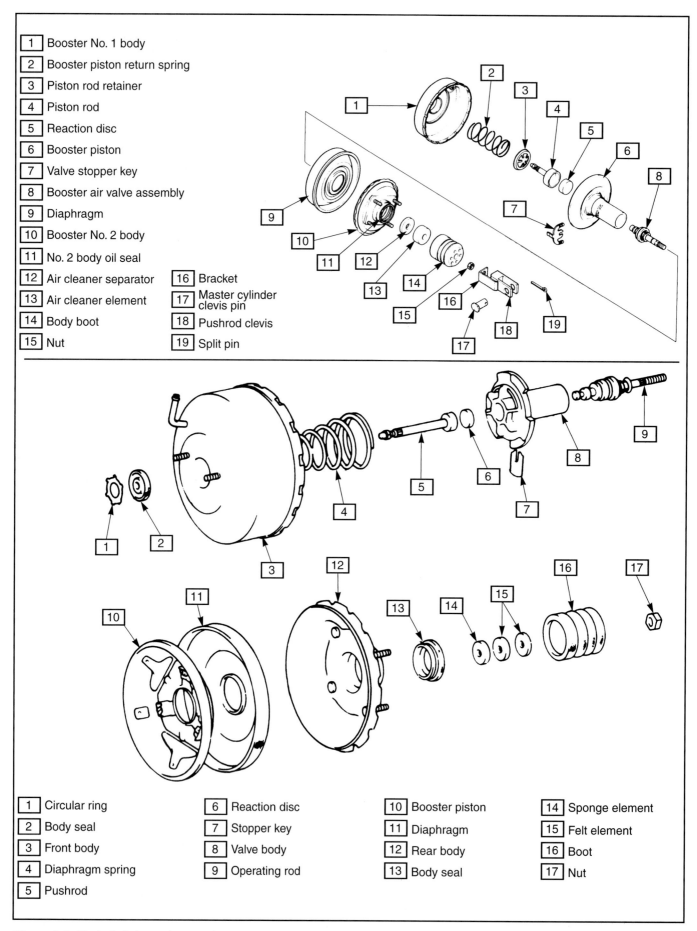

1	Booster No. 1 body
2	Booster piston return spring
3	Piston rod retainer
4	Piston rod
5	Reaction disc
6	Booster piston
7	Valve stopper key
8	Booster air valve assembly
9	Diaphragm
10	Booster No. 2 body
11	No. 2 body oil seal

12	Air cleaner separator	16	Bracket
13	Air cleaner element	17	Master cylinder clevis pin
14	Body boot	18	Pushrod clevis
15	Nut	19	Split pin

1	Circular ring	6	Reaction disc	10	Booster piston	14	Sponge element
2	Body seal	7	Stopper key	11	Diaphragm	15	Felt element
3	Front body	8	Valve body	12	Rear body	16	Boot
4	Diaphragm spring	9	Operating rod	13	Body seal	17	Nut
5	Pushrod						

Figure 8-8. *Exploded views of two typical brake boosters and their related internal parts. (Chevrolet)*

Note: A spanner wrench should be used to rotate the housings, since the two halves of the housing are tightly installed.

After the tabs are separated, allow the housings to separate. This should be done slowly to allow the diaphragm return spring to decompress. Then remove the internal parts of the assist unit. **Figure 8-8** shows the internal parts of two typical vacuum power assist units. Disassembly procedures vary depending on the type of assist unit. Notice in **Figure 8-8** that control valves and pushrods can be one-piece units, or can be separated into two or more parts. Diaphragms and support plates also vary widely. For this reason, you must obtain the proper service manual before starting repairs.

As you disassemble the unit, carefully handle parts that will be reused, such as the control valve, spring, and reaction unit. On some units, the pushrod and control valve assembly is held to the diaphragm assembly with a snap ring. Use the correct pliers to remove the snap ring. Carefully pull the control valve from the center of the diaphragm. Lay the parts in order as you remove them, this will make reassembly easier and reduce the possibility of making an error.

After all internal parts are removed, gently pull the diaphragm from the support. On most power assist units, the diaphragm is attached to a support plate, **Figure 8-9.** This plate should be carefully removed from the diaphragm to avoid damage. Check the edge of the old diaphragm for tears, this is where most leaks occur.

Note: Some vacuum boosters have a letter or number code on a label or stamped on them. You will need this code when ordering parts. Be sure to place all parts in the order of removal. This will make booster reassembly much easier.

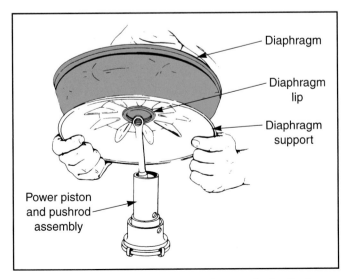

- Diaphragm
- Diaphragm lip
- Diaphragm support
- Power piston and pushrod assembly

Figure 8-9. *Removing a diaphragm from a single diaphragm booster assembly. (General Motors)*

Check all internal parts for damage, dirt, and corrosion. Note whether there is any buildup of gasoline, brake fluid, or moisture in the bottom of the unit. Small amounts of gasoline and moisture are considered normal, but brake fluid usually indicates a leaking master cylinder. The master cylinder should be carefully checked and replaced if necessary. Use crocus cloth or 400 grit sandpaper to remove any corrosion from moving parts. Then clean all components with denatured alcohol and dry with compressed air.

Note: Most control valves and/or control valve and power piston/pushrod assemblies are serviced as a complete unit.

To begin reassembly, lubricate the new diaphragm at all points where it will contact other parts of the unit. Most manufacturers recommend silicone grease or aerosol silicone lubricant. Install the new diaphragm on the support plate. Then reinstall the pushrod and control valve assembly into the center of the diaphragm. Place the diaphragm and return spring in the housing, using new seals where necessary. Carefully check the service manual for exact reassembly procedures.

Note: Some technicians file the leading edges of the locking tabs into a slope. This can make booster reassembly much easier. Be sure to apply some touch-up paint to the bare metal after booster reassembly.

Next, position the front and rear housings in the holding fixture, making sure the scribe marks are aligned. Make a final check that all parts are properly installed and aligned. Then tighten the fixture screw until both halves are flush. Rotate the front housing until the locking tabs engage. Stake the locking tabs to keep the housing from coming apart. Then loosen the fixture screw.

Install the check valve grommet and check valve, then install the front housing seal, silencer, and boot. If desired, you can use a vacuum pump to check unit operation before installing it on the vehicle.

Tandem Booster Overhaul

Overhaul of a tandem or dual diaphragm unit is similar to the process used to overhaul a single diaphragm unit. Start by removing the dust boot and silencer from the rear of the housing. Then remove the check valve and its grommet from the front housing. Scribe marks on the front and rear housings and mount the unit in a holding fixture. Tighten the fixture to hold the housing assembly together, then rotate the front housing counterclockwise using the spanner tool. After the housing tabs are unlocked, slowly back off the holding fixture and allow the housings to separate. This housing is also under strong spring tension, so care must be taken to ensure the housings do not fly apart.

Remove the internal parts of the assist unit. **Figure 8-10** shows the internal parts of a typical dual diaphragm vacuum power assist unit. Note the unit contains two diaphragms, control valves, and other components. Also note the internal separator is a stationary part. Be careful with parts that will be reused, such as the internal separator, control valves, return spring, and reaction unit. After all internal parts are removed, gently pull and remove the primary and secondary diaphragms from their related support plates.

Check all internal parts for damage, dirt, and corrosion. Note whether there is any buildup of gasoline, brake fluid, or moisture in the bottom of the unit. Small amounts of gasoline and moisture are considered normal. However, the presence of brake fluid usually indicates a leaking master cylinder. The master cylinder should be carefully checked and replaced if necessary. Use crocus cloth to polish minor corrosion from any moving parts. Then clean all components in denatured alcohol, and dry with compressed air.

To begin reassembly, lubricate both new diaphragms at all points where they will contact other parts of the unit. Most manufacturers recommend silicone lubricant. Then reassemble the diaphragms and other internal parts into the housing, **Figure 8-11.** Carefully check the service manual for exact reassembly procedures.

Position the front and rear housings in the holding fixture, making sure the scribe marks are aligned. Make a final check that all parts are properly installed and aligned. Then tighten the fixture screw until both halves are flush. Rotate the front housing until the locking tabs engage. Stake the locking tabs to keep the housing from coming apart, then loosen the fixture screw. Install the check valve grommet and check valve, then install the front housing seal, silencer, and boot.

Vacuum Booster Installation

To install the vacuum assist unit, place it over the studs in the firewall. While a helper holds the booster on the firewall, install and tighten the mounting nuts, and reconnect the pushrod.

 Note: If necessary, measure and adjust pushrod length as explained earlier in this chapter.

When adjustment is complete, reconnect the brake light switch, then install the master cylinder. Install the

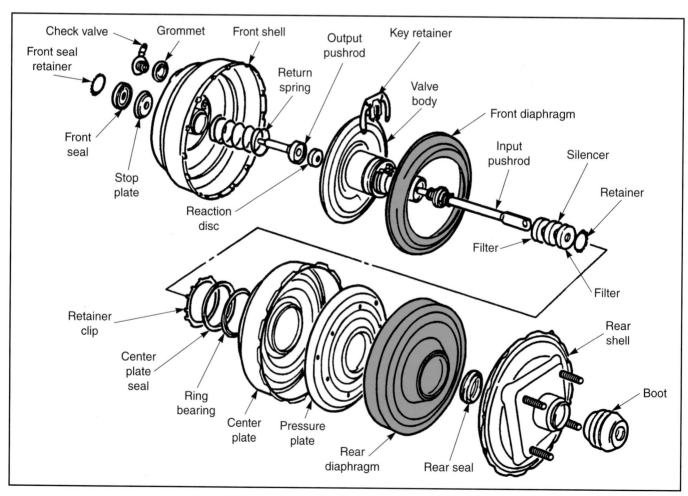

Figure 8-10. *An exploded view of a tandem diaphragm power brake booster assembly. (Corvette)*

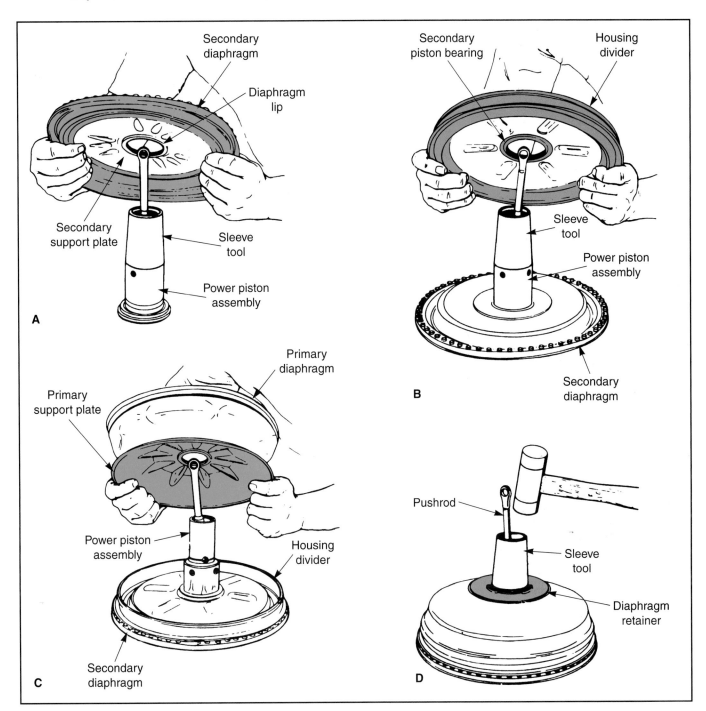

Figure 8-11. *Steps in tandem vacuum booster overhaul. Single diaphragm booster overhaul is similar. A—Assembling secondary diaphragm and support plate. B—Installing housing divider. C—Assembling primary diaphragm and support plate. D—Installing the diaphragm retainer. Exercise care as the diaphragms can be easily damaged. (Chevrolet)*

vacuum hose and low vacuum switch, if used. Bleed the master cylinder if necessary, then road test the vehicle and ensure that repairs are successful.

Vacuum Pump Service

Brake system vacuum pumps are driven by an electric motor or by a belt or gear from the engine. While it may be possible to service the vacuum pump, most are simply replaced.

Motor-driven Pump

To remove a motor-driven vacuum pump, locate the pump on the vehicle. A typical motor-driven pump is shown in **Figure 8-12.** Remove any covers or other components to reach the pump. Then remove the electrical connectors and vacuum hoses. Remove the fasteners and lift the pump from the vehicle.

To reinstall the pump, carefully place it in position and install and tighten the fasteners. Then reconnect the electrical connectors and vacuum hoses. Reinstall any covers or components, then start the vehicle and check pump operation.

138

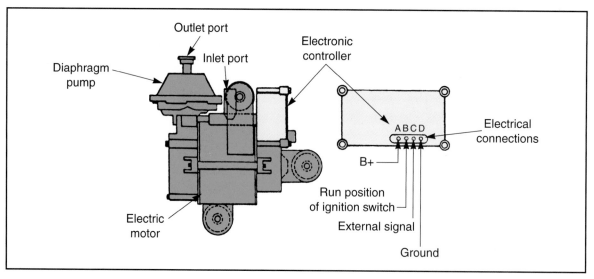

Figure 8-12. *One type of electrically-operated vacuum pump. The pump may be mounted in a variety of spots on the vehicle. This particular pump mounts underneath the battery, and may be enclosed in a metal or plastic housing. (General Motors)*

Belt-driven Pump

To remove belt-driven vacuum pumps, locate the pump on the engine, **Figure 8-13,** and remove the vacuum hose connections. Then loosen the pump adjuster bolts and remove the drive belt. Remove the pump fasteners and lift the pump from the engine compartment. To reinstall, place the pump in position and install and tighten the fasteners. Then install the belt and adjust it according to specifications. Reinstall the hoses, start the engine, and check pump operation.

Gear-driven Pump

To remove a gear-driven vacuum pump, first locate the pump. Gear-driven pumps are usually located in the spot formerly occupied by the distributor, **Figure 8-14.**

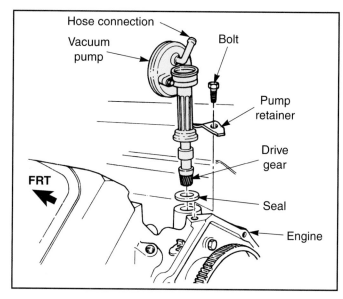

Figure 8-14. *An engine mounted, gear driven vacuum pump. These pumps are generally driven by the camshaft. (Chevrolet)*

Remove the hose and the hold-down bolt and clamp. Slowly pull the pump from the engine, ensuring the oil pump gear is not displaced. To install the replacement pump, place it in the distributor hole, making sure that it is fully seated and that the vacuum pump, oil pump, and camshaft gear engage. Then install the hold down clamp and tighten the bolt. Reattach the vacuum hose and start the engine. Check pump operation.

Hydro-boost Service

Hydro-boost systems are found on many light trucks, vans, and diesel-powered cars. The Hydro-boost is the most common type of hydraulic power assist unit. This section explains how to test and service Hydro-boost systems.

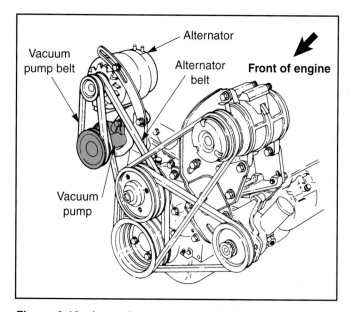

Figure 8-13. *An engine mounted, belt driven vacuum pump. (Ford)*

Checking Hydro-boost Condition

Before condemning the Hydro-boost unit, visually inspect and check the operation of the brake hydraulic system and friction elements. If these check out good, test the operation of the power assist by the following method.

First, check the fluid level and add as needed. If the fluid level is very low, check for leaks. Then pump the brake pedal at least 30 times with the engine off. This will bleed off any hydraulic pressure stored in the accumulator. While holding firm pressure on the brake pedal, start the engine. The brake pedal should move downward, then push up against foot pressure. If it does not, the system is not providing hydraulic assist.

To test the accumulator, turn the steering wheel in one direction against the stop (as far as it can turn) for about five seconds. Then shut off the engine and wait at least 30 minutes. After 30 minutes have passed, press on the brake pedal. There should be power assist for at least two to three stops. If the brake pedal is hard (no power assist) before the third or fourth application, the accumulator is defective or its check valve is leaking. **Figure 8-15** gives a list of Hydro-boost symptoms and their causes.

Common Hydro-boost Problems

The most common causes of Hydro-boost problems are internal and external leaks, **Figure 8-15**. External leaks may result in low fluid level, leading to erratic steering and brake booster operation. Internal leaks can cause fluid pressure to bypass the power unit, leading to loss of power

Hydro-Boost Diagnosis		
Condition	Cause	Correction
Excessive Brake Pedal Effort	Loose or broken power steering belt No fluid in power steering reservoir Hydro-Boost unit leaking	Replace or rebuild booster Check vacuum hose and check valve Check engine vacuum
Slow Brake Pedal Return	Excessive friction in Hydro-boost unit Restriction in power steering return line Pedal rod damage	Overhaul or replace Hydro-boost Replace return line Overhaul or replace Hydro-boost
Brakes Grab	System contamination.	Disassemble and clean Hydro-boost and steering system
Pedal Vibrates, Booster Chatters	Power steering pump slips Low power steering fluid level System contamination	Tighten or replace belt Fill reservoir, check for leaks Disassemble and clean Hydro-boost and steering system
Accumulator Leak	System contamination Internal accumulator leak	Disassemble and clean Hydro-boost and steering system Replace accumulator and power piston

A

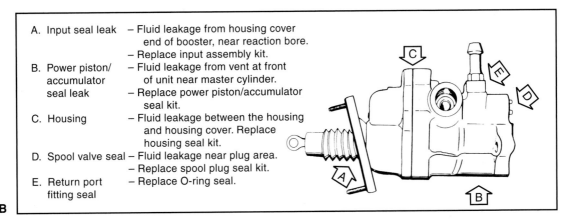

B

A. Input seal leak — Fluid leakage from housing cover end of booster, near reaction bore.
— Replace input assembly kit.
B. Power piston/ accumulator seal leak — Fluid leakage from vent at front of unit near master cylinder.
— Replace power piston/accumulator seal kit.
C. Housing — Fluid leakage between the housing and housing cover. Replace housing seal kit.
D. Spool valve seal — Fluid leakage near plug area.
— Replace spool plug seal kit.
E. Return port fitting seal — Replace O-ring seal.

Figure 8-15. A—Chart showing Hydro-boost conditions and their causes. B—Possible leak points on a Hydro-boost unit. (GMC)

braking assist. Other causes of lack of power assist are pump or control system failure. Pump failure may be caused by a slipping or missing belt, internal pump failure, and leaking or clogged hoses.

Many Hydro-boost problems are often preceded by unusual noises. Normal Hydro-boost noise will only occur for a short time. However, noises that are louder than normal or occur constantly may be an indication of system wear or air in the steering or Hydro-boost system.

Since the Hydro-boost is part of the power steering system, defects in the steering system can affect Hydro-boost operation. This also means a Hydro-boost problem can affect the steering system. Therefore, when diagnosing a Hydro-boost failure, make sure the steering system does not also have a problem. If both the power brakes and steering are weak, check the pump before condemning the brake unit. If the power steering belt squeals when the brakes are applied, the belt may be loose, or the accumulator may have failed. Adjust or replace the belt and check tension, **Figure 8-16.**

Hydro-boost Removal

Before removing a Hydro-boost unit, check the condition of the rest of the hydraulic system. On a Hydro-boost system, check for leaks and swelling at the high and low pressure hoses, **Figure 8-16.** Also check for the presence of metal or dirt in the hydraulic fluid. If foreign material is present or fluid appears to be overheated (will be dark and smell burnt), the system should be thoroughly flushed *before* the assist unit is removed.

Before taking any steps to remove the Hydro-boost unit, press and release the brake pedal at least 30 times (engine off) to depressurize the accumulator.

 Warning: Always depressurize the accumulator before disconnecting any hydraulic component.

Refer to **Figure 8-17.** Remove the master cylinder. Then disconnect the hydraulic hoses from the Hydro-boost unit. Plug all openings to prevent fluid loss and system contamination. This also minimizes fluid loss and contamination by air or dirt. Remove any electrical connectors.

Disconnect the pushrod from the brake pedal and remove the nuts holding the unit to the firewall studs. Remember to unplug the brake light switch so the battery will not be discharged. On some vehicles, the studs are part of the assist unit, and the nuts must be removed from under the dashboard. After all fasteners are removed, lift the unit from the engine compartment.

Hydro-boost Overhaul

The following sections explain how to overhaul Hydro-boost assist units. Designs and repair methods vary widely, so the proper service manual should be consulted. If you are replacing the Hydro-boost assembly, skip this section.

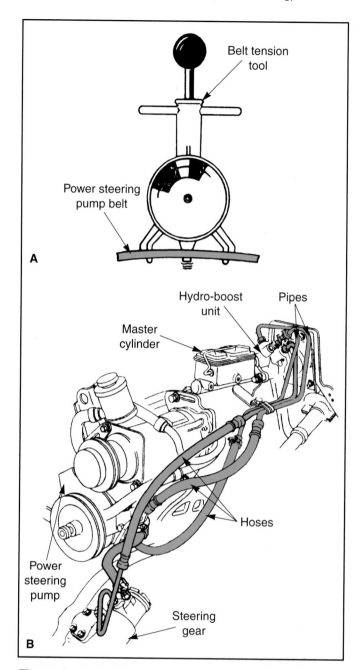

Figure 8-16. *A—Check belts for proper tension with a belt tension tool. If adjustment is necessary, do not move the pump by prying against the reservoir, pulling on the filler neck or hammering; pump damage may occur. On serpentine belts, make sure the tensioner is in good condition. B—Before removing any hydraulic assist units, check all the lines for signs of leakage, swelling, cracking, etc. (General Motors)*

To overhaul a Hydro-boost unit, begin by draining any excess oil and mounting the housing in a vise. Do not overtighten the vise. Next, place the **accumulator piston compressor** or a C-clamp over the end of the accumulator and install a nut onto the stud. See **Figure 8-18.**

Compress the accumulator cover with a C-clamp, and insert a small punch into the hole of the unit housing, **Figure 8-18,** and loosen and push out the snap ring. After the snap ring is removed, slowly loosen the C-clamp and remove the accumulator.

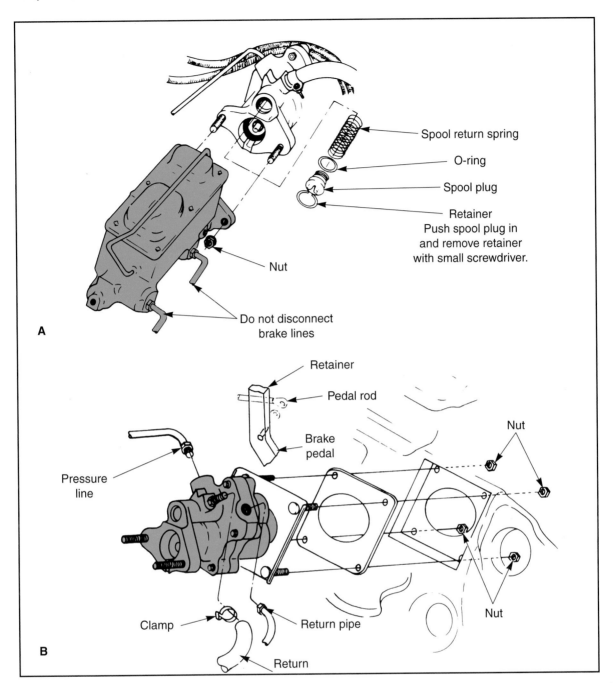

Figure 8-17. *A—Removing the master cylinder. B—Removing Hydro-boost unit. Be sure to position the hydraulic hoses to minimize fluid loss. (General Motors)*

Caution: The accumulator is under high pressure. Do not attempt to remove the snap ring without first compressing the accumulator cover. If you do not have the piston compressor shown in Figure 8-18, do not attempt to remove the accumulator.

Once the accumulator is removed, remove the other internal parts of the Hydro-boost unit. On most Hydro-boost units, the housing assembly must be split to remove some of the internal parts. **Figure 8-19** shows typical Hydro-boost internal parts.

Caution: Pushrod removal is not recommended. If the pushrod requires replacement or its seals are leaking, replace the Hydro-boost pushrod housing or the complete unit.

Clean all parts in approved solvent. Check all internal parts for cracks and damage. Carefully check the spool valve and valve bore for scratches, scoring, or wear. If the spool valve or bore show any signs of damage or the housing is leaking, replace the entire unit. Inspect the power piston. If it is deeply scratched, replace the Hydro-boost

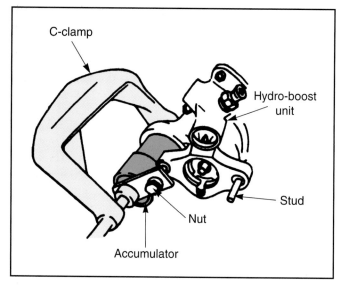

Figure 8-18. *Compressing the accumulator with a C-clamp. Use caution as these springs are under high tension. Wear safety glasses. (General Motors)*

housing or complete assembly. Other worn parts can be replaced as needed. If the accumulator uses a gas-filled bag, check the bag for leaks.

Start by lubricating all parts with the proper type of power steering fluid. Then install the internal parts. Be sure to use all the parts in the overhaul kit. Install the accumulator cover last, using the accumulator piston compressor and C-clamp.

 Caution: Be sure that the snap ring is fully seated before releasing the C-clamp.

Hydro-boost Installation

To install the Hydro-boost unit, place it over the studs in the firewall, or push the studs through the firewall and install the nuts. Install and tighten the mounting nuts, and reconnect the pushrod. Then install the hydraulic lines and master cylinder. Bleed the unit as needed. After all installation and bleeding procedures are complete, road test the vehicle and check for proper power assist operation.

Flushing and Bleeding Hydro-boost Units

The Hydro-boost unit is operated by pressure developed in the power steering pump. Therefore, any debris or contamination in the fluid, or air in the power steering system will affect Hydro-boost system operation. The most common reason for flushing is the fluid has become overheated and has lost its ability to lubricate and prevent corrosion. The most common reason for bleeding is that repairs or a leak have allowed air to enter the system.

If the power steering pump or steering gear has failed and is replaced, the entire hydraulic system, including the Hydro-boost unit, should be flushed and bled. If the pump has produced metal shavings, the Hydro-boost (and steering gear) should be disassembled to remove debris. Hoses should also be thoroughly cleaned and blown out with compressed air. If this is not done, metal particles from a damaged pump or steering gear may cause problems in the Hydro-boost system. If the problem is not as serious, the system can be satisfactorily cleaned by flushing.

System Flushing

To flush the Hydro-boost hydraulic system, begin by draining the system of as much fluid as possible. To do this, remove the fluid return hose from the power steering

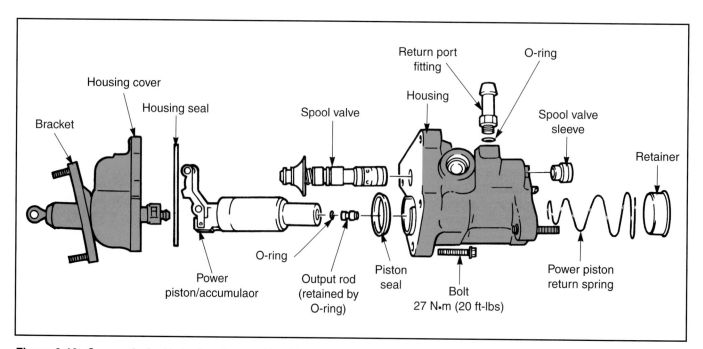

Figure 8-19. *One particular Hydro-boost assembly which has been separated to show the internal parts. (Cadillac)*

pump, **Figure 8-20.** This is the large low pressure hose. Place a pan under the hose and allow fluid to drain out.

Plug the return port at the pump reservoir and add fresh power steering fluid to the reservoir. Then start the engine and allow it to run at idle. Add new fluid at the reservoir as the old fluid drains from the return hose. Have an assistant slowly pump the brake pedal and turn the steering wheel.

 Note: Raise the vehicle while performing a flush or bleed operation. This will prevent flat spots on the tires.

When only fresh, uncontaminated fluid flows out of the return hose, reattach the hose and bleed the system. If the fluid continues to show particles of metal or other debris, it will be necessary to disassemble and clean the Hydro-boost unit, power steering pump, and power steering gear.

System Bleeding

Any time a Hydro-boost or power steering system component is removed, the system should be bled. Start the bleeding process by filling the power steering pump reservoir to the full mark. Then allow the vehicle to sit

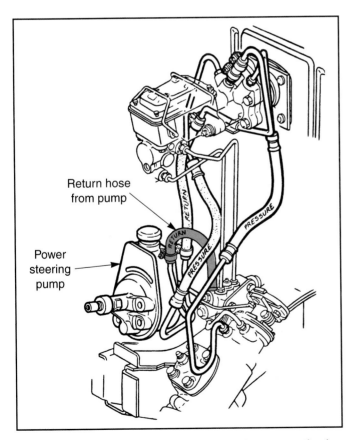

Figure 8-20. *To flush the Hydro-boost unit, remove the low pressure return hose and place in a drain pan. Dispose of the old fluid; do not reuse it. Dispose of properly. (General Motors)*

undisturbed for about five minutes. Start the engine and allow it to idle for about one minute; do not turn the wheel or apply the brake pedal. Stop the engine and add fluid to the reservoir if necessary. Repeat these steps until no further fluid is needed. Start the engine and turn the steering wheel from side to side while applying the brake pedal.

Check the power steering fluid level and add if needed. Repeat these steps until the fluid level stabilizes. If the power steering fluid is extremely foamy, restart the engine and apply the brake pedal while turning the steering wheel from side to side for about one minute. Then stop the engine and allow the vehicle to sit for about five minutes. Repeat the running and sitting steps until the fluid becomes clear.

If the fluid remains foamy and the level in the reservoir rises, a return hose may be sucking air, or there may be a defect in the power steering pump. Refer to the proper service manual for power steering troubleshooting and repair instructions.

Powermaster Service

Of the three power assist units discussed in this chapter, the Powermaster by far, can be the most difficult to diagnose and service. As you learned in Chapter 7, the Powermaster has more in common with anti-lock brake systems than either vacuum or Hydro-boost assist units. However, with a little patience, Powermaster systems can be successfully diagnosed and repaired.

 Note: Several revised service procedures have been issued for the Powermaster system via technical service bulletins. The most recent service information should be used.

Powermaster Diagnostics

Powermaster diagnostics has changed several times since the system was originally introduced. Most Powermaster problems can be classified into two areas:
- Hard pedal.
- Increased pedal travel.

Hard Pedal

A hard pedal complaint is usually caused by a defect in the Powermaster unit. The hard pedal is an indication the Powermaster is not providing sufficient boost. As with all brake repairs, a thorough visual inspection should be performed before any specific tests.

Note: Do not confuse a hard pedal complaint with the hard pedal achieved from depressurizing the system.

Depressurize the system by applying and releasing the brake pedal 10-20 times with the ignition key off. Clean off the reservoir cover and open. The fluid on the side away from the pressure switch, referred to as the booster side, should be close to the top. If the system is pressurized, the fluid will be just below the minimum mark.

Pressure Testing and Checking Warning Light

The Powermaster system has no self-diagnostics, unlike most of the anti-lock brake systems you will study later. Like a base brake system, the Powermaster uses the red Brake light as its trouble light. How this light performs can provide clues to the cause of a malfunctioning Powermaster. Use a pressure gauge to check Powermaster system operation.

Typical Powermaster pressures are as follows:
❑ Light on/Pump on: 355-435 psi (2479-2999 kPa).
❑ Light off/Pump on: 490-530 psi (3378-3654 kPa).
❑ Light off/Pump off: 540-770 psi (3723-5309 kPa).

During normal operation, the pump will run every 2-4 brake applications. The point at which the pump turns on should be at least 100 psi (689.5 kPa) less than the point at which the pump turns on. If pressure does not reach the 540-770 psi (3723-5309 kPa) range, check Ports A and B in the booster. If pressure exceeds 770 psi, replace the pressure switch.

Since the Powermaster operates at such pressures, the system must be depressurized when performing some tests and when servicing the unit. **Figure 8-21** shows typical warning light operation.

Warning Light Operation	
Action	**Condition**
With normal brake application, light does not come on.	No action required
Light comes on, then goes off.	Defective accumulator
Light comes on and stays on.	Defective pump motor, pressure switch, or Powermaster unit
Rapid brake pedal application causes light to go on and off.	Normal condition

Figure 8-21. *Chart outlining warning light diagnostics for vehicles equipped with the Powermaster system.*

Accumulator Testing

To test the accumulator, first depressurize the Powermaster system. Then, turn the ignition key on and note the time from the point the motor starts until the light goes out. Compare the results to the following:

0-2 seconds—Replace accumulator.
3-7 seconds—Accumulator ok.

If the light stays on for more than 8 seconds, continue timing until the motor shuts off (to a maximum of 2 minutes). If the motor fails to shut off after 2 minutes, check the pressure switch.

If the motor shuts off in less than 15 seconds, replace the accumulator and retest. A special pressure gauge is available for exact testing. If the motor runs for more than 7 seconds or fails to shut off after accumulator replacement, the pressure switch, pump motor, or the Powermaster unit itself is probably defective.

Pressure Switch Testing

A defective pressure switch will either prevent the pump from running or will fail to shut it off. Disconnect the plug-in terminal from the pressure switch and check the switch with an ohmmeter. All switch terminals should have continuity with each other. If not, the switch is defective. If the switch tests good and is not leaking, continue on to test the pump.

Pump Motor Testing

The first step in checking the pump motor is to make sure the circuit fuse is ok. Then check the connections at the pump and pressure switch, **Figure 8-22.** Disconnect the pressure switch and pump motor plug-in connectors. Perform the following tests with the ignition switch *on*:
❑ Switch connector terminal A should have continuity with pump terminal D.
❑ Switch connector terminal B should have 12 volts to ground (test w/voltmeter).
❑ Pump connection terminal B should have 12 volts to ground.
Perform the following tests with the ignition key *off*.
❑ Pump terminal A should have continuity to chassis ground.
❑ Pump terminal C should have 12 volts to ground. If this terminal does not show voltage, check fuse.

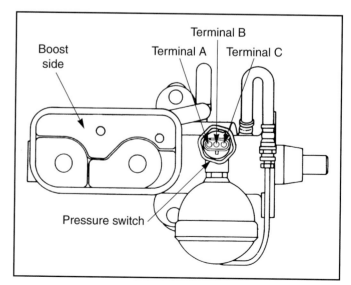

Figure 8-22. *Upper view of a Powermaster unit showing the terminal positions on the pressure switch. (Pontiac)*

An open circuit in any of these tests will be caused by a poor connection or wiring problem, **Figure 8-23.** Next, check the pump motor itself. Use an ohmmeter to make these tests.

❑ There should be resistance between motor terminals B and D.

❑ There should be an open circuit between motor terminals A and C.

If the pump is defective, the Powermaster unit should be replaced.

Master Cylinder/Booster Unit

Master cylinder and booster diagnosis is limited to inspecting the reservoir with the top off. Depressurize the system and remove the cover. Place a clear plastic tube over Port A and turn the ignition key to on, **Figure 8-24.** If fluid rises in the tube from Port A, the booster is leaking internally. Repeat this procedure on Port B. If fluid flows up the tube from Port B, the check valve in the pump is leaking. In either case, Powermaster unit replacement is needed.

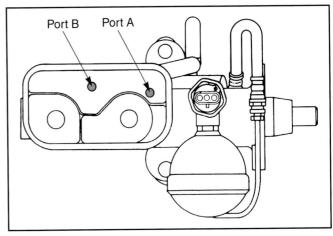

Figure 8-24. *Upper view showing Ports A and B in the reservoir. (Pontiac)*

Increased Pedal Travel

Increased brake pedal travel on vehicles equipped with the Powermaster system is usually caused by a malfunction in the Powermaster unit itself or in the hydraulic

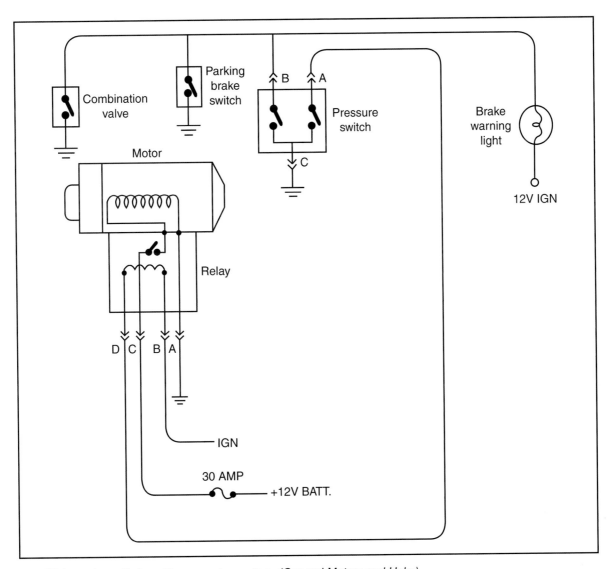

Figure 8-23. *Wiring schematic for a Powermaster system. (General Motors and Helm)*

system. One component that can cause increased pedal travel is the combination valve. To test the combination valve, bleed the left front brake only. If pedal travel stays the same or worsens after bleeding, a hydraulic leak is present somewhere in the brake system. If pedal travel improves after bleeding, the combination valve has an internal defect.

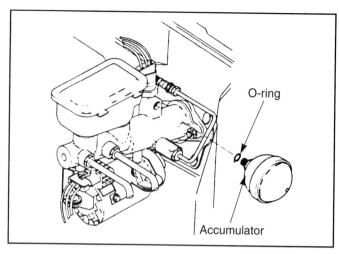

 Note: Powermaster equipped vehicles use a special brass combination valve. If the combination valve is defective, be sure to replace it with this brass valve.

To properly evaluate pedal travel, apply the brakes with the vehicle parked, ignition off, and without power assist (be sure to depressurize the accumulator with 10-20 brake applies). Pedal travel should be within 2.25" (57.2 mm) of maximum at 100% pedal force, assuming normally worn brake linings.

Powermaster Service

Due to problems encountered in rebuilding the power assist unit and parts availability, it is recommended the entire Powermaster unit be replaced if the power assist unit or pump is defective. While the pump can be replaced separately, the increased power of the new pump could cause any worn internal seals or check valves in the power assist unit to leak.

However, some portions of the Powermaster system can be replaced. These include the accumulator, pressure switch, hoses, and lines. In all cases, be sure to depressurize the system before beginning work.

Accumulator Replacement

To replace the Powermaster accumulator, shut off and remove the key from the ignition. Depress the brake pedal 10-20 times to depressurize the accumulator. Using a wrench, loosen the accumulator. Turn the accumulator counterclockwise to loosen and remove. Be sure the old O-ring comes off with the accumulator, **Figure 8-25.**

To install the new accumulator, first lubricate the new O-ring with clean brake fluid and place the O-ring on the accumulator. There is a space on the accumulator for the O-ring, just under the threads. After installing the O-ring, install the accumulator on the Powermaster by turning it clockwise by hand. After the O-ring has seated, use the wrench to tighten the accumulator to the Powermaster unit.

After installing the new accumulator, turn the ignition switch on and allow the Powermaster to pressurize. After the system has pressurized, shut off the ignition, pump the brake pedal 10-20 times, and repressurize the system again. Add fluid if needed.

Figure 8-25. *Powermaster accumulator replacement is fairly easy. Be sure to depressurize the system before removal. (Pontiac)*

Pressure Switch Replacement

If the pressure switch needs to be replaced, begin by depressurizing the system. Remove the plug-in wiring harness from the switch.

 Note: If the switch was leaking, examine the plug-in harness wire insulation for brake fluid intrusion by checking for swelling. If the wires in the switch harness are swollen, replace the plug-in connector and all related wiring.

Using an adjustable wrench, loosen the pressure switch. A special socket that makes this task easier is available, **Figure 8-26.** After removing the switch, make sure the old O-ring is on the switch. To install the new switch, lightly coat the new O-ring with clean brake fluid and

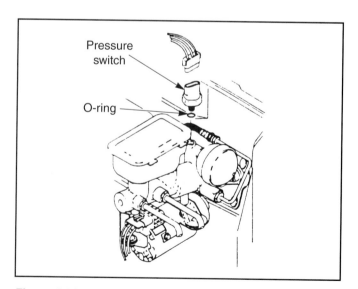

Figure 8-26. *Hydraulic switch replacement. Inspect the wiring to the switch, especially if the switch has been leaking. (Pontiac)*

install the new switch by hand. Reinstall the plug-in wiring harness. Bleeding is done in the same manner as with accumulator replacement.

Hose and Line Replacement

The Powermaster unit has two hoses and one metal line, which is part of the pressure hose from the pump. The return hose is a formed neoprene hose with two clamps, much like a fuel hose. Use a flare-nut wrench to remove the pressure hose and metal line.

Powermaster Replacement

To replace the Powermaster, begin by disconnecting all wiring plug-in connectors. Remove any wiring harnesses, brackets, accessories, etc., that could interfere with unit removal. Disconnect the brake pedal pushrod from the pedal arm. Disconnect the brake lines from the unit, remove the retaining nuts, and then, the Powermaster unit.

Remove the reservoir cover and drain the brake fluid into a container. Do not reuse the old fluid. Add fluid to the reservoir on the new unit and bench bleed. Installation of the new unit on the vehicle is the reverse of removal. Be sure to adjust the stoplamp switch after brake pushrod installation.

Bleeding the Powermaster

The Powermaster should bleed normally using the standard bleed procedure you learned in Chapter 6. A special adapter is available for pressure bleeding. Make sure the pump is operating properly. However, you may encounter a problem in bleeding if the pump has not been properly filled. If fluid is not circulating to the pump after 20 seconds, it may be necessary to reprime the pump.

To prime the pump, turn off the ignition and apply the brake pedal 10-20 times to depressurize the system. Clean and remove the reservoir cover. Then, disconnect the booster tube and nut assembly from the casting boss at the pump outlet. Once this is done, wait for brake fluid to gravity bleed from the disconnected tube. Once fluid begins to flow, reconnect the tube and refill the reservoir. The pump should now properly circulate fluid.

Summary

The first question to ask when servicing a defective power assist unit is whether it should be repaired or replaced. Vacuum boosters are usually replaced, but sometimes the unit must be overhauled. Make sure that parts are available, and do not try to overhaul any vacuum assist unit without the proper special tools. Before disassembling the vacuum booster, test its condition. Also test the vacuum hoses and check valve.

To remove the vacuum booster, remove the master cylinder, pushrod connection, and vacuum hose. Then unbolt the unit from the firewall and remove it from the vehicle. To disassemble and service any vacuum assist

unit, a special holding fixture and tools are needed. Separate the two halves of the housing and remove the internal parts. Clean and inspect the parts and replace as needed. Overhaul of dual diaphragm units is similar to those with single diaphragms.

After reassembling the unit, reinstall it on the vehicle. If necessary, check pushrod length. When installation is complete, road test unit operation. Brake system vacuum pumps are usually replaced instead of being serviced. Replacement methods vary, depending on whether the pump is driven by an electric motor or by a belt or gear from the engine.

Before beginning repairs to a hydraulic power assist unit, check the condition of the brake hydraulic friction components. Then test the operation of the power assist. When testing Hydro-boost systems, begin by checking the fluid level and adding fluid as needed.

To remove a Hydro-boost unit from the vehicle, always begin by depressurizing the accumulator. Then remove the pushrod, master cylinder, and hydraulic lines before removing the power assist unit.

To overhaul the unit, first disassemble it and clean all parts. Then check for damage and order needed new parts. Reassemble the unit and install it on the vehicle. Road test and check for proper operation.

Hydro-boost units must sometimes be flushed and bled. The return hose must be removed to drain fluid from the unit, after which new fluid is added as the unit operates. When all old fluid is removed, reinstall the return hose.

To bleed a Hydro-boost unit, fill the reservoir. Alternately run and stop the engine, checking fluid level each time. When the fluid is no longer foamy and the level is correct, the unit has been bled.

Powermaster systems can be tricky to diagnose and service. Some parts, such as the accumulator, pressure switch, lines and hoses can be serviced separately. However, a pump or booster problem requires unit replacement.

Review Questions—Chapter 8

Please do not write in this text. Write your answers on a separate sheet of paper.

1. Check the brake friction and hydraulic components _____ testing the booster.

2. *True or False.* It is often cheaper to replace a vacuum brake booster than to rebuild it.

3. *True or False.* A vacuum power assist can be rebuilt with simple hand tools.

4. *True or False.* If a hissing sound is heard only when the brakes are first applied, the booster is defective.

5. If the brake pedal has no power assist when the engine has been off for 30 minutes, the _____ is defective.

6. An overly long _____ can cause brake dragging.

7. Once the vacuum booster is disassembled and a defective pushrod assembly found what should the technician do to service it?

8. Before completing booster reinstallation by installing the master cylinder, check the _____ length.

9. Most vacuum pumps are _____ instead of _____.

10. What is the most common cause of problems in a hydraulic booster system?

11. When diagnosing any Hydro-boost system, the first step in diagnosis is to check the _____.

12. If dirt and metal are found in the fluid of a Hydro-boost system, the system should be flushed _____ the booster unit is removed.

13. Never remove the accumulator cover snap ring before _____.

14. If the hydraulic booster system uses an electric pump to produce pressure, test procedures are similar to those for a system using a _____.

15. Any time that any part is removed from a Hydro-boost system, the system must be _____ to remove _____.

ASE Certification-Type Questions

1. When a vacuum brake unit seems to have no power assist, what should the technician check *first*?
   (A) The intake manifold bolts.
   (B) The vacuum hose.
   (C) The power unit internals.
   (D) The pushrod length.

2. If the vacuum booster seems to have no reserve (power assisted stops after the vacuum source is lost), what is the most likely problem?
   (A) Defective diaphragm.
   (B) Defective control valve.
   (C) Defective check valve.
   (D) Loose pedal linkage.

3. If a special holding fixture is not used to hold the vacuum power assist as the two halves are separated, what could happen?
   (A) The two halves could fly apart with great force.
   (B) The diaphragm could stick to the front housing.
   (C) The return spring could be bent.
   (D) Nothing.

4. Technician A says that a small amount of gasoline inside of the power assist housing is a normal condition. Technician B says that a small amount of brake fluid inside of the power assist housing is a normal condition. Who is right?
   (A) A only.
   (B) B only.
   (C) Both A & B.
   (D) Neither A nor B.

5. When should pushrod length be checked?
   (A) When the vacuum booster is completely disassembled.
   (B) After vacuum booster overhaul, but before it is installed on the vehicle.
   (C) After the vacuum booster is installed on the vehicle, but before the master cylinder is installed.
   (D) After the master cylinder is installed on the vacuum booster, but before the engine is started.

6. Technician A says that an electric motor-driven vacuum pump may quit pumping if the low vacuum switch fails. Technician B says an engine driven vacuum pump can quit pumping if the low vacuum switch fails. Who is right?
   (A) A only
   (B) B only.
   (C) Both A & B.
   (D) Neither A nor B.

7. Technician A says that Hydro-boost units are pressurized by the power steering pump. Technician B says that Powermaster pump motors are similar to the motors used in an ABS system. Who is right?
   (A) A only.
   (B) B only.
   (C) Both A & B.
   (D) Neither A nor B.

8. Accumulators provide backup pressure when ___.
   (A) the engine is running
   (B) the hydraulic pump is running
   (C) hydraulic pressure to the assist unit is interrupted
   (D) braking up in reverse

9. Technician A says that accumulators are spring loaded. Technician B says that accumulators are charged with pressurized gas. Who is right?
   (A) A only.
   (B) B only.
   (C) Both A & B.
   (D) Neither A nor B.

10. Problems in the steering system can affect the _____ power assist system.
    (A) Hydro-boost
    (B) Powermaster
    (C) single diaphragm
    (D) tandem diaphragm

11. All of the following defects require Hydro-boost unit replacement, EXCEPT:
    (A) damaged spool valve.
    (B) damaged spool valve bore.
    (C) leaking housing cover seal.
    (D) leaking housing.

12. During a Hydro-boost flush operation, metal continues to show in the fluid. Technician A says this is caused by a defective power steering pump. Technician B says this is metal flaking off the pressure lines. Who is right?
    (A) A only.
    (B) B only.
    (C) Both A & B.
    (D) Neither A nor B.

13. A Powermaster system needs to be depressurized. How many times should the brake pedal be pressed to totally depressurize the system?
    (A) 10-20
    (B) 20-40
    (C) 1-2
    (D) Until the pedal sinks to the floor.

14. A vehicle equipped with the Powermaster system is leaking from Port A in the reservoir. Technician A says the pump is defective and must be replaced. Technician B says the booster unit is defective and the Powermaster must be replaced as a unit. Who is right?
    (A) A only.
    (B) B only.
    (C) Both A & B.
    (D) Neither A nor B.

15. A technician is attempting to manually bleed a new Powermaster unit. The pump will not circulate fluid. What should the technician do *first?*
    (A) Replace the Powermaster.
    (B) Check the battery.
    (C) Dump the fluid and refill with fresh.
    (D) Remove the booster tube from the Powermaster and allow to gravity bleed.

Brake hoses as they come off the manufacturing line. (Continental Teves)

Chapter 9

Hydraulic Valves, Switches, Lines, and Hoses

After studying this chapter, you will be able to:

❑ Explain how metering valves operate.
❑ Explain the purpose and construction of the proportioning valve.
❑ Discuss the reason load-sensing proportioning valves are used.
❑ Explain the purpose and construction of the residual pressure valve.
❑ Discuss the operation of a pressure differential valve and switch.
❑ Explain the purpose and construction of the low brake fluid switch.
❑ Explain the purpose and construction of the brake light switch.
❑ Identify the reasons for using steel hydraulic lines.
❑ Identify types of brake line flares.
❑ Discuss why flexible hoses are used to connect some brake parts.
❑ Identify various types of brake fittings and explain their function.

Important Terms

Pressure control valves
Brake switches
Metering valve
Proportioning valve
Load-sensing proportioning valve
Height-sensing proportioning valve

Residual pressure valve
Pressure differential valve
Combination valve
Low brake fluid switch
Brake light switch
Stoplight switch
Brake lines
Armored line

Tubing flares
Double-lap flares
Interference angle
ISO flare
Flare-nut
Flexible hoses
Banjo fitting
Tubing flares

This chapter covers the valves and electrical switches used to control and monitor the brake hydraulic system. It also covers the steel and flexible hydraulic lines used to connect the brake hydraulic devices, as well as fittings and flares necessary for proper brake system operation. This chapter also discusses the low fluid level and brake light switches.

Pressure Control Valves and Switches

All modern brake hydraulic systems contain *pressure control valves.* These systems also contain *brake switches,* which operate dashboard-mounted warning lights, or illuminate the rear brake lights. Pressure control valves affect the hydraulic pressure delivered to the wheel units, which can make brake action more efficient. Brake switches are used to illuminate warning lights on the dashboard, alerting the driver to the presence of a brake problem. Some valves and switches work together to sense problems and warn the driver. The following sections explain the construction and operation of these valves and switches

If the front brakes apply before the rear brakes, vehicle stability is affected. If the front wheels grip before the rear wheels, the back of the vehicle will have more momentum than the front. The back tries to pass the front, which can cause the vehicle to become unstable, possibly resulting in a skid.

Unfortunately, gravity and the design of modern brake systems makes this problem worse. When the brakes are first applied, much of the vehicle's weight is transferred to the front wheels. This gives the front wheels more traction, and therefore, more stopping power than the rear wheels. In addition, many modern vehicles are designed with front disc and rear drum brakes.

Metering Valve

As you learned in Chapter 5, the front brake pads are very close to the rotor and are not held in a retracted position. When the brakes are first applied, hydraulic pressure moves the front disc pads into contact with the rotor almost immediately. However, the rear brake shoes are held in the retracted position by return springs and must overcome spring pressure to move the shoes into contact with the drum. This means the front pads would apply much more quickly than the rear shoes. A metering valve is used to keep this from happening.

The *metering valve,* **Figure 9-1,** is installed in the brake line between the master cylinder and the front brakes. Vehicles with diagonally split braking systems will have two metering valves. The basic metering valve design contains a spring-loaded valve, which remains closed when no pressure is applied. When the brake pedal is applied, master cylinder pressure is sent to the front and rear wheels. The rear wheels receive full pressure immediately. Since fluid

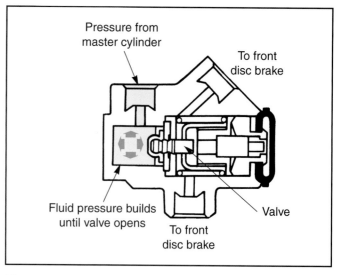

Figure 9-1. *A cross-sectional view of a metering valve. Shown with the valve closed; no fluid pressure is being transmitted to the brake calipers. (Niehoff)*

flows to the rear brakes immediately, system pressure remains relatively low and the closed metering valve keeps the front brakes from applying.

When the rear brakes have applied, there is no more fluid movement into the rear hydraulic units, and system pressure rises. When the system pressure reaches a certain point, the metering valve opens, **Figure 9-2.** Full pressure then applies the front brakes. When the brakes are released, fluid can bypass the main metering valve and return to the master cylinder. While most metering valves are mounted in the brake lines or in a combination valve, some metering valves are installed in the master cylinder, **Figure 9-3.**

After the front brakes are serviced, the front brake lines may contain air. This air may compress during brake

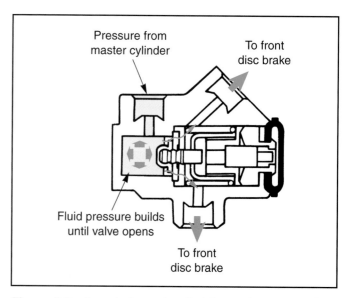

Figure 9-2. *A metering valve that is ready to open from hydraulic pressure. When it opens, brake fluid under pressure (arrows) will travel to the front brake calipers. (Raybestos)*

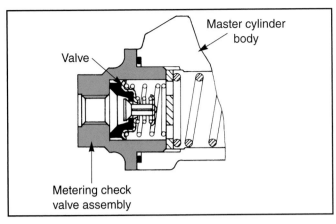

Figure 9-3. *A special metering valve installed in the master cylinder body. (ATE)*

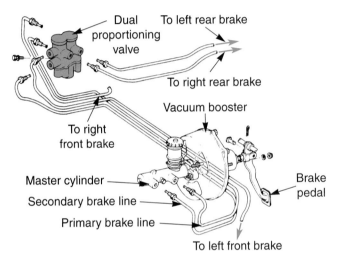

Figure 9-4. *A dual proportioning valve setup. Note that with this valve, a separate line is sent to each individual rear brake assembly. (Bendix)*

bleeding and keep the metering valve from opening. In this circumstance, it may not be possible to bleed the front brakes. Therefore, many metering valves have an external plunger which can be depressed during the bleeding operation to allow brake fluid to flow into the front wheel units.

Proportioning Valve

As mentioned earlier, when the brakes are applied, much of the vehicle's weight is transferred to the front wheels. If the brakes are applied hard during an emergency stop, so much vehicle weight is transferred to the front wheels that the rear wheels can easily lose traction and lock up. Rear wheel lockup can cause wear on the rear tires and can cause the vehicle to spin out of control, especially on wet or icy roads. The danger of lockup is worse on vehicles with rear drum brakes than on vehicles with rear discs. However, rear wheel lockup is a problem on many vehicles with four-wheel disc brakes, especially those with front-wheel drive.

To prevent rear wheel lockup, a ***proportioning valve,*** **Figure 9-4,** is installed in the rear brake line. Inside the proportioning valve assembly, a calibrated spring holds the valve away from the opening to the rear brakes. Diagonally split braking systems may have two proportioning valves. In some cases, the proportioning valves are installed at the master cylinder outlets, **Figure 9-5.**

During normal braking, the system pressure cannot overcome spring tension, and passes through the opening to the rear brakes. When the brake pressure rises past a certain point, such as during a panic stop, it will overcome spring tension and push the valve against the opening. This seals the rear brake system from further increases in pressure. Since the amount of pressure entering the rear brakes is limited, the brake friction elements cannot be applied any harder. Note that rear brake pressure is not decreased,

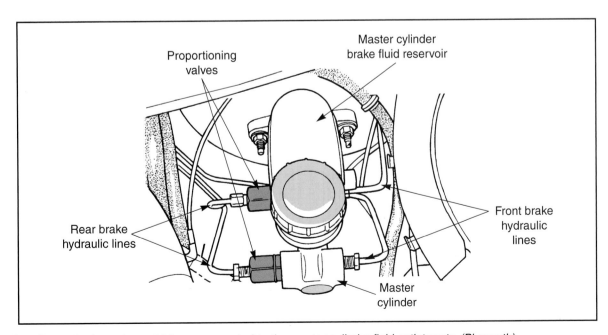

Figure 9-5. *Proportioning valves which are connected to the master cylinder fluid outlet ports. (Plymouth)*

it is merely kept from going any higher. When the brake pedal is released, the fluid pressure drops, and the spring opens the passage to the rear brakes. Fluid pressure can then return to the master cylinder.

Load-sensing Proportioning Valve

A variation of the proportioning valve is the **load-sensing proportioning valve** or *height-sensing proportioning valve.* This type of valve can vary the maximum pressure to the rear brakes according to the load placed on the vehicle. The load-sensing proportioning valve can be thought of as a standard proportioning valve with a mechanism that controls rear brake pressure. Load-sensing proportioning valves are used on pickup trucks, vans, and other vehicles designed to regularly carry heavy loads. The valve is usually installed at the rear wheels, mounted on the vehicle frame as shown in **Figure 9-6.** A link or lever from the valve is connected from the rear axle assembly.

 Note: Do not confuse this valve with a height sensor for a ride control suspension system, they are not the same.

In operation, the link moves an adjusting mechanism inside the proportioning valve assembly. See **Figure 9-7.** Loading the vehicle causes the body to compress the springs, which reduces the distance between the axle and frame. The reduced distance between the axle and frame causes the link to move the adjustment mechanism. This increases spring tension on the proportioning valve, which allows more pressure to enter the rear brakes before the valve closes. Therefore, the rear brakes take more of the braking effort when the vehicle is loaded.

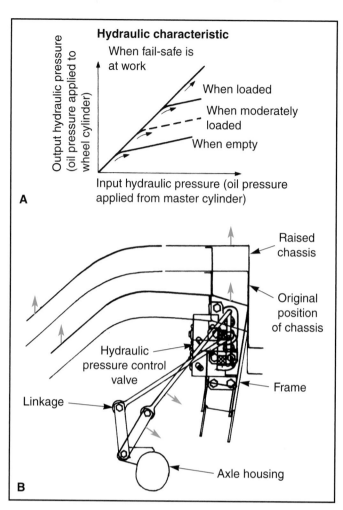

Figure 9-7. *A—Hydraulic pressure chart. B—Regulated (load-sensing) proportioning valve and linkage assembly depicting the linkage and valve changes in the original and raised frame positions. (Bendix/Chevrolet)*

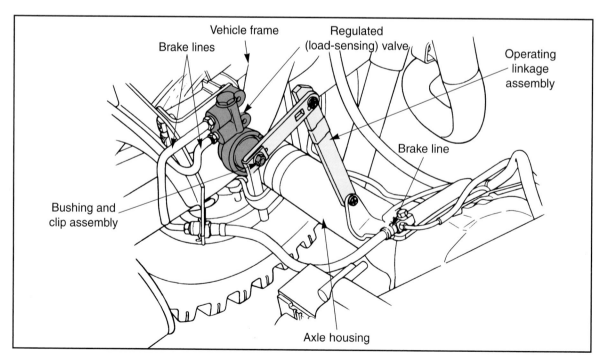

Figure 9-6. *Overall view of one particular regulated (load-sensing) proportioning valve, and its operating linkage. (GMC)*

When the vehicle is lightly loaded, the rear will rise. This increases the distance between the axle and body, causing the link to extend and lower the pressure sent to the rear wheels. A delay mechanism is built into the proportioning valve to keep quick axle movements (such as those caused by bumps) from affecting the valve.

Residual Pressure Valve

On any vehicle with drum brakes, a small amount of pressure must be maintained in the system to keep the wheel cylinder cup lips from collapsing, usually about 7 psi (47 kPa). All vehicles with drum brakes have a *residual pressure valve* to maintain this pressure. **Figure 9-8** shows a residual pressure valve installed in the end of the master cylinder bore.

Most modern vehicles with drum brakes have a residual pressure valve installed in the outlet to the rear wheels, **Figure 9-9.** Older vehicles have residual pressure valves installed in each outlet. While this leaves a slight residual pressure on the entire brake system, it is not enough to cause brake drag or accidental brake apply. Some manufacturers feel the small amount of residual pressure makes initial brake application slightly quicker.

The residual pressure valve consists of two check valves. The *apply check valve* allows fluid flow to the rest of the brake system whenever master cylinder pressure is greater than system residual pressure, **Figure 9-10.** The

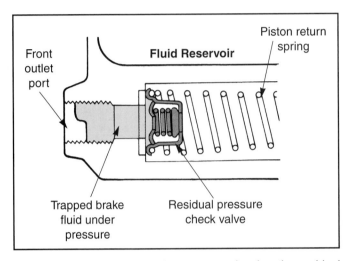

Figure 9-8. *A master cylinder cutaway showing the residual pressure check valve assembly and location. (Raybestos)*

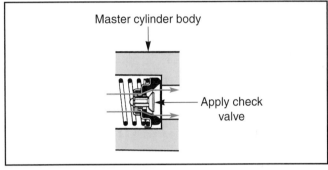

Figure 9-10. *Cutaway of one type of residual pressure valve showing the apply valve allowing fluid pressure to travel from the master cylinder to the entire system. (ATE)*

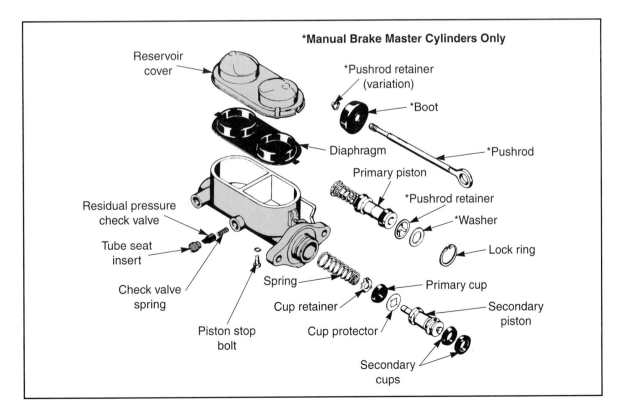

Figure 9-9. *A master cylinder with a residual pressure check valve for use with rear drum brakes. (Wagner)*

return check valve allows fluid to flow back to the master cylinder when the brakes are released and system pressure becomes higher than master cylinder pressure, **Figure 9-11.** However, the return check valve contains a spring which opposes returning fluid pressure. This spring is designed to close the valve when the system pressure goes below a certain point.

When the brakes are applied, the apply check valve unseats and fluid flows to the wheel units. When the brakes are released, the apply check valve closes, and fluid returns to the master cylinder through the return check valve. When the system pressure becomes lower than spring pressure, the spring closes the valve. This traps a small amount of pressure in the brake hydraulic system.

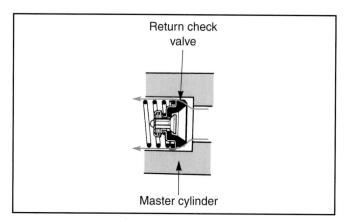

Figure 9-11. *The brakes have been released and the return check valve is being forced open. Fluid is now flowing back into the master cylinder. (ATE)*

Pressure Differential Valve

The ***pressure differential valve*** is a warning device used in all split brake systems. Remember that a split braking system contains two separate brake hydraulic systems or sides. The pressure differential valve is installed so that each side of the hydraulic system presses on one side of the valve, **Figure 9-12.** Attached to the valve is an electrical switch. This switch receives electrical power whenever the ignition is on. The switch's electrical circuit goes through a dashboard-mounted brake warning light. To complete the circuit, the switch must be grounded by the pressure differential valve.

When the hydraulic system is operating normally, pressure on both sides of the valve will be equal. With the pressure on both sides equal, the valve is centered and the switch is open. When one side of the hydraulic system fails, pressing on the brake pedal will result in normal pressure on one side, and much lower pressure on the other side of the system. The difference in pressure pushes the valve to move toward the side with less pressure, **Figure 9-13.** When the valve moves, it moves the switch plunger, causing the switch contacts to close. This grounds the switch, allowing current flow to turn on the dashboard warning light.

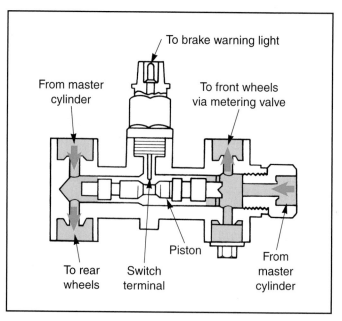

Figure 9-12. *A cross-sectional view of a pressure differential valve with equal pressure on both sides of the piston. (Niehoff)*

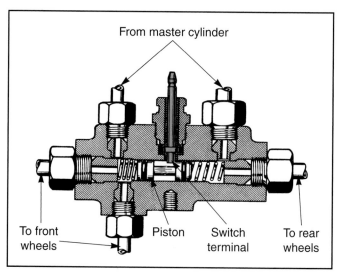

Figure 9-13. *Cutaway of a pressure differential valve and warning switch. Note that there has been a failure in the front brake system. The master cylinder pressure has moved the piston to the left, making switch contact and turning on the dash warning light. (Chevrolet)*

Combination Valve

For manufacturing ease and quicker service, two or more hydraulic valves are often combined into one assembly. This multiple valve assembly is called a ***combination valve.*** A combination valve may contain up to three different hydraulic valves. Common two-unit combination valves usually contain a pressure differential valve and either a metering valve or proportioning valve, **Figure 9-14.** A few two-unit combination valves contain the metering and proportioning valves. Three-unit combination valves are composed of a metering valve, proportioning valve, and pressure differential valve, **Figure 9-15.**

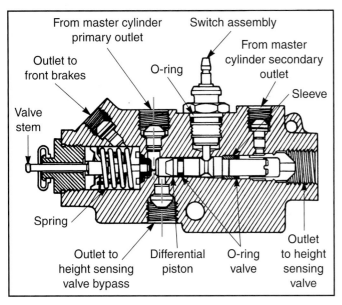

Figure 9-14. *A cross-sectional view of a two unit combination valve. The left half is a metering valve, and the right half is a pressure differential valve. Study the construction. (Dodge)*

Low Brake Fluid Switch

Master cylinders in many vehicles are equipped with a *low brake fluid switch.* This switch is usually installed in the reservoir cap, **Figure 9-16,** or in the side of the reservoir, **Figure 9-17.** This switch consists of a float and a set of contacts. The switch contacts have electrical power when the ignition is on. The contacts are attached to the float, and close when the float falls below a certain point. The switch electrical circuit is powered through a dashboard-mounted fluid level warning light. When the fluid level becomes too low, the float falls in the reservoir, causing the contacts to close. Closing the contacts grounds the circuit and illuminates the dashboard warning light.

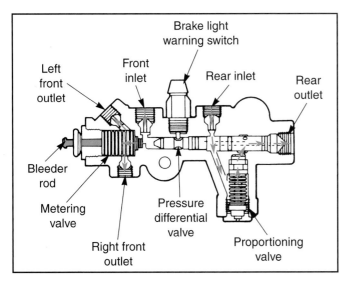

Figure 9-15. *Cross-sectional view of a three-way combination valve. Note that this valve incorporates a warning switch, proportioning and metering valve in one unit. Study the construction. (Raybestos)*

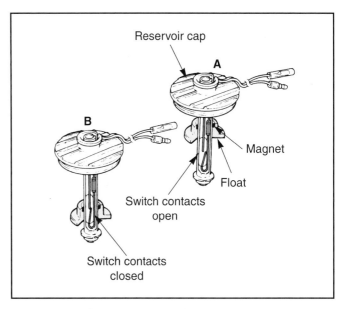

Figure 9-16. *Cutaway views of float level sensors that are a part of the reservoir cap. A—Switch contacts are open, circuit is ok. B—Switch contacts are closed. Warning light will come on if the sensor is functioning correctly. This will alert the driver to a low fluid level in the reservoir. (Honda)*

Figure 9-17. *This master cylinder has a brake fluid level switch located in the side of the reservoir.*

Brake Light Switch

To operate the brake lights, a **brake light switch** or **stoplight switch** is installed on the brake pedal linkage. One end of the brake pedal switch is mounted to the vehicle body under the dashboard. The other end is installed on the brake pedal linkage. The two most common types of brake switches are the plunger type, **Figure 9-18,** or the pin actuated type.

Some vehicles use a **sliding contact brake light switch, Figure 9-19.** This switch can be easily recognized since it is much larger than other types of brake switches. The sliding contact brake light switch also contains the brake switches for the cruise control and anti-lock brake systems when they are used.

Figure 9-18. *A plunger type stoplight switch assembly. 1-Brake pedal mounting bracket. 2-Switch retainer. 3-Stoplight switch. 4-Stoplight switch mounting bracket. 5-Switch actuator lever. 6-Brake pedal. 7-Plunger. 8-Normal brake pedal travel. (General Motors)*

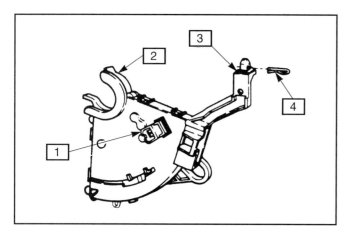

Figure 9-19. *A slide brake light switch is used with some anti-lock brake systems. 1-Cruise control terminal. 2-Switch assembly. 3-Washer. 4-Retainer clip. (General Motors)*

With either type, switch operation is the same; pressing on the brake pedal causes the end of the switch assembly to move in relation to the end mounted on the body. This movement causes the switch contacts to close, completing the electrical circuit between the vehicle battery and the brake lights. Completing the circuit causes the lights to illuminate.

On some vehicles, the brake switch has battery current at all times, and the brake lights will come on whenever the pedal is depressed. On very old vehicles, the brake switch power is routed through the ignition switch and the brake lights can illuminate only when the ignition switch is in the on position.

Brake Lines and Hoses

For the hydraulic pressure developed in the master cylinder to reach the wheel brakes, they must be connected by some sort of hydraulic tubing. This tubing consists of rigid steel lines and flexible rubber hoses.

Steel Brake Lines

Whenever possible, the hydraulic system uses rigid steel **brake lines,** sometimes called *tubing,* to transmit hydraulic pressure. Steel lines are resistant to collision damage and vibration, can stand up to high brake system pressures, and are relatively inexpensive. For added safety, the steel used in the lines is double-wall (double thickness) welded steel made from copper-coated sheet steel. Common steel line sizes range from 1/8-3/8" (3.25-9.5 mm).

Steel lines are often coated with tin, zinc, lead, or Teflon™ to reduce damage from corrosion. The lines are

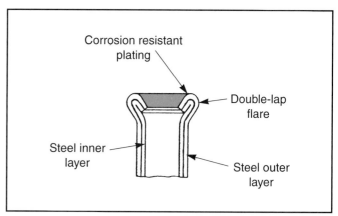

Figure 9-20. *Typical steel brake line (tubing) construction. (Ford)*

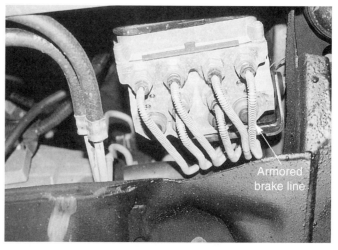

Figure 9-21. *Armored brake line is used in location where the tubing could be subjected to flying debris and other minor abuse. These are connected to an anti-lock brake hydraulic actuator.*

clamped to the vehicle unibody or frame at close intervals to reduce damage from vibration and road debris. The clamps often contain rubber bushings to reduce the possibility of normal vehicle vibration rubbing a hole in the line. **Figure 9-20** shows typical steel line construction.

Another type of line is the **armored line.** Armored lines are standard steel tubes around which steel wire is coiled, **Figure 9-21.** The steel spring protects the line from impact damage. Armored lines are usually used under the vehicle.

Line Installation

Factory steel lines are pre-bent to the shape needed to snugly fit the frame and other parts of the vehicle. The bends are relatively gentle to reduce the possibility of kinks. **Figure 9-22** shows the routing of brake lines on a typical vehicle. Aftermarket steel lines are available in pre-cut lengths, but must be bent and flared to be installed on the vehicle. These procedures are covered in Chapter 10.

Tubing Flares

To make a solid connection to the hydraulic units, the ends of all steel lines have **tubing flares.** A tubing flare is an expanded, or flared section at the end of the line. The flare fits firmly against a seat, forming a leak-proof connection.

To connect to other hydraulic units, many steel brake lines use **double-lap flares** at all connecting points. The double-lap flare will withstand vibration and high pressures. Notice in **Figure 9-23A** the flare angle is 45° while the angle of the seat is 44°. This is called an **interference angle.** When the nut is tightened into the fitting, the interference angle causes the seat and the flare to wedge together. This wedging effect provide an effective seal against leakage.

Another type of flare used for brake line connections is the **ISO flare.** ISO stands for **International Standards Organization,** a group which sets standards for international trade and manufacturing. Similar to the double-lap flare, the ISO flare provides a secure crack-resistant joint. An ISO flare is shown in **Figure 9-23B.** Note that both flares use a threaded nut, called a **flare-nut,** to hold the flare against its seat.

Flexible Hoses

Since the wheel brake units are independently suspended (they can move in relation to the frame and body), hydraulic connections between the wheels and body cannot be rigid. **Flexible hoses, Figure 9-24,** are used at the wheels to allow for movement. As the wheels rise, fall, and turn, the flexible hose will transmit high pressure without

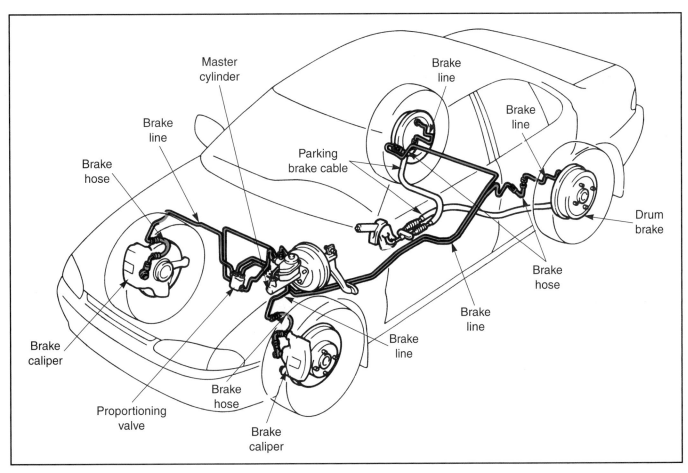

Figure 9-22. *The layout of brake lines and hoses as used on one particular vehicle. (Chevrolet)*

breaking. Most flexible hose lines are made from natural rubber and synthetic fabric. There are usually two-plies (layers) of rubber and two-plies of braided fabric material for added pressure resistance. The hose is usually covered by a synthetic rubber sheath to reduce damage from atmospheric pollutants. See **Figure 9-25.** Some flexible brake hoses use braided steel mesh between the rubber layers for added strength. Braided mesh may also be used over the outside of the hose to protect it from road debris.

Hose fittings are threaded to accept the flared tubing of steel brake lines. The flexible hoses are firmly mounted at each end, and all movement occurs in the hose itself.

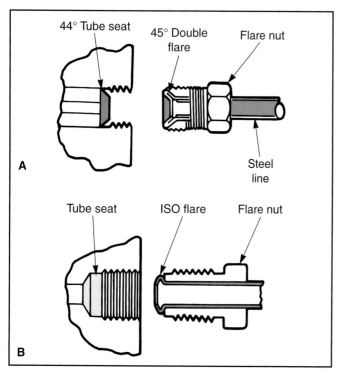

Figure 9-23. A—A cutaway view of a double-lap tubing flare, flare nut, and the tubing seat. B—A cutaway view of an ISO (International Standards Organization) type tubing flare. (Niehoff)

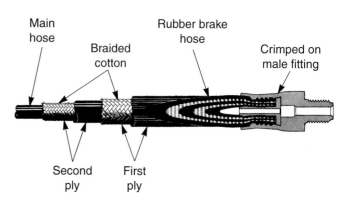

Figure 9-25. A sectioned view of a flexible rubber brake hose. Study the construction. (Wagner)

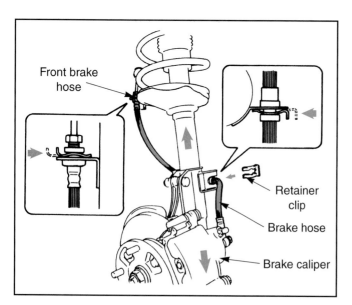

Figure 9-24. A front suspension which uses a flexible rubber brake hose. The hose will allow the suspension to travel up and down. A steel line would flex a few times, then break off, causing a loss of braking pressure to that wheel and eventual loss of fluid to that portion of the hydraulic circuit. (Chevrolet)

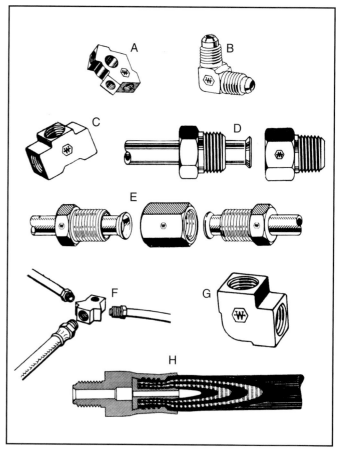

Figure 9-26. An assortment of typical tubing and hose fittings and connectors. A—Distribution block for four fittings. B—Male elbow. C—Female branch tee. D—Adapter for reducing size. E—Inverted union to connect both pipes. F—Distribution block for three fittings. G—Female elbow. H—Crimped on hose fitting. (Weatherhead)

Brake Fittings

Brake fittings are threaded connectors made of steel or occasionally brass. Fittings form junctions or connections for the brake lines and hoses. The fitting can be as simple as the female pipe fitting at the end of a flexible hose, or a complex block which splits a single hydraulic line into several separate lines. Brake fittings are always solidly mounted on the frame or axle. Some common brake fittings are depicted in **Figure 9-26.** All brake fittings are threaded to accept the pipe fittings used on hoses and lines. These fittings can be standard or metric pipe fittings. On almost all brake systems, the fitting is a female pipe thread designed to accept the male nut on the end of the brake line.

Note that in addition to the fittings discussed above, the pipe thread connections to the major hydraulic devices such as the master cylinder, calipers, wheel cylinders, and control valves, are also known as fittings. These fittings are female pipe threads.

Another type of fitting, shown in **Figure 9-27,** uses two washers, a drilled bolt, and a specialized type of compression fitting called a *banjo fitting.* This fitting is installed by placing one washer on each side of the banjo fitting, then installing and tightening the bolt. Tightening the bolt causes the washers to seal the fitting.

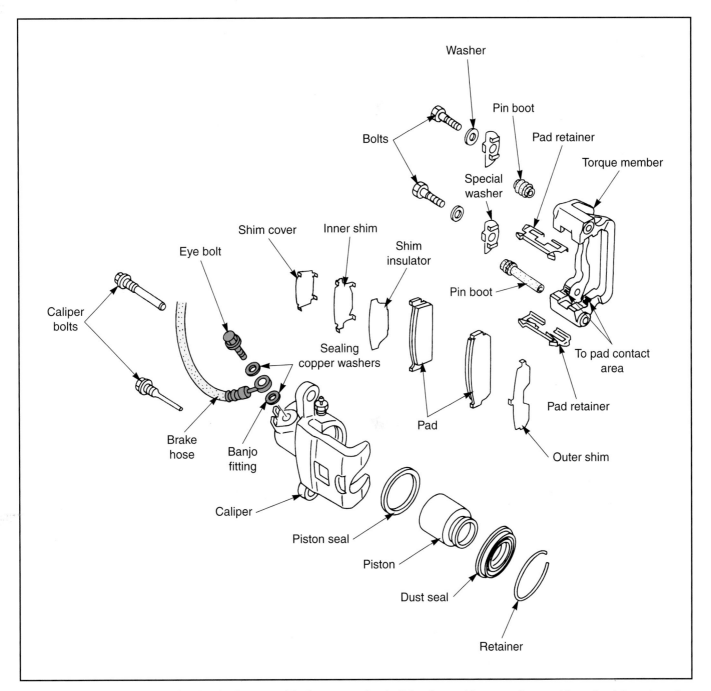

Figure 9-27. *Exploded view of a disc brake assembly that uses a banjo fitting (named because it resembles a banjo), and sealing washers for brake hose attachment to a caliper. (Infiniti)*

Summary

The modern brake hydraulic system contains various valves which control hydraulic pressure to parts of the brake system. One valve is the metering valve, used to prevent application of the front brakes until the rear brakes are partially applied. Another valve is the proportioning valve which controls pressure to the rear wheels during hard braking. A variation of the proportioning valve controls pressure according to the weight placed on the vehicle. The pressure differential valve warns the driver if one side of the hydraulic system looses pressure. All of these valves may be combined into one unit called a combination valve. Combination valves can contain two or three valves.

Electrical brake switches are also used in various ways. In addition to the switch attached to the pressure differential valve, switches also control a dashboard light which warns the driver that fluid level in the master cylinder reservoir low. Another switch is installed on the brake pedal linkage to illuminate the brake lights when the brake pedal is depressed.

Hydraulic pressure from the master cylinder reaches the wheel brakes by way of rigid steel lines or flexible rubber hoses. Steel line hydraulic units that do not move in relation to each other, while hoses connect the moveable wheel assemblies with the frame or body lines.

Brake fittings are threaded connectors which connect brake lines and hoses. Fittings can be standard or metric pipe threads. Double-lap or ISO flares are used to ensure a solid connection between steel lines and fittings with no leaks. The two types of flares are the double-lap and the ISO flare. One of these two types must be used in brake system flares.

Review Questions—Chapter 9

Please do not write in this text. Write your answers on a separate sheet of paper.

1. A _____ responds to changes in vehicle load.

2. _____ may contain 2-3 valves and an electrical switch.

3. The low brake fluid switch is installed on the master cylinder _____.

4. How does the low brake fluid switch respond to a low fluid level?

5. The rear brake light switch is installed on the _____ linkage.

6. Rigid brake lines are made of _____.

7. Brake lines are usually coated with another metal or Teflon to reduce _____.

8. Define a tubing flare.

9. What are the two types of brake tubing flares?

10. Flexible hoses are firmly mounted at _____.

ASE Certification-Type Questions

1. Technician A says that the metering valve keeps the rear wheels from locking up. Technician B says that the proportioning valve keeps the front wheels from applying too quickly. Who is right?
   (A) A only.
   (B) B only.
   (C) Both A & B.
   (D) Neither A nor B.

2. When is the metering valve closed?
   (A) When no pressure is applied to it.
   (B) When heavy pressure is applied to it.
   (C) When the driver releases the brake pedal.
   (D) When the front disc pads are applied.

3. The load-sensing proportioning works because a load placed in the vehicle trunk _____.
   (A) compresses the frame bushings
   (B) compresses the shock absorbers
   (C) compresses the rear springs
   (D) None of the above.

4. Technician A says that during light braking, the proportioning valve is open. Technician B says that during light braking, the residual pressure valve is closed. Who is right?
   (A) A only.
   (B) B only.
   (C) Both A & B.
   (D) Neither A nor B.

5. A properly functioning residual pressure valve keeps the _____ seals from collapsing.
   (A) caliper
   (B) wheel cylinder
   (C) master cylinder
   (D) Both A & B.

6. Under which of the following conditions will the pressure differential valve illuminate the dashboard light ?
   (A) The engine is first started.
   (B) The metering valve sticks closed.
   (C) A severe leak develops in a rear wheel line.
   (D) A front hose swells shut.

7. Which of the following switches will *not* illuminate a dashboard light?

    (A) Pressure differential switch.

    (B) Brake fluid level switch.

    (C) Brake light switch.

    (D) None of the above.

8. All of the following statements about brake lines and hoses are true, EXCEPT:

    (A) brake lines are double-walled and single flared.

    (B) steel lines are often coated to reduce corrosion damage.

    (C) brake hoses are combinations of rubber and fabric.

    (D) brake hose ends are solidly mounted.

9. Brake fittings are threaded connectors sometimes made of _____.

    (A) copper

    (B) aluminum

    (C) phenolic plastic

    (D) None of the above.

10. A banjo fitting is a special type of _____.

    (A) double-lap flare

    (B) ISO flare

    (C) tubing nut

    (D) compression fitting

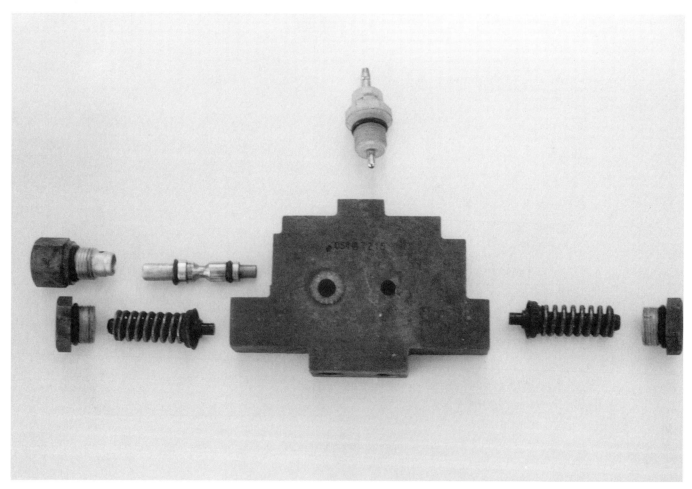

Exploded view of a combination valve. The top valve is the pressure differential switch. The bottom valves are proportioning valves. This valve is used on a diagonal split brake system.

Chapter 10

Hydraulic Valve, Switch, Line, and Hose Service

After studying this chapter, you will be able to:

- ❑ Diagnose braking problems caused by hydraulic system valves.
- ❑ Test the operation of hydraulic system valves.
- ❑ Test the operation of low brake fluid switches.
- ❑ Test the operation of brake light switches.
- ❑ Remove and replace valves and switches.
- ❑ Check steel hydraulic lines, hoses, and fittings for defects.
- ❑ Bend, cut, and flare replacement steel hydraulic lines.
- ❑ Replace flexible hoses.
- ❑ Replace brake fittings.

Important Terms

Hydraulic pressure gauges	ISO flare	Collet
Coil spring bender	Flaring bar	Forming mandrel
Lever bender	Flaring yoke	Clamping nut
Double-lap flare	Tool body	Forcing screw

In this chapter, we will discuss the diagnosis and repair of hydraulic valves and switches. Common valve problems will be covered, as well as adjustment and replacement procedures. In addition, service procedures for electrical switches connected with the hydraulic system will be discussed. In this chapter, you will also learn to service hydraulic lines, hoses, and fittings, including flaring steel brake lines.

Hydraulic Valve Service

The brake system hydraulic valves are relatively trouble free, but can stick or leak. Valve failure can cause internal hydraulic system malfunctions, leading to various complaints, some of which may be difficult to diagnose without pressure gauges. The following sections identify problems caused by the hydraulic valves, and explain how to perform hydraulic valve diagnosis and service procedures.

Hydraulic Valve Problems

There are several kinds of hydraulic valve problems, varying with the type of valve, and how it has failed. The metering and proportioning valves can be the source of several problems, including **wheel lockup** (skidding), failure of the brakes to apply, and uneven braking. The most common problem caused by residual pressure valves is leaking wheel cylinders and poor performance when the brakes are first applied. The pressure differential valve and switch assembly is probably the most frequent cause of driver complaints, since it can illuminate the dashboard light. Any of the valves can develop external **fluid leaks.**

Front wheel skidding when the brakes are first applied can be caused by a metering valve that is stuck open. If the front brakes do not apply at any time, the cause may be a metering valve that is sticking closed. In some cases, the driver may not notice the lack of front brake application until the rear brakes wear out prematurely.

Rear wheel skidding can be caused by the proportioning valve sticking open. On a vehicle with a diagonally split hydraulic system, failure of one of the proportioning valves to close will cause the related rear wheel to skid while the other rear wheel operates properly. If the proportioning valve sticks closed, or closes too quickly, the rear brakes will not operate. This may cause front wheel lockup or premature front brake wear.

In rare cases, the metering or proportioning valves can stick open or become clogged, holding pressure in the hydraulic system between the valve and the wheel units. This can cause brake drag or complete wheel lockup. A valve that is slow to release can cause temporary lockup when the brakes are first released.

Residual pressure valve failure can cause wheel cylinder leakage since the cups are not held tightly against the wheel cylinder bore. In extremely rare cases, the residual pressure valve will stick closed, causing brake drag. If

the outlet check valve sticks closed, the affected brakes will not apply. This condition is also rare.

If the pressure differential valve moves to one side and sticks, the valve will ground the electrical switch, causing the dashboard light to stay on at all times. If the valve sticks in the centered position, it will not illuminate the dashboard light when one side of the system loses pressure.

Note: The master cylinder and other hydraulic and friction parts should be ruled out as the cause of a problem before checking the hydraulic valves.

Testing the Operation of Hydraulic System Valves

Some hydraulic valve operations can be checked by simple observation. In some cases, a fitting must be loosened to observe fluid flow. In other cases, the valve can only be tested by using one or more pressure gauges. Remember that modern vehicles often have several or all of their hydraulic control valves built into a common unit called a combination valve.

Testing Metering Valve Operation

To test the metering valve, loosen but do not remove one of the front brake bleeder screws. Then have an assistant press gently on the brake pedal while you observe the bleeder screw.

Note: Do not depress the metering valve bypass plunger.

If fluid squirts out of the bleeder screw as soon as the pedal is pressed, the valve is stuck open. If no fluid exits, have your assistant slowly exert more pressure on the pedal. At a certain point, fluid should begin to exit from the bleeder. If it does, the metering valve is working properly. To check the exact pressure when the metering valve opens, refer to the section on pressure checking valves. If fluid does not exit the bleeder screw under high pedal pressure, the valve is stuck closed.

Testing and Adjusting Load-sensing Proportioning Valves

Many load-sensing proportioning valves are adjustable. Load-sensing proportioning valves do not require periodic adjustment. However, damage, sagging springs, or suspension part replacement can often change the vehicle's ride height. Adding helper springs or air shocks can also change the ride height, making valve adjustment necessary.

To check the adjustment of a load-sensing proportioning valve, the vehicle must be on a flat surface with a normal load, **Figure 10-1.** The best way to assure the vehicle is on a flat surface is to place it on an alignment rack or drive-on hoist. Do not raise the vehicle by the frame.

After ensuring the vehicle suspension is at its normal height, obtain a length of rubber or plastic vacuum hose with a 3/8″ (9.3 mm) outside diameter and a 1/4″ (7 mm) inside diameter. Cut a piece of this hose to a length of 5/8″ (16.3 mm) and slice it lengthwise as shown in **Figure 10-2.**

 Note: This piece of vacuum hose will be used as a gauge. Measure and cut accurately.

After cutting the vacuum hose, loosen the valve adjuster setscrew and place the piece of hose around the valve operating rod. Make sure the adjuster sleeve is resting on the lower mounting bracket, then tighten the setscrew. This positions the valve for normal operation. Remove the piece of hose.

If further adjustment is necessary, the adjuster sleeve can be moved up or down. Begin by loosening the adjuster setscrew. To decrease pressure, move the adjuster sleeve upward, making it shorter. To increase pressure, move the adjuster sleeve downward, making it longer. The movement needed is very small, and should be carefully

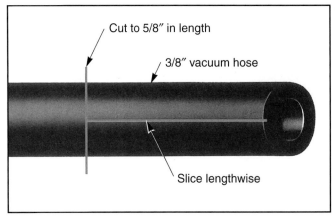

Figure 10-2. *A section of vacuum hose. When cutting the hose, make sure to measure it accurately, as it will be used as a gauge. (Gates)*

checked and recorded. Moving the adjuster sleeve up or down .039″ (1 mm) changes pressure by 60 psi (413 kPa). After making the adjustment, tighten the setscrew.

Some load-sensing proportioning valves must be set by using a special one-time gauge, **Figure 10-3.** The gauge is installed on the proportioning valve with the slotted tank over the correct spot on the valve body. The gauge is then used to correctly position and install the control link. After the gauge is used, the technician cuts off the tang, and the adjustment procedure is finished.

Figure 10-1. *Pickup truck parked on a level shop floor. Do not take measurements with cargo or passengers in the vehicle. A full tank of fuel should be part of the vehicle's normal load. Air the tires to the recommended inflation. (Toyota)*

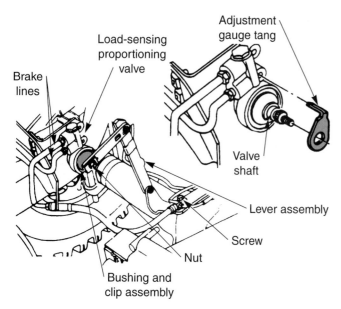

Figure 10-3. *A load-sensing proportioning valve assembly with its special adjustment gauge. (General Motors)*

Note: To make accurate adjustment to a load-sensing proportioning valve, install and use pressure gauges.

Testing and Centering Pressure Differential Valves

To test the operation of the pressure differential valve, have an assistant press firmly on the brake pedal with the ignition switch in the *on* position. Then open any bleeder screw. The dashboard brake light should come on. If it does not illuminate, the valve, switch, or wiring is defective. Close the bleeder and have the assistant press on the brake again. The light should go out after one or two applications.

Most pressure differential switches are *self-centering.* Once a pressure problem has been corrected, the switch will center itself. If the pressure differential valve is an older type that does not center itself, have an assistant press firmly on the brake pedal with the ignition switch in the *on* position. Then, open an outlet hydraulic line fitting at the valve. If the light goes out, the valve has been centered. If the light stays on, tighten the fitting and loosen the other outlet line fitting. If the light still does not go off, the valve is stuck and should be replaced. If the light flickers, it may be possible to center the valve by alternately loosening both outlet lines as the brake pedal is applied.

Testing Valves With Pressure Gauges

The metering and proportioning valves can be checked with *hydraulic pressure gauges.* In some cases, the only way to isolate a problem is to install pressure gauges and determine what is happening on either side of the valve. To perform this test, you will need a gauge with a capacity of at least 1000 psi (6985 kPa).

Caution: Never use an undercapacity gauge. Not only will the gauge be damaged, but it can fly apart with great force, causing personal injury.

Testing Metering Valves

To test a metering valve, install two gauges at the valve locations shown in **Figure 10-4.** One gauge will measure the pressure being placed on the metering valve, while the other gauge will indicate the pressure leaving the valve for the front brakes. After installing the gauges, press on the brake pedal. Observe the gauges. If small amounts of pressure immediately register on both gauges, the valve is stuck open and must be replaced.

Note: Do not depress the metering valve plunger.

When the gauge on the input side reaches a certain pressure, usually 150-175 psi (1034-1206 kPa), the pressure gauge on the outlet side should begin rising. This indicates the metering valve has opened. Check the opening pressure against specifications. If the pressure at opening is much higher or lower than specified, the valve is sticking and requires replacement.

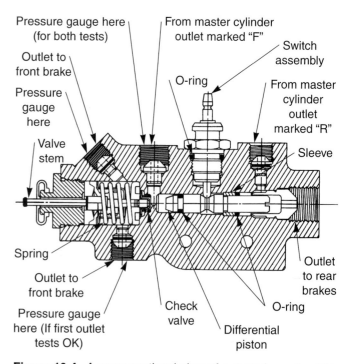

Figure 10-4. *A cross-sectional view of a metering valve. This valve also houses a warning light switch. Connect the pressure testing gauges, one to the fluid outlet ports, and in this case, one between the master cylinder and the metering valve inlet port. (Dodge)*

When the brake pedal is released, both gauges should quickly drop to zero. If the output gauge remains at a higher value than the input gauge, the return check valve is sticking and the valve should be replaced.

Testing Proportioning Valves

Pressure testing a proportioning valve, including load-sensing proportioning valves, also requires two gauges, as shown in **Figure 10-5.** One gauge measures the pressure entering the proportioning valve, while the other gauge indicates the pressure leaving the valve for the rear brakes. After installing the gauges, press on the brake pedal. Pressure should rise on both gauges until the specified cut-off pressure is reached. Cutoff pressures range from 550-610 psi (3790-4203 kPa). After this point, input pressure should continue to rise while output pressure remains constant. If the pressure continues to rise, the valve is not closing and should be replaced.

When the brake pedal is released, both gauges should drop to zero. If the output gauge remains at a higher value than the input gauge, the proportioning valve is sticking and the valve should be replaced.

To pressure check the proportioning valves on a vehicle with a split diagonal brake system, install one gauge into a valve input line, and the other gauge at one of the output lines past the proportioning valve. See **Figure 10-6.** Then test the proportioning valve in the same manner as the single valve system. To check the other proportioning valve, the gauge must be installed into the other line. Special adapters are often needed to pressure test proportioning valves that are installed on the master cylinder, **Figures 10-7** and **10-8.**

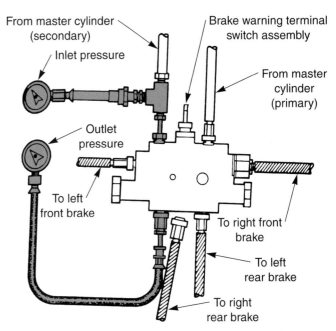

Figure 10-6. *Proportioning valve test setup for a vehicle with split diagonal brakes. (Chrysler)*

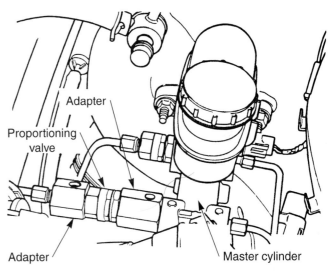

Figure 10-7. *When testing proportioning valves installed on or in the master cylinder, special adapters are often required. (Chrysler)*

Testing Residual Pressure Valves

To test a residual pressure valve using a pressure gauge, install the gauge at the master cylinder outlet or other convenient fitting on the hydraulic system. If only one side of a dual system uses a residual pressure valve, be sure you attach the gauge to the proper side.

Have an assistant press on the brake pedal while you observe the gauge. Pressure should rise with pedal application. Then release the pedal and observe the gauge. If the gauge drops until it reaches about 7-10 psi (48.26-68.95 kPa) and holds at this pressure, the residual pressure valve is working properly. If pressure drops to zero, the valve is not working.

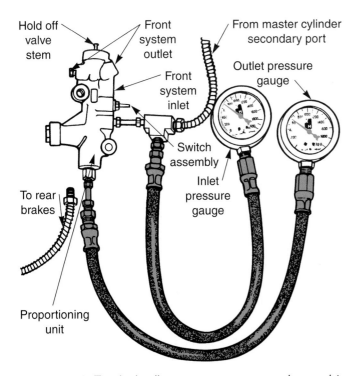

Figure 10-5. *Two hydraulic pressure gauges can be used to test proportioning valves. This setup is used to test a rear-wheel drive vehicle. Handle the gauges with care. (Chrysler)*

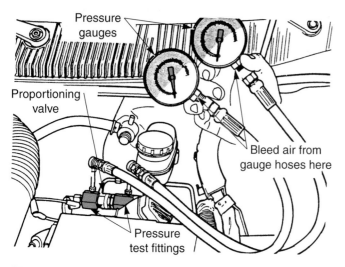

Figure 10-8. *Proportioning valve test setup using adapters. (Chrysler)*

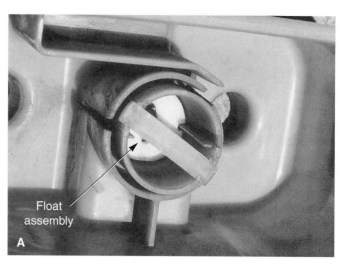

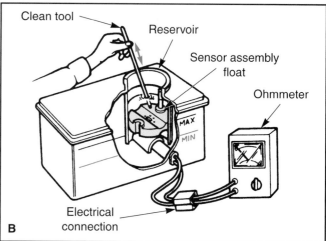

Figure 10-9. *A—Brake fluid level sensor. B—Testing a brake fluid level sensor by pushing the float down past the "minimum" fluid level to turn the switch on. If the circuit is ok, the ohmmeter needle will move. (Mazda)*

Electrical Switch Service

The electrical switches used in the brake system (low brake fluid, brake light, pressure differential.) are on-off switches. Therefore, testing these switches is relatively simple. Before assuming that an electrical problem is caused by the brake system, check the vehicle's fuses and wiring. Also check for corrosion at connections and disconnected or open wiring.

Testing Low Brake Fluid Switches

Before checking the low brake fluid switch, check the brake fluid level to ensure it is not low. Add fluid if necessary and observe whether the light goes out. If the fluid is low, check for a leak or worn brake linings.

The simplest test of a low brake fluid switch is to manually operate the float and determine whether the light comes on. Turn the ignition switch to the *on* position and have an assistant observe the light as you move the float up and down. The light should come on and off as the float is moved up and down. If the light does not go off when the float is raised, the switch may be defective, or the wiring may be shorted. If the light does not come on when the float is at its lowest position, the switch may be defective, the bulb or fuse may be burned out, or the wiring may be disconnected. Testing a fluid level switch with an ohmmeter is shown in **Figure 10-9**.

Testing Brake Light Switches

The easiest test of the brake light switch is to have an assistant step on the brake while you observe the brake lights. Note that some brake lights are wired through the ignition switch, and the lights will not illuminate unless the ignition is on.

If only one brake light does not work, the problem is usually a burned out or ungrounded lightbulb. On most older vehicles, the brake wiring passes through the turn signal switch, and a defect in the switch may cause one side of the vehicle to have no brake lights.

If no brake lights come on, check for power at the brake switch. Most brake light switches are located on the brake pedal linkage under the dashboard. A few older vehicles have hydraulic brake switches installed on the master cylinder. If the brake switch has power, make sure it is properly adjusted. If adjusting the switch does not cause the brake lights to operate, check for disconnected wiring or bad grounds at the taillight bulbs. If everything else checks out, test the switch with an ohmmeter, **Figure 10-10**. The switch should have continuity (zero resistance) when the plunger is fully extended, and no continuity (infinite resistance) with the plunger fully retracted. If the switch does not operate as specified, it should be replaced.

If the brake lights are on at all times, check the brake switch adjustment. If the adjustment is correct, or if adjusting the switch does not cause the lights to go off, disconnect the switch's electrical connector. If the lights go off, the switch is defective. If the lights stay on, there is a shorted wire energizing the stoplight circuit at all times.

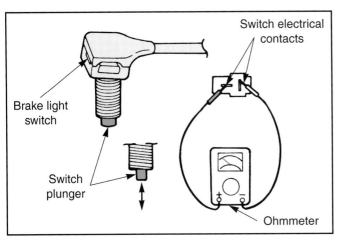

Figure 10-10. *With the ohmmeter leads touching the switch terminals, check the brake light switch for continuity (completed circuit). Follow the vehicle manufacturer's testing procedures. (Hyundai)*

Note: If you find a defective brake light switch on a vehicle with anti-lock brakes, check the ABS control module for codes. This is discussed in Chapter 22.

Adjusting Brake Light Switches

To adjust a brake light switch, first determine the switch adjustment method. Most brake switches are self-adjusting. Install the switch as far at it will go in the holder. Then pull up on the brake pedal. A series of clicks will be heard as the switch moves in its holder, **Figure 10-11.** After making this adjustment, recheck brake light operation.

Note: A few brake light switches are not adjustable. They must be replaced if they do not operate correctly.

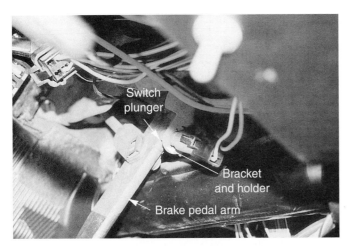

Figure 10-11. *Self-adjusting brake light switch. The switch body is installed in a stationary bracket and the plunger contacts the brake pedal arm.*

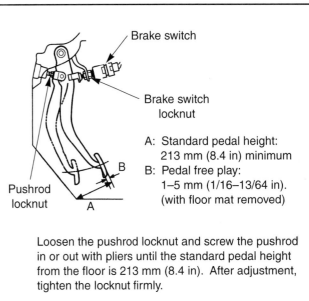

A: Standard pedal height: 213 mm (8.4 in) minimum
B: Pedal free play: 1–5 mm (1/16–13/64 in). (with floor mat removed)

Loosen the pushrod locknut and screw the pushrod in or out with pliers until the standard pedal height from the floor is 213 mm (8.4 in). After adjustment, tighten the locknut firmly.

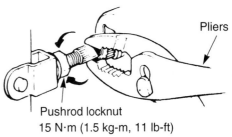

Pushrod locknut
15 N·m (1.5 kg-m, 11 lb-ft)

Screw in the brake switch until its plunger is fully depressed (threaded end touching the pad on the pedal arm). Then back off the switch 1/2 turn and tighten the locknut firmly.

Figure 10-12. *Some brake light switches use a locknut to secure the switch. (Honda)*

If the switch is a threaded type, loosen the locknut and adjust it so the brake lights are off when the pedal is in the released position. The brake lights should come on when the pedal is lightly depressed (no more than about 1/2″ or 13 mm), **Figure 10-12.**

Some switches use a spacer for adjustment. Adjust the switch plunger so it is fully bottomed out against the spacer with the pedal in the fully released position. Then remove the spacer and check brake light operation. Some switches require the brake pedal to be depressed during adjustment, **Figure 10-13.**

Removing and Replacing Valves and Switches

In almost every modern vehicle, hydraulic valves are replaced instead of being repaired. Therefore, no information is given concerning the overhaul of hydraulic valves or switches.

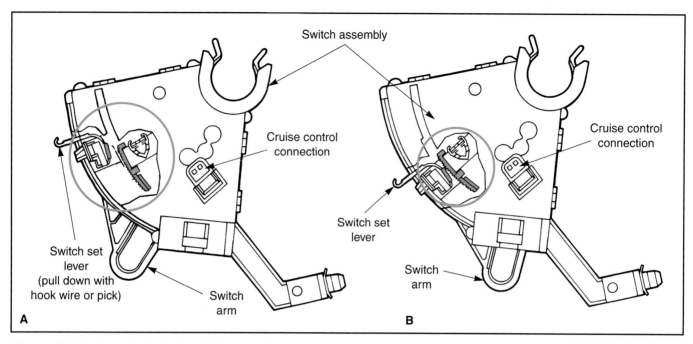

Figure 10-13. *Some brake light switches require a special adjustment procedure. A—After switch installation, depress the brake pedal. Then, using a wire hook or mechanics pick, pull the switch set lever down until you hear a click. B—Release the set lever and then, the brake pedal. The switch should adjust properly as shown. Follow the vehicle manufacturer's set and installation guidelines. (Chevrolet)*

 Caution: Be sure to use a pan to catch dripping fluid and be sure to wipe down all surfaces touched by brake fluid as soon as possible.

Removing and Replacing Hydraulic Valves

Removing and replacing brake hydraulic system valves is relatively easy. Begin by disconnecting any electrical connections. Then loosen and remove the fittings holding the lines to the valve assembly. Place a pan under the valve to catch any dripping fluid. If necessary, cap the lines to reduce fluid loss and entry of air and moisture.

After all electrical connections and lines are removed, remove the fasteners holding the valve to the vehicle. Compare the replacement part to the old unit. This ensures the replacement is the correct part. Install the valve assembly on the vehicle, but do not tighten the fasteners. Start all hydraulic fittings and snugly tighten them. Then tighten the valve to vehicle fasteners. Finish tightening the line fittings and either manual or power bleed the system as explained in Chapter 6. Install any electrical connectors and recheck valve and system operation.

Removing and Replacing Electrical Switches

To replace an electrical switch, remove the battery negative cable. Then unplug the electrical connector and remove the old switch. Most low fluid level switches are

part of the reservoir cap assembly and the entire cap is replaced if the switch is defective. Pressure differential switches are threaded into the pressure differential valve assembly and can be removed for replacement. In most cases, it is not necessary to bleed the hydraulic system after replacement. Most brake light switches are installed on the brake pedal assembly and can be loosened to remove them.

After installing the new switch, reconnect the battery, and adjust the switch if necessary. Then recheck switch and system operation.

Brake Line, Hose, and Fitting Service

The following sections explain how to check and service the steel hydraulic lines, hoses, and fittings. In most cases, repair procedures are simple. However, they should never be done haphazardly, since without the lines and hoses, the other system parts are useless. A small leak or kink, left uncorrected, can lead to complete brake failure.

Line, Hose, and Fitting Defects

The most common defect found in lines, hoses, and fittings are leaks. Leaks can usually be spotted by visual observation. Any leakage from a line, hose, or fitting must be corrected.

The steel lines may be damaged by road debris, or may clog internally. If any steel line shows signs of damage, especially kinking or crushing, it should be replaced. Clogging is harder to detect, but can usually be isolated by loosening a fitting past the suspected clog point. Then have an assistant press on the brake while observing the fitting. If no fluid exits or fluid flow is slow, the line has a restriction.

Flexible hoses can develop swelled spots on the outer covering. This is a sign of an internal leak, and the hose should be replaced. If the outer covering shows any signs of road damage or cracking, the hose should also be replaced.

Occasionally, a hose will swell internally, blocking off the fluid passage. This can cause the related wheel brake to apply slowly or not at all. Pulling in one direction is often caused by a swelled hose on the opposite wheel. Some hoses swell to the point they not only impede the flow of fluid into the wheel hydraulic unit, but keep it from moving out when the brakes are released. The vehicle will pull in one direction when the brakes are applied, and pull in the opposite direction when the brakes are released.

Fittings are essentially trouble free, but can be damaged by road debris or become internally clogged. If a fitting is damaged or leaking, it should be replaced. Do not attempt to repair a cross-threaded or stripped fitting by installing an oversized nut or undertightening the nut.

Bending and Cutting Brake Lines

When a brake line must be replaced, the new line must closely match the length and shape of the old line. The new line must be bent and if needed, cut and flared. First, obtain the proper length of double-wall steel line. Preflared brake lines of standard lengths are available. Do not attempt to repair an old brake line by shortening the line and reflaring, or by installing a compression union.

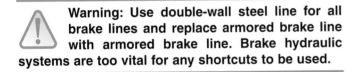 **Warning: Use double-wall steel line for all brake lines and replace armored brake line with armored brake line. Brake hydraulic systems are too vital for any shortcuts to be used.**

It is usually best to bend the line into the proper shape first, then cut any excess line from the ends. Cutting first may result in a brake line that is too short.

Bending New Brake Lines

It is sometimes possible to make slight bends in steel lines by hand. However, trying to make a sharp bend in a steel line will inevitably kink the line, leading to flow restrictions and leaks. Therefore, a tubing bender should always be used to make any sharp bends. Remember to place the bend in exactly the right place. If the old line is available, use it as a template for making the new line.

If the new line is relatively small and bends will not be more than about 45°, a *coil spring bender* can be used, **Figure 10-14.** Begin by selecting the proper coil spring. The right spring will slip firmly over the line.

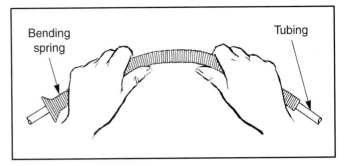

Figure 10-14. *A coil spring bender can be used to bend steel brake line.*

Then slowly bend the line into the proper shape. The spring will prevent kinking. Bend the line slightly further than the finished bend, then release. Make sure the angle of the bend is correct, then remove the bender from the line. If the line is large, or if the finished angle will be greater than 45°, use a *gear bender* or *lever bender,* **Figure 10-15.** Install the bender on the tubing, and work the lever to bend it to the needed shape. Remember to slightly overbend the line so that it will spring back to the proper angle.

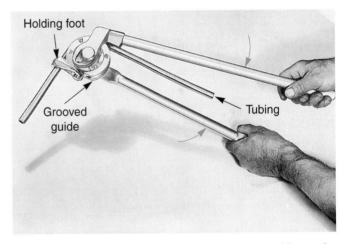

Figure 10-15. *Lever tube bender used to fashion tubing to the correct angle. (Imperial-Eastman)*

Cutting Brake Lines

Once the new line is bent into the proper shape, it can then be cut to the proper length. To cut any brake line, always use a tubing cutter, not a hacksaw, **Figure 10-16.** A hacksaw cannot make an accurate cut, and the ragged edges may interfere with the flaring process. A tubing cutter makes cuts that are perfectly square, and the line will remain perfectly round. When using the tubing cutter, slightly tighten the cutter and make at least one complete revolution before tightening again. Tightening the cutter too much will bend the line. Once the line is cut, inspect the end to make sure the line was cut squarely. Then, ream the inside of the tubing to remove any burrs and blow out with compressed air, **Figure 10-17.**

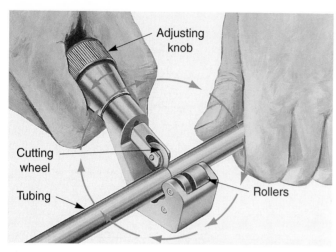

Figure 10-16. *Cutting tubing. Tighten the cutter adjusting knob after each full revolution of the cutter around the tubing. Do not overtighten; this will tend to crush the tubing, making the cutting procedure difficult.*

Flaring Brake Lines

There are two types of flares used on modern vehicles, the **double-lap flare,** and the **ISO flare.** The following sections explain how to make these flares. Note that special tools are required to make both types of flares.

> ⚠ **Warning: Check the original brake line to determine what type of flare was originally used. Fitting designs are different, and double-lap and ISO flares will not interchange. Do not use a single-lap flare on any brake line.**

Making a Double-lap Flare

The making of a double-lap flare is a two-step process. A special flaring tool, such as the one shown in **Figure 10-18,** is needed. It consists of a *flaring bar,* adapter, and a screw-operated *flaring yoke.* Before going further, make sure the line has been cut square and has no burrs.

> 🧰 **Note: Always place the flare nut over the line before flaring the end. Once the end is flared, the flare nut will not fit over the end of the line. Make sure the flare nut is facing in the proper direction when installed.**

Insert the line through the proper size hole in the flaring bar. Determine what size adapter is needed, and use it as a height gauge to determine how far to extend the line from the flaring bar, **Figure 10-19.** Tighten the flaring bar to hold the line securely, then place the adapter over the end of the line. Use the yoke assembly to push the adapter until it is flush with the flaring bar, **Figure 10-20.** This is the first step of the flare.

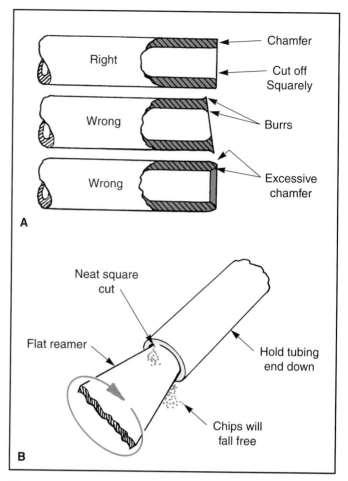

Figure 10-17. *A—Make sure the tubing is cut square. B—Lightly chamfer the tubing with a flat reamer. Thoroughly clean the inside of the tubing with air pressure to remove any metal particles. Wear safety glasses.*

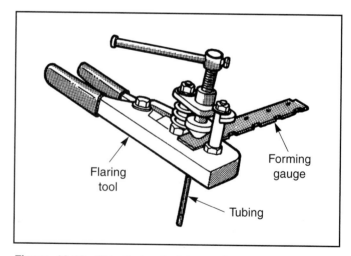

Figure 10-18. *This flaring tool is used to create double-lap flares.*

Remove the adapter and use the swivel of the yoke assembly to finish the flare by folding the metal in on itself, **Figure 10-21.** Then remove the yoke and loosen the flaring bar. Inspect the flare to ensure it is made correctly. If the flare is not correct, cut off the flare and try again.

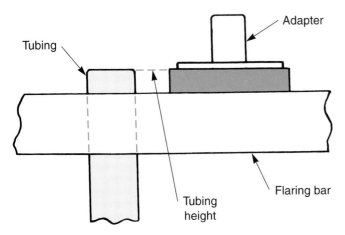

Figure 10-19. *Using the correct size adapter to adjust the tubing height in the tool. (Snap-On Tools)*

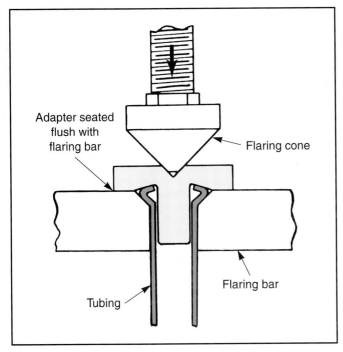

Figure 10-20. *The flaring cone has pushed the adapter into flush contact with the flaring bar. Note the shape of the tubing flare at this stage. (Snap-On Tools)*

Making an ISO Flare

To make an ISO flare, you will need a special flaring tool containing a *tool body*, which works with a *collet*, *forming mandrel*, *clamping nut*, and *forcing screw*. See **Figure 10-22.** Begin making the ISO flare by placing the clamping nut over the line and clamping the flaring tool in a bench vise.

Note: The nut used with an ISO fitting is different from the flare nut used on a double-lap flare fitting, Figure 10-23. Do not confuse the two.

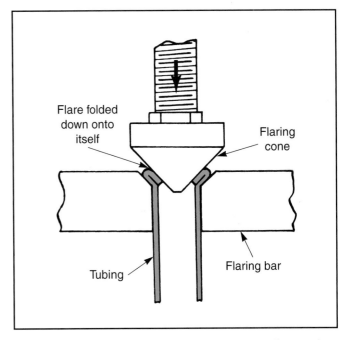

Figure 10-21. *A flare being folded down onto itself to produce the finished double-lap flare (double metal thickness in the seat area). (Snap-On Tools)*

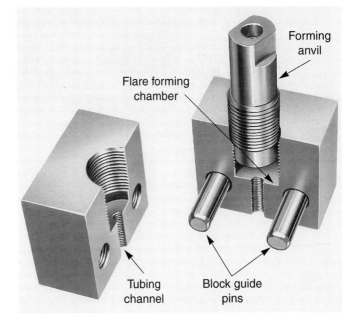

Figure 10-22. *An ISO flare forming tool. Follow the tool manufacturer's operating instructions. (Kent-Moore)*

Select the proper size collet and forming mandrel for the tubing diameter being used. See **Figure 10-24.** Then insert the tubing into the correct collet, leaving about 3/4″ (19.05 mm) of tubing extending out of the collet, **Figure 10-25.** Insert the line and collet assembly into the tool body. The end of the tubing should contact the forming mandrel. Tighten the clamping nut into the tool body, ensuring it is tight enough to keep the tubing from moving during the forming process.

Turn the forcing screw until it bottoms out against the forming mandrel. Once the screw bottoms out, stop tightening. Overtightening may cause the flare to become oversized. Then disassemble the flaring tool and check the flare to ensure that it is correctly formed.

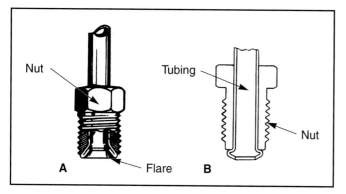

Figure 10-23. *A—Completed double-lap flare with nut. B—ISO flare with nut. Notice that they are not the same.*

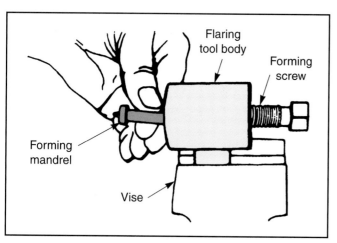

Figure 10-24. *Placing the correct size forming mandrel in the tool body. Note that the tool body is securely held in a bench vise. (Pontiac)*

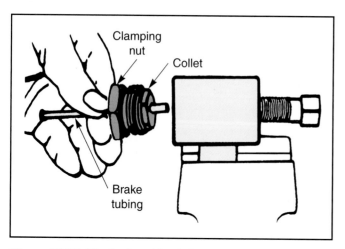

Figure 10-25. *The brake tubing is inserted to the correct depth in the collet and is being placed in the flaring tool body. (Pontiac)*

Hose Replacement

To replace a flexible hose, loosen the tubing fittings. Then remove the hose clip, **Figure 10-26,** where used. Match the old and new hoses to ensure a proper fit. Place the new hose in position and lightly install the tubing nuts and other fittings. Then install the hose clips.

 Caution: The hose should be fastened securely at both ends. Do not allow a hose end to vibrate or move around. If clips or other fasteners become damaged, locate and obtain replacements.

After the clips or other mounting devices are tight, tighten the tubing fittings and bleed the hose. Then recheck system operation. When installing compression fittings, be sure to install new sealing washers, if used.

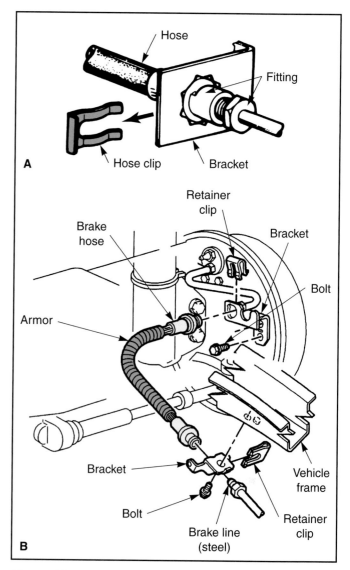

Figure 10-26. *A—Removing a hose retaining clip. This will allow the brake hose to be removed from its mounting bracket. B—Hose removal. Make sure the new hose is properly routed to prevent damage. (Toyota and Pontiac)*

Fitting Replacement

Many hydraulic brake system fittings are part of another assembly. If they are defective, the entire assembly must be replaced. However, some fittings, such as those which divide brake lines into individual lines, **Figure 10-27,** can be replaced without affecting other hydraulic system parts. To replace these fittings, loosen and remove the tubing nuts holding the lines to the fitting. Cap the lines to reduce fluid loss and entry of air and moisture. Then remove the fasteners holding the fitting.

Compare the new fitting to the old unit to ensure it is the correct part. Install the new fitting, but do not tighten the fasteners. Start all hydraulic line tubing nuts and snugly tighten them. Then tighten the fasteners holding the fitting to the vehicle. Tighten the line tubing nuts and bleed the system as explained in Chapter 6. Before releasing the vehicle, check for leaks.

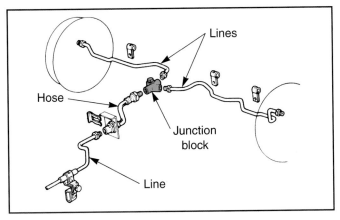

Figure 10-27. *A junction block used to split a single brake line into two separate lines. (Toyota)*

Summary

Hydraulic valve problems vary with the type of valve and include wheel lockup, failure of the brakes to apply, uneven braking, leaks, and illumination of warning lights. Hydraulic valve operation can be checked by observation or with pressure gauges. Valves which can be checked with pressure gauges are the proportioning, metering, and residual pressure valves. Some load-sensing proportioning valves are adjustable. Some pressure differential valves can be centered manually.

Low brake fluid, brake light, and pressure differential switches are all on-off switches and can be checked easily. Before assuming the switch is bad, check fuses, connections, wiring, and corrosion. Most brake light switches can be adjusted. Removing and replacing hydraulic valves and electrical switches is relatively easy. Always recheck system operation after replacing any valve or switch.

Lines, hoses and fittings can develop external leaks which can be spotted visually. Lines can become kinked or clogged, and hoses can swell internally or externally. Any damaged line, hose, or fitting must be replaced.

To replace lines they must be bent and cut to the proper configuration. Line ends must also be properly flared. The two types of flares are the double-lap flare and the ISO flare. These flares must be made carefully. Fittings should be replaced if they become damaged. Replacement steps should be followed carefully.

Review Questions—Chapter 10

Please do not write in this text. Write your answers on a separate sheet of paper.

1. A defective _____ can cause front wheel skidding.

2. The proportioning valve prevents _____ skidding.

3. A failed pressure differential valve may cause the _____ to be on constantly.

4. A defective metering valve can cause _____ when the brakes are first released.

5. A _____ may require adjustment.

6. If an assistant presses on the brake pedal while you open a bleeder screw, the dashboard brake light comes on. What is wrong with the pressure differential valve?

7. To test the metering and proportioning valves, what kind of test equipment is used?

8. If a brake light switch cannot be adjusted, it should be _____.

9. If any hydraulic lines are removed, the hydraulic system must be _____.

10. A brake hose that is swelled shut could cause the vehicle to do what during braking?

11. When can the technician safely install a compression union on a leaking brake line?

12. Replacement brake lines should never be cut with a _____.

13. Place the flare nut over the new line _____ the line is flared.

14. Making a double-lap flare is a _____ step process.

15. If a hydraulic fitting is part of another brake system unit, what should be done to replace it?

ASE Certification-Type Questions

1. Technician A says that brake system hydraulic valves can stick. Technician B says that brake system hydraulic valves can cause hydraulic system malfunctions. Who is right?
   (A) A only.
   (B) B only.
   (C) Both A & B.
   (D) Neither A nor B.

2. All of the following could cause front wheel lockup, EXCEPT:

   (A) metering valve sticking open.
   (B) metering valve sticking closed.
   (C) proportioning valve sticking closed.
   (D) residual pressure valve sticking open.

3. If the pressure differential valve sticks in the centered position, which of the following will happen?

   (A) Skidding of one wheel at the rear.
   (B) Skidding of one wheel at the front.
   (C) Constant illumination of the dashboard brake warning light.
   (D) No illumination of the dashboard brake warning light.

4. Which of the following valves sometimes has a provision for adjustment?

   (A) Metering valve.
   (B) Proportioning valve.
   (C) Residual pressure valve.
   (D) Pressure differential valve.

5. To check the operation of a pressure differential valve, have an assistant press on the brake pedal while you open the _____ bleeder valve.

   (A) left front
   (B) right rear
   (C) left rear
   (D) Any of the above.

6. Technician A says that pressure testing a metering valve requires two gauges. Technician B says that pressure testing a proportioning valve requires two gauges. Who is right?

   (A) A only.
   (B) B only.
   (C) Both A & B.
   (D) Neither A nor B.

7. If one brake light does not work when the brake pedal is depressed, the *most likely* problem is _____.

   (A) a burned out bulb
   (B) a blown fuse
   (C) a defective brake switch
   (D) low fluid level in the master cylinder

8. When installing a new hydraulic valve, which of the following should the technician do *first*?

   (A) Bleed the brake system.
   (B) Tighten the tubing nuts.
   (C) Loosely install the valve fasteners.
   (D) Tighten the valve fasteners.

9. Technician A says that replacing most electrical switches does not require bleeding the brake system. Technician B says that most brake light switches are installed on the master cylinder. Who is right?

   (A) A only.
   (B) B only.
   (C) Both A & B.
   (D) Neither A nor B.

10. Which of the following would be the *most likely* result of a hose swelling shut internally?

    (A) Pulling to one side.
    (B) Leaks.
    (C) Excessive pedal travel.
    (D) Dragging brakes.

11. When should the technician repair a brake line by shortening the line and reflaring?

    (A) When there is extra line available.
    (B) If a new line must be ordered.
    (C) If ISO flaring tools are available.
    (D) Never.

12. Technician A says that a new brake line should be cut to length, then bent. Technician B says that a new brake line should be bent to the proper shape, then flared. Who is right?

    (A) A only.
    (B) B only.
    (C) Both A & B.
    (D) Neither A nor B.

13. Which of the following flares can be used on modern vehicles?

    (A) Double-lap.
    (B) ISO.
    (C) Single-lap.
    (D) Both A & B.

14. Technician A says that double-lap and ISO flares will not interchange. Technician B says that double-lap and ISO flares can be interchanged only in an emergency. Who is right?

    (A) A only.
    (B) B only.
    (C) Both A & B.
    (D) Neither A nor B.

15. Under which of the following situations should a fitting be repaired instead of being replaced?

    (A) Stripped threads.
    (B) Leaks.
    (C) Cracks.
    (D) None of the above.

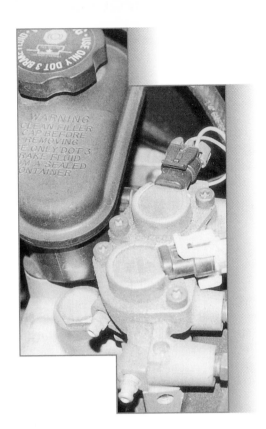

Chapter 11

Friction Brake Theory

After studying this chapter, you will be able to:

❑ Define friction.
❑ Explain the relationship of weight and speed to momentum.
❑ Define coefficient of friction.
❑ Define static and kinetic friction.
❑ Define brake fade.
❑ Explain the relationship of friction to heat development.
❑ Identify and describe the factors affecting stopping power.
❑ Identify outside factors affecting coefficient of friction.
❑ Describe brake lining materials and construction.

Important Terms

Friction	Static friction	Heat dissipation	Semi-metallic
Momentum	Kinetic friction	Weight transfer	Metallic
Kinetic energy	Brake horsepower	Inertia	Rotor
Coefficient of friction	Brake fade	Non-metallic	Drum

Before studying the operation of the brake friction members, you must have an understanding of the basic principles of friction. These principles explain what friction is, how it is used to overcome vehicle motion, and various factors affecting brake material construction. This chapter is an introduction to brake friction principles which will prepare you for the disc and drum brake theory and service chapters.

What is Friction?

When two things move against each other, there is a resistance to the movement between them. This is caused by microscopic imperfections (high spots) that exist on even the smoothest surfaces, **Figure 11-1**. The imperfections on one surface contact the imperfections on the other surface as they move against each other. The resistance caused by this contact is called *friction.*

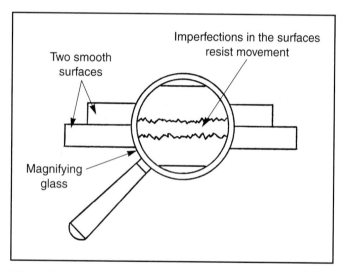

Figure 11-1. *A magnifying glass showing the surface imperfections which cause friction between the two moving surfaces. These imperfections act like two pieces of sandpaper being rubbed together.*

If there was no friction, tires would have no traction against the road. Without friction, bolts would not tighten, and doors would not stay closed. However, in many instances, it is desirable to minimize friction as much as possible. This is why engines and drivetrain parts have elaborate lubrication systems.

In the brake system, friction is put to work to overcome the vehicle's momentum, in other words, to stop the vehicle. What momentum is, and how it is overcome is explained in the following paragraphs.

Vehicle Momentum

When a vehicle is stopped, it has weight, but no momentum. *Momentum* is the combination of the vehicle's weight and speed. See **Figure 11-2**. It is often referred to as *kinetic energy,* and is measured in foot-pounds (ft-lbs.). As a technician, you will not be asked to calculate momentum. However, for your own understanding of the basic principles of brake operation, you should remember the following examples.

At any given speed, momentum increases with the weight of the vehicle. For instance, a 4000 lb. (1812 kg) vehicle has twice the momentum of a 2000 lb. (906 kg) vehicle, when both of them are traveling at the same speed. For example the 4000 lb. (1812 kg) vehicle traveling at 30 mph (48 kph) has a momentum of 124,000 ft-lbs. (168 640 N•m). The 2000 lb. (906 kg) vehicle at 30 mph has a momentum of 62,000 ft-lbs. (84 320 N•m). Note that the momentum of the 4000 lb. vehicle is exactly twice that of the 2000 lb. vehicle.

Speed, however, has much more effect on momentum. For any vehicle, momentum increases by the square of the speed. For example, a 3000 lb. (1359 kg) vehicle has four times the momentum at 40 mph (64 kph) as it does at 20 mph (32 kph). At 60 mph (92 kph), momentum is nine times higher. At 20 mph, momentum of the 3000 lb. vehicle is 41,000 ft-lbs. (55 760 N•m). At 40 mph, momentum rises to 166,000 ft-lbs. (225 760 N•m). At 60 mph, momentum is 372,000 ft-lbs. (505 920 N•m).

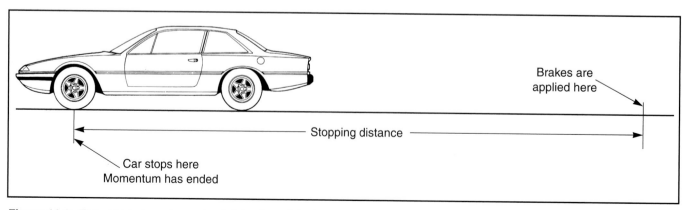

Figure 11-2. *The vehicle has stopped, retaining its weight, but now has lost its forward momentum (kinetic energy). (Ferrari)*

Brakes perform more work stopping a light vehicle going at a high rate of speed than a heavier vehicle going slowly. You would think that a 4000 lb. vehicle moving at 25 mph (40 kph) would have the same amount of momentum as a 2000 lb. (906 kg) vehicle moving at 50 mph (80 kph). However, the 4000 lb. (1812 kg) vehicle has only 86,000 ft-lbs. of momentum at 25 mph, while the 2000 lb. vehicle has over 172,000 ft-lbs. of momentum at 50 mph.

about .005. While this seems like a very small coefficient of friction, it is occurring during every revolution of the brake rotors, and is sufficient to stop the vehicle.

Note: The coefficient of friction will always be less than one. If the coefficient was greater than one, the brakes would lock up the wheels.

Putting Friction to Work

The job of the brake system is to overcome momentum and stop the vehicle. To do this, it uses hydraulics, pneumatics, mechanical leverage, and friction. Hydraulics, pneumatics, and mechanical leverage were discussed in earlier chapters. The following sections explain how friction is put to work to stop the vehicle.

Coefficient of Friction

As stated earlier, friction is always present between two materials that slide against each other. The **coefficient of friction** is the amount of friction that can be produced as two materials move against each other. The coefficient of friction is determined by a simple calculation. In the example shown in **Figure 11-3,** the coefficient of friction is calculated by measuring the force required to slide a block over a surface and then dividing it by the weight of the block.

If it takes 10 lbs. (16 kg) of force to slide a 10 lb. (16 kg) block over a flat surface, the coefficient of friction for the block is 1. Another 10 lb. (16 kg) object, such as a block of ice. may only require 5 lbs. (8 kg) of force to slide it across the same surface. In this case, the coefficient of friction for the block of ice is .5. See **Figure 11-4.**

To determine the coefficient of friction for a vehicle braking system, change weight to pressure and change the sliding force to momentum. For example, a 4000 lb. (1812 kg) vehicle moving at 25 mph (45 kph) has a momentum of about 86,000 ft-lbs (116 960 N•m). If the brake pads are pressed against the rotors with a steady force of 500 psi (3450 kPa) and the vehicle is brought to a stop, the coefficient of friction is determined by dividing 500 by 86,000. This results in a coefficient of friction of

Static and Kinetic Friction

The two basic types of friction are *stationary* or **static friction** and **kinetic friction,** sometimes called *sliding* or *dynamic friction.* Keep in mind that static friction is a holding action that keeps a stationary object in place, while kinetic friction slows a moving object by converting momentum to heat. Note that static friction is always higher than kinetic friction.

The most obvious use of static friction is the parking brake, **Figure 11-5.** When the parking brake is applied, static friction between the applied brake components resists movement. To move the vehicle, static friction must be eliminated by releasing the brakes. Since the vehicle has no movement, there is no momentum to overcome, and no heat is generated.

As discussed earlier, vehicle weight times vehicle speed equals momentum. Applying the brakes on a moving vehicle causes the stationary friction members (pads or shoes) to be forced into contact with the rotating friction members (rotors or drums). This contact, **Figure 11-6,** causes friction and heat, which results in the rotating parts slowing and eventually stopping. Since the momentum of the rotating parts is called kinetic energy, the friction used to stop the rotating parts is called kinetic friction.

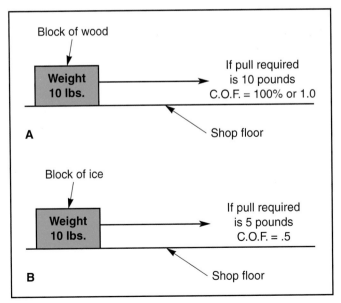

Figure 11-4. *Two examples illustrating the coefficient of friction. A—C.O.F. at 10 pounds of pull. B—C.O.F. at five pounds. (TRW Inc.)*

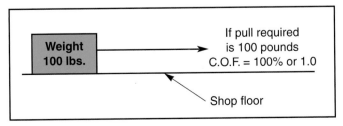

Figure 11-3. *A 100 pound (45.359 kg) weight being pulled across the shop floor. The coefficient of friction (C.O.F.) is 1.0. (TRW Inc.)*

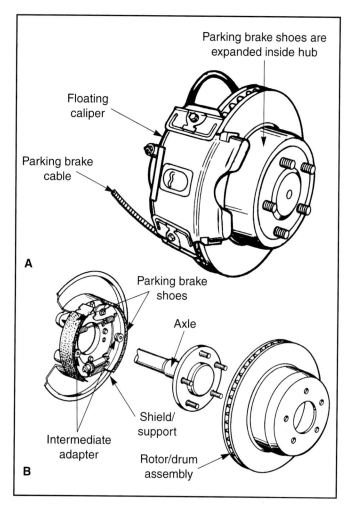

Figure 11-5. *Static friction (not moving). A—With the parking brake shoes expanded into contact with the inside hub (drum), the rotor is prevented from turning. B—Exploded view of the various brake components. (Chrysler)*

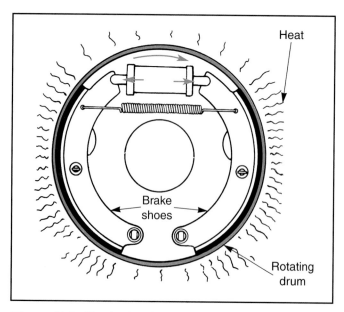

Figure 11-6. *The brake shoes are being forced into contact with the revolving brake drum producing kinetic friction and heat. This will cause the rotating parts to slow down and come to a stop. (Bean)*

Brake System Horsepower

Ordinarily, we think of horsepower in connection with the engine. However, the horsepower developed by the engine to move the vehicle from 0-60 mph (0-95 kph) must be absorbed by the braking system when slowing the vehicle from 60 mph to 0.

Brake horsepower is the amount of power needed to stop a vehicle and is much greater than engine horsepower. The engine in a modern vehicle is often capable of moving the vehicle from 0-60 mph in one-eighth of a mile, or 660 feet. The brake system of the same vehicle can take the vehicle from 60 mph to 0 in about 150 feet. The distance taken to get from 60 to 0 is about one-quarter the distance taken to get from 0-60 mph, **Figure 11-7.** Therefore, if the engine in the above example has 150 horsepower, the brake system has over 600 brake horsepower.

Factors Affecting Friction Development

Many factors affect the development of friction by the brake system. While some may seem obvious, others will seem unlikely. However, they all have their part to play in the proper development of friction.

Apply Pressure

The more pressure applied to the brake friction members, the more they resist movement, and the more friction is developed. More friction means more braking action. Pressure is created by a combination of mechanical leverage and hydraulic pressure, plus the action of the power assist unit. These factors of pressure development were discussed in earlier chapters.

Friction Material Temperature

The temperature of the friction materials has a great effect on the amount of friction developed. As the friction material gets hotter, its ability to stop the vehicle is reduced. Not only must friction materials be designed to operate under greatly varying temperatures, they must have roughly the same coefficient of friction, both cold and hot. Too much variation means the brake pedal feel and required pressure would change drastically as the brakes become heated.

Friction Material Contact Area

Although a small braking surface could produce as much friction as a larger surface by being applied harder, it would quickly overheat and become useless. Brake friction members must be large enough to absorb and spread the frictional heat out. For this reason, the larger the vehicle, the larger the brake friction components.

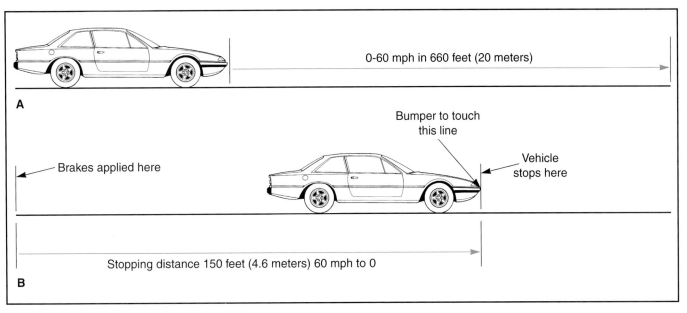

Figure 11-7. Brake horsepower. A—Vehicle accelerates from 0 to 60 mph in 660 feet. B—The same vehicle being brought to a stop with its brake system, while traveling from 60 mph to 0 in 150 feet. (Parker Automotive)

Friction Material Finish

The finish, or smoothness of the friction materials has an effect on the vehicle's braking ability. A rough brake surface would have a higher coefficient of friction, but would grab and wear quickly. Modern brake lining material develops a smooth finish as it is used. As the top layer of material wears away, the underlying surface maintains the smooth finish.

Type of Friction Material

The materials used in a brake friction unit have a great effect on its stopping ability. More force is needed to move some materials over a surface than others, even when apply pressure, contact area, and finish are the same. The friction characteristics of a brake material make up its coefficient of friction. If the coefficient of friction is too high, the brakes will work too well and cause the wheels to lock up. If the coefficient of friction is too low, the brake pedal would require excessive force to stop the vehicle. Types of friction materials are discussed in more detail later in this chapter.

Heat Removal

Friction always causes heat. The more friction needed to stop the vehicle, the greater the amount of heat generated during braking. Therefore, the temperature of the brake components rises as the brakes are applied. If a vehicle weighs 4000 lbs (1812 kg), one emergency stop from 60 mph (95 kph) can raise brake lining temperatures by 160°F (71°C). Repeated hard stops can continue to raise the temperature by equal amounts, **Figure 11-8.**

The build-up of heat can lower the coefficient of friction in the brake pads or shoes to the point where the brakes begin to fade. **Brake fade** is the term given to gradual brake failure caused by brake overheating. As the brakes begin to fade, it takes more pedal pressure to stop the vehicle. After a certain point, the brakes will not have any effect, regardless of the pressure applied to the brake pedal.

Sometimes, when low quality brake fluid is used, or the fluid is water contaminated, the fluid will boil due to excessive frictional heat. When this happens, pressing on the brake pedal will compress the vaporized fluid instead of applying the brakes.

Not only is brake fade dangerous, the heat generated by excessive braking can wear the linings, overheat and warp the drums and rotors, and cause premature failure of the hydraulic system, wheel bearings, and seals.

Brakes are Designed to Fade

Brakes are designed to fade at a certain temperature. Once this temperature is passed, the frictional materials melt and the brake linings will no longer stop the vehicle. While this seems to be a poor brake design, it makes sense when considering the potential damage from allowing brake temperatures to continuously rise. If temperatures were allowed to rise without dissipation, the wheel bearing seals, lubricating grease, CV axle boots, and any other nearby parts would be destroyed. In extreme cases, nearby combustible components, including the tires, could catch fire.

To prevent this, a built-in fade point is designed into the lining material. This point is high enough to allow for most braking situations. However, this emphasizes the importance of removing heat as efficiently as possible.

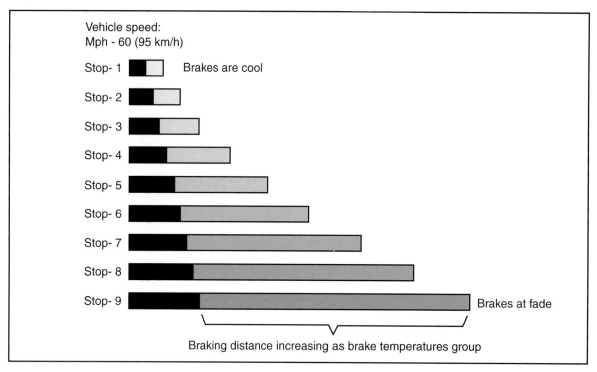

Figure 11-8. *Chart illustrating a brake temperature range from cool to the point of brake fade. At Stop 9, the brakes have begun to fade severely. Notice as the brake temperatures go up, stopping distances become longer. (FMC)*

Heat Dissipation

Heat is removed from the brake friction surfaces by direct transfer to the surrounding air. This process is called **heat dissipation, Figure 11-9.** Modern brake systems are designed to provide the best possible heat dissipation. The size of the friction surfaces is designed for maximum heat dissipation. The larger the friction area, the better the heat dissipation. Rotors and drums are made from cast iron, which can do an efficient job of absorbing heat and transferring it to the outside air.

The disc brake rotor is exposed to outside air throughout its diameter, except where it contacts the brake pads. In addition, the dust shields and sometimes the wheel covers are designed to direct air over the rotor. Disc brake rotors on heavier vehicles contain internal fins to allow air to flow through the rotor. The design of the fins actually pumps air through the rotor internals. In some cases, the rotor braking surface is drilled to increase airflow.

Brake drums used on heavier vehicles are finned on their outside diameter. Air is directed over the drum by vehicle movement. Some brake drums are made of aluminum, with a cast iron lining that contacts the shoes.

Other Factors Affecting Braking

Other factors affect the operation of the brake system, and therefore vehicle braking ability. While these are not parts of the brake friction system, they directly affect its operation.

Weight Transfer

When a vehicle is at rest, most of the vehicle weight is over the front wheels, since the engine and transmission are at the front of the vehicle. As much as 60% of the total weight of a rear-wheel drive vehicle is supported by the front wheels. On a front-wheel drive vehicle, the figure is closer to 80%.

In addition, more weight is placed on the front wheels during braking, due to **weight transfer.** When the brakes are applied, extra vehicle weight is transferred from the rear of the vehicle to the front. This is caused by inertia. **Inertia** is the resistance to any change in momentum, in this case, the tendency of the moving vehicle to keep moving. Although the brakes are slowing the wheel assemblies, the body, drivetrain, and frame try to continue moving forward. See **Figure 11-10.**

The vehicle's original weight distribution and the effects of weight transfer cause the rear wheels to have less weight and the front wheels to have more weight. Since more braking must be done by the front wheels than the rear wheels, the front brake systems has a larger friction area than the rear brake system.

Tire and Road Conditions

If the tires do not grip the road properly, the brake system will not work. Traction between the tires and the road must be maintained for proper stopping. If the brake system works so well that the tire stops rotating, it is said to be skidding. When the road is wet or icy, the tire is skidding on a layer of water or ice. On dry pavement, a skidding tire causes so much frictional heat that the tire rubber melts. The tire then skids on a layer of liquid rubber.

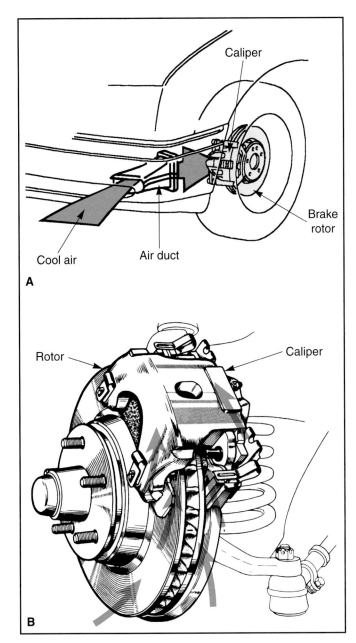

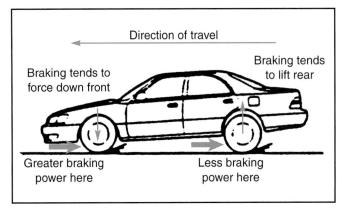

Figure 11-10. *Braking forces create weight transfer when stopping the vehicle. Note how the front end is going down and the rear end is rising.*

the closed throttle plate. This creates a drag on the drivetrain which helps to slow the vehicle. The lower the transmission gear and final drive gear ratios, the faster the engine is turning in relation to road speed, and the more the engine braking effect. This is why drivers are advised to switch to a lower gear when descending a steep hill.

Older vehicles with large, high compression engines and lower gearing were able to get considerable engine braking to assist the wheel brakes. Newer vehicles with small engines and overdrive gears have only a small amount of engine braking in top gear, and the wheel brakes must do more of the braking. A vehicle with an automatic transmission also has less engine braking capacity, since some frictional drag is lost inside of the torque converter when the lockup clutch is not applied.

Water on Brake Linings

In many parts of the world, it is not uncommon to travel through roads that are partially or completely covered with water. In addition to the increasingly common use of off-road vehicles, passenger cars sometimes enter areas where the pavement is covered by water. If water reaches the brake linings, it acts as a lubricant, causing brake fade. There is no way to design brake linings to overcome brake fade due to water. The only cure, other than avoiding flooded roads, is to allow the brakes to dry out. Disc brake systems dry out more quickly, since the exposed rotor will spin off water once the vehicle is on dry pavement. Drum brakes are enclosed, and water takes a longer time to exit the assembly.

Figure 11-9. *A—Cool air flowing through an air duct and being directed onto the brake rotor and other parts. B—Arrows indicate the airflow through and around a typical disc brake assembly. Heat, generated by braking, is being carried away by the airflow. (Bendix & Mercedes-Benz)*

In addition to loss of braking ability, one or more skidding tires can cause loss of steering control. Even if skidding is not severe, it can cause flat spots on the tire surface. Therefore, both the tires and brakes must be in good condition for adequate braking. On modern vehicles, anti-lock brake systems are able to assist in braking and preventing skids. ABS is covered in Chapters 21 and 22.

Engine Size and Drive Train Gearing

The size and compression ratio of the engine has an effect on braking. When the accelerator pedal is released, the engine is trying to draw air into the cylinders against

Friction Members

There are two friction surfaces in any brake assembly. The brake pad or shoe lining, which are made of a mixture of heat resistant materials, and the cast iron rotor or drum. These two frictional surfaces are discussed in the following paragraphs.

Pad and Shoe Linings

Materials used on modern brake shoes and pads must have about the same coefficient of friction when either hot or cold. They must also resist fading at high temperatures, but fade when there is enough heat to cause damage to other vehicle parts. Brake lining materials must also be able to stop when the linings are wet, and recover quickly when dried out. They must stop the vehicle smoothly and quietly, last for tens of thousands of miles, and not cause excessive wear to the rotor or drum.

Lining Materials

Brake linings use various metals and high temperature synthetic fibers such as Kevlar and other heat resistant compounds. Asbestos is no longer used in new brake linings, but may be found in some brake linings currently installed on vehicles.

Some vehicles use linings made from organic or completely **non-metallic** materials. These materials are mixed from various compounds and molded into the proper shape. Non-metallic lining materials are quieter (less prone to squeaking) and do not damage the cast iron drums or rotors. However, they provide the lowest coefficient of friction, and therefore, the least amount of braking power.

Most brake linings, especially those sold as lifetime guaranteed brakes, are **semi-metallic** linings. Semi-metallic materials are made from a combination of non-metallic materials and iron, mixed and molded into the proper shape. Semi-metallic linings are harder and last longer. They are also more fade resistant than completely non-metallic linings. However, semi-metallic linings increase brake pedal effort, may squeak on application, and cause some wear to the rotors and drums.

Metallic brake linings are used on high performance and competition vehicles only. They are made from sintered metal (powdered metal that is formed into linings). Metallic brake linings resist brake fade very well, but require high pedal pressure, are noisy, and severely wear rotors and drums.

Temperature Markings

It is very important that brake lining materials have about the same coefficient of friction when cold as when hot. Brake pads and shoes have makings on the edges that give the approximate coefficient of friction at low and high temperatures. The markings are in letters, with lining materials marked A being the lowest, and improving through the alphabet. Typical letter markings and their relationship to temperature are shown in **Figure 11-11.** The best linings have markings that indicate the coefficient of friction is the same at high and low temperatures. The more variation in the high and low temperature letters, the poorer the quality.

Lining Attachment Methods

Disc brake linings are bonded (glued) or riveted to a metal backing pad, **Figure 11-12.** The attachment method is not an indicator of quality, since both methods are widely used. As a general rule, most pads used on heavier cars and trucks are riveted to the backing plate, while pads used on smaller vehicles are bonded.

Drum brake linings are also bonded or riveted to a metal shoe, **Figure 11-13.** Attachment methods vary, but modern shoes tend to be bonded rather than riveted.

Friction Material Code	Coefficient of Friction
C	<.15
D	.15-.25
E	.25-.35
F	.35-.45
G	.45-.55
H	>.55

Figure 11-11. *A—A disc brake pad illustrating the various edge code markings. B—Chart showing friction material codes used on brake linings. (Jack Klasey)*

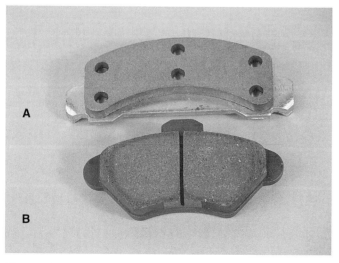

Figure 11-12. *Brake lining to shoe attachment methods. A— Disc brake pad that has been riveted to the shoe. B— Disc brake pad which is bonded (glued) to the metal shoe. (Jack Klasey)*

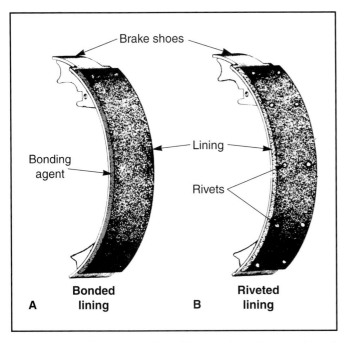

Figure 11-13. *A—Brake lining which has been bonded (glued) to the shoe. B—Lining that was riveted to the shoe. Both attachment methods are currently used. (General Motors)*

Rotor and Drum Construction

The basic job of drums and rotors is to provide a contact surface for the brake linings, and to absorb heat. Modern brake rotors and drums are always made from cast iron. A few vehicles are equipped with aluminum drums having cast iron liners.

Figure 11-14. *A brake rotor displaying a smooth braking surface. (General Motors)*

Rotors

Both sides of the brake **rotor** are machined smooth where they contact the brake pads, **Figure 11-14.** This contact area is extremely smooth for smooth stops. The rotor's cast iron is designed to last through several sets of linings. Rotors are heavy for maximum heat absorption, and have enough extra metal so they can be resurfaced if they become slightly damaged.

Drums

The inner surface of the brake **drum** is machined smooth where it contacts the brake shoes, **Figure 11-15.** The weight of the drum helps to absorb heat and transfer it to the outside air. Drums are also built with extra metal so that they can be resurfaced if they become slightly damaged. A drum should outlast several sets of shoes.

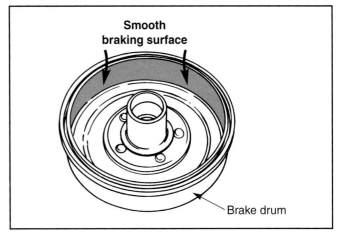

Figure 11-15. *A brake drum depicting a smooth braking surface. (FMC)*

Summary

When the microscopic imperfections of two moving surfaces contact each other, they create resistance. This resistance is called friction. Friction should be avoided in some vehicle systems, but is vital to the operation of the brakes.

Momentum is a combination of vehicle weight and speed. The job of the brake system is to overcome momentum and stop the vehicle. The brakes use friction to stop the vehicle. Friction is always present between two materials that slide against each other. The coefficient of friction is the amount of friction that can be produced as the two surfaces slide against each other. The coefficient of friction is determined by dividing sliding force by the weight of the object.

The two kinds of friction are static friction and kinetic friction. Static friction is friction that keeps a stationary object in place. Kinetic friction slows a moving object by converting movement to heat. Static friction is always higher than kinetic friction.

The brake system can absorb much more horsepower than the vehicle engine can produce. For this reason, stopping distances from a given speed are much shorter than the distances needed to accelerate to the same speed.

Many factors affect the ability of a frictional material to function. Brake apply pressure, temperature, contact area, and type of material all affect the braking ability of a material.

Friction always causes heat, and this heat must be removed to prevent fading. Brake materials are designed to fade at a certain temperature to avoid starting a fire. Heat is absorbed by the metal of the drums and rotors and dissipated to the outside air. Other factors that affect braking ability are weight transfer, tire and road conditions, engine braking ability, and water on the brake linings.

All brake assemblies have two frictional surfaces. The matching frictional surfaces are brake pads and rotors and brake shoes and drums. Brake pad and shoe linings are a mixture of heat resistant composition materials. The linings are marked with a letter series indicating their hot and cold braking performance. Rotors and drums are made of cast iron, and always turn with the wheel and tire.

Review Questions—Chapter 11

Please do not write in this text. Write your answers on a separate sheet of paper.

1. Define friction.

2. Define momentum.

3. The coefficient of friction must always be less than _____, or the brakes would cause the wheels to _____.

4. The type of friction that keeps a stationary object in place is _____ friction.

5. Friction that stops a moving object is _____ friction. It does this by converting _____ into _____.

6. Does a vehicle's brake system have more or less horsepower than its engine?

7. What five factors affect friction development?

8. When heat causes the coefficient of friction to reach zero, this is called brake _____.

9. Excessive brake heat may cause substandard brake fluid to _____.

10. How is brake heat removed?

11. Which set of brakes do the most braking, front or rear?

12. After driving through water, _____ brakes dry off sooner than _____ brakes.

13. Linings are attached to the pad or shoe by _____ or _____.

14. Modern rotors and drums are made from _____.

15. A drum or rotor should outlast several sets of _____.

ASE Certification-Type Questions

1. Technician A says that a vehicle that has weight has momentum, even when it is stopped. Technician B says that momentum is a combination of speed and weight. Who is right?
   - (A) A only.
   - (B) B only.
   - (C) Both A & B.
   - (D) Neither A nor B.

2. If it takes 2 pounds of force to move an 8 pound weight across the floor, the coefficient of friction is _____.
   - (A) 4
   - (B) .5
   - (C) .25
   - (D) .125

3. Technician A says that static friction turns movement into heat. Technician B says that static friction is a form of kinetic friction. Who is right?
   - (A) A only.
   - (B) B only.
   - (C) Both A & B.
   - (D) Neither A nor B.

4. Technician A says that it is not unusual for a vehicle with a 200 horsepower engine to have an 800 horsepower brake system. Technician B says that the horsepower of a vehicle brake system has an effect on stopping distances. Who is right?
   - (A) A only.
   - (B) B only.
   - (C) Both A & B.
   - (D) Neither A nor B.

5. Which of the following affect the braking ability of a brake material?
   - (A) Its temperature.
   - (B) What it is made of.
   - (C) How hard it is applied.
   - (D) All of the above.

6. Brake linings are designed to fade a certain _____.
   - (A) temperature
   - (B) humidity
   - (C) speed
   - (D) momentum

7. During braking weight transfer, in which direction is the vehicle weight transferred?
   - (A) Front to back.
   - (B) Back to front.
   - (C) Left to right.
   - (D) Right to left.

8. If a tire skids on dry pavement, it is skidding on a layer of _____.

   (A) melted asphalt.
   (B) melted rubber.
   (C) water vapor.
   (D) gas vapor.

9. Technician A says that non-metallic linings make the most noise. Technician B says that non-metallic linings will quickly wear rotors and drums. Who is right?

   (A) A only.
   (B) B only.
   (C) Both A & B.
   (D) Neither A nor B.

10. A great variation in the letter markings on a brake shoe lining indicate that the lining will stop well when _____.

    (A) cold
    (B) hot
    (C) hot or cold
    (D) None of the above.

Disc brakes are used on the front and rear axles of many vehicles. This is one of the front disc brake assemblies on a vehicle with four-wheel disc brakes. (Chrysler)

Chapter 12

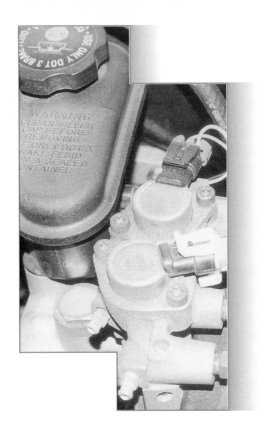

Disc Brake System Components and Operation

After studying this chapter, you will be able to:

- ❑ Identify the components of a disc brake.
- ❑ Identify the two main types of rotors.
- ❑ Identify the three types of front caliper piston arrangements.
- ❑ Identify and explain the operation of fixed and floating calipers.
- ❑ Identify floating caliper mounting methods.
- ❑ Identify and explain the operation of rear calipers.
- ❑ Identify brake pad materials and construction.

Important Terms

Rotor	Caliper	Bushings	Riveting
Machining	Dust boot	Fixed caliper	Bonding
Turning	Floating calipers	Disc brake pads	Anti-rattle clip
Solid rotors	Guide pins	Lining	Anti-squeal compounds
Ventilated rotors	Slider pins	Shoe	Wear indicator
Splash shield			

This chapter is designed to provide a clear understanding of front and rear disc brake operation. Caliper hydraulic components were covered in Chapter 5. This chapter covers all major types of disc brake rotor and caliper designs and related parts. Studying this chapter will prepare you for the service and repair information in Chapter 13.

Disc Brake Rotor

The job of the disc brake *rotor* is to provide a smooth braking surface for the pads to contact. When the stationary pads contact the spinning rotor, the resulting friction slows the rotor and stops wheel rotation. Much of the resulting frictional heat is absorbed and dissipated by the rotor. **Figure 12-1** shows a typical rotor and pad assembly.

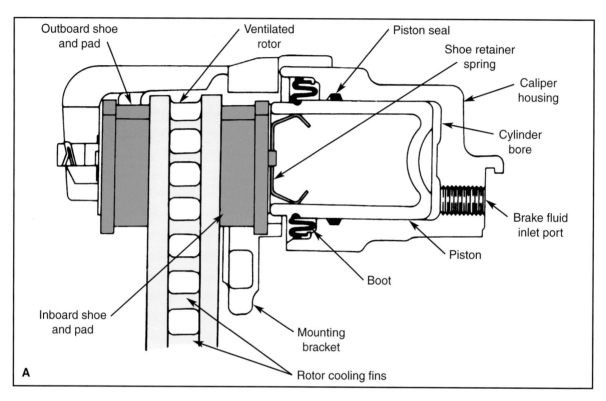

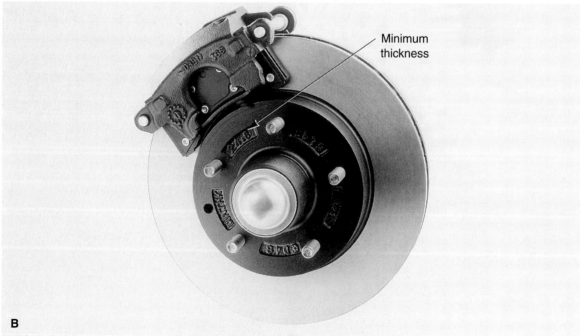

Figure 12-1. *A—Cross-sectional view of a typical single piston, sliding (floating) disc brake assembly. Note the ventilated rotor (disc). B—Front view. Note the minimum thickness numbers stamped on the hub portion of the disc. (General Motors)*

Note: In this and other chapters, we will be referring to *inboard* and *outboard* components. Inboard components are those installed on the inside of the wheel assembly. Outboard components are installed on the outside of the wheel assembly.

Rotor Construction

All disc brake rotors are made of cast iron and are similar in appearance. Rotors are manufactured in many different diameters and thicknesses. Since the linings clamp against the rotor to stop it, both sides are machined to provide a smooth braking surface. The rotor must also be the same thickness throughout its diameter. Variations in thickness will cause pulsation when the brakes are applied. As well as its diameter and thickness, rotors can be either ventilated or solid.

Rotor Diameter

The rotor diameter directly affects the system's braking ability. A larger diameter rotor provides more braking area and exposes more area to the air, which aids in heat dissipation. Rotor diameter is chosen depending on the brake system and the vehicle's intended use. Most disc brake rotors are 9-11" (22.86-27.94 cm) in diameter. Larger rotors are used on sports cars, trucks, and vans.

Rotor Thickness

To absorb heat, rotors must have sufficient metal. The rotor is built with enough metal so the surface finish can be restored by removing a layer of metal. This removal process is called **machining** or **turning.** The minimum thickness of the rotor is usually stamped on the hub or the inner part of the rotor body, **Figure 12-1.** If the rotor is below the minimum thickness before or after machining, it must be replaced.

Solid Rotors

Solid rotors have no openings between the machined surfaces. The rotor is cooled by air passing over the outside surfaces of the rotor. Solid rotors are smaller than ventilated rotors and are used on lighter vehicles.

Ventilated Rotors

Ventilated rotors have internal fins between the two friction surfaces. See **Figure 12-2.** The fins are arranged to create a centrifugal air pump inside the rotor. As the rotor spins, the fins draw air into the center of the rotor and discharge it from the edges. This extra air circulation causes the rotor to give up frictional heat much more rapidly than would otherwise be possible. This type of rotor is always used on larger cars and light trucks and on many smaller vehicles.

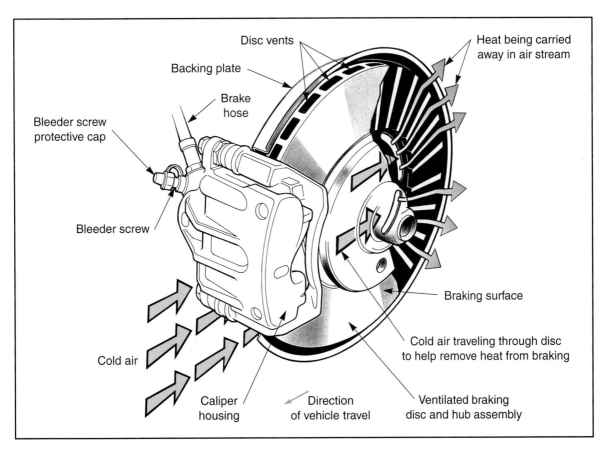

Figure 12-2. *Brake rotors are either solid or ventilated. Note that with the ventilated rotor, air can pass between the braking surfaces to remove heat. This provides a more rapid cooling. (Saab)*

A variation of the ventilated rotor has both internal fins and holes drilled into the braking surfaces from one side to the other. This type of rotor is shown in **Figure 12-3.** These are used on high performance and competition vehicles. They are not installed on most production automobiles because they quickly wear out the brake pads.

Rotor Attachment

Rotors are attached to a wheel hub. The hub is attached directly to the wheel spindle, knuckle, or axle through wheel bearings, discussed in Chapter 16. For proper braking ability, rotors must run true without any scoring, lateral runout, or wobble. For this reason, the rotor must be accurately attached to the hub. Most rotors are separate from the hub, and are held in place by the lug nuts and/or wheel studs when the tire and rim are installed, **Figure 12-4.** Some rotors, especially those used in the front of rear-wheel drive vehicles, are integral with the hub (the hub and rotor are formed as one piece of cast iron). An integral rotor is shown in **Figure 12-5.**

Splash Shield

To keep as much water and debris as possible away from the rotor, most disc brake assemblies have a *splash*

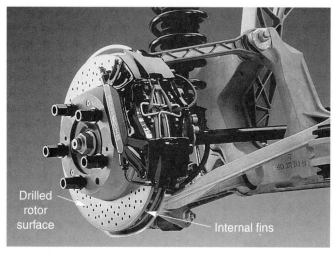

Figure 12-3. *A rear disc brake rotor which incorporates internal fins and cross-drilled (side-to-side) air holes. These air holes aid in cooling the rotor. (Porsche)*

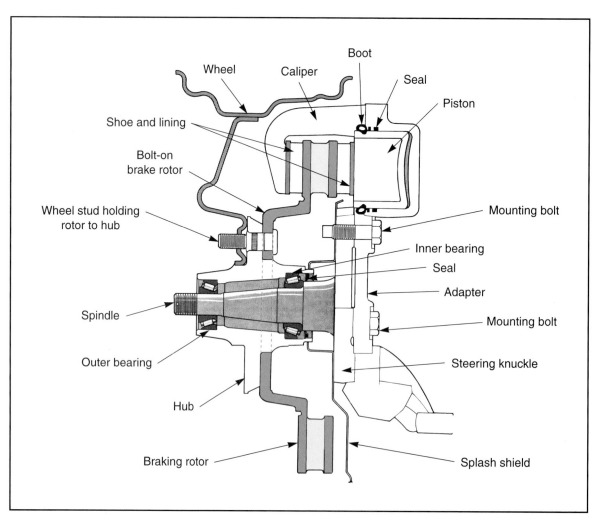

Figure 12-4. *A brake rotor which is secured to the hub with wheel studs. Study the construction. (Dodge)*

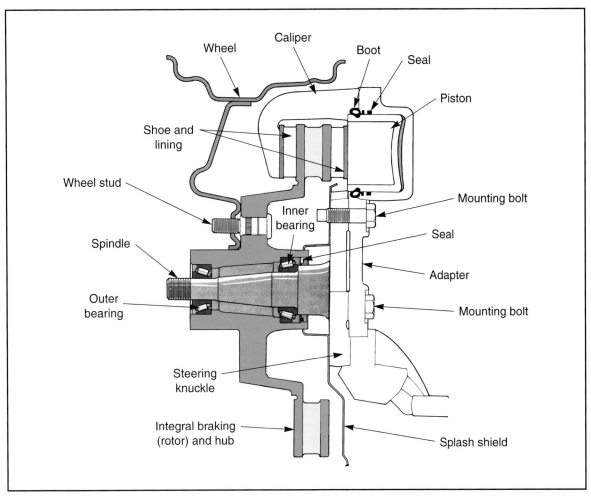

Figure 12-5. *A disc brake assembly that incorporates the brake rotor and hub into an integral (one-piece) unit. Compare this to the sectional rotor/hub assembly shown in* **Figure 12-4.** *(Dodge)*

shield installed over the inner surface of the rotor. The rotor's outside surface is protected by the wheel itself. The splash shield is a sheet metal stamping or a molded plastic assembly that covers part or all of the rotor's inner surface. The splash shield deflects water and debris from the inside rotor surface. The shield is usually fastened to the spindle assembly, and is far enough away from the rotor to allow air to circulate. Some splash shields are designed as air scoops, directing air over the rotor when the vehicle is moving. **Figure 12-6** illustrates the use of a splash shield and wheel for protection.

On some rear disc brake assemblies, the shield also acts as a backing plate to hold the emergency brake shoes. The shield may be held by the caliper assembly on some four-wheel drive vehicles' front brakes.

Front Disc Brake Caliper

The function of the disc brake *caliper* is to provide a housing for the hydraulic components and the brake pads. Calipers are made of cast iron or aluminum. The caliper is usually attached by bolts to the spindle assembly on the

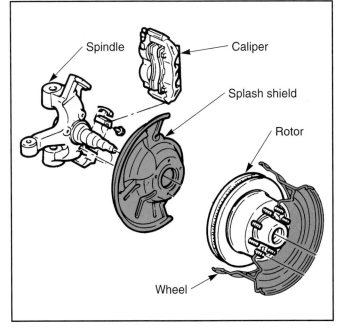

Figure 12-6. *An exploded view of a front brake assembly that uses a splash shield and vehicle wheel to protect the brake assembly from dirt, water, stones, etc. (Ford)*

front wheels or to the axle on rear wheels, **Figure 12-7.** Calipers have a *dust boot* over every piston to keep dust and water away from the inner seals, to retract the piston, and to reduce the possibility of sticking.

Caliper Operation

As discussed in Chapter 5, all calipers operate in roughly the same manner. When the brake pedal is pressed, the master cylinder develops hydraulic pressure in the brake system. This pressure travels to the caliper and causes the piston to move outward. Piston movement forces the pads into contact with the rotor, causing friction which slows the rotor and the vehicle. The piston seal prevents leaks and retracts the piston slightly when hydraulic pressure is removed. This allows the linings to just clear the rotor.

Higher Pedal Effort

Unlike drum brakes, disc brakes have no servo action. Servo action means that wheel rotation helps the brakes to apply. With no servo action, disc brakes require more pedal force to achieve the same braking effect as drum brakes. For this reason, power assist units are used on most modern vehicles.

Pad-to-Rotor Clearance

Modern disc brakes have almost no clearance between the lining and rotor when the brakes are released. Because residual pressure check valves are not used with disc brakes, no static pressure is present to apply the pads. Therefore, even with very low clearances, there is almost no brake drag or wear.

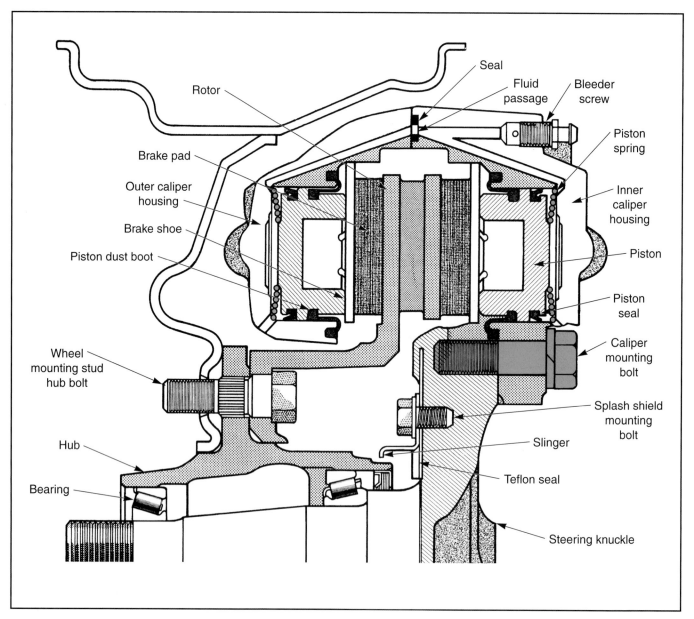

Figure 12-7. *A brake caliper which is attached to the spindle (steering knuckle) with mounting bolts. (Dodge)*

Low-drag calipers have greater clearance. This clearance can be as much as .006″ (.15 mm) when the brakes are released. This is designed to remove all brake drag, which aids in fuel economy. Calipers using this design are always used with quick take-up master cylinders, discussed in Chapter 5.

Brake pad-to-rotor clearance adjustment is automatic. As the pad linings wear, the piston moves further past the seal. The seals can retract the piston enough to eliminate brake drag, but no more. Therefore, as the pads wear, the piston moves out further. Extra brake fluid enters the piston chamber to compensate for piston travel. This is illustrated in **Figure 12-8.**

Caliper Types

There are two methods of mounting disc brake calipers, depending on the hydraulic piston arrangement. Calipers are grouped as floating and fixed calipers. On modern vehicles, floating calipers are the most common. Fixed calipers are used on a few modern high performance vehicles. The two types of calipers are discussed in the next sections.

Floating Calipers

Floating calipers use one or two pistons, located on the same side of the caliper. The caliper can move or "float" back and forth in relation to the rotor. This floating action is made possible by attaching the caliper in a way that allows it to slide on the mounting hardware.

Floating Caliper Construction

Floating calipers are commonly made of cast iron or aluminum. The most common method of allowing the caliper to slide is by the use of mounting bolts that allow the caliper to move over them. This is shown in **Figure 12-9.**

These bolts are called *guide pins* or *slider pins.* To reduce wear of the caliper metal, the guide pins contact *bushings,* sometimes called *sleeves.* The actual movement between the bolts and caliper is through these bushings.

On other vehicles, particularly light trucks and vans, the sliding takes place between machined ways. A *way* is a flat surface designed to allow smooth movement between itself and another component that slides over it. Both the caliper and the spindle mounting surfaces are machined smooth so they can move against each other with a minimum of friction. See **Figure 12-10.**

Caliper movement over the ways is controlled by the use of supports, sometimes called *keys.* These supports may be held in place by a single mounting screw, **Figure 12-11.** A leaf spring, usually referred to as a *clip,* cuts down on caliper rattling. On other calipers, the sliding action is controlled and damped by one-piece combination metal and rubber bushings, **Figure 12-12.** Some disc brake calipers have only one pin, and the side away from the rotor is installed on a machined way, **Figure 12-13.**

One and two piston calipers may be installed on an integral cast extension of the spindle, as shown in **Figure 12-14.** Other vehicles have a special adapter or bracket, which bolts to the spindle. The caliper is then bolted to the adapter. A caliper with a bolt-on adapter is shown in **Figure 12-15.**

Floating Caliper Operation

One or two piston floating calipers operate in the same manner. As the brakes are applied, hydraulic pressure builds up in the cylinder bore behind the piston and their seals. Pressure forces the piston outward. When the piston moves enough to force the inner brake pad into contact with the rotor, resistance to outward piston movement is increased. It now becomes easier for hydraulic pressure to push the caliper backward instead

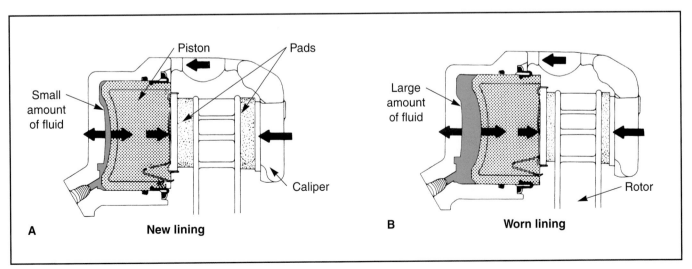

Figure 12-8. *A—Illustrates brake pads that are new. Note the small amount of brake fluid behind the piston. B—The pads have worn down, causing piston to travel farther out in its bore to maintain the correct rotor-to-pad clearance. There is now a larger volume of brake fluid behind the piston. (Delco)*

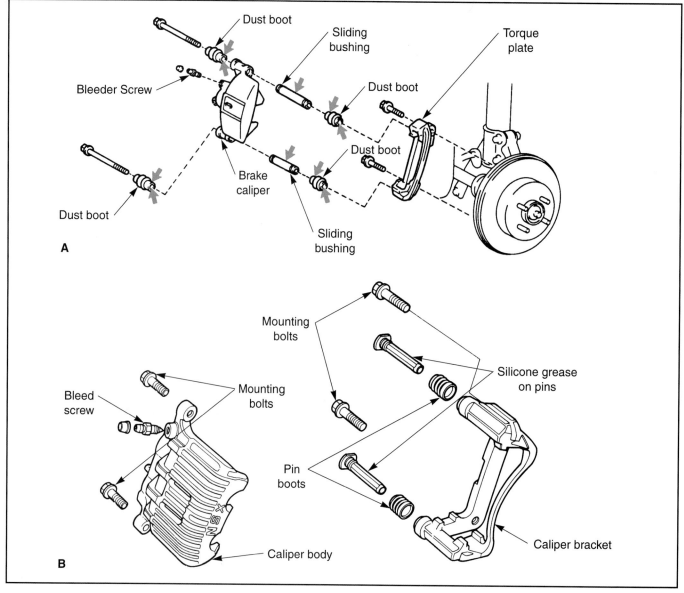

Figure 12-9. *Two styles of floating (sliding) caliper brake assemblies. A—One piston setup. B—Two-piston arrangement. Compare both. (Nissan & Toyota)*

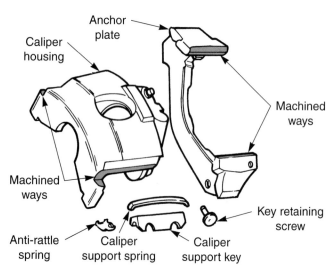

Figure 12-10. *One type of floating caliper and anchor plate showing the machined ways. (Ford)*

of pushing further on the inner pad. The caliper then "floats" inward, bringing the outer brake pad into contact with the rotor. Pad pressure is then equal on both sides, which slows the rotor.

When the brakes are released, the outer seal returns the piston to its normal position. This movement is very slight as it is done by the seal trying to straighten itself, which tends to draw the piston in. As the piston returns to its unapplied position, the pads are moved away from the rotor. The loss of pad pressure allows the pads and caliper to "float" back to the normal resting position. This action is shown in **Figure 12-16.**

Fixed Calipers

The *fixed caliper* is rigidly bolted to the spindle and does not move during braking. Basic fixed caliper operation is different since hydraulic pistons are installed on both sides of the rotor.

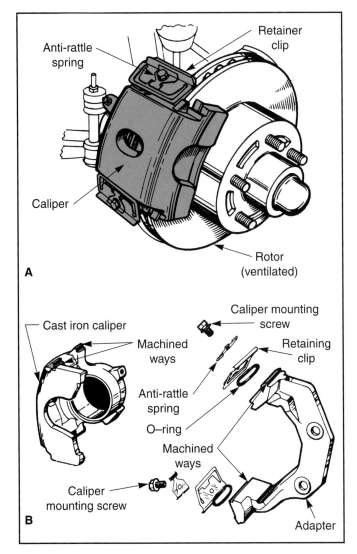

Figure 12-11. *A single piston floating caliper. A—Overall view. B—Exploded view. Note the caliper mounting screws, retaining clips and O-rings, which hold the caliper and adapter plate together at the proper tension. (Chrysler)*

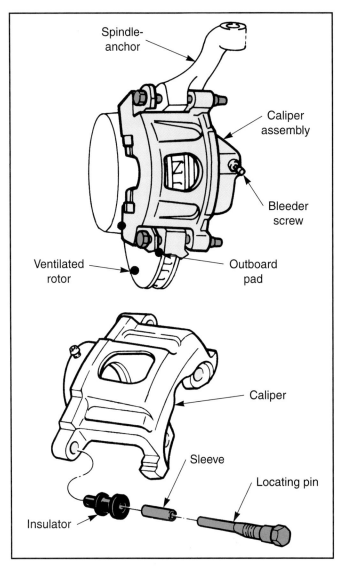

Figure 12-12. *A single-piston, floating caliper assembly which slides (travels) from side-to-side on locating pins, sleeves, and insulators. (Ford)*

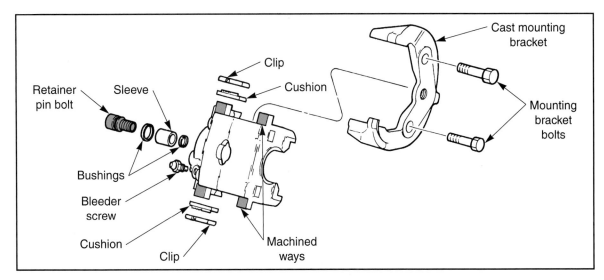

Figure 12-13. *An exploded view of a floating caliper assembly using only one retainer pin bolt and machined ways to secure it to the cast mounting bracket. (TRW Inc.)*

Fixed Caliper Construction

Fixed calipers are usually made of cast iron. Some fixed calipers are a one piece design, **Figure 12-17,** while others are split so they can be disassembled for easier piston removal and honing, **Figure 12-18.** Older fixed caliper designs used two pistons, one on each side. All modern fixed calipers have a total of four pistons, two on each side.

Fixed Caliper Operation

When the brakes are applied, hydraulic pressure enters the caliper assembly, **Figure 12-19.** Pressure flows

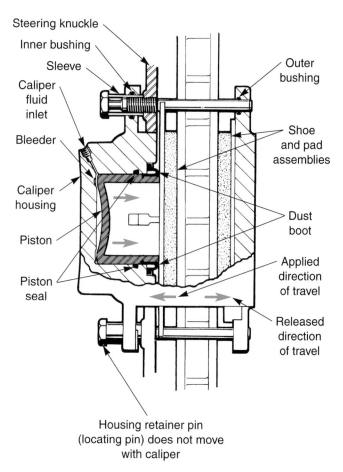

Figure 12-16. *This shows the operation of a single-piston, floating caliper. As the piston is forced outward, hydraulic pressure (which is equal in all directions), forces the caliper to float (slide) inward. This movement applies the outboard pad. (Cadillac)*

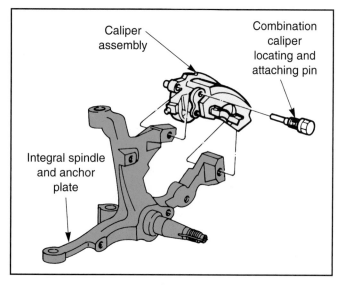

Figure 12-14. *A single-piston, sliding caliper assembly. This caliper bolts onto an integral cast extension of the spindle. (Ford)*

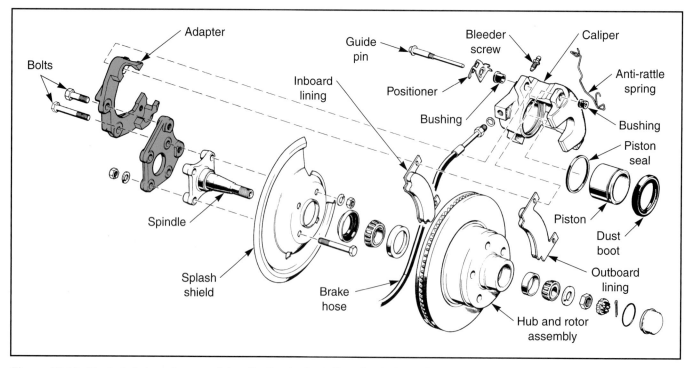

Figure 12-15. *Exploded view of a one-piston, floating brake caliper that bolts to the adapter. (Chrysler)*

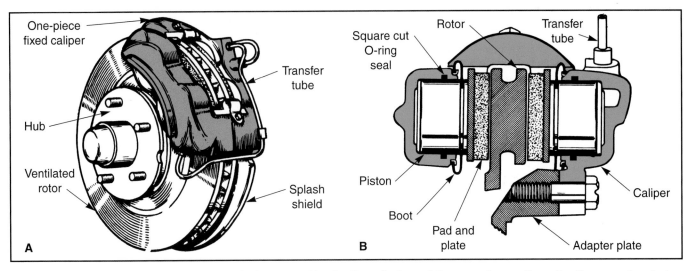

Figure 12-17. *A four-piston, fixed caliper brake assembly. A—Overall view of the one-piece caliper. B—Cross-sectional view showing the internal parts. (Kelsey-Hayes)*

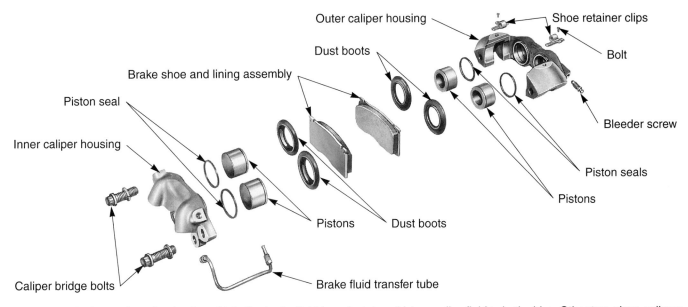

Figure 12-18. *A two-piece fixed caliper. Note the brake fluid transfer tube which supplies fluid to both sides. Other two-piece calipers use internal fluid passages. (Plymouth)*

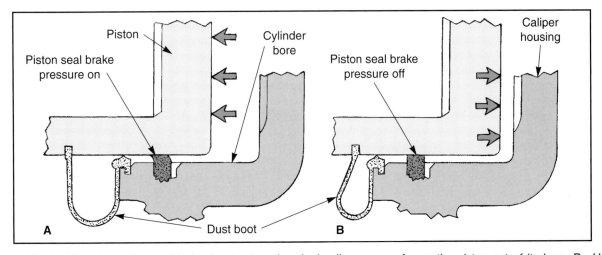

Figure 12-19. *A—This shows the seal being bent outward as hydraulic pressure forces the piston out of its bore. B—Here the square-cut seal is in a normal resting position and shape after the piston has returned. (Plymouth)*

through internal passages or external transfer tubes, pressurizing all four pistons at the same time. The pistons move outward, forcing the brake pads into contact with the spinning rotor, slowing or completely stopping it. When the brakes are released, hydraulic pressure decreases and the pistons are pulled clear of the rotor by the action of the piston seals.

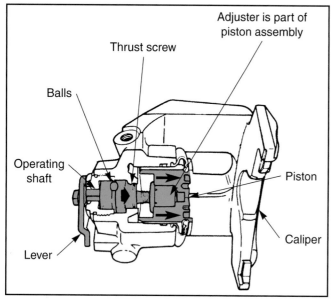

Figure 12-20. A cutaway view of a ball-and-ramp parking brake assembly and clamp unit.

Rear Disc Brake Caliper

Rear disc brake calipers do the same job and have roughly the same components as front brake calipers. However, they are usually much smaller since the rear brakes perform less of the braking work than the front brakes. Virtually all rear brake calipers are single piston floating designs.

Rear disc brakes must have a provision for a parking brake. This brake can be an internal device which applies the pads, **Figure 12-20,** or a separate set of small shoes, **Figure 12-21.**

Rear Caliper Construction

There are three caliper designs used on rear brakes. They are the screw type, which uses an integrated screw to actuate the parking brake. The second is the ball-and-ramp, which contains three steel balls between a shaft and screw mechanism. The third is the cam-type that uses a lever operated cam. Information on each of these can also be found in Chapters 6 and 18.

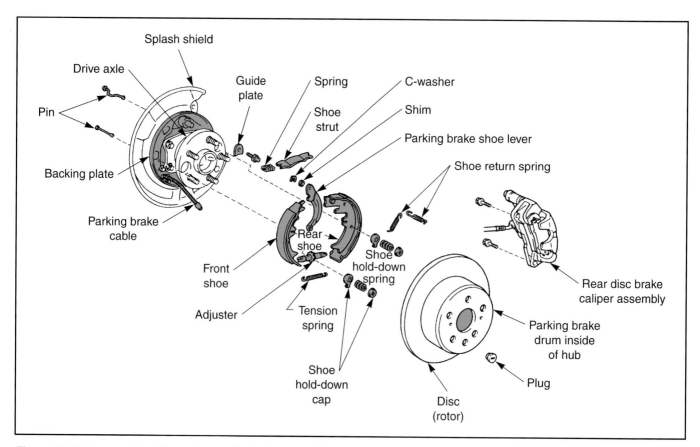

Figure 12-21. An exploded view of a rear disc brake assembly which uses a set of drum brakes for parking. The parking brake cable will force the brake shoes into contact with the drum section located inside the rotor hub. (Toyota)

Disc Brake Pads

The **disc brake pads** are designed to create friction when forced into contact with the spinning rotor during braking. The disc brake pad consists of a **lining** (the actual brake material) and a **shoe** (the metal lining support), **Figure 12-22.** The lining is designed to give the best coefficient of friction with acceptable wear, smoothness, and quiet operation. The lining absorbs heat and transfers it to the shoe assembly. The shoe provides a rigid area for the piston(s) to push against. Disc brake pads are made in many shapes and sizes.

Figure 12-22. *Brake pads come in all sizes, shapes, and materials. (Jack Klasey)*

Lining Construction

Linings are made of heat resistant materials, as discussed in Chapter 11. The lining is constructed by mixing the various temperature resistant and bonding materials. This mixture is carefully controlled to create the proper coefficient of friction. The mixture is then placed in a mold, compressed, and heated. The finished pad is machined to give it a smooth surface. Some pads are grooved to aid in self-cleaning, air transfer for cooling, and noise reduction. The groove also acts as a wear indicator. When the pad is worn until the groove is gone, the lining should be replaced.

Shoe Construction

The shoes to which the linings are attached are made of flat, heavy gauge steel, punched and formed to match the caliper mounting and lining. Some shoes have alignment dowels stamped into the metal. Anti-rattle springs or clips which reduce brake squeal may be riveted or welded to the shoe. Some clips are snapped on the shoe, and must be installed or transferred from the old shoes when new shoes are installed.

Lining-to-Shoe Attachment

The lining is attached to the shoe in one of two ways. One method is **riveting.** Brass rivets are used to attach the lining to the shoe. The lining and shoe are made with a number of matching holes, with the rivets installed through the holes. The rivets are then secured tightly using the proper riveting tools. Brass rivets are used since they will not cause as much damage if the lining wears down until they contact the rotor. A riveted shoe can have anywhere from two to eight rivets. A riveted brake pad is shown in **Figure 12-23A.**

The second method of lining attachment is called bonding. **Bonding** is a process that glues the lining to the shoe. A special high temperature cement or epoxy resin is placed between the lining and shoe. Then the lining and shoe are pressed together until the cement is thoroughly dried or cured. Some cements require heat to properly dry and hold. A typical bonded lining is illustrated in **Figure 12-23B.**

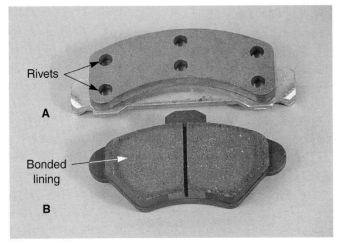

Figure 12-23. *A—Brake lining-to-shoe attachment using rivets. B—Bonded brake lining-to-shoe. (Bendix)*

Noise Reduction Devices

Common brake noises are squeaks and squeals from brake shoe and rotor contact, and clicks and knocks from loose pad to caliper contact. Various clips, springs, and other devices are attached to the brake pad shoe to reduce noise. When the pad is installed on the vehicle, these form a tension fit with the pad and other parts of the brake assembly. The springs and clips absorb some of the vibrations and noise from the pad as it rubs against the rotor.

Other designs use a piece of flexible heat resistant material, sometimes called an insulator, installed on the back of the pad. This material also absorbs noise and vibration. When used, they are installed on the outboard pad, against the caliper housing. Some designs have insulators on both pads, while others use a flat metal plate called an **anti-rattle clip,** plate, or shim. **Figure 12-24** shows different noise reducing clips, springs, and shims.

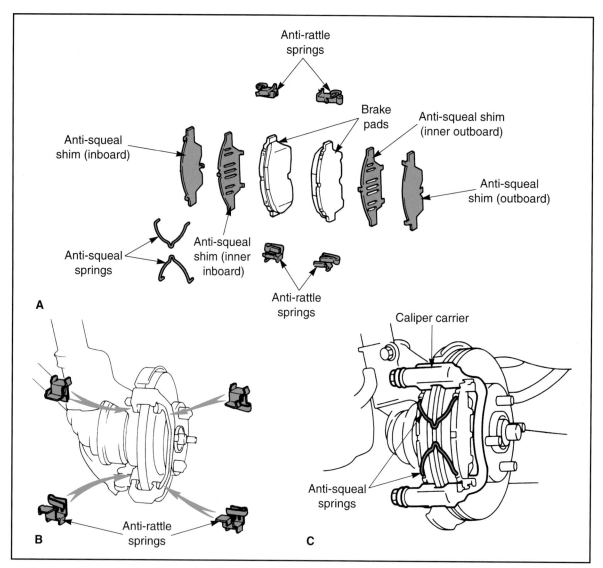

Figure 12-24. *An assortment of noise reducing items. A—Shims and springs. B—Four anti-rattle springs. C—Anti-squeal springs in their mounting position. (Chevrolet)*

Anti-Squeal Compounds

Companies that make gaskets and adhesives for automotive use also produce chemically-based **anti-squeal compounds,** **Figure 12-25.** Most are designed to be installed on the back of a brake pad before installation, in

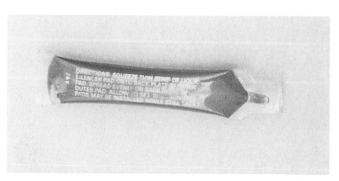

Figure 12-25. *One type of anti-squeak compound. This type is meant to be applied to the back of the pad.*

the same location as an insulator. A few graphite-based compounds are designed to be placed on the friction material itself. Read the directions on the compound before use.

Lining Wear Warning Devices

A small piece of flat spring steel called a **wear indicator** is frequently attached to one of the shoes. As the pad linings wear down to their minimum thickness, the end of the spring contacts the spinning rotor. A high pitched squeal is produced whenever the brakes are *not applied*, alerting the driver that it is time for brake service. Operation of a typical wear indicator is depicted in **Figure 12-26.**

Another type of wear indicator consists of an electrical lead attached next to or inside the pad, **Figure 12-27.** When the ignition switch is in the on position, an electrical circuit exists through a dashboard light to this sensor. If the pads wear down far enough, the electrical lead contacts the rotor when the brakes are applied. This grounds the lead and completes the circuit, illuminating the dashboard light.

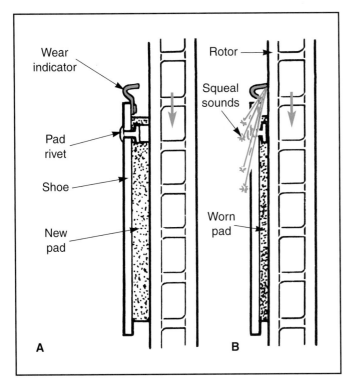

Figure 12-26. *A spring steel lining wear indicator. A—Pads are new; the wear indicator is not able to contact the rotor. B—Pads have worn down permitting the wear indicator to contact the rotor. A high pitched squeal sound is being produced. This alerts the driver that brake service is necessary. (Chevrolet)*

Summary

Disc brakes use a rotor or disc that provides a braking surface for the pads. The rotor also aids in cooling. Rotors are round and come in many different diameters and thickness. They are constructed from cast iron and are either solid or ventilated with fins between braking surfaces. The rotor must be thick enough to absorb heat, and flat enough to provide for smooth stops. The rotor is protected by a splash shield.

Calipers are constructed of cast iron or aluminum. They are either floating or fixed. The floating type can move in relation to the spindle. Fixed types are solidly bolted onto the spindle. Rubber dust boots protect the pistons from water and dirt.

Floating calipers have one or two apply pistons. Modern fixed calipers are fixed types. The pistons are moved outward by hydraulic pressure. The piston seals are used to return the piston to its unapplied position when brake pressure is released.

Brake pad linings are made from various heat resistant materials. They are bonded or riveted to the steel shoe. The shoe is a backing plate for the lining and a rigid meeting place for the apply piston(s).

To reduce brake noise, shoes are sometimes equipped with special insulators. Some brake shoes also incorporate warning devices to alert the driver that brake work is necessary.

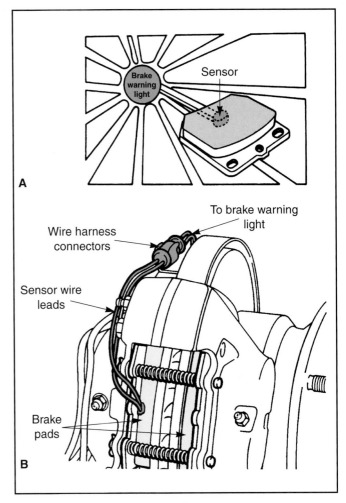

Figure 12-27. *An electric type pad wear sensor. A—Pad with the sensor inside the lining and a dashboard warning light. B—Brake pads in position and sensor wire lead harness with connector. When the worn pads allow the sensor to contact the rotor, the electrical circuit is complete. The dashboard light will come on alerting the driver. (Bendix)*

Review Questions—Chapter 12

Please do not write in this text. Write your answers on a separate sheet of paper

1. A ventilated rotor has internal _____ to remove heat.

2. Solid rotors are cooled by air passing over the _____ surfaces of the rotor.

3. Rotors are attached to the wheel _____, which is attached to the spindle through _____.

4. On some vehicles, an extra _____ holds the caliper to the spindle.

5. Removing brake drag improves _____.

6. A caliper which moves in relation to the spindle mounting is called a _____ caliper.

7. All modern fixed calipers have _____ pistons, _____ on each side.

8. All rear disc brake calipers are equipped with a _____ brake.

9. What two parts make up a disc brake pad?

10. The two disc brake pad parts in Question 9 are attached to each other by what two methods?

ASE Certification-Type Questions

1. Technician A says that all rotors are cast as an integral part of the hub. Technician B says that all rotors spin with the wheel. Who is right?
   (A) A only.
   (B) B only.
   (C) Both A & B.
   (D) Neither A nor B.

2. On a ventilated rotor, the fins are used for _____.
   (A) cooling.
   (B) cleaning.
   (C) increasing airflow.
   (D) Both A & C.

3. Technician A says that every floating caliper has at least one piston. Technician B says that every piston has a dust boot. Who is right?
   (A) A only.
   (B) B only.
   (C) Both A & B.
   (D) Neither A nor B.

4. Splash shields are installed on the _____ side of the rotor.
   (A) inboard
   (B) outboard
   (C) inboard and outboard
   (D) None of the above.

5. Most disc brake calipers are made of _____.
   (A) aluminum
   (B) cast iron
   (C) steel
   (D) Both A & B.

6. Technician A says that a floating caliper has one or two pistons. Technician B says that a fixed caliper has one or two pistons. Who is right?
   (A) A only.
   (B) B only.
   (C) Both A & B.
   (D) Neither A nor B.

7. The piston is forced outward in the cylinder bore by _____.
   (A) rotor movement
   (B) seal twisting
   (C) hydraulic pressure.
   (D) servo action

8. Four-piston calipers are usually constructed in _____ sections.
   (A) two
   (B) three
   (C) four
   (D) None of the above.

9. Technician A says that on some disc brakes, the linings are constantly touching the rotor. Technician B says that piston adjustment is automatic on all disc brakes. Who is right?
   (A) A only.
   (B) B only.
   (C) Both A & B.
   (D) Neither A nor B.

10. Guide pins are used on _____ disc brake calipers.
    (A) single piston floating
    (B) dual piston floating
    (C) four piston fixed
    (D) Both A & B.

11. Technician A says that the parking brake on a rear caliper system consists of small drum brakes. Technician B says that the parking brake on a rear caliper system uses the disc brake pads to clamp the wheel. Who is right?
    (A) A only.
    (B) B only.
    (C) Both A & B.
    (D) Neither A nor B.

12. By which of the following methods is a disc brake pad lining manufactured?
    (A) Cutting the lining material from a large block.
    (B) Placing a mixture in a mold, then freezing it.
    (C) Placing a mixture in a mold, then heating it.
    (D) Building up layers of material in a form.

13. The bonded disc brake pad uses _____ to hold the shoe to the lining.
    (A) rivets
    (B) cement
    (C) screws
    (D) Both A or B.

14. Technician A says that clips used on disc brake pads are anti-squeal devices. Technician B says that clips used on disc brake pads are wear reducing devices. Who is right?

(A) A only.

(B) B only.

(C) Both A & B.

(D) Neither A nor B.

15. The pad wear indicator can warn the driver of excessive wear by _____.

(A) noise

(B) a light

(C) smell

(D) Both A or B.

Measuring rotor thickness with a disc brake micrometer. This can be done on or off the car, but in all cases, before any estimate is given or parts ordered.

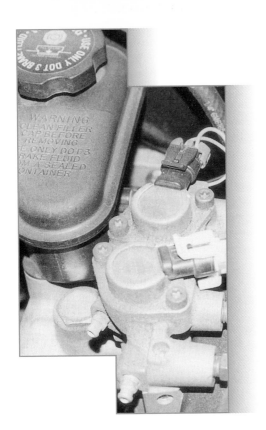

Chapter 13

Disc Brake Service

After studying this chapter, you will be able to:
- ❑ Identify common disc brake problems.
- ❑ Diagnose disc brake problems.
- ❑ Remove and replace front disc brake calipers.
- ❑ Remove and replace rear disc brake calipers.
- ❑ Remove and replace disc brake pads.
- ❑ Refinish disc brake rotors.

Important Terms

Noise	Grooving	Silencer band	Hot spots
Pulsation	Heat-checking	Collar	Swirl grinding
Loaded calipers	Bluing	Rough cut	Non-directional finish
Staking	Arbor	Finish cut	Bedding-in
Scoring	Adapters	Oscillating	

In Chapter 6, you studied the service of disc brake caliper hydraulic systems. In this chapter, you will learn how to diagnose and service the disc brake system friction components. Pad and rotor service is very similar for every kind of disc brake system, varying only according to size, mounting method, and whether the caliper has a provision for a parking brake. Variations, where they occur, will be noted in the text.

Common Disc Brake Problems

The most common disc brake problems are noise and pulsation. Common brake **noise** includes squeaks and squeals from brake shoe and rotor contact. Disc brakes commonly produce high pitched squeals or squeaks when the brakes are applied. This is often caused by glazed or worn pads, but may be the result of polished (overly smooth) rotors, excessively hard pad material, or the wear indicator contacting the rotor. A grinding or rubbing noise when the brakes are applied may indicate the pad linings are worn and the metal shoes are contacting the rotor. Clicks and knocks are produced by loose pad-to-caliper contact.

Pulsation is a type of vibration. It is usually felt as a side-to-side motion in the steering wheel, or an up-and-down motion in the brake pedal, or both, when the brakes are applied. Pulsation is usually caused by variations in the rotor's surface. Long use or excessive heat can cause the rotor to develop thickness variations, high spots, or warping. Pulsation is also caused by hard spots (places in the rotor which have become overheated and lost their original finish).

If heavier than normal pedal pressure is needed for braking, this may be caused by worn or excessively hard brake pads. Another cause of a hard pedal is an overheated brake system. Overheated rotors and pads have a poor coefficient of friction, meaning the pedal must be applied much harder to have the same braking effect. However, before assuming the cause of a hard pedal is the disc brakes, check the brake hydraulic system and any power assist units first.

A spongy pedal can be caused by caliper and mounting hardware flexing. This is usually not a problem unless the vehicle is designed to operate with high hydraulic system pressures. Extreme caliper or bracket wear, cracks at the mounting points, or loose bolts can also cause a spongy pedal.

Pads and rotors that wear out ahead of time are often caused by driver habits, or severe usage, such as mountain driving or trailer towing. If the pads are wearing unevenly, check for a sticking piston or slide pins, misaligned caliper, or flexing. **Figure 13-1** lists common disc brake problems and their causes.

Caliper and Pad Service

While calipers and brake pads are similar in basic components and operation, there are many differences in design. These differences are in the areas of mounting, noise reduction clips and insulators, and fasteners. These are addressed where applicable.

Disc Brake Problems (All)	
Condition	**Possible Causes**
High-pitch squeal only when brakes applied	Glazed linings or polished rotor.
High-pitch squeal only when brakes released	Pad wear sensor contacting rotor, replace pads.
High-pitch squeal at all times	Splash shield contacting rotor.
Metallic grinding when brakes applied	Pads worn down to metal. Sticking caliper.
One pad worn more than the other	Caliper piston or slides sticking. Caliper misaligned.
Brake pedal pulsates	Excessive rotor runout. Normal ABS operation.
Excessive pedal effort	Glazed linings. If linings OK, problem is in power booster system.
Rear Disc Brake Problems	
Vehicle rolls with parking brake applied	Rear caliper pistons sticking. Parking brake cable misadjusted.
Vehicle rolls when in gear with parking brake applied	If resistance felt when moving, rear brakes OK. If resistance is not felt, rear caliper pistons sticking or parking brake cable misadjusted.

Figure 13-1. *Troubleshooting chart listing problems that can occur with disc brakes.*

Performing Disc Brake Service on ABS/TCS Equipped Vehicles

Many of the most common brake service procedures, such as pad replacement, rotor service, and wheel bearing replacement, are not affected by the presence of anti-lock brakes (ABS) or traction controls (TCS). If the lining or friction service replacement procedures involve the wheel speed sensors, treat them gently, and recheck the gap where applicable. Do not drop or hammer on the sensor rings, or use them to pry on other components. Do not replace any system hoses with standard (non-ABS) hoses. The higher pressures in these systems can rupture a standard brake hose.

Checking Caliper and Pad Condition

Calipers are usually trouble free. However, they can occasionally develop external brake fluid leaks or sticking apply pistons. Hydraulic system problems were discussed in Chapter 6. If there are no hydraulic system defects, the caliper is usually not a source of problems. However, the caliper should be checked for wear, cracks (especially at the mounting points), and torn dust boots.

Observing the pad thickness and rotor condition is the quickest way to determine whether the pads should be replaced. Uneven wear between the inboard and outboard pads is a sign of a sticking caliper piston or slide. Uneven wear on one or both pads indicates that the caliper is misaligned with the rotor.

Note in **Figure 13-2** that pad thickness can be visually checked by observing the pad lining thickness through openings in the caliper. However, this provides only a general idea of pad thickness. If the pads have a wear sensor, you can check the amount of clearance between the sensor and the rotor. If there is any doubt as to pad condition, remove the caliper and check by measuring the pads against specified minimum thickness. Note the condition of the rotor. If the rotor is scored or appears to have been overheated, the pads need replacement. Also, make sure the wheel turns freely. If the wheel will not turn easily, the caliper piston may be sticking, or there may be a problem with the wheel bearings.

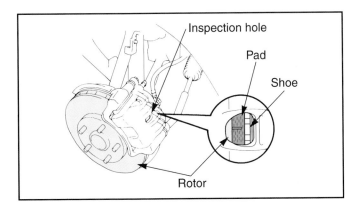

Figure 13-2. *Brake pad thickness can be checked on this assembly by looking through the inspection hole in the caliper. (Toyota)*

Front Caliper Removal and Pad Replacement

 Note: The following procedure is for replacing the pads without overhauling the caliper. If the caliper must be overhauled, refer to Chapter 6.

To remove the caliper, first raise and support the vehicle in a safe manner. If a lift is not available, support the vehicle at the frame with jackstands. Mark the wheel stud closest to the tire valve stem with crayon to ensure the tire is reinstalled in the same position. Then remove the tire and rim.

 Warning: Before proceeding, carefully check the temperature of the hub and rotor assembly. If the assembly feels hot, allow it to cool, or use gloves to protect yourself from burns.

If the pads are going to be replaced and no other service is needed, it is not necessary to remove the caliper hose. If the caliper uses an electrical pad wear sensor, disconnect the sensor electrical connector, **Figure 13-3.**

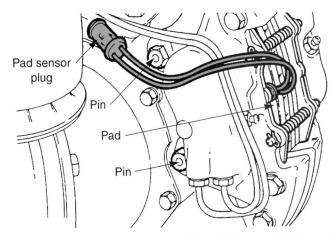

Figure 13-3. *A disc brake assembly which uses electric pad wear sensors. (Land Rover)*

Use a small prybar, C-clamp, large adjustable pliers, or other tool to lightly push the pads away from the rotor, **Figure 13-4.** This will make caliper removal easier. If the rotors and pads are badly grooved, the pads may need to be moved back a considerable distance before they can be removed.

On vehicles equipped with ABS systems, some manufacturers recommend opening the bleeder screw to allow fluid to escape, rather than pushing it back into the hydraulic actuator and master cylinder. This minimizes the chance of contamination, which could cause problems. Check the service manual before proceeding.

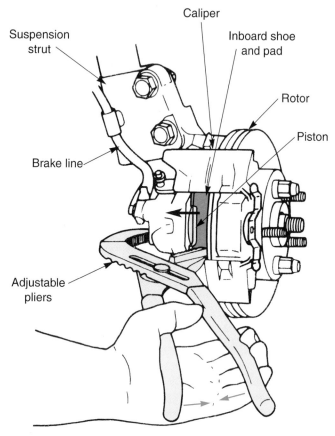

Figure 13-4. *Using pliers to lightly force the inboard pad and piston away from the rotor. This allows for easier caliper removal. (Pontiac)*

 Caution: If the pads will be reused, do not damage them by prying. It is usually possible to pry on the metal shoe portion of the pads to move them. On some calipers, the pads can be loosened by pushing the caliper housing forward with a large C-clamp.

Remove the fasteners holding the caliper to the spindle. There are several methods of attaching the caliper to the spindle:

❑ On some vehicles, the caliper is held by bolts which thread into the caliper and slide on the spindle through steel sleeves or bushings, **Figure 13-5.** These bolts can be standard capscrews, or may have round heads with an internal Allen or Torx® fitting.

❑ Other brake systems use bolts which thread onto the spindle assembly. The caliper slides on these bolts through hardened steel sleeves and/or bushings. See **Figure 13-6.**

❑ On some vehicles, the caliper is held in place by rubber or metal clips or bushings. The bushings, also called support keys, are in turn held by screws or bolts, **Figure 13-7.** The fasteners can be removed and the clips or bushings lightly tapped out to remove the caliper. The fasteners shown in **Figure 13-8** are removed and the clips are lifted off to free the caliper.

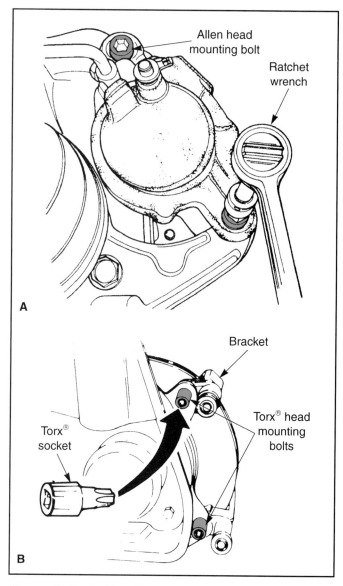

Figure 13-5. *A—This illustrates Allen head mounting bolts used on a brake caliper. B—Torx® head bolts and the bit needed for removal and installation. (General Motors)*

Figure 13-6. *A brake caliper that uses caliper guide pin bolts that, when installed, thread into the spindle.*

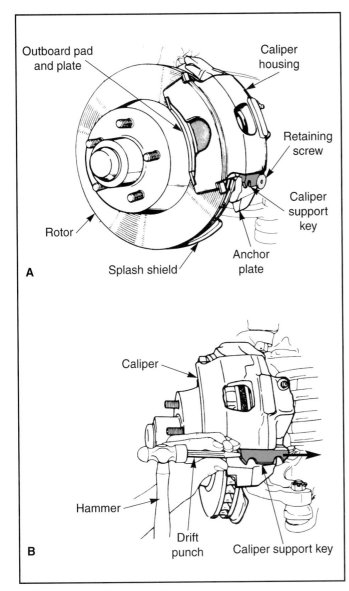

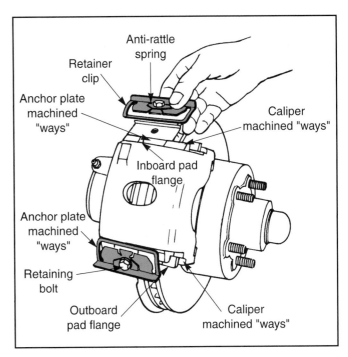

Figure 13-8. *This caliper can be freed from the anchor plate by removing the retainer clips. (FMC)*

Figure 13-7. *A—Caliper which is held in place with a caliper support key and retaining screw. B—After the retaining screw has been removed, the support key is driven from its slot with a punch and hammer. (Bendix)*

❑ On a few vehicles, an adapter bracket must be removed before the caliper can be removed. On other vehicles, the caliper and bracket can be easily removed as a unit, and separated later.

After the caliper fasteners are removed, lift the caliper from the rotor. In some cases, it may be necessary to twist the caliper slightly for removal. On some systems, the pads will remain with the rotor, **Figure 13-9,** while on others, the pads will come away from the rotor with the caliper assembly, **Figure 13-10.** If the caliper will not be over-hauled, place it on the frame, or use a piece of wire to attach it to the vehicle.

 Caution: Do not allow the caliper to hang by the hydraulic hose. Hose damage may result. Use a wire hook, Figure 13-11.

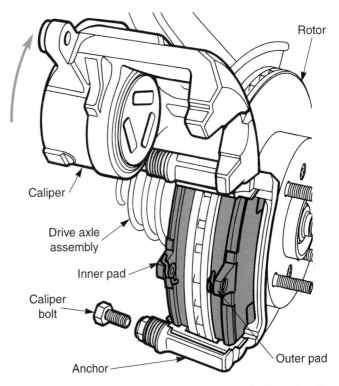

Figure 13-9. *Caliper assembly being removed. Note that the brake pads have remained with the rotor and anchor unit. (Sterling)*

Pad Removal

Remove the pads from the caliper, or from around the rotor as necessary. Some calipers have outer pads that are held by clips or have been clinched (metal tabs on the outer pad shoe clamped by force against the caliper). In

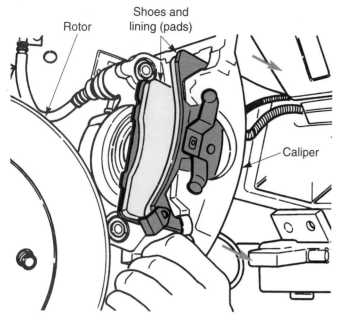

Figure 13-10. *A caliper assembly being removed. Note the pads have come off with the caliper. (Chrysler)*

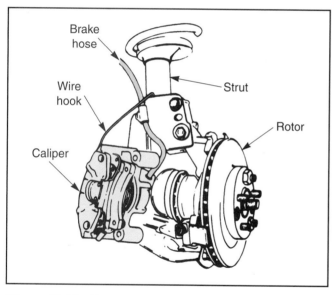

Figure 13-11. *Always support the caliper with a wire hook. Never let it hang by the brake hose. Hose damage may result. (Chevrolet)*

these cases, the outer pads must be pried to release them from the caliper. Most pads will come off easily after the caliper is removed from the rotor.

If necessary, check the pad thickness with a micrometer or caliper, and compare against service manual specifications. In most cases, however, the pads are obviously worn enough to require replacement.

With the pads out of the way, carefully check the caliper for damage or leaking. Also check the rotor as explained later in this chapter. If the caliper shows any signs of leakage or damage, it should be overhauled or replaced. Many technicians prefer to install *loaded calipers,* which are new calipers with the pads already installed.

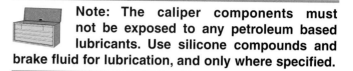

Note: The caliper components must not be exposed to any petroleum based lubricants. Use silicone compounds and brake fluid for lubrication, and only where specified.

Installing New Pads on the Front Calipers

Before installing the new pads, loosen the bleeder screw and use a large C-clamp to push the caliper piston into its bore. This is shown in **Figure 13-12.** Place the old inner pad, a metal bar, or a block of wood on the piston surface. This will minimize the chance of piston damage from the C-clamp. Place a pan under the caliper to catch brake fluid from the bleeder.

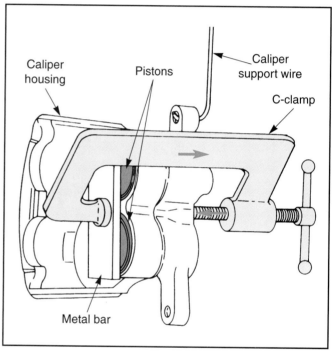

Figure 13-12. *Using a C-clamp and a metal bar to force the caliper pistons back into their bores. This will provide the necessary pad-to-rotor clearance when reinstalling the caliper. (Pontiac)*

As soon as the piston is seated, stop turning the C-clamp and tighten the bleeder screw. Continuing to turn the C-clamp after the piston is seated may damage the caliper. After the bleeder screw is tight, remove the C-clamp.

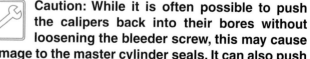

Caution: While it is often possible to push the calipers back into their bores without loosening the bleeder screw, this may cause damage to the master cylinder seals. It can also push dirt and any debris into and through the hydraulic actuator on ABS/TCS equipped vehicles. If this dirt becomes trapped in the valves, it can cause the hydraulic actuator to malfunction.

Compare the new pads with the old ones to ensure they are correct. Be sure to note whether the mounting surfaces (metal shoe) portion is correct, and the linings are not too thick to prevent the caliper from fitting over the rotor. After ensuring the pads are correct, install any clips and anti-squeal insulators on the new pads. If the vehicle has a separate pad wear sensor, install it in the proper position, **Figure 13-13**. If desired, place *anti-squeal compound* on the pad shoes where they contact the caliper. See **Figure 13-14**.

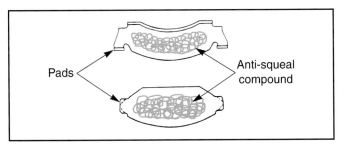

Figure 13-13. *Installing one type of outboard brake shoe and lining. Note the retaining clip and audible wear sensor position. (Chrysler)*

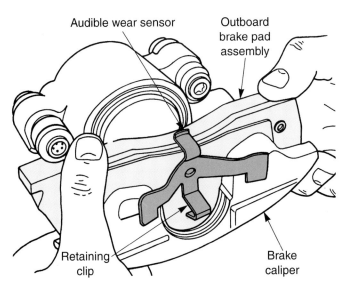

Figure 13-14. *Two different styles of brake pad shoes showing the correct placement of anti-squeal compound. Do not allow this compound to touch the friction material. (Bendix)*

 Note: Some technicians prefer to slightly bevel (grind) the edges of the pads to reduce the likelihood of brake squealing. Some newer pads come with beveled edges from the manufacturer.

Place high temperature lubricant on the parts of the caliper that move in relation to the spindle assembly. Typical lubrication points would be the sliding surfaces of the caliper and spindle, guide pins, sleeves and bushings, and any related moving parts. Do not use motor oil, wheel bearing grease, or chassis lubricant. The high temperatures of the brake system will cause it to burn off almost immediately, leaving a sticky residue which will interfere with brake operation.

 Caution: Do not allow non-graphite anti-squeal compounds or lubricants to touch the pad friction surfaces or rotor.

Caliper Reinstallation

Caliper reinstallation is relatively simple, but must be done correctly if the brakes are to operate properly. Install the rotor if it was removed. Before reinstalling the caliper, check the spindle assembly, splash shield, and other related parts for damage. Place high temperature lube on any sliding surfaces of the spindle or adapter bracket.

 Note: If the calipers were removed from the vehicle, be careful not to switch calipers between the left and right sides of the vehicle. On many vehicles, reversing the calipers will place the bleeder screws in a position that makes it impossible to completely remove all air from the caliper.

Place the caliper over the rotor, **Figure 13-15.** If the caliper does not slide easily over the rotor, do not force it into place. Remove it and check to see if the piston is fully retracted and the pads are correct. Some vehicles require that one end of the caliper be installed first. On other vehicles, the caliper must enter straight into the mounting bracket. After the caliper is in position, install the attaching hardware and the brake hose if necessary.

 Caution: Start all fasteners by hand before using a hand or air tool to tighten.

After the fasteners are tight, ensure the rotor can turn freely with the caliper installed. While turning the rotor, listen for scraping noises that indicate the caliper or another stationary part is contacting the rotor. Bleed the brakes if necessary, then reinstall the tire and rim.

Staking the Brake Pads

Some brake caliper designs require the technician to pinch a portion of the outer pad shoe against the caliper. This is referred to as *clinching* or **staking** the brake pads. Staking must be performed to prevent the outer pad from moving. This procedure is done using a large pair of adjustable pliers (channel lock) or a hammer. Have an assistant pump the brake pedal to bring the pads against the rotor. Then, while the assistant presses the brake pedal, stake the outer pads to the caliper, **Figure 13-16.**

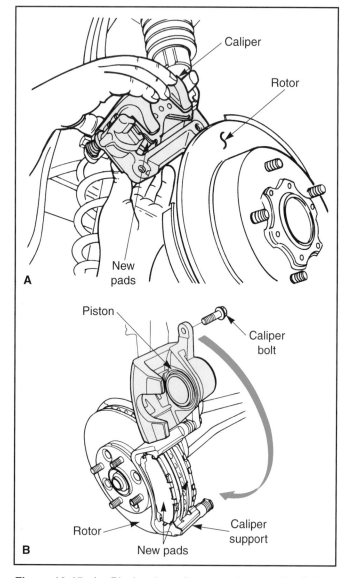

A—Placing the caliper over the rotor.

Figure 13-15. *A—Placing the caliper over the rotor. B—Caliper is being installed over pads and rotor after one side (end) has been attached to the caliper support. (Chevrolet & Honda)*

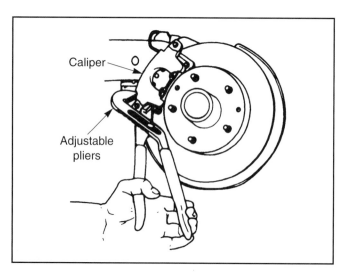

Figure 13-16. *Some brake pads require that metal tabs on the pad shoe be staked to the caliper. Use a pair of adjustable pliers to stake the pads. (Chevrolet)*

Note: Some technicians stake pads by wedging a small chisel between the bottom edge of the outer pad and the hub, then staking the pad using a second chisel and hammer. This practice is not recommended.

Rear Caliper and Pad Removal

Rear disc brakes with the parking brake assembly built in require slightly different procedures, **Figure 13-17.** In some cases, special tools are needed to retract the piston into the bore. When servicing pads and rotors on a rear disc brake assembly, the first step is to disconnect the parking brake cable from the caliper. Once the parking brake cable is disconnected, loosen the caliper bolts and carefully remove the caliper. You may need to remove other components before caliper removal can be performed.

Three common parking brakes are used with rear wheel disc brakes. They are the screw, ball and ramp, and cam. Special procedures for each type is discussed in the following paragraphs. See Chapter 6 for additional information on caliper service.

Note: Always service one rear caliper at a time.

Screw Type Caliper Service

On disc brake calipers with screw type parking brake mechanisms, the cable actuating lever is connected to an actuating or **high lead screw.** The screw passes through a splined nut cast in the piston. When the cable moves the lever, the screw rotates and moves the piston outward, tightening the pads against the disc.

When servicing this type of caliper, there are two ways to compress the piston in the caliper. The first is to turn the piston back into the caliper using a spanner wrench or special tool once the caliper is removed, **Figure 13-18.** The second method allows you to push the piston back in, similar to front calipers. The first step is to remove the parking brake actuator lever from the caliper after the cable has been removed. *Do not* turn the caliper high lead screw. Once this lever has been removed, compressing the piston in is much easier, **Figure 13-19.** Once the pads are removed, carefully push the piston back into the caliper. The piston can usually be pushed into the caliper without damage.

When installing the pads, make sure the "D" shaped locator lines up with the "D" shaped projection on the back of the inner brake pad, **Figure 13-20.** The two-way check valve should also be replaced whenever the caliper is serviced. If the actuating lever was removed, replace the lever seal and make sure the high lead screw is all the way out after caliper installation.

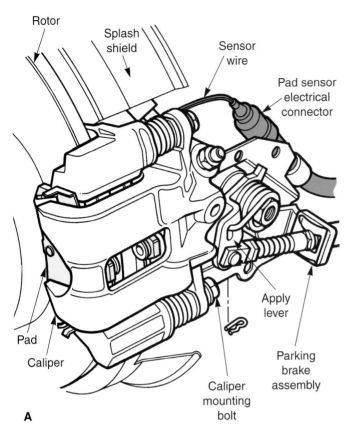

A

B

Figure 13-17. *A—A rear wheel disc brake assembly, with the parking brake cable and hook disconnected from the caliper apply lever. B—A screw-type rear disc caliper assembly. Remove the parking brake cable and lever to push in the piston. (Chevrolet)*

Ball-and-Ramp Caliper Service

The ball-and-ramp caliper assembly works by using three steel balls along matching tapered ramps to apply the brake pads. To replace the pads without disassembling the caliper, a special tool, **Figure 13-21,** must be used to turn the piston into the caliper bore. Carefully push the piston

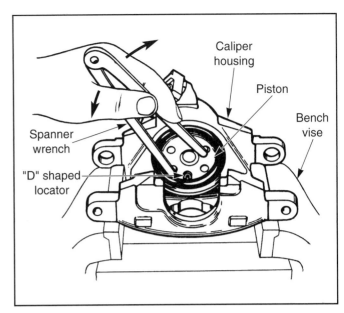

Figure 13-18. *A spanner wrench is sometimes needed to turn screw type rear caliper pistons into the body. (Bendix)*

back into the caliper using the special tool to bottom the piston in the caliper bore. There is no way to bottom the piston without using this tool.

Cam Type Caliper Service

The cam type caliper is a simple design in which an eccentric cam moves a rod. The rod pushes the piston and pads into contact with the rotor. The rod is threaded and mates with an adjusting nut in the piston. The rod and adjusting nut remove any slack caused by pad wear. To retract the piston when new pads are installed, the piston should be turned clockwise as shown in **Figure 13-22.**

Rear Caliper Reinstallation

Rear caliper reinstallation is the reverse of installation. If it was removed, reinstall the rotor. Place the new pads in the caliper or on the rotor assembly. Reinstall the caliper and reconnect the parking brake linkage if it was removed. If the vehicle uses an electrical wear sensor, attach the electrical connector. If needed, bleed the system. If the vehicle has a drum-in-disc system, be sure to adjust the parking brake. Be sure to stake the pads if necessary.

Adjusting Rear Disc Brake Calipers

After rear disc brake service, you may need to adjust the caliper pistons to the rotors. In some cases, the brake pedal only needs to be pumped several times to bring the pads in adjustment. However, a special adjustment procedure is sometimes needed to bring the pads into position and to obtain a good pedal.

Using a flat-head screwdriver, carefully position the tip against the top lip of the caliper piston, **Figure 13-23.** Clamp a pair of vice grips on the caliper's parking brake actuating mechanism.

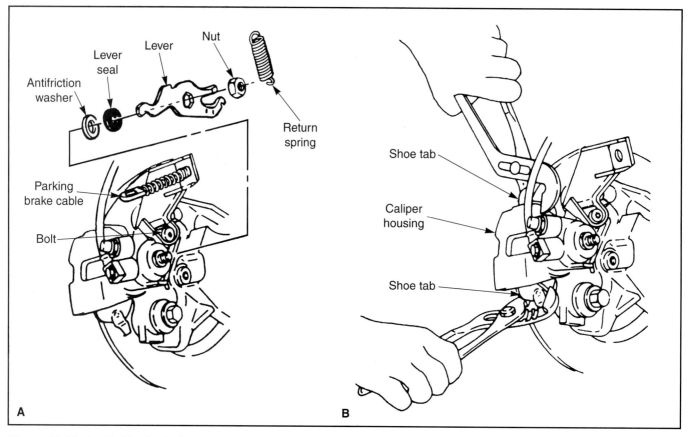

Figure 13-19. *A—Parking brake lever removal can allow a screw caliper's piston to be pressed in. B—Pressing the caliper piston in with two pairs of adjustable pliers. (General Motors)*

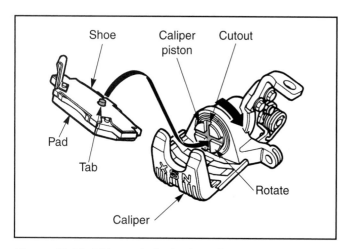

Figure 13-20. *This particular brake caliper setup requires the tab on the brake shoe to be placed in one of four caliper piston cutouts (slots). (Nissan)*

 Note: In some cases, this is easier to perform using a box-end wrench.

While carefully prying against the piston, use the vice grips to apply and release the parking brake. The caliper piston should adjust the pads until they just rest against the rotor. If the caliper piston fails to move, try tapping the caliper piston area with a ball peen hammer. Be careful not to damage the bleed screw. If the piston continues to stick, remove the caliper and perform an overhaul.

Rotor Service

The following sections discuss the service of disc brake rotors. The condition of the rotor is as important as pad condition. Many common disc brake problems, such as noises and pulsation, are caused by the rotor. Therefore, it is very important that you carefully check the rotors when the pads are replaced.

Sometimes the rotor is not refinished when the pads are replaced. However, the usual procedure in most shops is to refinish the rotor to allow it to wear into the new disc pads. Whether it is refinished or not, the rotor should be checked as explained in the following paragraphs.

 Caution: Some manufacturers recommend that rotors be refinished only if they are scored or out-of-round, which would produce a pulsation. In some cases, extensive driving is required after refinishing to burnish the pads properly. Some rotors cannot be machined and must be replaced if they are scored or out-of-round.

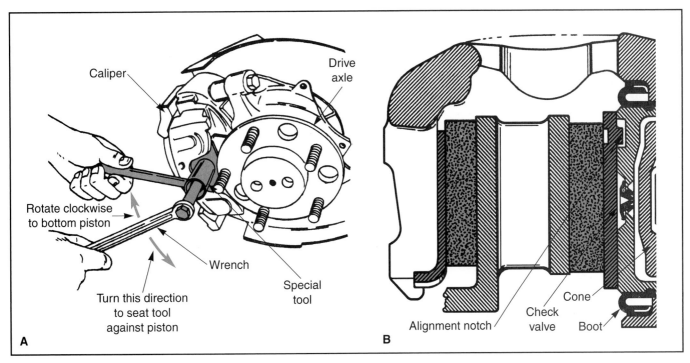

Figure 13-21. *A—Installing new brake pads on ball and ramp calipers requires the piston be screwed back into its bore. A special tool is needed for this procedure. Position the tool and rotate the handle counterclockwise while holding the shaft. Continue until the tool is seated against the piston. Loosen the tool handle about one-quarter of a turn. Now hold the tool handle and turn the shaft until the piston is completely bottomed. Even though the inward travel of the piston has stopped, it will continue to rotate after bottoming. B—Make sure the tabs on the pad are installed in the alignment notches in the piston. (Wagner)*

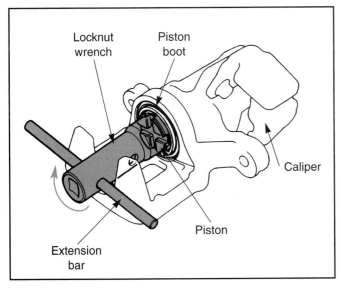

Figure 13-22. *Retracting the piston by turning in a clockwise direction with a special locknut wrench. This will provide the necessary clearance for installing the new brake pads. (Honda)*

Checking Rotor Condition

Before removing the rotor from the vehicle, check it for damage, warping, and proper thickness. Normal rotor wear patterns consist of small scratches and a slight polishing of the braking surfaces. They do not greatly affect braking and can usually be removed by light sanding instead of machining.

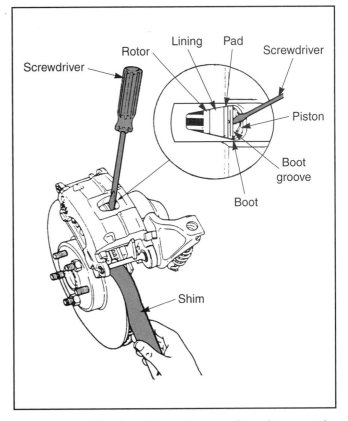

Figure 13-23. *After installation, the rear caliper piston must be adjusted so the pads rest just against the rotor. This is a special adjustment procedure that can be used to adjust rear brake calipers. (Chevrolet)*

Visually inspect the rotor for *scoring* or *grooving* on the braking surface, **Figure 13-24.** Scoring and grooving are deep cuts in the rotor surface. They always follow the rotor's curve of rotation. If the pads have worn to the rivets or the metal shoe surface, the rotor will be badly scored. Sometimes, a rotor will be lightly scored by long usage, especially in sandy or dusty areas. Do not assume the inboard rotor surface is good if the outboard surface shows no damage. Scoring can exist on one or both braking surfaces of the same rotor. Machining (turning) the surfaces is required if the rotor shows any scoring or grooves.

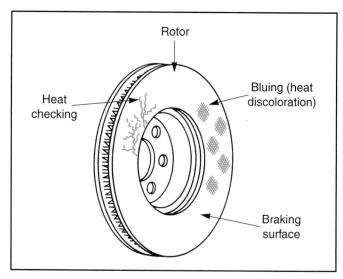

Figure 13-25. *This illustrates heat-checking and bluing caused by heavy braking, severe service, etc. A heat-checked rotor should be replaced. (EIS)*

Figure 13-24. *Scoring or grooving on the braking surface is caused by dirt, exposed rivets, etc. Rotor machining will be needed to reestablish the proper brake surface. Some minor scoring is considered normal.*

Excessive brake heat can cause *heat-checking* (tiny surface cracks) or *bluing.* There can be a combination of heat-checking and bluing or just one condition by itself. These patterns can also form on one or both sides of the rotor, **Figure 13-25.** Bluing can sometimes be removed by machining. If heat-checking is present or bluing cannot be removed without excessively reducing rotor thickness, the rotor should be replaced.

 Caution: Never attempt to machine a heat-checked or cracked rotor. Replace any rotor that appears to be heat-checked or cracked.

Checking Rotor Thickness and Runout

To check for rotor thickness (parallelism), you will need a micrometer, **Figure 13-26.** Before checking thickness, calibrate the micrometer. Then install the micrometer on the rotor. If the rotor is grooved, place the point of the

micrometer in the deepest groove. Then read the thickness on the micrometer and compare it to the specified minimum thickness. The thickness should be more than the minimum if the rotor will not be turned. If the rotor must be turned, there should be enough metal remaining to be at or above the minimum thickness after the turning process.

Turn the rotor about one-quarter turn and repeat the thickness measurement procedure. Measure at least four places on the rotor. If the thickness variation between parts of the rotor is more than about .01″ (.254 mm), the rotor should be turned to prevent brake pulsation.

To check for excessive runout (warping), a dial indicator should be used. Before checking runout, eliminate any looseness in the rotor and hub assembly. If the rotor is separate from the hub, install at least three of the wheel bolts onto the lugs. If possible, the flat (non-tapered) side of the nuts should contact the rotor. Lightly tighten the nuts until there is no play between the rotor and hub. If the rotor is integral with the hub, make sure there is no play in the wheel bearings. If play is evident, tighten the wheel bearings until all play is removed. Then proceed with the runout checking procedure.

Place the dial indicator over the rotor so that the pointer is contacting the rotor about two-thirds of the way to the edge of the braking surface. The pointer should be on a flat spot, not over any grooves. Then tighten the mounting clamp. A typical dial indicator installation is shown in **Figure 13-27.** Then set the micrometer dial to zero. Turn the rotor slowly and observe the movement of the dial indicator needle. If the needle indicates runout of more than .01″ (.254 mm), the rotor should be turned or replaced.

 Note: Mark the point of maximum runout for later reference.

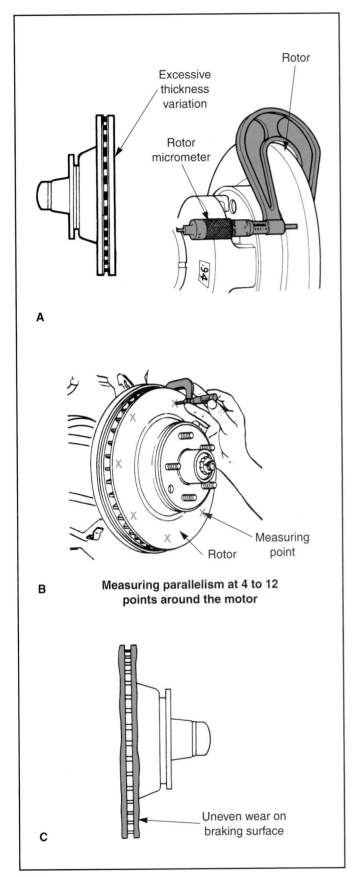

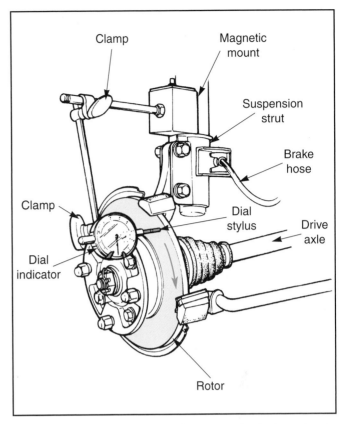

Figure 13-27. *Using a dial indicator to check the rotor for lateral runout. Note the magnetic mount used for holding the instrument assembly to the suspension strut. (Chevrolet)*

Rotor Removal and Installation

To remove the rotor, remove the caliper as explained earlier in this chapter. Some rotors are held in place by screw or bolt fasteners, or the wheel nuts when the tire and rim are installed. Most can be taken off the hub once the rim and caliper are removed. Other rotors are integral (part of the hub assembly), and the wheel bearings must be removed to remove the rotor, **Figure 13-28.** To remove these rotors, remove the dust cap and cotter pin holding the spindle nut in place. Remove the nut and pull the rotor and hub assembly from the vehicle. Be careful not to damage any bearing parts during removal. On some vehicles, you may have to remove a caliper mounting bracket or other components to remove the rotor.

 Note: Some front-wheel drive vehicles require a puller to remove the hub before the rotor can be removed. Unless the rotor is to be replaced, determine if the rotor needs machining or use an on-car brake lathe.

Turning Rotors

The rotor can be turned on or off the vehicle, depending on the type of rotor installation, and the lathe available. On late-model vehicles, some manufacturers recommend the rotors be turned on the vehicle.

Figure 13-26. *Measuring a brake rotor. A—The micrometer is placed on the rotor. B—Measure at 4 to12 spots around the rotor for the most accurate reading. C—Side view of a brake rotor illustrating uneven wear of the braking surface. (TRW, Inc. & Wagner)*

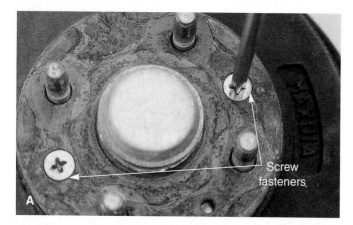

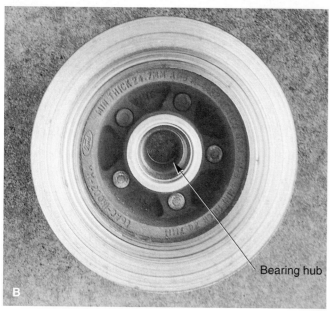

Figure 13-28. *Rotor removal. A—Some rotors are attached to the hub with screw fasteners. B—Integral rotors house the wheel bearings.*

Turning Rotors Off the Vehicle

After the rotor is removed from the vehicle, bring it to the brake lathe. If the rotor is an integral type, remove the grease seal and inner bearing and clean all grease from the interior of the hub. Check the brake lathe *arbor* and rotor *adapters* for dirt and metal, and clean as needed. Make sure the cutters are sharp and tightly attached. Inspect the lathe and be certain that all other parts, including shields, are solidly attached and in good working condition.

Note: Some front-wheel drive rotors require the use of a special hub adapter, which has lugs, much like the vehicle's wheel hub. Check the service manual.

Install the rotor on the brake lathe using the proper adapters, **Figure 13-29A.** Consult the lathe makers manual if you have any doubts as to how this should be done. If you have not checked the rotor's thickness or have any questions as to the thickness, check the rotor with a micrometer, **Figure 13-29B.** Then install the silencer band, **Figure 13-29C,** on the outer edge of the rotor.

The *silencer band* reduces noise, but more importantly eliminates vibration in the rotor as it is being cut. Vibration will cause *tool chatter* (a rapid bouncing of the tool against the rotor surface) as the bit cuts. This will produce a rough "wood grain" surface on the rotor. The rotor will need to be recut, unnecessarily removing metal. If you forget to install the band, the rotor will chatter as soon as the bits start cutting. Turn the lathe off immediately and install the silencer band. Then begin the cut again.

Start the brake lathe and check that the rotor is turning smoothly. Then turn the feed dials of each cutter until they are near the rotor surface. Be sure the cutters are directly across from each other. Slowly turn the cutter assembly until the cutters are in approximately the center of the rotor braking surface. Then slowly turn one cutter feed dial until the cutter tip just touches the rotor surface. Hold the feed dial and turn the numbered *collar* to zero, **Figure 13-30.** Repeat this operation for the other cutter.

Move the cutters to the innermost part of the rotor; do not ground the tip in the corner between the rotor hub and braking surface. Then adjust the cut and feed speed. A cut made to remove a great deal of metal, with the speed set relatively high is a *rough cut.* A cut made to remove a small amount of metal at slow feed speeds is a *finish cut.*

The amount and speed of cutting will be governed by the total amount to be removed and the finish desired, **Figure 13-30.** If the rotor is deeply grooved, a great deal of metal must be removed to obtain a smooth finish on the rotor. In this case, it will be necessary to make several rough cuts before making the finish cut. If the rotor is only lightly damaged, or just requires that a shiny surface be removed, it may be refinished by a single finish cut.

Caution: Remove only enough metal to clean up imperfections. Careless cutting may make an otherwise machinable rotor too thin and it will require replacement. If the imperfections are too deep, replace the rotor. Cutting too much at once can also cause the tips to wear prematurely or break.

Rough Cut

To make a rough cut, set the cutters to the maximum cutting depth and set the speed to a relatively fast setting. As a general rule, take no more than .006" (.152 mm) from each side on any single cut. Check the collars on the lathe cutters to determine whether they are scribed in thousandths of an inch or in millimeters.

After cutting depth and feed speed are established, engage the feed lever and watch the rotor as it cuts. Allow the cutting blades to cut the entire braking surface and exit the outer edge of the rotor. Then disengage the feed lever and inspect both rotor surfaces. If the first cut left grooves or shiny spots, repeat the rough cut as needed. After all damaged areas are removed, make a finish cut.

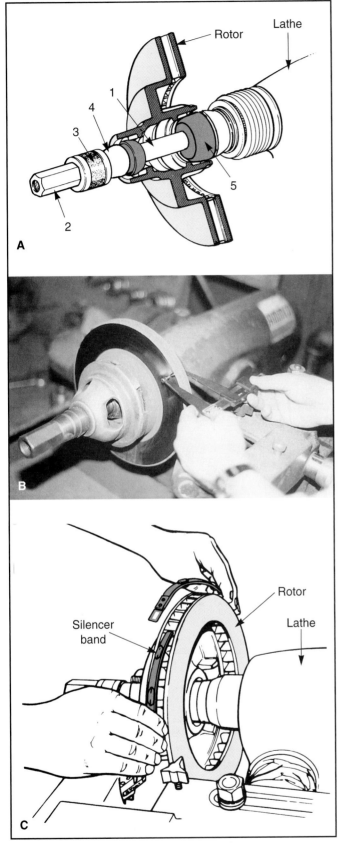

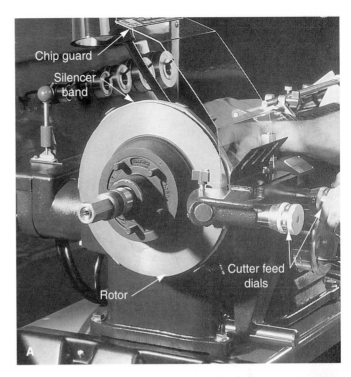

Figure 13-29. *A—A brake rotor with integral hub installed on a lathe. 1—Shaft. 2—Arbor nut. 3—Self-aligning spacer. 4—Tapered cone adapter. 5—Tapered cone adapter. B—You can measure the rotor for thickness before cutting. C—A silencer band being placed around the rotor to help reduce noise and vibration as the rotor is machined. (Ammco)*

Figure 13-30. *A—An ASE certified technician turning a brake rotor. Always follow the tool manufacturer's machining procedures and safety guidelines. B—As the tool moves over the rotor surface, watch for the presence of any damaged areas that are not removed. C—The line on the inside of this rotor is a groove that was below the cutting bit. More metal must be removed from the rotor to eliminate this spot. (Ammco)*

Note: If the rotor is warped, observe the mark that you made when checking maximum runout. At the mark, one side should be cutting deeply, while the other side is cutting lightly or not at all. This indicates that the rotor is properly installed on the lathe arbor.

Finish Cut

To make a finish cut, set the cutters to a small cutting depth, no more than .002" (.050 mm), and set the feed speed to a low setting. Engage the feed lever and observe the rotor as it is cut. Allow the cutting blades to cut the entire braking surface and exit the outer edge of the rotor. Then disengage the feed lever and inspect both rotor surfaces. After making the finish cut, check the rotor thickness with a micrometer. If the rotor is now too thin, it must be discarded.

Turning Rotors On the Vehicle

On some vehicles, the rotor is pressed into the CV axle shaft in such a way that removal is very time-consuming and difficult. In these cases, it is much easier to turn the rotor on the vehicle. To turn a rotor on the vehicle, a special on-vehicle lathe, **Figure 13-31,** must be used. Both front and rear rotors can be turned using this lathe.

Setup and cutting instructions are similar to the process for a bench-mounted lathe. The cutters are set to just touch the rotor surface, then brought to the middle of the rotor. The depth of cut is set and the feed turned on. As with any type of machining operation, watch the rotor carefully as it is being cut.

Some on-vehicle lathes require the rotor be turned by engine power as the cutters move across the rotor braking surface. A sequence of installing and using this type of lathe is shown in **Figure 13-32.** Other on-vehicle lathes are equipped with a drive motor which turns the rotor and CV axle assembly with the transmission in neutral. Some motor driven on-vehicle lathes have a provision for changing drive speed to make rough and final cuts.

When using an on-vehicle lathe, it is very important to set all cutters and drive mechanisms very carefully. This is because the design of the on-vehicle lathe is less rigid than the bench lathe, and slight misalignment can cause the rotor to be cut improperly. If the lathe and rotor appear to **oscillating** (wobbling) excessively when the cut is started, turn off the lathe immediately and recheck all adjustments.

Removing Hot Spots

Hot spots, sometimes called *hard spots,* are rotor sections that have been overheated by severe brake operation and become much harder than the surrounding metal.

Figure 13-31. *An ASE certified brake technician using an on-car brake rotor lathe to turn a rotor. This particular vehicle is front-wheel drive. When using these lathes, carefully follow all the tool manufacturer's installation and machining procedures. (Hunter Engineering Company)*

Figure 13-32. *A vehicle powered, on-car brake lathe operating sequence. The procedures will vary between the various lathe manufacturers. 1—The vehicle has been raised to a comfortable height. 2—Wheel and tire are removed. 3—Lug bolts are reinstalled (if needed) to secure the rotor to the hub. 4—Remove the caliper mounting bolts. 5—Support the caliper on a wire hook. 6—Clean the caliper mounting area. 7—Bolt on the lathe mounting legs. 8—Mount the lathe head on the vehicle. 9—Carefully align the carbide cutter bits to the centerline of the rotor. 10—Place the vibration dampener on the rotor, remove clip before machining. 11—Adjust the cutter bits. 12—Manually turn the lathe head in as far as the cutter bits will allow. Then, engage the lathe head drive. Repeat steps 11 and 12 as needed. (Kwik-Way Mfg. Co.)*

These spots cannot be removed by cutting bits. After turning is complete, these spots will remain as raised places on the finished surface.

To remove hot spots, a special motor driven grinder, **Figure 13-33,** must be used. This grinder is installed in place of the cutting bit and rotates a grinding stone, or wheel, against the braking surface as the rotor turns. To use this grinder, set clearances in the same manner as when setting the cutting bits. Then start the grinder and set the feed to low speed. As the grinding wheel moves over the hard spot, it will grind it down to match the other areas of the rotor.

If the grinder cannot remove all the hot spots, the rotor should be replaced. Ideally, you should replace any rotor that has hot spots.

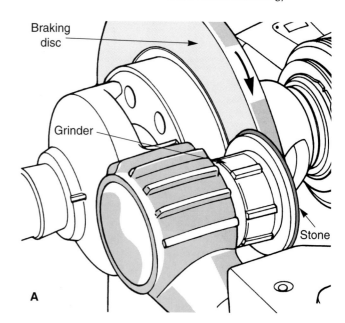

A

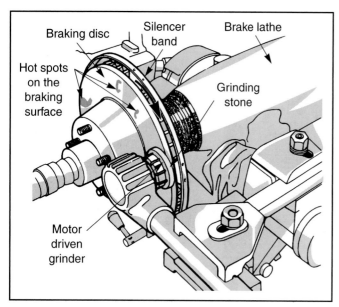

Figure 13-33. *Removing hot spots from the braking disc (rotor) with a motor driven grinder. The grinder is also handy for removing rust and lining deposits. Wear your safety glasses. (Dodge)*

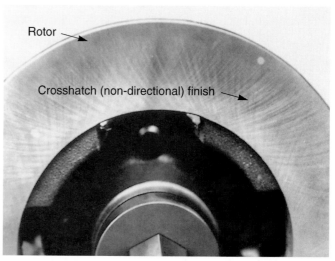

Figure 13-34. *A—Lathe mounted grinder being used to apply the proper non-directional crosshatch (swirl) pattern to the brake rotor. B—Finished rotor. Note the grinding pattern. (Chrysler)*

Swirl Grinding Rotors

The **swirl grinding** process is used to make a final **non-directional finish** on the rotor. A non-directional finish eliminates the microscopic tool marks made by the cutting bits, replacing them with a series of extremely fine random scratches. This finish helps to eliminate noises and aids in pad break-in.

There are essentially two ways to perform swirl grinding: the **lathe-mounted grinder, Figure 13-34,** and the **hand-held grinder, Figure 13-35.** With either design, the basic operation is to hold a spinning sanding disc against the rotor surfaces as the lathe turns. The combination of lathe and grinder rotation creates a swirl pattern that prevents the development of vibrations that take the form of squeals or other noise.

To perform swirl grinding, follow the manufacturer's instructions to attach the grinder, if necessary. Then rotate

the grinder against the rotor as it turns. It is not necessary to operate the swirl grinder for a long period; 30-60 seconds on each side is sufficient.

Once the swirl grinding operation has been performed, remove the rotor from the lathe and clean it thoroughly to remove all chips. This is especially important if the rotor is an integral type with bearings installed in the hub. After cleaning, the rotor can be reinstalled on the vehicle.

Rotor Installation

Rotor installation is the reverse of removal. Most rotors simply fit back on over the hub. Once installed, reinstall the rotor screw or bolt, caliper, and wheel. To reinstall an integral rotor, clean the spindle assembly of all old grease and dirt. Install the inner bearing and seal, place the rotor over the spindle, and install the outer bearings. Install

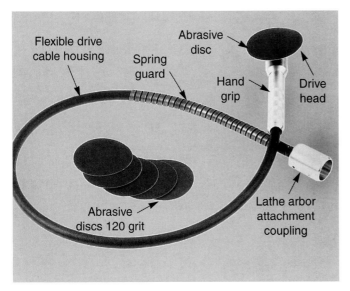

Flexible drive cable housing
Spring guard
Abrasive disc
Hand grip
Drive head
Abrasive discs 120 grit
Lathe arbor attachment coupling

Figure 13-35. One type of hand-held swirl grinder used to produce a "non-directional" rotor finish. This unit is powered by the lathe arbor, which turns a flexible drive shaft (not shown). The grinder spins at approximately twice the speed of the lathe arbor shaft. (Ammco)

and adjust the spindle nut and install a new cotter pin. Check that the rotor turns freely. Bearing service will be covered in more detail in Chapter 17.

Final System Check and Road Test

Reinstall the tires and torque the lug nuts to specifications using a torque wrench or torque sticks, if available. Simply tightening the lug nuts with an impact can lead to uneven torque, which could distort rotors and lead to pulsation. Check the master cylinder reservoir and add fluid if necessary. After adding fluid, pump the brakes a few times to set the pad to rotor clearances. If necessary, bleed the brake system.

⚠ **Warning: Do not attempt to move the vehicle until the brake pedal has been pumped several times. If brake clearances are excessive (common after the pads have been replaced), the brake pedal will go to the floor without applying the brakes.**

Bedding-in Brake Pads

Road test the vehicle, making at least ten gentle stops to seat and burnish the linings. This process is referred to as *bedding-in.* Accelerate to about 35-40 mph (56-64 kph), then apply the brakes with light to moderate pressure to reduce speed to approximately 5 mph (8 kph), do not come to a full stop. Allow at least 1/3 mile between stops. Allow the brakes to cool completely before releasing the vehicle to the customer.

While making these stops, be alert for noises, hard pedal, and pulsation. If the rear brakes were serviced, make sure that the parking brake works properly. It is a good idea to check the parking brake even if the rear brakes were not serviced. After you are sure that the vehicle stops properly, return it to the owner. Caution the owner to avoid hard prolonged braking for at least 200 miles (320 km).

Summary

Common disc brake problems are noise, pulsation, and a hard pedal. The caliper hydraulic system is usually trouble-free. Disc brakes can be checked by observing the thickness of the pads and the condition of the rotor.

The caliper can be removed by first removing the wheel and tire to gain access. Then lightly pry the pads away from the rotor and remove the caliper attaching hardware. Then lift the caliper from the rotor and remove the pads if necessary. Check the caliper for leaks and damage. If necessary, check the pad thickness.

Push the piston into the bore with a C-clamp after loosening the bleeder screw. Then install the new pads with all hardware. Apply anti-squeal and high temperature lubricant if necessary, then reinstall the caliper over the rotor. Install and tighten the mounting hardware.

Common rotor problems include thickness variations, warping, and heat damage. To remove the rotor, determine what kind it is. Some rotors can be taken off the hub after the caliper is removed, while others are removed by removing the wheel bearing cotter pin and nut. The rotor and hub can then be pulled from the spindle.

Rotors can be turned on or off of the vehicle. Correctly mount the rotor in the arbor using the proper adapters. Then adjust the cutters to take off needed material. Make rough cuts to remove a great deal of metal, and finish cuts to produce a smooth final surface. Hot spots can be removed by grinding. Swirl grind to place a non-directional finish on the turned rotor.

After turning, install the rotor, add fluid to the reservoir if necessary, and road test the vehicle. Check brake operation and seat the brake pads, then return the vehicle to the owner.

Review Questions—Chapter 13

Please do not write in this text. Write your answers on a separate sheet of paper.

1. Why should you use wire to hang the caliper to the body, or place it where it will not hang?

2. Before pushing the piston into the caliper body on some ABS vehicles, you should loosen the _____.

3. Anti-squeal compound, when used, should be installed between the disc brake pads and the _____ mounting surfaces.

4. Before attempting to turn a rotor, visually check it for _____ and _____.

5. As a minimum, rotor thickness should be checked at _____ spots on the rotor.

6. Excessive rotor runout can be checked with a _____.

7. To remove a(n) _____ rotor from the vehicle, the wheel bearings must be removed.

8. Place the following rotor turning steps in order.
   ___ (A) Set the feed speed.
   ___ (B) Install the rotor using the proper adapters.
   ___ (C) Clean the arbor.
   ___ (D) Set the rotor cutting depth.
   ___ (E) Install the silencer band.
   ___ (F) Start the lathe motor.
   ___ (G) Determine whether the rotor can be turned.
   ___ (H) Swirl grind the rotor.
   ___ (I) Inspect the turned rotor surface.

9. A motor driven grinder must be used to remove _____ from the rotor.

10. A non-directional finish is a series of fine _____ scratches.

ASE Certification-Type Questions

1. All of the following statements about are true, EXCEPT:
   (A) disc brake pad and rotor service is similar for every kind of disc brake system.
   (B) the most common disc brake problems are noise and hard pedal.
   (C) disc brakes commonly produce high pitched squeals or squeaks.
   (D) pulsation is usually caused by rotor problems.

2. Technician A says that observing pad thickness is a good way to determine pad condition. Technician B says that a grooved rotor is a sign of wheel bearing wear. Who is right?
   (A) A only.
   (B) B only.
   (C) Both A & B.
   (D) Neither A nor B

3. Pushing the pads away from the rotor makes _____ removal easier.
   (A) pad
   (B) caliper
   (C) rotor
   (D) dust boot

4. Loaded calipers are new calipers with the _____ already installed.
   (A) pads
   (B) pins
   (C) rotor
   (D) All of the above.

5. All of the following statements about pushing the caliper piston into its bore are true, EXCEPT:
   (A) before pushing on the piston, loosen the bleeder screw.
   (B) push on the piston with a large C-clamp.
   (C) place a pan under the caliper to catch brake fluid.
   (D) tighten the bleeder only after the C-clamp is removed.

6. Do not allow any anti-squeal compound or lubricant to touch the _____.
   (A) pad friction surfaces
   (B) pad shoe
   (C) rotor
   (D) Both A & C.

7. A rotor will not turn freely after the caliper is installed. Technician A says to drive the vehicle and apply the brakes a few times to loosen the rotor. Technician B says to loosen the wheel bearings until the rotor turns freely. Who is right?
   (A) A only.
   (B) B only.
   (C) Both A & B.
   (D) Neither A nor B.

8. To check a rotor for runout, use a _____.
   (A) dial indicator
   (B) brake micrometer
   (C) feeler gauge
   (D) Any of the above.

9. Technician A says that all bearing play should be removed from an integral rotor before runout is checked. Technician B says that a non-integral rotor should be bolted to the hub before the runout is checked. Who is right?
   (A) A only.
   (B) B only.
   (C) Both A & B.
   (D) Neither A nor B

10. A brake lathe can be used to remove _____ from the rotor.
    (A) scoring
    (B) bluing
    (C) high spots
    (D) All of the above.

11. Technician A says that a rough cut should be used to remove a small amount of metal from the rotor. Technician B says that a rough cut should not be used to make a final finish on the rotor surface. Who is right?
    (A) A only.
    (B) B only.
    (C) Both A & B
    (D) Neither A nor B.

12. The cutting depth is set by the _____.
    (A) cutter feed
    (B) speed feed
    (C) numbered collar
    (D) cutting bits

13. Hot spots are caused by _____ overheating.
    (A) brake
    (B) lathe
    (C) grinder
    (D) swirl grinder

14. The swirl grinder puts a non-directional finish on the _____ surface.
    (A) pad lining
    (B) rotor
    (C) hub
    (D) Both A & B.

15. When would the technician want to use an on-car rotor lathe?
    (A) When the rotors are too hot to remove.
    (B) When the bench lathe is being used.
    (C) When the rotors are pressed to the CV shaft.
    (D) On a non-drive axle only.

A drum brake assembly used on a smaller car. The solid anchor at the bottom always indicates a non-servo brake assembly.

Chapter 14

Drum Brake System Components and Operation

After studying this chapter, you will be able to:

- ❏ Identify major drum brake components.
- ❏ Explain the purpose of backing plates.
- ❏ Identify the construction and purpose of brake shoes.
- ❏ Identify and explain the purpose of brake springs.
- ❏ Identify types of brake drums.
- ❏ Explain self-energizing, leverage, and servo action.
- ❏ Identify servo and non-servo brakes and explain their operation.

Important Terms

Drum brakes	Star wheel adjuster	Cross ribs
Backing plate	Floating adjuster	Maximum diameter
Anchors	Automatic adjuster	Discard diameter
Support pads	Cable adjusters	Brake shoe
Brake shoes	Link adjuster	self-energizing
Primary shoe	Lever adjuster	Brake shoe leverage
Secondary shoe	Ratchet adjusters	Servo brakes
Drum brake springs	Brake drum	Non-servo brakes
Eccentric cams	Axial ribs	

Drum brakes have been used on automobiles for many years and are still in widespread use. While drum brakes are simple in design and operation, they must be thoroughly understood before attempting to service them. This chapter covers the construction, major components, and operation of drum brakes. The construction and operation of wheel cylinders was covered in Chapter 5.

Drum Brake Components

Modern **drum brakes** are sometimes called *internal expanding brakes,* since the brake components are internal (inside a drum), and the shoes must expand, or move outward, to contact the drum. Drum brake components vary only slightly between manufacturers. Most design differences are in the methods of brake shoe anchoring and spring placement.

While they have been completely replaced by disc brakes on the front axles, most vehicles use drum brakes on the rear axle. Some vehicles use a combination disc/drum brake assembly which will be discussed later. Drum brakes will continue to see use because of their relative low cost.

 Note: Most cars and trucks built before 1974 were equipped with front drum brakes. While this chapter concentrates on rear drum brakes, many of the principles also apply to front drum brakes.

Backing Plate

To provide a foundation for the drum brake components and to act as a splash shield against water and road debris, a **backing plate** (sometimes called a support plate) is used. The wheel cylinder, brake shoes, and springs are attached to the backing plate. A typical backing plate is shown in **Figure 14-1.**

The wheel cylinder is usually solidly attached to the top of the backing plate. Most wheel cylinders are bolted to the backing plate. On some smaller cars, the wheel cylinder is held by a clip. A few designs place the wheel cylinder at the bottom of the backing plate. One design allows the wheel cylinder to move slightly to aid in shoe centering.

Anchors, which are solid attaching points for the brake shoes and springs, are also installed on the backing plate. Anchors are often called *anchor pins* or *anchor plates,* depending on their construction. **Figure 14-2** shows some typical anchor designs. Some anchors support only the shoes, while others are connecting points for the shoes and return springs. The flat spots on the backing plate are called **support pads.**

Most backing plates are discs of stamped steel, bolted or riveted to the spindle assembly on front brakes and to the axle flange on rear brakes, **Figure 14-3.** A few backing plates are made of aluminum or plastic. These plates are used primarily as splash shields with the wheel cylinder and anchors mounted on a steel support plate.

There is a very close fit between the backing plate and the rotating drum, **Figure 14-4.** This close fit keeps most water and dirt out of the drum brake assembly. However, if the vehicle is driven through deep water, the close fit traps water and the vehicle must be driven for a considerable distance before the brakes will operate properly.

Brake Shoe Assemblies

Commonly referred to simply as **brake shoes,** brake shoe assemblies consist of two major parts, the steel shoe, and the friction material lining. These parts are discussed in the following sections.

Shoe Construction

The steel brake **shoe** holds and supports the brake lining. Wheel cylinder movement is transmitted to the lining by the shoe. The two parts of a brake shoe are the **table** and **web.** They may be welded together or stamped from a single piece of steel. The table and web vary in thickness and the web may or may not be reinforced. A typical brake shoe is shown in **Figure 14-5.**

Nibs and Support Pads

Some shoes have raised spots on the table which contact raised spots on the backing plate when the shoe is installed. These are called **nibs.** The nibs and support pads provide a bearing surface for the shoe to slide on. Some shoes do not have nibs and the edge of the shoe table slides on the backing plate support pads.

Figure 14-1. *The backing plate provides a building foundation for the various brake components.*

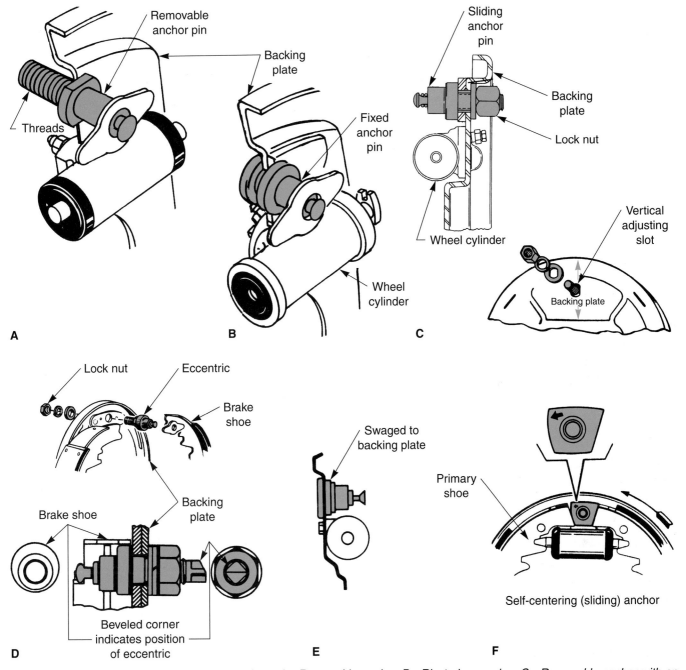

Figure 14-2. *An assortment of brake shoe anchors. A—Removable anchor. B—Riveted-on anchor. C—Removable anchor with an adjusting slot. D—Removable eccentric type. E—Swaged-on. F—Self-centering (sliding) anchor. (General Motors, Wagner, Bendix and Raybestos)*

Shoe and Wheel Cylinder Connections

There are two methods of connecting the wheel cylinder piston to the shoe web. One method is by using a **link,** sometimes called a *strut* or *pin,* between the piston and web. The link usually has a rounded end which contacts the piston. The other end is slotted or grooved to fit securely against a matching cutout on the web. The link passes through the wheel cylinder dust boot to contact the piston, **Figure 14-6A.**

On the second design, the wheel cylinder piston directly contacts a projection on the shoe web, **Figure 14-6B.** Piston movement moves the shoe directly. The wheel cylinder dust boot has a large center opening to allow the shoe projection to touch the piston without damaging the boot.

Shoe Anchoring Methods

The end of the web opposite the wheel cylinder is made to match the fixed anchors attached to the backing plate. Some webs have a rounded end to fit the round anchors as shown in **Figure 14-7.** Other web ends are flat to match a flat tapered anchor. This anchor design is called a **keystone** anchor. The design of the anchor allows the shoes to center themselves in relation to the drum by sliding up or down on the anchor.

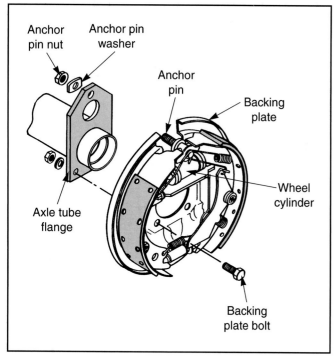

Figure 14-3. *This backing plate bolts to the tube flange. Note the brake shoe anchor also bolts to the flange. (Bendix)*

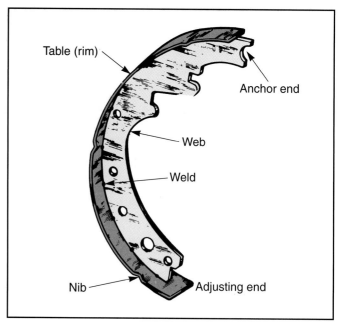

Figure 14-5. *Typical brake shoe construction. The web and table are secured together with welds. (Bendix)*

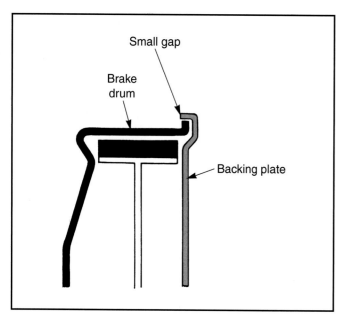

Figure 14-4. *Backing plate to brake drum fit. This close fit helps to prevent water, dirt, etc., from entering. (FMC)*

Some shoes on older vehicles are solidly anchored to the backing plate by pins which pass through the shoe pivot hole and are solidly mounted on the backing plate. The web is held to the pin by a snap ring or a cotter pin.

Parking Brake Attachment

Rear drum brakes are used as parking brakes on most vehicles. The parking brake is applied by moving the shoes into contact with the drum. This is done by a mechanical

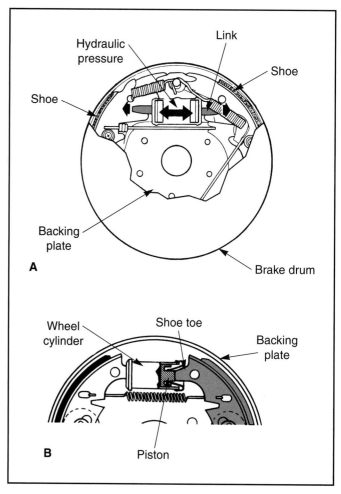

Figure 14-6. *A—Wheel cylinder to brake shoe connection method using links. B—In this brake assembly, the shoe toe is in direct contact with the wheel cylinder pistons. (Niehoff and FMC)*

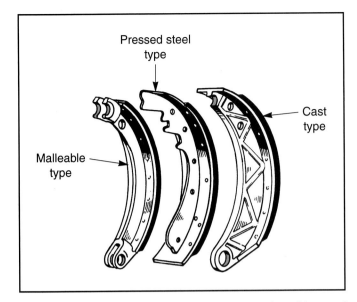

Figure 14-7. Several styles of brake shoe webs with round ends. (Bean)

connection between the parking brake linkage and the rear brake shoe web. Parking brakes are discussed in more detail in Chapter 18.

Linings

The brake *lining* is the friction material which contacts the drum. Brake linings must be able to produce friction in order to properly stop the vehicle. Some of these factors are the ability to stop well in cold and wet weather, and resistance to compression and swelling when hot. The lining should not fade under normal braking conditions. Lining materials were discussed in Chapter 11.

Lining Size and Attachment

Brake lining thickness and surface area varies with vehicle size and anticipated demands. Older vehicles with front drum brakes have linings with a large surface area for more braking power. These linings are relatively thin to reduce heat buildup. Smaller brake shoes are used on the rear. On modern front- and rear-wheel drive vehicles, the rear brake linings have a smaller surface area, since they carry a smaller portion of the braking load. These linings are relatively thick for longer wear.

To perform their job, the brake linings must be securely attached to the shoe table. They must also be attached tightly enough to allow heat transfer between the lining and the shoe. The brake lining is attached to the shoe table in one of two ways.

Some linings are attached with *rivets,* **Figure 14-8.** The rivets are usually made of brass or copper to reduce drum damage should the lining wear down to the rivets. Some rivets are tubular, with a hole in the center. This allows brake dust to exit the pad and drum surface.

Other linings are attached by bonding. *Bonding* is a method of gluing the lining to the shoe table by pressure and heat. A high temperature cement is placed on the back

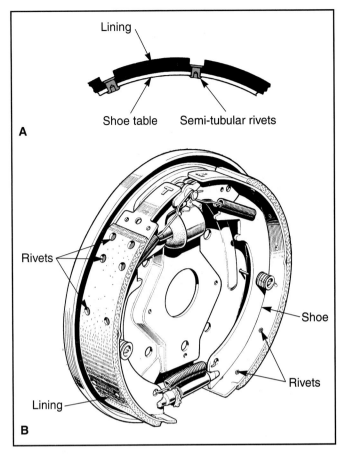

Figure 14-8. Brake lining attached to the shoe table with rivets. A—Cross-sectional view of rivets. B—Assembled brake unit which uses riveted lining. (Girling & FMC)

of the lining. Then the lining is tightly held against the shoe table as heat is applied. This creates a tight bond, securely holding the two pieces together. A typical bonded shoe is shown in **Figure 14-9.**

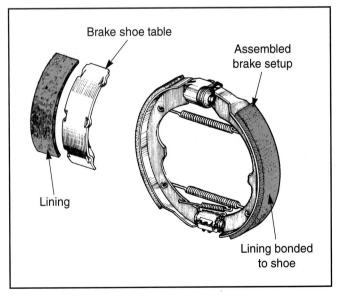

Figure 14-9. Brake shoes with bonded (glued on) linings. No rivets are used with this method. (Toyota & Nissan)

Primary and Secondary Shoes

There are two types of brake shoes in every drum brake assembly. The *leading* or **primary shoe** faces the front of the vehicle, and the *trailing* or **secondary shoe** faces the rear of the vehicle. Since the secondary shoe performs more of the braking effort, the primary shoe is often smaller. A set of primary and secondary shoes are shown in **Figure 14-10.**

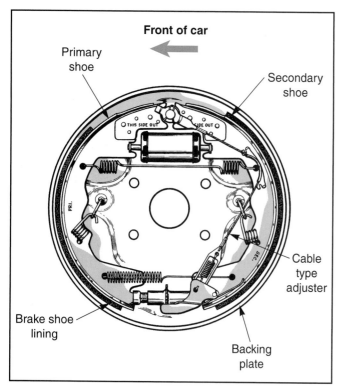

Figure 14-10. *A brake assembly showing the primary and secondary shoes. (Dodge)*

 Note: Most newer drum brake designs have shoes with linings the same size. This is done to simplify manufacturing and installation.

Lining Contact Methods

Lining-to-drum contact is very important for proper brake function in any given design. The contact can be designed by *arc grinding* the shoes. Arc grinding is actually a sanding process done at the factory which removes small amounts of the lining's top layer until it has the proper curve. Brake shoes can be arc ground in one of two ways.

The *eccentric* or **contour** ground lining is used on most vehicles. The contour lining is thicker at the center. When the brakes are first applied and the lining moves into contact with the drum, the thicker center section contacts the drum first. See **Figure 14-11A.** At each end of the lining, there is a small gap or clearance of about .01″ (.25 mm) between the lining and the drum.

As brake pressure increases, the brake shoe, lining, and drum will distort, or flex, to allow full lining contact against the brake drum. This is shown in **Figure 14-11B.** Eccentric grinding compensates for drum and shoe flexing, allowing high and even lining-to-drum pressure around the full circumference of the lining.

The *full contact* or **concentric** ground lining is used on large trucks or other medium- and heavy-duty vehicles. On these vehicles, the drum and shoes are too large and rigid to allow much flexing. A concentric brake shoe contacts the drum evenly. There are no high spots that can prevent proper lining to drum contact or adjustment. A concentric shoe and drum are shown in **Figure 14-12.**

Brake Lining Grooves

On older vehicles with very large brake shoes, **grooves** were often cut or formed in the brake lining. These

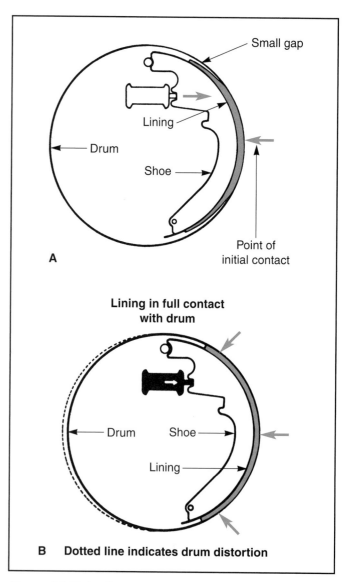

Figure 14-11 *A—Contour ground lining. Note that the thicker center section contacts the drum first, as the brakes are applied. B—Brake pressure has increased. The brake shoe (and some drum) have flexed or distorted, allowing full lining-to-drum braking surface contact. (Bendix)*

grooves helped to relieve stress on the shoe web, reducing uneven contact pressure. The grooves helped to dry out linings that had become wet by channeling out water and allowing air circulation around the lining. One type of grooved lining is illustrated in **Figure 14-13.** Most modern brake linings with smaller contact areas usually do not have grooves. However, some aftermarket replacement shoes have grooved linings.

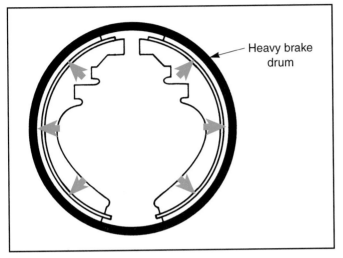

Figure 14-12. Concentric shoes and brake drum. This heavy-duty drum and shoes will not flex. The shoes contact the drum braking surface evenly. (Ammco)

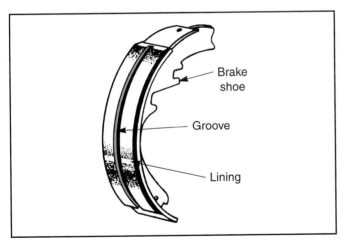

Figure 14-13. A brake shoe with grooved lining. The two are bonded together. (Wagner)

Brake Shoe Springs

The two major types of **drum brake springs** are hold-down springs and return springs. The holddown springs hold the shoes against the backing plate, while at the same time allowing them to move when the brakes are applied. The return springs, sometimes called *retracting springs,* bring the shoes to the unapplied position when hydraulic pressure is removed.

Holddown Springs

Earlier in this chapter, we discussed the brake shoe nibs and the corresponding support pads. The job of the **hold-down springs** is to keep the nibs against the support pads without holding them so tightly they cannot move or wear excessively. The two types of hold-down springs and methods of attachment are discussed in the following paragraphs.

Coil Spring Hold-down

The **coil spring hold-down** assembly consists of a round metal pin, spring, and washer. One end of the pin is formed into a disc shape and the other end is flattened. This pin is installed through a hole in the backing plate and a hole in the brake shoe web. The pin passes through a coil spring and washer installed over the shoe web. Note the washer contains a slot. When the washer is depressed and moved one-quarter turn (90°), the pin's flattened area cannot pass though the slot. When the washer is released, spring tension holds the pin in place. This type of hold-down is illustrated in **Figure 14-14.**

Another version of the coil spring hold-down is shown in **Figure 14-15.** This type of coil spring uses with a hooked end. The hooked end is attached to a clip installed into the backing plate.

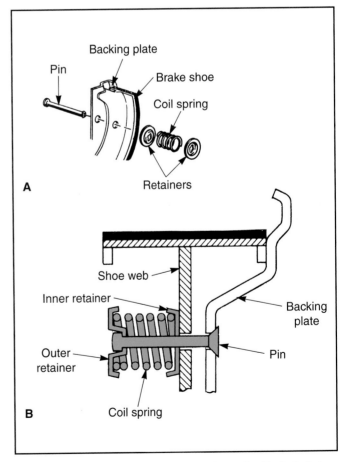

Figure 14-14. A coil spring shoe hold-down. A—Exploded view. B—Cross-sectional view of the assembled parts. (Raybestos)

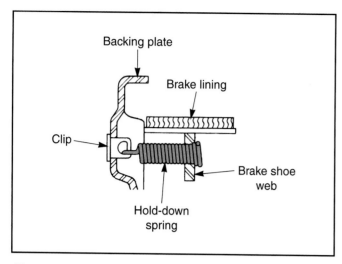

Figure 14-15. *A coil spring and clip brake shoe hold-down.* *(Wagner)*

Spring Clip Holddown

Another brake shoe holddown is the *spring clip holddown.* This design also makes use of a pin. The clip is a flat piece of spring steel, bent in a U-shape with a hole at each end. The pin passes through holes in the backing plate and brake shoe web and then into both holes in the spring clip. The pin is turned one-quarter turn, locking it in the retaining notch on the spring clip. A spring clip hold-down is shown in **Figure 14-16.**

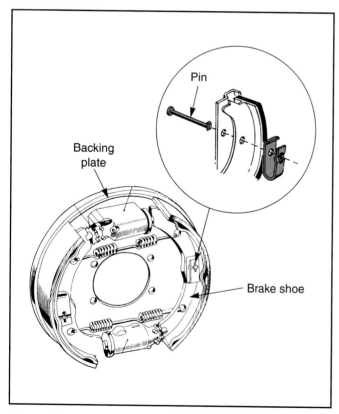

Figure 14-16. *A spring clip brake shoe retainer. Note the U-shape of the clip. The clips are made of spring steel, which when installed, provide constant tension. (Honda)*

Shoe Return Springs

Shoe *return springs* are always coil springs. Return springs are connected between the brake shoe and a stationary support or from one shoe to another. The function of shoe return springs is to return the brake shoes to the unapplied position when hydraulic pressure is removed from the wheel cylinder. **Figure 14-17A** shows two return springs attached to the front (primary) and rear (secondary) brake shoes and a stationary support. Most spring ends are made to be inserted into a hole in the shoe web. The other end of the spring is hooked to fit over the stationary anchor.

Wheel cylinder hydraulic pressure pushes the shoes outward to contact the drum. At the same time, the return springs are stretched by shoe movement. When the brake pedal is released, the wheel cylinder loses hydraulic pressure. With hydraulic pressure gone, the springs return to their original position, retracting the brake shoes and pushing the wheel cylinder pistons to the unapplied position. Note that some newer brake assemblies use a single spring to retract the shoes, **Figure 14-17B.**

Some brake assemblies have additional return springs attached between two shoes as in **Figure 14-17C.** The springs help to return the shoes and also maintain the proper alignment between the shoes and anchors. Return springs are sometimes color-coded to indicate model changes or to identify them for proper reinstallation.

Anti-Rattle Springs

Anti-rattle springs are used in the brake assemblies to reduce rattle and clicking noises. They do this by providing a slight amount of spring tension between two parts. This tension eliminates play, keeping the parts from striking each other. Most anti-rattle springs are coil springs, such as the brake lever spring shown in **Figure 14-18.** Sometimes flat metal strips are used as springs to create tension between parts.

Drum Brake Adjusting Devices

To properly apply the brakes, the clearance between the shoes and the drum must be as small as possible without actual contact. Too much clearance will cause a low brake pedal and low frictional pressure. Too little clearance will cause the brakes to drag, overheating and ruining the linings and drum.

All drum brakes have a provision for adjusting clearance. The two major types of adjusters are the star wheel and the ratchet. The star wheel is the most widely used, while the ratchet is used only on the rear brakes of a few vehicles.

On some older vehicles, the brake adjusters consist of *eccentric cams* attached to each shoe web. Turning each cam adjusted the related shoe. On a few older vehicles with non-servo brakes, the bottom anchor could be moved to adjust brake clearance. These adjusters are not used on modern vehicles.

Star Wheel Adjusters

Almost all modern drum brakes use some form of the *star wheel adjuster.* On many systems, the star wheel adjuster is placed at the bottom of the shoe assembly, with the wheel cylinder at the top, as shown in **Figure 14-19.** On some drum brake systems, the adjuster is placed at the top of the assembly, directly under the wheel cylinder, with the bottom of the shoes anchored, **Figure 14-20.**

The star wheel design is called a *floating adjuster,* since it is not anchored to the backing plate and can move with the shoes. The star wheel adjuster can be manually turned to adjust the clearance during brake service. All star wheels used on modern vehicles are also operated by automatic adjuster linkage. The automatic linkage adjusts the brake clearance as the vehicle is driven.

A star wheel assembly is comprised of three main parts—star wheel, pivot nut, and socket. Washers and

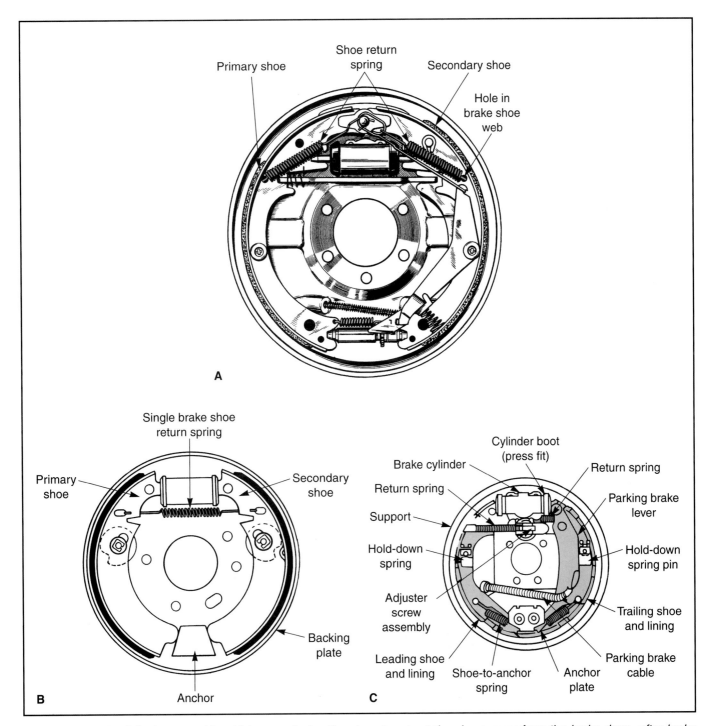

Figure 14-17. *A—A brake assembly which uses dual coil springs to retract the shoes away from the brake drum, after brake hydraulic pressure has been released. B—Brake assembly using a single brake shoe coil return spring to retract the shoes. C—This brake setup uses several additional coil springs. These help to retract the shoes and/or maintain the proper alignment between the shoes and anchors. (Bendix, Bean)*

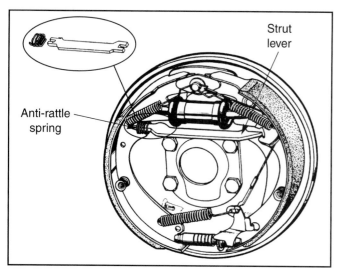

Figure 14-18. A brake strut lever which uses a coil spring for an anti-rattle device. (Oldsmobile)

other parts are used to ensure there is no friction between the star wheel, pivot nut, and socket. An exploded view of a typical star wheel adjuster is shown in **Figure 14-21.** The

star wheel has external threads that mesh with the pivot nut's internal threads. Turning the star wheel causes it to thread in or out of the pivot nut. This makes the star wheel assembly longer or shorter, depending on which way the star wheel is turned. The socket has a machined hole that matches a machined projection on the star wheel. This allows the star wheel to be turned with a minimum of friction.

A washer is installed between the star wheel and socket. This washer reduces friction and wear between the moving parts. To keep the adjuster assembly from coming apart, a spring is used to place tension on the assembly. The spring may be part of the star wheel assembly, attached to both shoes on opposite ends of the star wheel, or attached to one shoe and the self-adjuster lever.

Star wheel adjusters are threaded according to the side of the vehicle they belong, and are marked with the letters *L* or *R.* These indicate whether the adjuster should be installed on the left or right side of the vehicle. Adjusters should never be placed on the wrong side of the vehicle. If this is done, thread rotation will cause the adjuster to back off instead of tightening.

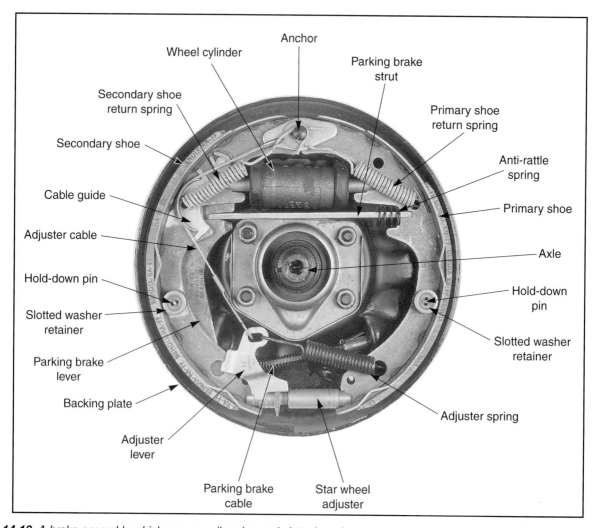

Figure 14-19. A brake assembly which uses a coil spring and slotted washer retainer as shoe hold-downs. (Chrysler)

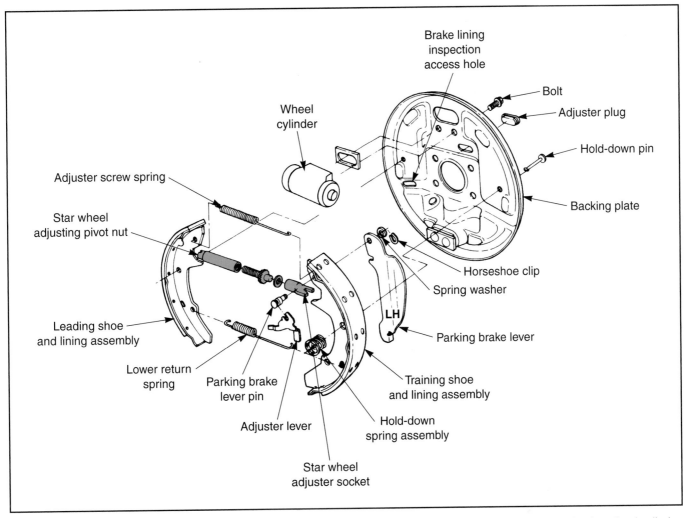

Figure 14-20. *Exploded view of a drum brake assembly which uses a star wheel adjuster located directly under the wheel cylinder. (Ford)*

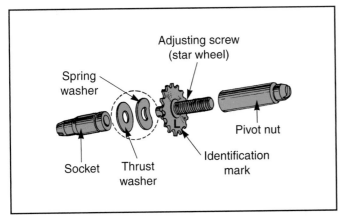

Figure 14-21. *Exploded view of a star wheel adjuster. Note the L identification mark. This mark indicates the adjuster is to be used on the left side of the vehicle. Never install on the wrong side. (Wagner)*

Automatic Star Wheel Adjusters

On all modern vehicles, the star wheel is turned by an ***automatic adjuster.*** Some adjusters are operated by a cable. Others are operated by a lever or a link. The basic operating mechanism for all these adjusters is the same. As the vehicle moves backwards, when the brakes are applied, movement of the top of the rear shoe away from the anchor causes the adjuster to move as needed.

Cable Adjusters

Cable adjusters use a small braided cable and associated parts to maintain proper lining-to-drum clearance as the brake lining wears. It operates by using brake shoe movement to pull on an adjusting lever, which then turns the star wheel. Refer to **Figure 14-22** as you read the following paragraph. Notice that the cable is connected to the backing plate and passes through a guide attached to the rear shoe before connecting to the adjuster lever.

When the brakes are applied with the vehicle moving in reverse, they cause the rear shoe to leave its anchor. Since the cable is attached to the adjuster lever and the stationary backing plate through a pivot point, shoe movement tightens the cable. As the cable tightens, it pulls on the adjuster lever that is touching the star wheel. If the lining is worn enough, the shoe will move enough to pull the lever above the next tooth on the star wheel. The lever falls into engagement with the tooth.

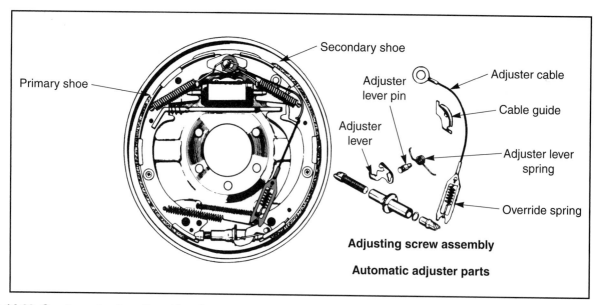

Figure 14-22. One type of automatic cable adjuster as used with duo-servo brakes. (Bendix)

When the brakes are released, the operating shoe returns to its anchor, and an adjuster return spring pulls the lever down. The falling lever causes the star wheel to turn one notch and become slightly longer. This expands the shoes to maintain the proper drum clearance. This adjustment procedure goes on whenever the brakes are applied in reverse when there is enough wear to allow the lever to move over a star wheel tooth.

Link Adjuster

The *link adjuster,* **Figure 14-23,** works in the same manner as the cable type. Linkage rods are used in place

of the cable. Instead of a cable guide, a lever or crank is mounted near the center of the brake shoe web. When the rear shoe moves away from its anchor, it tightens the link and lever. This pulls up the adjuster lever. When the brakes are released, the return spring moves the adjuster lever to turn the star wheel.

Lever Adjuster

On the *lever adjuster,* a lever and link are used. This assembly is illustrated in **Figure 14-24.** The adjusting lever is attached to the web at the center of the shoe, and can pivot at the attaching point. The lower end of the lever,

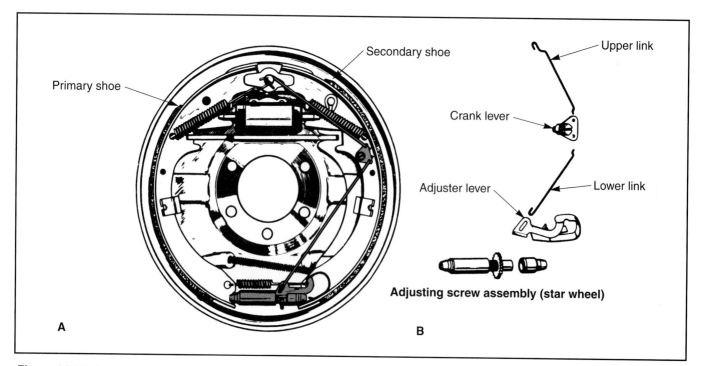

Figure 14-23. An automatic link-type brake shoe adjuster. Note that metal links (rods) are used in place of the cable. Compare to Figure 14-22. (Bendix)

called the *pawl,* contacts the star wheel. The top of the lever is connected to a metal link. The link is attached to the stationary backing plate anchor. At the bottom of the lever, a coil spring is mounted between the lever and the brake shoe table.

Applying the brakes in reverse causes the rear shoe to move away from the anchor. Since the top of the lever is attached by the link to the stationary anchor, it is pulled inward by the brake shoe movement. This causes the lever to pivot on its attaching point, and the pawl moves downward. The downward movement of the pawl turns the star wheel to adjust the brakes. If the brake linings are not sufficiently worn, the pawl cannot move enough to adjust the brakes.

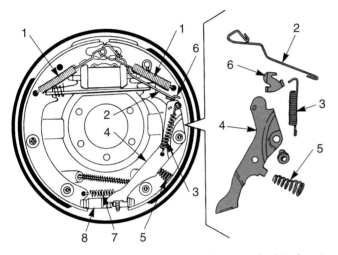

Figure 14-24. *A lever-type of brake adjuster and related parts. 1—Return springs. 2—Pivot link. 3—Override spring. 4—Lever pawl. 5—Lever return spring. 6—Pivot lever. 7—Adjuster spring. 8—Star wheel assembly. (FMC)*

When the pawl moves downward, it compresses the coil spring. When the brakes are released, the spring decompresses and returns the pawl as the adjuster lever is returned with the brake shoe.

> **Note: Most self-adjusters have an override mechanism to prevent lever damage should the star wheel stick and be unable to turn. The lever and pawl assembly is made in two pieces and tensioned by a spring. If the linkage cannot turn the star wheel, the spring allows the two parts of the lever to compress. When the brakes are released, the spring returns the lever assembly to its original shape.**

Manual Adjustment Provision

All star wheel adjusters have a provision for manual adjustment. On floating adjuster systems, the adjustment can be made through an access hole in the backing plate.

On some older cars, the access hole is in the drum. On some vehicles, a metal plug is installed on the backing plate or the brake drum. The plug must be knocked out to gain access to the star wheel. A special tool can then be inserted through the hole to turn the star wheel. When adjustment is complete, the hole made by removing the metal plug must be sealed by a rubber plug. These plugs can be obtained from any parts store.

On non-servo brake systems, the adjuster is a non-floating type located directly under the wheel cylinder. The backing plates of shoe assemblies with non-floating adjusters have no access hole. The drum must be removed to adjust brake clearance.

Ratchet Adjusters

Ratchet adjusters are used on rear non-servo brakes of some smaller cars. Common ratchet adjusters consist of a *quadrant* and a *wheel,* **Figure 14-25A,** or small and large ratchets, **Figure 14-25B.** The quadrant and wheel have interlocking teeth. The teeth slide over each other in one direction and lock when movement is reversed. A spring maintains constant contact between the sets of teeth.

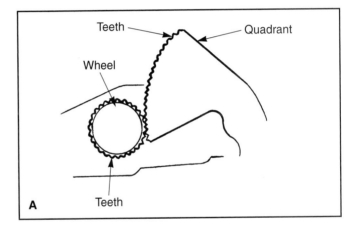

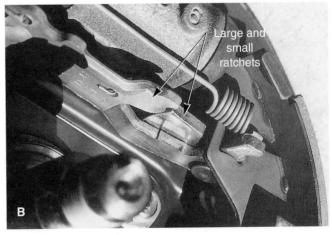

Figure 14-25. *A—Non-servo type brake adjuster which consists of a toothed wheel and a toothed quadrant. B—Another style of ratchet adjuster used with non-servo brakes. This mechanical adjuster will keep the proper lining-to-drum clearance. (EIS)*

Brake shoe movement causes the teeth on one ratchet to push against the teeth on the other ratchet. If the brakes are worn enough, the ratchet teeth travel over each other and take up the clearance. When the brakes are released, the teeth lock together and maintain the adjustment. The action of the teeth in both directions is shown in **Figure 14-26.**

Most ratchet adjusters are operated when the vehicle is braked while traveling in reverse. However, some adjusters are operated by the application of the parking brake. A common cause of loose brake adjustment is the failure of the driver to apply the brakes in reverse (such as placing the transmission in forward instead of applying the brakes), or not using the parking brake.

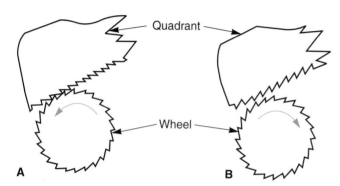

Figure 14-26. *A wheel and quadrant ratchet brake adjuster. A—Wheel is free to turn; brakes are being adjusted. B—Brakes are now released. The wheel and quadrant teeth are meshed together maintaining the correct lining-to-drum clearance.*

Manual Adjustment Provision

Ratchet adjusters can be operated manually to adjust the brakes. However, the brake drum must be removed to make the adjustment. Since these brake designs are used on the rear brakes of small cars, manufacturers feel that precise adjustment with the drum installed is not necessary.

Brake Drum

The **brake drum** provides a cast iron braking surface for the shoes to work against. The drum also absorbs and dissipates large amounts of heat. Brake drums are simple in construction and operation. The design differences between various brake drums are discussed in the following paragraphs.

Types of Drums

The external drum was one of the first types used on automobiles. The outer surface of the drum was machined to provide a smooth, finished area for the band lining to contact. This type can sometimes be found today as a parking brake for large trucks. All modern brake drums are the internal type, with the inner drum surface machined

smooth to provide a braking surface. Refer to **Figure 14-27.** The drum outer surface is heavy to absorb heat while the center section is relatively thin.

Drum Attachment Methods

Drums used on modern rear-wheel drive vehicles are attached to the hub or axle flange by the wheel lugs when the tire and rim are installed, **Figure 14-28.** The lug bolts are pressed into the hub flange and pass through holes in the brake drum. If the brakes have not been serviced, the drum may be held against the flange with two or three flat, spring steel speednuts. These nuts are for vehicle assembly line purposes only and do not require reinstallation when the drum is reinstalled.

A variation of this design uses 1-3 tapered screws or bolts to hold the drum to the axle flange, **Figure 14-29.** The wheel lugs pass through the drum, and the drum can be pulled from the hub once the screws are removed. The screws should be reinstalled after the brakes are serviced.

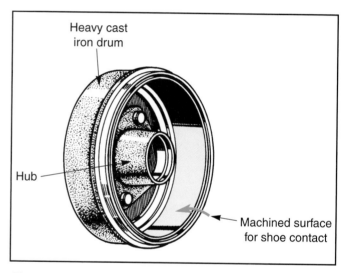

Figure 14-27. *A cast iron brake drum with a machined internal braking surface for lining contact. (Nissan)*

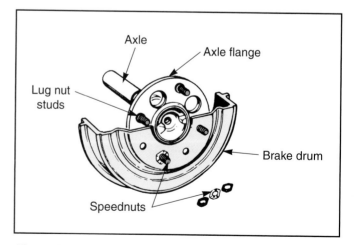

Figure 14-28. *A brake drum which is held onto the axle flange lug bolts (studs). Note the factory installed speednuts. (EIS)*

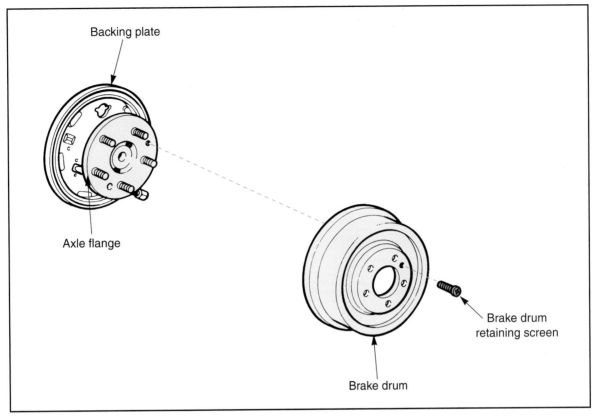

Figure 14-29. *An exploded view of a rear brake assembly. The brake drum is held to the axle flange with a retaining screw. (Toyota)*

Other drums are integral with the hub, as shown in **Figure 14-30.** These drums are removed by removing the wheel bearing nut and pulling the hub from the spindle or by pulling the axle assembly. Integral hubs are used on the front wheels of some older rear-wheel drive vehicles, the rear wheels of many modern front-wheel drive vehicles, and on the rear axles of large trucks and vans.

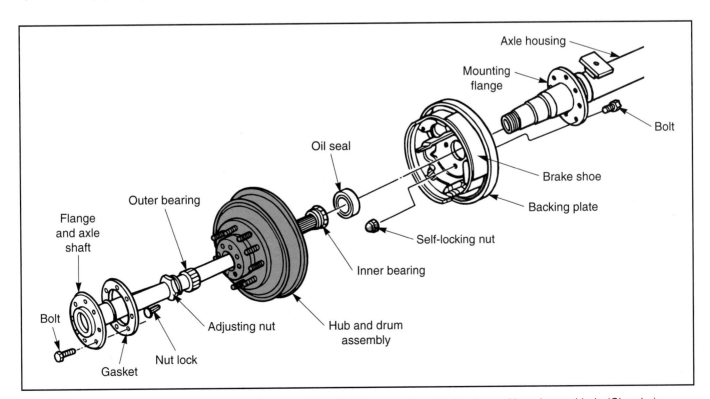

Figure 14-30. *An exploded view of a rear brake assembly which incorporates a brake drum with an integral hub. (Chrysler)*

Drum Materials

Almost any metal can be used to make a brake drum, and many metals have been tried in the past and discarded. Brake drums manufactured in the last 30 years have been made of cast iron, or aluminum with a cast iron liner. Each type is discussed in the following paragraphs.

Cast Iron Drums

Cast iron drums are durable since cast iron wears and withstands heat very well. Cast iron contains **graphite** (a soft form of the element carbon), which adds flexibility and corrosion resistance, and helps retard scoring caused by grit. Cast iron drums are usually thick and heavy to absorb the heat generated during braking.

The weight of cast iron makes the drums rigid, which reduces flexing under extreme brake pressures. A steel backing plate is cast into the center section of some large drums. The steel backing plate gives the drum extra strength. A cross-section of a cast iron drum is shown in **Figure 14-31.**

Aluminum Drums

An aluminum drum is shaped like a cast iron drum. It consists of a cast iron **liner** (braking surface) surrounded by an aluminum housing. The drum is made by casting melted aluminum around the cast iron liner. The casting process creates a permanent bond between the cast iron and aluminum. Heat can travel quickly between the liner and the aluminum housing. This design gives the braking surface life of a cast iron drum with more rapid dissipation of heat through the aluminum housing. Increasing use of front disc brakes has reduced the use of this design. An aluminum drum with a cast iron liner is shown in **Figure 14-32.**

Drum Cooling Provisions

To dissipate more heat to the surrounding air, the external surface area of a brake drum can be increased by the use of fins or ribs. These ribs increase the contact between the drum metal and the surrounding air, allowing heat to be removed quickly. **Axial ribs,** or fins, are ribs which run parallel around the drum. **Cross ribs** run across the drum at right angles to the braking surface. Axial ribs and fins are illustrated in **Figure 14-33.** Not all drums have fins or ribs.

Drum Wear Marks

As the drum is used, some of the metal is removed. More metal is removed when the drum is machined (turned) to remove scoring. Most brake drums are marked with a dimension indicating the maximum amount of metal that can be removed. This wear limit is called the **maximum diameter** or the **discard diameter.** The wear limit can be given in thousandths of an inch or in millimeters. The dimension is cast or stamped on when the drum is formed. Wear limits are usually located on the face (front) of the drum. See **Figure 14-34.** A drum that exceeds

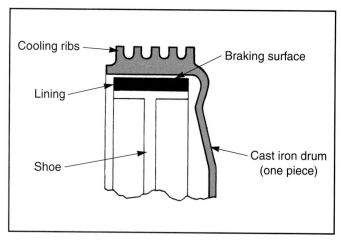

Figure 14-31. *A cross-section of a cast iron brake drum. Note the use of cooling ribs to help reduce heat buildup during braking. (FMC)*

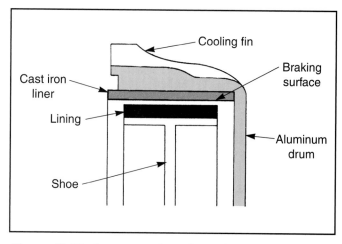

Figure 14-32. *A cross-section of an aluminum brake drum which incorporates a cast iron ring for the braking surface. (Bean)*

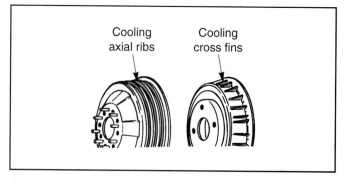

Figure 14-33. *To aid in cooling, brake drums can use axial or cross ribs, as well as fins. (FMC)*

the maximum allowable diameter following machining must be discarded. If the wear limits are exceeded, the drum will be too thin to absorb heat properly, and brake fading may occur even under normal braking. In addition, a thin drum can warp, crack, or even shatter during hard braking.

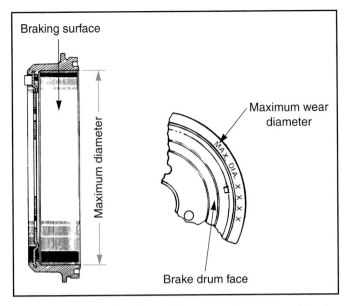

Figure 14-34. *A brake drum with cast maximum wear diameter tolerances. (Nissan & Ford)*

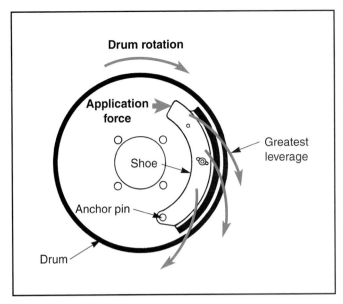

Figure 14-35. *A brake shoe being self-energized (pulled into) as it contacts the spinning brake drum. (Bean)*

Drum Brake Operation

Basic drum brake operation is simple: the wheel cylinder pushes the shoes into contact with the rotating drum, and the resulting friction slows the drum. However, other forces assist wheel cylinder apply pressure. The exact effect of these forces depends on the brake design.

Two types of drum brake designs are used on modern automobiles and light trucks: the servo brake and the non-servo brake. Servo brakes are more efficient, but cost more to manufacture and wear faster than non-servo brake designs. As a general rule, larger vehicles use servo brakes, while smaller vehicles have non-servo brakes. Non-servo brakes are common on front-wheel drive vehicles. Both types of brakes make use of the tendency of drum brakes to assist the driver's foot pressure when the brakes are applied.

Braking Assist

Pressing harder on the brake pedal does not increase braking effect by the exact amount of pressure. Factors such as shoe anchor pin location and the direction of drum rotation can increase or decrease the effects of more pedal pressure. The first of these factors is called *brake shoe self-energizing,* and the second is *brake shoe leverage.*

Brake Shoe Self-Energizing

When the shoe is anchored, **Figure 14-35,** as the drum rotates, friction at the point where the lining contacts the drum will try to pull the lining into the drum. This wedges the lining into the drum with more force than was originally applied. The faster the vehicle is moving, the greater the drum rotation, and the greater the wedging effect. This is the self-energizing feature of drum brake systems.

The farther the shoe anchor pin is located from the drum, the greater the wedging effect. However, once hydraulic pressure is released, brake lining contact will have a tendency to force the lining away from the drum. This is called de-energizing. This action is shown in **Figure 14-36.** If the brake assembly uses two shoes, one shoe will always be self-energized and one de-energized depending on the direction of brake drum rotation. Other design factors can cause both shoes to be self-energizing.

Brake Shoe Leverage

The actuating force that is applied to one end of a brake shoe produces leverage when the other end of the

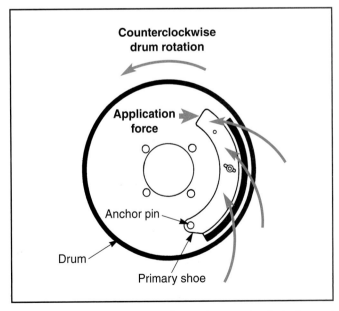

Figure 14-36. *The brake drum is rotating counterclockwise and is trying to force the shoe away from the braking surface. Compare this action to **Figure 14-35.** (FMC)*

shoe is solidly anchored. This force turns the shoe into a lever. Lever effect assists wheel cylinder pressure and self-energizing force to press the lining into tighter contact with the brake drum. This leverage increases as drum and shoe sizes increase. The amount of leverage is designed into the shoe by the manufacturer. The effect of shoe leverage is illustrated in **Figure 14-37.**

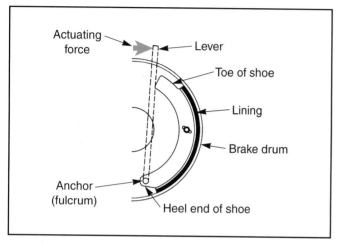

Figure 14-37. The effect of brake shoe leverage. The farther the actuating force moves up the lever from the pivot point, the greater the leverage will become. A longer brake shoe will produce more leverage than a short shoe, if both receive the same amount of actuating force. (Wagner)

Servo Brakes

Servo brakes are sometimes called *duo-servo brakes* since both the primary and secondary shoes contribute to the normal braking process. The servo brake system uses a dual piston wheel cylinder at the top of the backing plate. The bottom of the shoe assembly is not attached to the backing plate. Instead, the shoes are connected through a floating star wheel adjuster.

When the brake pedal is pressed, both shoes are forced out against the rotating brake drum by the wheel cylinder pistons. When the primary shoe comes into contact with the drum, it tries to rotate with the drum. This rotation is transferred to the secondary shoe through the floating star wheel adjuster. This force transfer is called **servo action.** Servo action causes the secondary shoe, after it contacts the rear shoe anchor, to be jammed up and out at the same time. Both shoes try to rotate with the drum, but are held stationary by the anchor. The self-energizing effect of the shoes wedges them against the drum. Note that in this design, both shoes are self-energizing. This compounds the secondary shoe's self-energized braking.

If the vehicle is braked when moving in reverse, **Figure 14-38,** the primary shoe acts as the secondary shoe, and brake effectiveness remains almost as good as when braking while going forward.

In a servo system, the primary shoe performs less of the braking job than the secondary shoe. As a result, it does not wear as fast, and most primary shoes have a shorter lining with less frictional area. The secondary shoe does more of the braking job. The lining is longer (more frictional area) than the primary shoe.

Non-Servo Brakes

Non-servo brakes are used on smaller vehicles, often with front-wheel drive. On these vehicles, the rear brakes carry only a small portion of the braking load and can be designed more simply.

Non-servo brakes use the same type of wheel cylinder as servo brakes. Wheel cylinder pressure pushes both brake shoes outward. The front shoe lining is self-energizing as it presses against the fixed anchor point. However, the rear lining is pushing against the rear piston of the wheel cylinder and has no self-energizing action. The rear brake shoe is applied only by hydraulic pressure behind the wheel cylinder piston. This is enough to cause the rear lining to solidly contact the drum and provide braking action. **Figure 14-39** shows the action of a non-servo brake assembly. In reverse, the rear shoe is self-energizing while the front shoe is applied by wheel cylinder pressure only.

Combination Disc/Drum Brakes

Some vehicles have combination disc and drum brakes. The drum brake assemblies are located in the center of the rotor and are used as parking brakes. The drum brakes are operated mechanically and there is no wheel cylinder. Chapters 18 and 19 cover these brakes in more detail.

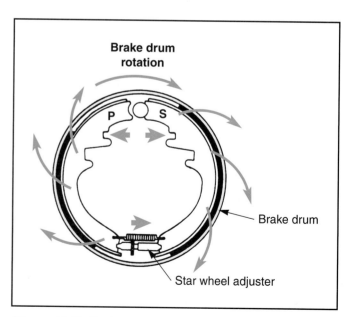

Figure 14-38. Brake shoe action with the vehicle moving in a reverse direction. (Wagner)

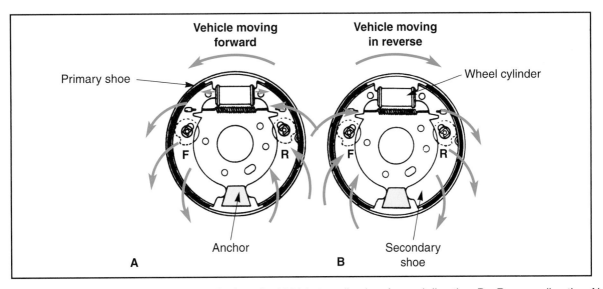

Figure 14-39. *Brake shoe action with non-servo brakes. A—Vehicle traveling in a forward direction. B—Reverse direction. Note that in reverse, the secondary shoe acts as the primary shoe. (Wagner)*

Summary

Modern drum brake components are always located inside the drum. Drum brakes are sometimes called internal expanding brakes, since the shoes must expand to contact the drum. Drum brake components are similar between all manufacturers.

The backing plate is the stationary foundation for the brake shoes and related parts. The backing plate contains anchors which secure the shoes and springs. The wheel cylinder is usually mounted at the top of the backing plate. Brake shoes hold the lining and move it in and out of contact with the brake drum. The main brake shoe components are the table and web.

Brake linings are attached to the shoe with rivets or by bonding. Lining to drum contact is important for good brakes. To maintain this contact, linings are ground eccentric (thicker in the middle) or concentric (the same thickness over the entire lining). The most common method on cars and light trucks is the eccentric grind.

Brake shoes are held in position by the hold-down springs. When wheel cylinder pressure is released, the shoes are retracted by return springs. Other springs are used to prevent rattles.

Shoes are adjusted with cams, toothed ratchets, or star wheels. Ratchet and star wheel adjusters can be operated by hand or automatically by self-adjusters. All modern vehicles have self-adjusters.

Drums are made from cast iron or aluminum with a cast iron braking surface. Cooling fins or ribs are molded onto the external surface of most brake drums. These can be axial ribs or cross ribs. Drums are marked with maximum wear and discard numbers.

Some drums are held in place by the wheel rim when the lug nuts are installed. Other drums are integral with the hub, and are held on by the wheel bearing assembly.

Brake shoes are actuated by the wheel cylinders. Brake pedal force is assisted by self-energizing action of the brakes. Self energizing is the tendency of the brake lining to wedge itself into the rotating drum. Braking force is also assisted by leverage created by the placement of the shoe anchors.

Brakes can have either servo or non-servo action. Servo action uses the rotation of the drum to assist braking. Non-servo brakes are usually used on the rear brakes of smaller vehicles where braking force is not as critical.

Review Questions—Chapter 14

Please do not write in this text. Write your answers on a separate sheet of paper.

1. Why are modern drum brakes sometimes called internal expanding brakes?

2. The close fit between the _____ and drum keeps dirt and water out of the drum.

3. The table and web are parts of the brake _____.

4. On most vehicles, the rear drum brakes are used as the _____ brakes.

5. Linings should not fade under _____ braking conditions.

6. Large linings are relatively _____ to reduce heat buildup.

7. The primary shoe faces the _____ of the vehicle.

8. Does the primary or secondary shoe perform more of the braking effort?

9. Eccentric grinding of the lining compensates for shoe and drum _____ during hard braking.

10. Match the type of spring with its function. Some answers are used more than once.

_____ Coil spring hold-down   (A) Returns shoes to unapplied position.
_____ Spring return
_____ Spring clip hold-down   (B) Holds shoes to backing plate.
_____ Anti-rattle spring   (C) Reduces noise.
_____ Retracting spring

11. For more heat dissipation, the outer surface of the drum is _____.

12. What does the maximum diameter of the drum refer to?

13. Self-energizing shoes are _____ into tighter contact with the drum when they are applied.

14. The amount of brake shoe leverage depends on where the shoe is _____.

15. Why are non-servo brakes used on the rear wheels of some vehicles?

ASE Certification-Type Questions

1. The backing plate is a stationary support for all of the following, EXCEPT:
   (A) the anchor pins.
   (B) the anchor plates.
   (C) the drum.
   (D) the brake shoes.

2. Technician A says that the primary shoe always faces the front of the vehicle. Technician B says that the primary shoe performs more of the braking job. Who is right?
   (A) A only.
   (B) B only.
   (C) Both A & B.
   (D) Neither A nor B.

3. The brake shoe _____ holds and supports the lining.
   (A) table
   (B) web
   (C) rivets
   (D) cement

4. Technician A says that eccentric ground brake shoes are used on large trucks. Technician B says that eccentric ground brake shoes allow for shoe and drum flexing. Who is right?
   (A) A only.
   (B) B only.
   (C) Both A & B.
   (D) Neither A nor B.

5. Shoe return springs are always _____ springs.
   (A) coil
   (B) clip
   (C) flat strip
   (D) torsion

6. The keystone anchor is used to allow the brake shoes to centralize against the _____.
   (A) drum
   (B) backing plate
   (C) wheel cylinder pistons
   (D) hold-down springs

7. Technician A says that adjusting cams are seldom used on modern brakes. Technician B says that there is one adjusting cam for each shoe. Who is right?
   (A) A only.
   (B) B only.
   (C) Both A & B.
   (D) Neither A nor B.

8. A floating star wheel assembly is not attached to the _____.
   (A) primary shoe
   (B) secondary shoe
   (C) pivot nut
   (D) backing plate

9. All of the following are types of star wheel self-adjusters, EXCEPT:
   (A) cable.
   (B) bar.
   (C) lever.
   (D) link.

10. All of the following statements about brake drums are true, EXCEPT:
    (A) brake drums can be made of cast iron or aluminum.
    (B) modern brake drums are machined on their internal surface.
    (C) brake drums have openings in the lining to aid in cooling.
    (D) most brake drums have markings to indicate maximum allowable wear.

11. The _____ links the shoes together at the bottom on a servo brake assembly.
    (A) keystone anchor
    (B) star wheel adjuster
    (C) hold-down spring(s)
    (D) tensioner clip

12. Technician A says that the self-energizing principle works only on non-servo brakes. Technician B says that on a non-servo brake, one shoe is self-energizing and the other shoe is not. Who is right?

    (A) A only.
    (B) B only.
    (C) Both A & B.
    (D) Neither A nor B.

13. Technician A says that manual adjusters cannot be used with non-servo brakes. Technician B says that a manual adjuster is turned by reaching through an access hole in the backing plate. Who is right?

    (A) A only.
    (B) B only.
    (C) Both A & B.
    (D) Neither A nor B.

14. Which of the following terms describes a brake design in which neither shoe assists in the application of the other?

    (A) Servo.
    (B) Duo-servo.
    (C) Self-energizing.
    (D) Non-servo.

15. On a servo brake assembly, the _____ shoe does most of the job of braking.

    (A) primary
    (B) secondary
    (C) front
    (D) leading

Fluid
seepage

Inspect all drum brake assemblies carefully before giving an estimate. The wheel cylinder on this drum brakes shows evidence of seepage.

Chapter 15

Drum Brake Service

After studying this chapter, you will be able to:

- ❏ Identify common drum brake problems.
- ❏ Diagnose drum brake problems.
- ❏ Remove and replace brake drums.
- ❏ Check brake shoe condition.
- ❏ Check brake springs for damage.
- ❏ Check condition of brake shoe adjuster mechanisms.
- ❏ Check drum diameter and compare with maximum wear limits.
- ❏ Remove and replace brake shoes.
- ❏ Refinish brake drums.
- ❏ Adjust drum brake clearance.

Important Terms

Integral	Taper	Cutting bar	Cutter feed
Non-integral	Arc grinding	Bit assembly	Bar feed dial
Rubber plugs	Arbor	Cutting bits	Collar
Knockout plug	Adapters	Silencer band	

Although disc brakes are now used on the front wheels of all modern vehicles, drum brakes are often found on the rear wheels of even the newest cars and trucks. Therefore, thorough knowledge of drum brake service is necessary in order to properly repair modern vehicles. This chapter covers the inspection and repair of drum brakes.

Common Drum Brake Problems

Typical problems caused by the drum brake system include poor braking, fading, dragging, pulsation, noises, and leaks. Before disassembling the wheel brakes, make a brief road test to check brake operation.

Warning: Do not attempt a road test if the brakes are completely inoperable. If the vehicle does not have good brakes, carefully drive the vehicle into the shop and inspect the brake system.

Worn and glazed linings are the most common drum brake defects. Poor braking or a hard pedal are commonly caused by glazed or worn linings. If the linings are completely worn, applying the brakes may cause squeaks, chatter, and/or vibration. A low pedal is another sign of worn linings.

Brake system noises consist of squeaks, chatter, rubbing noises, or clicks. Squeaking drum brakes are usually the result of worn or glazed linings, combined with an overly smooth drum surface. Brake chatter is usually caused by glazed linings, grease, or brake fluid on the linings. Rubbing noises are the result of metal-to-metal contact when the linings are completely worn down. Clicking noises are caused by broken or incorrectly installed hardware.

If the brakes operate normally when cold, but lose braking efficiency during hard stops, the brakes are fading. Fading is usually the result of thin drums, caused by long use or improper turning, combined with glazed and possibly thin linings. If the brakes fade only when they are driven through water, water is entering the drums.

Brake pedal pulsation is almost always caused by an out-of-round drum. Occasionally, oil on the linings, worn wheel bearings, or loose brake hardware can cause pulsation. Brake grabbing is usually caused by brake fluid or rear differential lube on the linings. Weak retracting springs, missing holddown springs, or backing plate damage can also cause grabbing.

Dragging brakes can be caused by improper manual adjustment, or a defective self-adjuster. A weak or broken retracting spring may cause the associated brake shoe to drag when the brakes are released. A very tight brake can cause wheel lockup. The shoes freezing to the backing plate in very cold weather due to poor lubrication can also cause brake drag.

If any of these problems occur erratically, (on and off), the problem could be a loose or cracked backing plate, loose anchors or wheel cylinders, or linings contaminated by brake fluid or oil. If any problems are noted during the road test, the brake drums must be removed and the brakes inspected and if necessary repaired. These problems and their possible causes are listed in **Figure 15-1.**

 Note: Before deciding that a problem is in the drum brakes, thoroughly check the front disc brake assembly (if used) hydraulic system, power assist, (when used), steering and suspension parts, and tires.

Drum Brake Problems	
Problem	**Possible Causes**
Poor braking or hard pedal	Power booster malfunction. Glazed or worn linings.
Low pedal	Linings worn. Linings out of adjustment. Air in hydraulic system.
Brakes fade	Thin drums. Glazed or worn linings. Water entering drums.
Rubbing or grinding noise	Linings worn metal-to-metal.
Brakes pulsate when applied	Out-of-round drum (if pulsation felt in seat of the pants). Oil on linings, wheel cylinder or axle seal leaking. Normal ABS operation.
Drum brakes drag	Improper manual adjustment. Defective self-adjuster. Weak retracting springs.
Brakes grab	Brake fluid or axle lubricant on linings. Weak retracting springs. Backing plate damage.
Brakes chatter	Glazed linings. Brake fluid or grease on linings.
Brakes click	Broken or incorrectly installed hardware.
Front Drum Brakes	
Front drum brakes pull to one side	Worn or misadjusted linings. Oil or brake fluid on linings. Clogged hose or sticking wheel cylinder. Weak retracting spring. Water entering drum.

Figure 15-1. *Chart listing drum brake problems and their causes.*

Brake Component Inspection

The following sections explain how to remove the brake drum and inspect the brake shoes, drums, springs, and other brake system components. This section briefly reviews the methods of checking for wheel cylinder leaks and other damage.

 Warning: Brake dust may contain asbestos. Be sure to wear a respirator when working on drum brakes.

Removing the Brake Drum

If you find any evidence of improper drum brake operation during the test drive, remove the drums and inspect the brakes. While disc brakes can be quickly inspected once the wheel is removed, the components of a drum brake assembly can only be checked by removing the drum. Drums are *integral* (one piece with a hub) or *non-integral* (separate from the hub). Removal procedures for each type are covered in the next sections.

 Note: Before removing the drum, use a hammer to deliver 1-2 light to medium blows. This helps to loosen the drum and causes loose brake dust to settle to the bottom of the drum.

Removing Non-Integral Drums

If the drums are separate from the hub or axle flange, they can be removed after removing the tire. Speednuts may be pressed over 2-3 studs if the drum is being removed for the first time. They can be removed using wire cutters or a screwdriver and do not have to be replaced when the drum is reinstalled. A few drums are held in place by 1-2 tapered screws. Remove these screws or bolts with the proper size Phillips screwdriver or socket and ratchet before attempting to remove the drum. Refer to **Figure 15-2.**

Note: An impact driver with bit may be needed to loosen drums held by Phillips screws.

If the drum will not slide off the hub, the shoes may be in close contact with the drum's inner surface. Sometimes the linings and drum are scored, and the interlocking scoring is preventing drum removal. To remove the drum in these cases, you must retract the linings by backing off the adjuster.

Begin by removing the *rubber plug* from the adjuster access hole. The plug may be located at the bottom or top

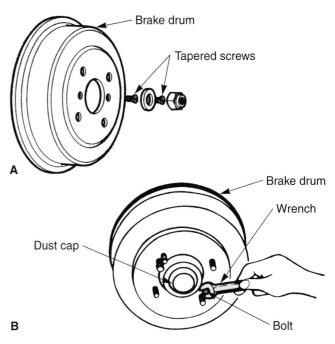

Figure 15-2. *A—Brake drum held in place with tapered head screws. B—Drum held to the hub with a bolt.* *(Bendix & British-Leyland)*

of the backing plate, directly under the wheel cylinder, or in the drum itself, depending on the brake design. **Figure 15-3** shows a typical plug location. If the hole contains a *knockout plug*, which is a metal slug in the backing plate, knock it out with a small punch. Then reach through the hole with an ice pick or small screwdriver and disengage the self-adjuster lever.

Note: On some vehicles, there is no provision for manually adjusting the brakes. If the drum and shoes are scored so the drum cannot be removed, first check for a knockout plug in the drum. If no plug is found, you may have to use diagonal pliers to clip off the holddown retainer pins, use a jaw puller to remove the drum, or in severe cases, cut an access hole in the drum using a torch. Before deciding that the adjuster must be backed off, make sure the parking brake has been released.

With the lever out of the way, insert a brake adjusting tool through the hole and engage the teeth in the star wheel, **Figure 15-4.** If the lever was under the star wheel, move the free end of the adjusting tool downward to turn the star wheel and increase the shoe-to-drum clearance. If the lever was on top of the star wheel, move the free end of the adjusting tool upward. Turn the star wheel until there is no brake shoe-to-drum contact when the drum is rotated. Once the linings are backed off, tap gently on the drum to loosen it, then pull it from the hub. It is acceptable to lightly pry between the drum and backing plate to loosen the drum.

Note: In many cases, the drum will stick to the flange at the center of the axle due to corrosion. Do not pound on the centering flange to remove the drum, as this will flatten and deform the flange, making drum removal extremely difficult.

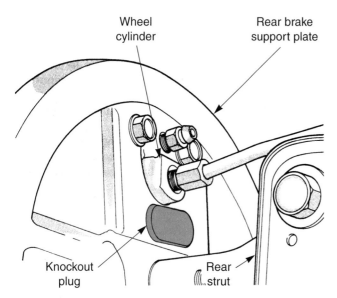

Figure 15-3. *Adjusting hole plug located under the wheel cylinder. (Chrysler)*

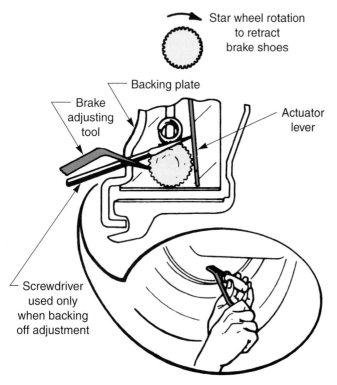

Figure 15-4. *Backing off the star wheel by using a screwdriver and brake adjusting tool. The screwdriver is not used when adjusting brakes. (Chevrolet)*

Removing Integral Drums

To remove an integral drum assembly, remove the grease cap, **Figure 15-5A,** to expose the wheel bearing retainer nut. Note that the nut is threaded onto the spindle. The nut is held by a cotter pin, either through slots in the nut itself, or through a slotted cover over the nut. Remove this cotter pin and unthread the nut from the spindle. Once the nut is removed, remove the washer and place both in the grease cap so they can be located later. Discard the cotter pin; do not reuse.

Carefully pull the drum and hub unit from the spindle, **Figure 15-5B.** Be careful not to drop the outer wheel bearing after the drum is removed from the spindle. If the drum will not slide off easily, back off the adjuster using the same procedure as for non-integral drums.

Warning: Do not step on the brake pedal with a brake drum removed. Hydraulic pressure will cause the wheel cylinder pistons to pop out of the cylinder.

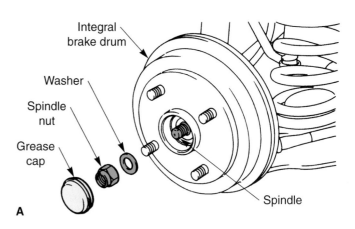

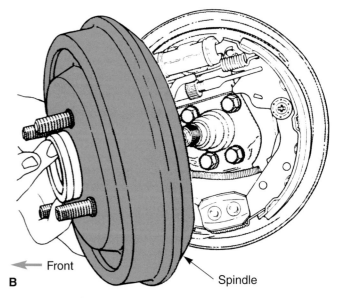

Figure 15-5. *A—Before the integral drum can be removed, the grease cap, spindle nut, washer, etc., must first be taken off. B—When removing the drum, do not drag the rear seal on the spindle threads. Seal damage will result. (Chevrolet)*

Cleaning Drum Brake Assemblies

After the drum has been removed, you should clean the brake assembly to remove the brake dust. Do not use compressed air to remove brake dust from the brake components. Use an approved HEPA vacuum cleaner or a closed liquid cleaner system, **Figure 15-6.**

Figure 15-6. *Cleaning drum brakes using a closed liquid cleaner. A—Start at the top of the brake assembly. B—Work down the assembly, removing as much brake dust as possible. Allow the assembly to air dry.*

Clean the brake assembly, starting at the top and working down. If shop towels or rags are used to remove brake dust, store them in a closed container until they can be professionally cleaned, or dispose of them with other toxic waste.

When cleaning the drum brakes on a rear-wheel drive axle, check for leaking axle seals. A leaking axle seal will allow gear oil to leak onto the backing plate and the brake linings. A severely leaking seal or one that has been leaking for a long time will coat the entire brake assembly with grease. Gear oil has a distinctive odor and feel, and can be easily distinguished from brake fluid. If the seal is leaking it must be replaced, along with the brake shoes if they are coated. Often the seal fails because the bearing is defective. Refer to Chapter 17 for bearing and seal service.

Checking Lining, Shoe, and Hardware Condition

Once the drum has been removed, the brakes can be inspected. Some of these checks can be made visually, while others require the use of measuring tools.

Checking Linings and Shoes

Check the linings for wear and damage. If the linings are badly scored, glazed, cracked, or worn down to the shoe table or rivets, they must be replaced. **Figure 15-7** illustrates some common lining defects. Darkening of the lining is a normal condition.

If the linings appear to be in good condition, they should be checked for sufficient thickness. The linings can

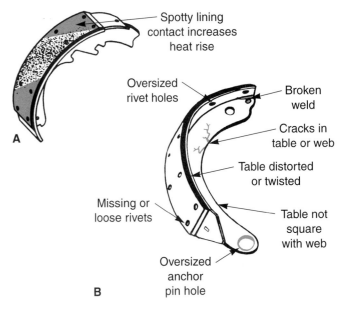

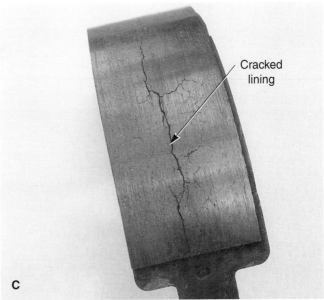

Figure 15-7. *Brake lining and shoe damage. A—Spotty lining-to-drum contact. B—An assortment of shoe damage. C—Lining that has cracked. (FMC, Atlas & World Bestos)*

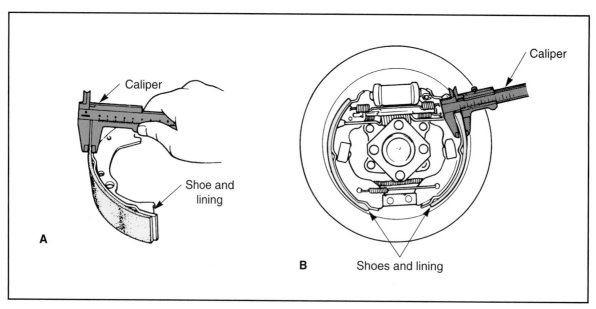

Figure 15-8. *Measuring lining thickness with a caliper. A—Measurement with shoe and lining off the vehicle. B—Measurement being obtained with the shoe and lining on the vehicle. (Chevrolet & Mazda)*

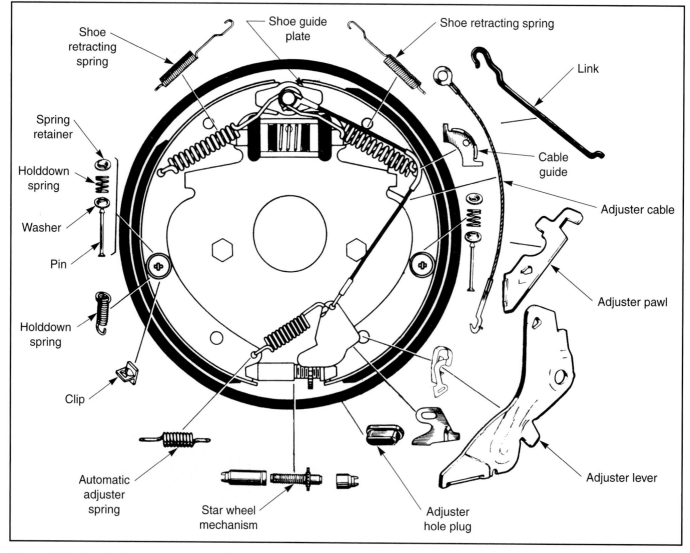

Figure 15-9. *Check all springs and other hardware for signs of damage. Replace as needed. (EIS)*

be measured for thickness using a caliper, **Figure 15-8.** Then compare the thickness against specifications. Many manufacturers consider the shoe linings to be acceptable if they extend about .030″ (.8 mm) above the rivets on riveted shoes, or about .060″ (1.6 mm) above the shoe table. If specifications are not available, ensure the linings are at least the thickness of a dime above the shoe table or rivet heads when the lining is riveted. A dime is approximately .060″ (1.6 mm) thick.

These specifications are the absolute minimum at which the brake linings will function adequately. It is considered good practice to replace the shoes when the linings are worn to the point they will need replacement within the next few thousand miles or kilometers. The small amount of extra use will not be worth the cost and inconvenience to the owner in bringing the vehicle back or replacing a drum damaged as a result of metal-to-metal contact with the shoes.

As part of the lining inspection, check the shoe web and table for damage. While rare, shoe damage can be the cause of hard to find problems such as noises. Typical shoe damage includes cracks, distortion (bent areas), looseness between the table and web, and excessive wear at contact points. If the shoe metal is contacting the drum, the shoes should be replaced.

Checking Springs and Brake Hardware

Most springs and other brake hardware can be reused, **Figure 15-9.** However, they should be closely inspected for stretching and overheating. Self adjusters, links, holddown springs, and washers should be checked for wear and distortion. This may require partial disassembly of the shoe mechanism. If the brake assembly has been severely overheated, the springs may have lost their tension and should be replaced even if they look good. Any bent or worn parts should be replaced.

Checking Wheel Cylinder Condition

Check the wheel cylinders for leaks by pulling the dust boot away from the end of the cylinder as shown in **Figure 15-10.** If fluid leaks from the cylinder when the dust boot is pulled back, the wheel cylinder should be rebuilt or replaced. Slight moisture under the dust boot is normal. If water leaks from the cylinder when the dust boot is removed, the dust boot is defective. If water is present inside the wheel cylinder, rusting has occurred, and the wheel cylinder should be rebuilt or replaced. Wheel cylinder removal and overhaul was covered in detail in Chapter 10.

Check the wheel cylinder mounting and hose connections. Mounting fasteners should be tight and there should be no cracks in the cylinder. The hose connection should be tight with no leaks. Any damage means the wheel cylinder should be replaced. Finally, loosen the bleeder screw. If the bleeder cannot be loosened, the wheel cylinder cannot be bled and should be replaced.

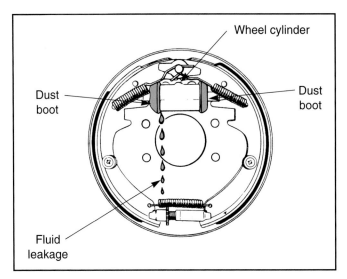

Figure 15-10. *A leaking wheel cylinder. Boot has been pulled away from the cylinder, allowing the leaking brake fluid to drip out. (FMC)*

Checking Brake Drum Condition

The drums should be inspected for evidence of scoring. If they are severely scored, they should be replaced. If the drums are smooth or only slightly scored, use a drum micrometer, **Figure 15-11,** to check diameter. Set the micrometer to the proper size. Then place the micrometer so the pointers contact the drum's inner braking surface. Slightly move the micrometer in the drum as shown until the maximum diameter is recorded.

When checking the drum size, keep in mind that turning the drum will increase its diameter. Therefore, a drum that is already close to its maximum diameter will become unusable if it needs to be turned to remove scoring. If the diameter of a drum is too large, or turning it makes it too large, it should be replaced.

Check the drum size at several places around the drum. See **Figure 15-11.** Ideally, the drum diameter should be checked every 45° (1/8 turn). Variations in drum diameter at different positions indicates the drum is out of round. This can be corrected by turning, as long as the drum minimum thickness is not exceeded.

Also check the drum for *taper.* Taper occurs when the outer diameter of the drum is greater than the inner diameter. Checking the diameter with the micrometer with the pointers at different depths will check for taper.

If the drums are integral types, make sure they are solidly mounted to the hub. If there is any play between the drum and hub, discard the drum. Check that the lugs are solidly mounted in the hub, and not damaged. Damaged lugs can be replaced without replacing the drum assembly. This is discussed in Chapter 17. On a non-integral drum, check the holes through which the lugs pass are not elongated or damaged, and the center of the drum is not cracked or badly deformed. Some common drum defects are shown in **Figure 15-12.**

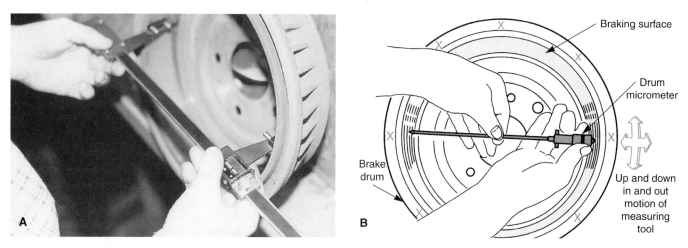

Figure 15-11. *A—Using one particular type of brake drum micrometer to check for wear on the braking surface. B—Check the drum braking surface at several points. This will provide the most accurate reading. (General Motors)*

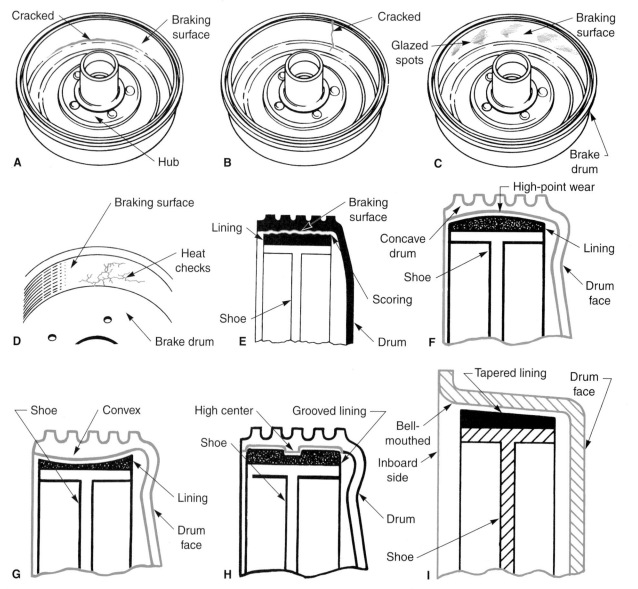

Figure 15-12. *Some common brake drum conditions. A—Cracked drum—excessive stress or too thin. B—Broken across the braking surface. C—Glazed spots. D—Heat checks in the metal caused by excessive heat. E—Scoring from dirt or sand. F—Concave drum. G—Convex drum. H—High center wear from grooved lining. I—Bell-mouthed or tapered drum. This condition is normally caused by a drum which is not rigid and/or is too wide. (Bean, Wagner, Bendix & EIS)*

Shoe and Hardware Service

All modern drum brakes use a single wheel cylinder and two shoes, and are similar in other basic components and operation. However, there are many minor differences, especially in attachment methods and whether the brake design is servo or non-servo. These differences can affect service procedures and are addressed where they apply.

Performing Drum Brake Work on ABS Equipped Vehicles

Many of the most common drum brake service procedures, such as shoe replacement, drum service, and wheel bearing replacement, are not affected by the presence of anti-lock brake (ABS) components. If the lining or friction service replacement procedures involve the wheel speed sensors, treat them gently, and recheck the gap where applicable. Remember not to drop or hammer on the sensor rings, or use them to pry on other components.

Removing the Brake Shoe Assemblies

The removal of brake shoes and associated hardware varies according to the brake design. The following are general brake removal steps. Always consult the proper service manual for complete removal and installation procedures.

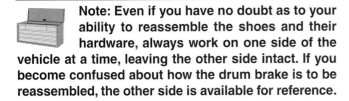

Note: Even if you have no doubt as to your ability to reassemble the shoes and their hardware, always work on one side of the vehicle at a time, leaving the other side intact. If you become confused about how the drum brake is to be reassembled, the other side is available for reference.

Servo Brake Shoe Removal

On most servo brake assemblies, both shoes can be removed as a unit with the star adjuster. To begin removal of the brake shoes, remove the shoe retracting springs from the backing plate anchor using a brake spring tool, **Figure 15-13.** The retracting springs are not interchangeable from front to rear and must be kept separate. Most retracting springs are color coded to aid in proper reassembly. If the self-adjuster link or cable eyelet is installed on the spring anchor, remove it now.

Once the retracting springs are removed, remove the holddown springs. **Figure 15-14** illustrates a typical holddown spring removal tool. You will probably need to hold the pins at the rear of the backing plate to prevent them from turning. Some holddown springs can be removed without a special tool. Place the springs, washers, and pins in a safe place. Many holddown springs are not interchangeable between the front and rear shoes. Non-interchangeable springs are usually painted different colors, and should be kept separate. Note that some

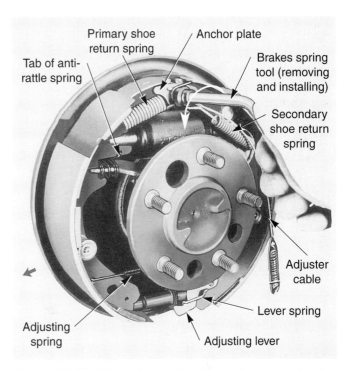

Figure 15-13. *Removing brake shoe return or retracting springs using a special spring tool. Keep all springs and parts in their proper order. (Chrysler)*

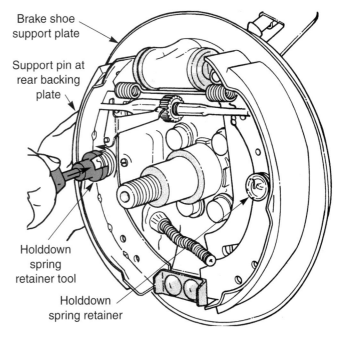

Figure 15-14. *Using a holddown spring retainer tool to remove the spring retainer. (Chrysler)*

rear shoe holddown springs also retain the self-adjuster lever, and use slightly different washers than the front springs.

Once the holddown springs are removed, remove the self-adjuster mechanism, **Figure 15-15.** It is not necessary to remove the star wheel at this time. If the brake assembly incorporates the parking brakes, it may be necessary to remove the parking brake lever from the rear shoe. Some parking brake levers are held to the rear shoe with a clip. This clip must be pried off before the shoes can be removed. On other vehicles, the lever fits into a slot on the rear shoe, and the lever can be removed after the hold-down springs are detached.

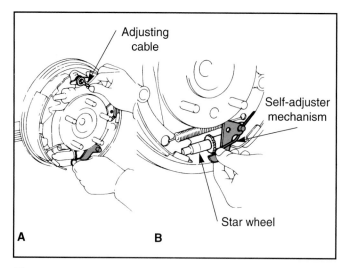

Figure 15-15. A—Removing the adjusting cable. B—Once the adjuster cable is free, the self-adjuster mechanism may be removed. (Toyota)

Once all attaching hardware has been removed, remove the shoes and star wheel assembly as a unit from the backing plate, **Figure 15-16.** On some vehicles, you may need to place a clip over the wheel cylinder to keep the pistons from popping out, **Figure 15-17.** Place the shoes and star wheel assembly on a workbench and remove the adjuster tensioning spring and adjuster from the shoes. If the parking brake assembly uses a strut rod between the front and rear shoes, remove it from the backing plate if necessary.

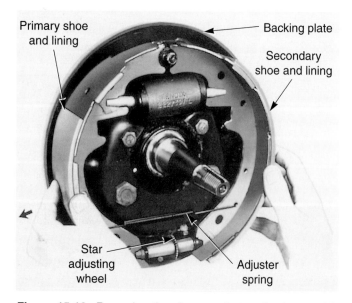

Figure 15-16. Removing the shoes and star wheel assembly from the backing plate. (Chrysler)

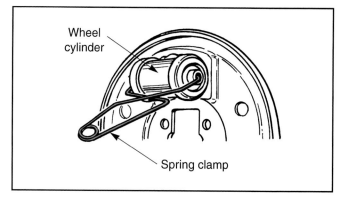

Figure 15-17. Spring type clamp is used to keep some wheel cylinder pistons from popping out of their bores from static (remaining) hydraulic line pressure. (FMC)

Non-Servo Brake Shoe Removal

Many non-servo brakes have a single retracting spring with an end attached to each shoe, as in **Figure 15-18A.** If the brakes are the non-servo type, each shoe can be removed separately, **Figure 15-18B.** Begin by removing the shoe retracting springs. Then remove the holddown spring from the rear brake shoe and remove the parking brake linkage clip if necessary. If the bottom of the shoe is anchored by a pin, remove the pin retainer. Remove the shoe from the backing plate. If necessary, place the shoe on a workbench and remove the self-adjuster linkage.

Remove the holddown spring from the front brake shoe, the anchor pin retainer if used, and remove the shoe from the backing plate. If necessary, remove the adjuster mechanism from the shoe.

Removing the Wheel Cylinder

Many wheel cylinders can be overhauled without removing them from the backing plate. This is possible after the shoes, springs, and other hardware have been removed. However, in most cases, the wheel cylinder can be easily removed from the backing plate. Remove the line or flexible hose from the wheel cylinder and cap the fitting. Then remove the attaching bolts or clips and pull the wheel cylinder from the backing plate, **Figure 15-19.**

 Note: If the wheel cylinder must be over-hauled, refer to Chapter 6 for procedures.

Removing the Backing Plate

The backing plate does not require removal unless it is damaged. To remove the backing plate, first remove the brake system parts, then loosen and remove the fasteners. On some rear-wheel drive vehicles, the axle must be removed to allow clearance for backing plate removal. Pull the backing plate away from the spindle or axle flange. Save any shims placed between the backing plate mounting surfaces.

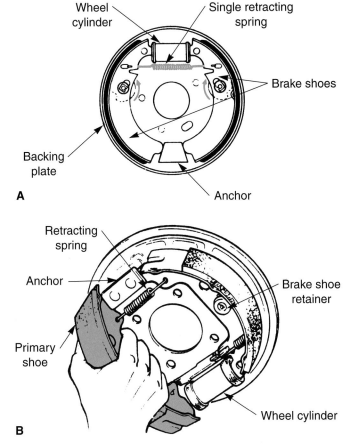

Figure 15-18. *A—A non-servo brake assembly which uses only one shoe retracting spring. B—Removing brake shoes separately on a non-servo type brake assembly. (Wagner)*

Cleaning and Inspection

Use a liquid cleaner or parts cleaning vat to clean all parts and the backing plate thoroughly. Do not use compressed air to dry the parts, allow them to air dry. Check the self-adjuster mechanism. Ensure the star wheel, when used, turns freely and the self-adjuster parts are not worn or binding. If necessary, the star wheel can be loosened by soaking in penetrating oil. If the star wheel cannot be made to turn easily, it should be replaced. If the self-adjuster is cable-operated, check the cable for fraying. If the cable has any broken strands, it should be replaced. If the ratchet adjuster's teeth appear to be worn or chipped, the adjuster should be replaced.

All brake springs should be checked for wear, **Figure 15-20.** A simple test often used by technicians to check springs after they are removed is to drop them on the shop floor. If a dropped spring makes a tinny sound when it strikes the floor, the spring has lost its tension and should be replaced. Note if paint has been burned off by excessive heat. Springs that show this sort of wear should be replaced. Some replacement brake shoes come with new springs.

Check the holddown springs and retainers for distortion and damage. Make sure the pins are not bent and replace any damaged parts. Check the retracting springs and other hardware for damage that might not have been

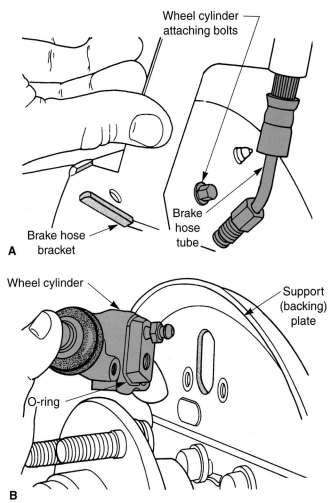

Figure 15-19. *Removing a wheel cylinder. A—Disconnect the brake hose and remove attachment fasteners. B—Remove the wheel cylinder from the backing plate. Note the O-ring seal on this particular type. The seal helps to keep water and dirt out of the brake assembly. (Chrysler)*

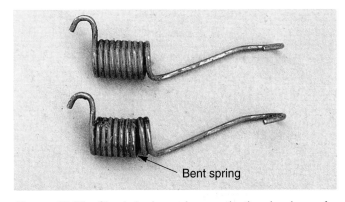

Figure 15-20. *Check brake springs and other hardware for damage. The top spring is ok, however, the lower spring is bent and should be replaced.*

visible with the shoes installed. If any parts are bent, worn, overheated, or otherwise damaged, replace them. Also check the backing plate for damage or wear where the shoes ride against the support pads, **Figure 15-21.**

Reinstalling the Brake Shoe Assemblies

Reassembly and reinstallation of the brake shoe assemblies varies somewhat, but can be performed by following the procedures in the following sections. Always consult the proper service manual for exact reinstallation procedures. If the backing plate was replaced, begin reassembly by installing the new backing plate.

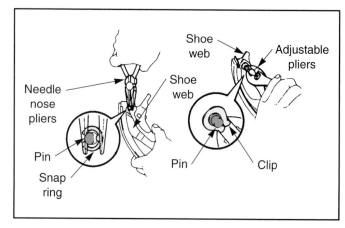

> ⚠️ **Warning: Modern brake shoes are designed to be installed without the need for arc grinding. Arc grinding creates large amounts of brake dust, which can be hazardous.**

Figure 15-22. *Installing pins and retainers to the brake shoe webs. (Toyota)*

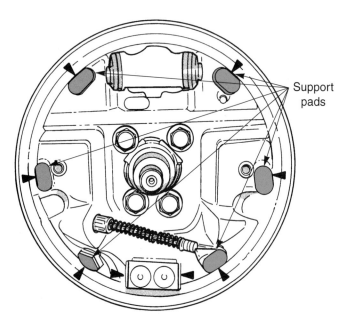

Figure 15-21. *Check the backing plate for damage and wear. Pay close attention for damage to the shoe support pads. (Plymouth)*

Reinstalling the Wheel Cylinder

Place the new or rebuilt wheel cylinder in place on the backing plate. Apply sealer around the wheel cylinder mounting area, if needed. Install the bolts or clips as applicable. Then install and tighten the flexible brake hose fitting. Finally, replace the piston to shoe links, if used. Refer to Chapter 6 for other information on wheel cylinder service.

Assembling the Brake Shoes and Related Parts

Before installing the brake shoes, compare them against the old shoes to ensure they are correct. If the shoes have different size linings, make sure the shoe with the smaller lining goes toward the *front* of the vehicle. Placing the shoe with the smaller lining toward the *rear* is a common mistake made by technicians.

Some brake shoes contain parts, such as pins and springs, which must be installed before the shoe is reinstalled on the backing plate. **Figure 15-22** shows a pin being installed on a brake shoe web.

Reassemble and lubricate the star wheel adjuster when necessary. Place a small amount of high temperature lubricant in the socket cavity, **Figure 15-23.** Do not lubricate the threads of the adjuster, as the lubricant will attract dirt and possibly jam the adjuster. Ratchet adjusters should be checked for free movement and lubricated when necessary.

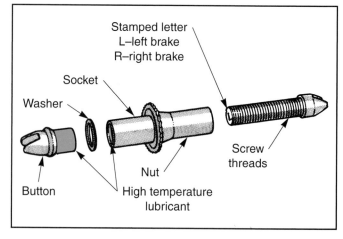

Figure 15-23. *Place a small dab of high temperature lube on the star wheel button and socket cavity. Wipe off excess that may come out when the star wheel is assembled. (Dodge)*

Reinstalling Servo Brakes

Many servo brakes can be partially assembled on the bench by placing the shoes and star wheel together as shown in **Figure 15-24.**

> **Note: Be sure that the primary (smaller) shoe faces front and that the adjuster is the correct one for that side of the vehicle.**

Install the adjuster tensioner spring by moving the shoes together. Place the holddown spring pins through the holes in the backing plate and lightly lubricate the support pads with high temperature grease. Then expand the shoes and place them on the backing plate, allowing the pins to pass through the proper holes in the shoe webs. If the parking brake lever is installed on the rear shoe, place it in position now. Align the shoes with the wheel cylinder links and shoe anchors. Self-adjuster spring tension will hold the shoes in place. If a parking brake strut is installed under the wheel cylinder, place it and its anti-rattle spring in position. See **Figure 15-25.**

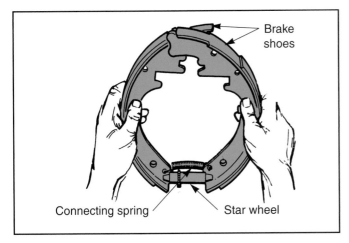

Figure 15-24. *Brake shoes, star wheel, and connecting spring assembled on a workbench before installation. (General Motors)*

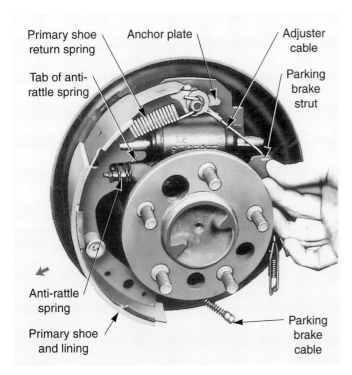

Figure 15-25. *Placing the parking brake strut and its anti-rattle spring into position, before installing the secondary shoe. (Chrysler)*

Install the self-adjuster lever and links as necessary. Place the holddown springs and washers over the pins, making sure the springs and washers are correctly positioned. Using the holddown spring tool, **Figure 15-26,** compress and turn the spring washer to lock it to the pin.

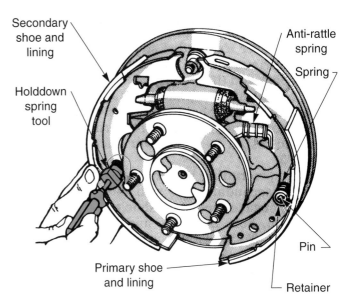

Figure 15-26. *Using a holddown spring tool to reinstall the spring retainer. (Dodge)*

Then install the retracting springs in the slots in the shoe web, and stretch them over the backing plate anchor using the spring installation tool, **Figure 15-27.**

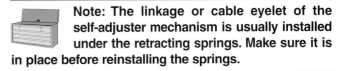

Note: The linkage or cable eyelet of the self-adjuster mechanism is usually installed under the retracting springs. Make sure it is in place before reinstalling the springs.

After the retracting springs are installed, you might want to use pliers to close the spring ends. Although this is not necessary in most cases, it reduces the chance of the spring ends coming loose during brake operation. Make sure the self-adjuster operates properly, then adjust the shoes, which is explained later in this chapter.

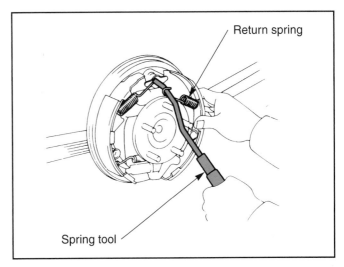

Figure 15-27. *Reinstalling a secondary shoe retracting spring with a special spring tool. Do not use standard pliers, they can damage the springs, causing premature failure. (Toyota)*

Reinstalling Non-Servo Brakes

Most non-servo brakes are installed one shoe at a time. Begin by placing the holddown spring pins through the holes in the backing plate and lightly lubricating the backing plate support pads. With the shoes on the bench, install the self-adjuster mechanism over the proper shoe web as necessary. Then install the front shoe over the holddown spring pin and attach the holddown spring, **Figure 15-28.** Install the retracting spring, if used, in the shoe and stretch it over the backing plate anchor.

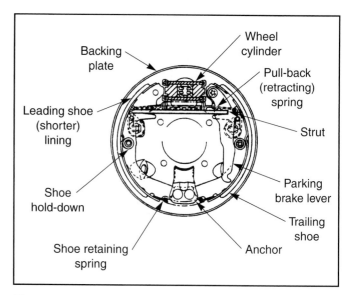

Figure 15-29. *Brake assembly which uses a single retracting spring setup. (EIS)*

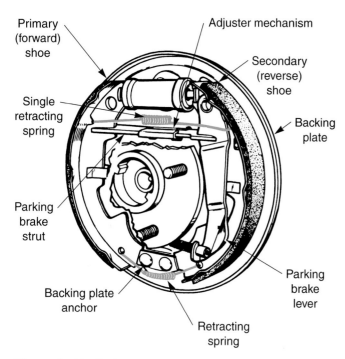

Figure 15-28. *Brake shoe retracting spring carefully placed over the backing plate anchor. (Bendix)*

Place the rear shoe over the holddown spring pin and attach the holddown spring. If the bottom of the shoe is anchored, place the shoe opening over the anchor pin and install the retainer. Install the rear retracting spring. If the brake assembly uses a single retracting spring between the two shoes, as in **Figure 15-29,** install it now. After all components are installed, make sure the self-adjuster operates properly. Then adjust the shoes as explained later in this chapter.

Caution: Never attempt to machine a cracked drum. Replace any drum that is cracked or heat-checked.

Lathe Preparation

Refer to **Figure 15-30** as you read the following paragraphs. Check the brake lathe *arbor* (the revolving spindle the drum is mounted to) and drum *adapters* (devices for fitting the drum to the arbor) for dirt, grease, and metal chips, and clean if needed. Check that the *cutting bar* (the long steel bar that holds the cutting assembly) and the *bit assembly* are tightly installed on the lathe. Make sure the *cutting bits* are not chipped or burnt. Inspect the lathe and be certain that all other parts are solidly attached and in good working condition.

Machining Drums

Unlike rotors, drums must always be removed from the vehicle for turning. After the drum is removed from the vehicle, remove any excess brake dust using a vacuum or liquid cleaner. If the wheel cylinder or axle seal was leaking, be sure to remove all brake fluid or oil residue from the drum. If the drum has an integral hub, remove the inner bearing and clean any grease from the interior of the hub. After these steps are taken, the drum can be turned.

Figure 15-30. *Check the brake lathe components before mounting the drum. Make sure the adapter setup is appropriate for the drum's size. Always follow the lathe manufacturers operating and safety instructions.*

Drum Installation

Install the drum on the brake lathe arbor using the proper adapters. Stagger the adapter feet, this will ensure a straighter drum installation. Consult the lathe user's manual if you have any doubts as to the proper drum mounting method. Then install the wide *silencer band,* **Figure 15-31,** on the outside of the drum.

 Caution: Failure to install the silencer band will cause the bit to chatter as it cuts. This will result in a rough finish that must be recut.

Cutter Adjustment

Start the brake lathe and check that the drum is turning evenly. If the drum appears to be wobbling, stop the lathe and reposition the drum. Then turn the *cutter feed* until the bit is near the drum surface. Slowly turn the *bar feed dial* until the bit is in roughly the center of the drum braking surface. Then slowly turn the feed dial until the cutting bit just touches the drum surface. Hold the feed dial and turn the numbered *collar* to zero and tighten, **Figure 15-32.**

⚠ **Warning: Remove only enough metal to remove scoring or out of round sections. Excess cutting may make an otherwise reusable drum too thin, and it will require replacement.**

Move the bar assembly until the cutting bit is at the innermost part of the drum. Be careful not to jam the cutting bit against the drum's inside corner. Then adjust the depth of cut using the feed dial. Adjust the feed speed using the speed dial.

After all adjustments have been made, the actual drum turning operation can be performed. A cut made to remove a great deal of metal, with the speed set relatively high is a rough cut. A cut made to remove a small amount of metal at slow feed speeds is a finish cut.

The amount and speed of cutting will be governed by the total amount of metal to be removed and the finish desired. If the drum is deeply grooved, a great deal of metal may need to be removed to obtain a smooth finish on the drum. In this case, it may be necessary to make several rough cuts before making the finish cut. If the drum is only lightly scored, or a glazed surface needs to be removed, it may be refinished by a single finish cut. The following sections explain how to make rough and finish cuts.

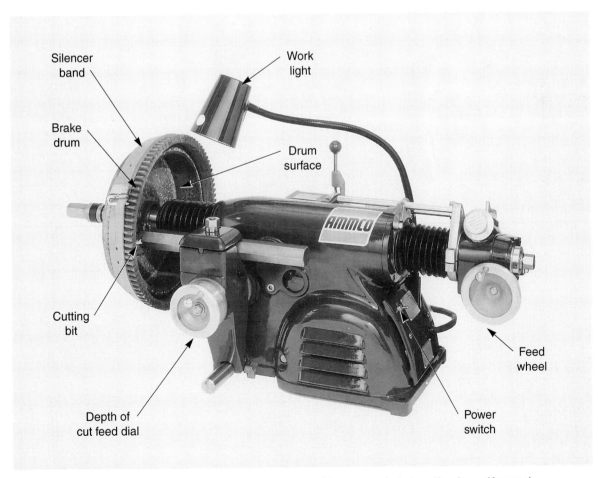

Figure 15-31. *A brake drum lathe that has been correctly setup, and is now ready to turn the drum. (Ammco)*

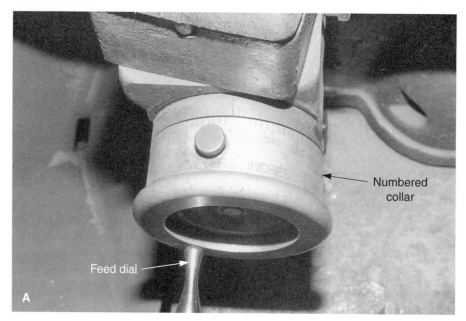

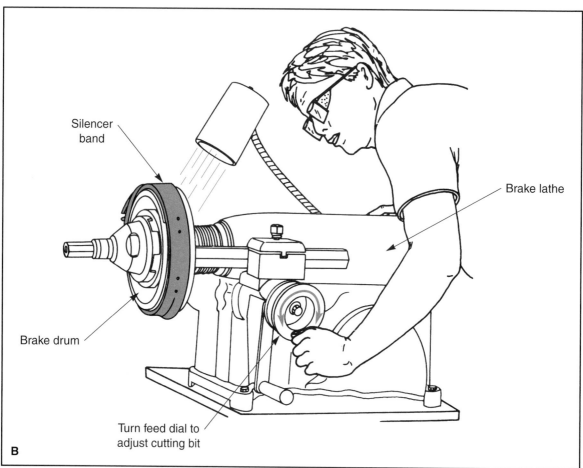

Figure 15-32. *A—Turn and set the feed dial and collar to cut the drum. Turn the feed dial slowly so as not to jam the bit into the drum surface. B—A technician turning a brake drum. (Ammco)*

Rough Cut

To make a ***rough cut,*** set the cutting bit to the maximum cutting depth and the speed to a relatively fast setting.

 Note: As a general rule, take no more than .008″ (.2 mm) from the drum surface on any single cut. Check the collar on the adjuster to determine whether it is marked in thousandths of an inch or millimeters.

After the cutting depth and feed speed are established, engage the feed lever and watch the drum as it cuts. Allow the bit to cut the entire braking surface and exit the outer edge of the drum. Then disengage the feed lever and inspect the drum surface. If the first cut left grooves or shiny spots, repeat the rough cut as needed. After all damaged areas are removed, make a finish cut.

Finish Cut

To make a *finish cut,* set the cutting depth to a small amount, usually about .002″ (.05 mm) and set the feed speed to a slow setting. Engage the feed lever and observe the drum as it is cut. Cutting time will be much longer than the rough cut. Allow the bit to cut the entire braking surface and exit the outer edge of the drum. Then disengage the feed lever and inspect the drum surface. After making the finish cut, recheck the drum thickness with a drum micrometer. If the drum is now too thin, it must be discarded.

 Warning: After cutting both drums of a single axle, check that the drum diameters are within .030″ (.75 mm) of each other. Too much variation in drum size may cause pulling.

Removing Hot Spots

Hot spots, sometimes called *hard spots,* are drum sections that have been overheated by severe brake operation and become much harder than the surrounding metal. Hot spots were more common in the past on vehicles with front drum brakes, but are still found occasionally. These spots cannot be removed by cutting bits. After the drum is turned, these spots will be visible as raised places on the finished surface.

To remove hot spots, a special motor driven grinder, **Figure 15-33,** must be used. This grinder is installed in place of the cutting bit and rotates a grinding stone, or wheel, against the braking surface as the drum turns. To use this grinder, set clearances in the same manner as when setting the cutting bit. Then start the grinder and set the feed to low speed. As the grinding wheel moves over the hard spot, it will grind it down to match the other areas of the drum. If the grinder cannot remove all of a hot spot, the drum should be replaced.

Drum Cleaning

Before the drum is reinstalled, remove all metal particles from the drum. This is especially important when the drum is an integral type. Metal shavings on the bearing races will destroy the bearings in a few miles. Also, metal shavings will be attracted to and could cover ABS wheel speed sensors on vehicles that have the sensors located inside the brake assembly.

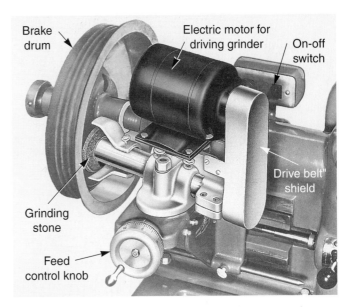

Figure 15-33. *Using a motor driven, lathe mounted stone grinder for removing drum hot or hard spots. (Kwik-Way)*

To clean the drum, remove all loose metal particles with a shop towel. Then use a liquid cleaner to finish removal. If the drum was cleaned with a petroleum based solvent (such as those found in parts cleaners), the braking surfaces should be sprayed with brake cleaner to remove petroleum residues.

Brake Drum Installation

After the brakes shoes and hardware are installed and the drum is turned, the drum can be reinstalled. Before this is done, the shoes should be adjusted to approximately the correct clearance. The drum can then be reinstalled. Reinstallation procedures vary according to whether the drum is an integral or non-integral design. Removal procedures for each type are covered in the next section.

If a new brake drum is being installed, you may have to prepare the drum before installation. Most new drums are coated with a rust prevention compound. These compounds can usually be removed with brake cleaners. Some new drums must be turned to remove the coating or to provide a finished surface. If the drum must be turned, perform a finish cut.

Making Initial Brake Adjustment

Before the drum is reinstalled, the brake shoes should be adjusted to approximately the proper clearance. This makes the final adjustment easier and quicker. Make this initial adjustment using a measurement tool such as the one in **Figure 15-34.** Insert the tool in the drum as shown and set the size so there is a light drag between the tool and drum. Then place the other side of the tool across the widest point of the shoe assembly. Adjust the brake shoes inward if the tool will not fit over the shoes.

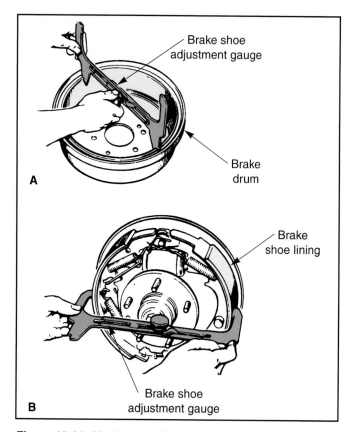

Figure 15-34. *Making an initial brake adjustment with a special adjusting gauge. A—Gauge is set to the drum inside braking surface diameter and locked. B—The measurement is transferred to the brake shoes, which are adjusted in or out with the star wheel until the gauge's shoe side just slips over the shoes. The brake drum may now be installed. (Ford)*

 Note: Most drum brakes with ratchet adjusters, and some with star wheel adjusters, can only be adjusted with the drum removed. Therefore, systems with ratchet adjusters should be adjusted as carefully as possible before installing the drum. If readjustment is necessary, the drum must be removed to gain access to the ratchet adjuster.

Adjust the brakes outward until the shoes meet the measuring tool dimension. On star wheel and ratchet adjusters, disengage the self-adjuster lever. Turn the star wheel or quadrant until the shoes lightly contact the measuring tool. This will make an adequate initial adjustment.

Installing Non-Integral Drums

Drums that are separate from the hub or axle flange can be reinstalled by sliding them over the shoe assembly, **Figure 15-35.** It is not necessary to reinstall any speednuts originally placed over the lugs. However, if the drum is held in place by tapered screws, they should be reinstalled and tightened. After the drum is reinstalled, the wheel and tire assembly can be reinstalled. Then make the final brake clearance adjustment.

 Note: On some vehicles, the brakes can be more easily adjusted with the wheel and tire removed. If clearances with the wheel and tire installed will be close, do not install them until brake adjustment is complete.

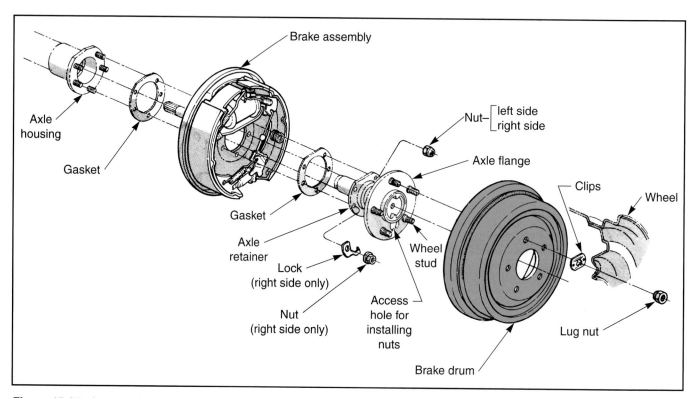

Figure 15-35. *An exploded view of a non-integral brake drum assembly. During installation, the drum will slide over the shoes and wheel studs. The drum will be held in place by the wheel and lug nuts. (Dodge)*

Installing Integral Drums

Before reinstalling an integral drum assembly, remove all grease and brake dust from the spindle assembly. Then reinstall the inner bearing and grease seal in the hub. Bearing cleaning, lubricating, and installation are discussed in Chapter 17. After installing the bearing and seal, slide the bearing and hub assembly over the spindle, **Figure 15-36.** Install the outer bearing, washer, and nut. Tighten the nut and adjust bearing preload.

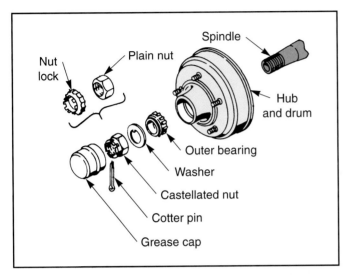

Figure 15-36. *Carefully slide the integral drum and hub onto the spindle. Be careful not to drag the grease seal over the spindle threads. This can lead to grease leaks. (Delco Moraine)*

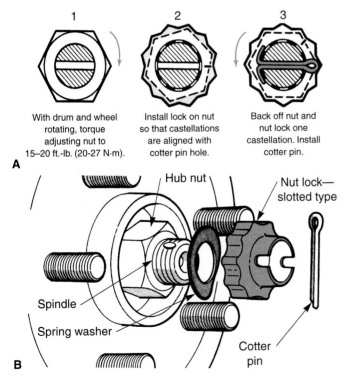

Figure 15-37. *A—Adjustment for one type of slotted nut retainer. B—This illustrates the hub nut, spring washer, slotted retainer, and cotter pin before assembly. (Chrysler & Ford)*

Install the slotted nut retainer, if used. Then install a new cotter pin, cutting and bending it to hold the nut in position, **Figure 15-37.** Then install the grease cap. Turn the wheel to ensure the cotter pin does not contact the cap.

Brake Shoe Adjustment with the Drum Installed

Brake shoe adjustment is a critical final step in successful brake service. The star wheel adjuster is the most common adjustment device on vehicles where the brakes can be adjusted with the drums installed. Locate the adjuster access hole in the backing plate or drum. Remove the rubber plug from the access hole. If the knockout plug is still in place, knock it out to gain access to the star wheel.

 Caution: Do not allow the knockout plug to fall into the drum brake assembly. If the plug does fall into the brake assembly, remove the drum and discard the plug.

Reach through the access hole with an ice pick or small screwdriver and disengage the self adjuster lever. Once the lever is disengaged, insert a brake adjusting tool through the hole and engage the teeth in the star wheel. This procedure is shown in **Figure 15-38.**

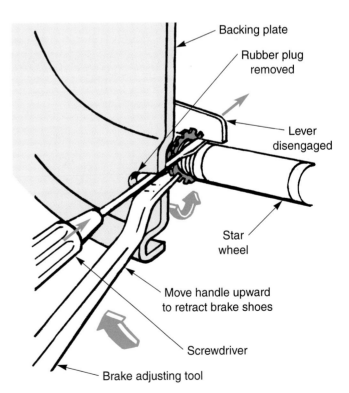

Figure 15-38. *Adjusting a star wheel to obtain proper brake shoe-to-drum clearance. Reinstall the rubber plug after final adjustment. (Delco)*

Note: If the star wheel is always moved in the direction that tightens the brake shoes against the drum, it is not necessary to disengage the lever. However, if the shoes are adjusted too tight, it will be necessary to disengage the lever to loosen them. If you are very careful not to overtighten the brakes, you do not have to disengage the lever.

If the lever is under the star wheel, move the free end of the tool upward to turn the star wheel and reduce the shoe to drum clearance. If the lever was on top of the star wheel, move the free end of the tool downward. Turn the star wheel until there is light brake shoe contact when the drum is rotated. When turning the wheel, you should feel no more than a light drag. After adjusting, apply the brake pedal several times to center the brake shoes on the drum. Then readjust clearance if necessary.

Install a rubber plug in the access hole when adjustment is completed. If a plug is not installed in the hole, water will enter the brake assembly and cause brake pulling or loss of braking if the vehicle is operated in wet weather. A typical rubber plug is shown in **Figure 15-39**.

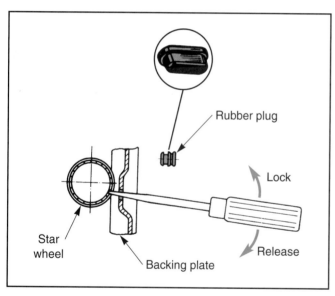

Figure 15-39. *A common brake adjustment access hole rubber plug. Some brake assemblies will use two of these plugs. Metal spring steel plugs are also used. (ATE & Bendix)*

Bleeding the Brakes

If necessary, bleed the brakes at this time. Brake bleeding was covered in Chapter 6. Note that brake bleeding may further center and seat the shoes. Further adjustment may be necessary after bleeding.

While bleeding, make sure the brake lights operate when the pedal is lightly depressed. If the brake lights do not operate, refer to the diagnosis procedures in Chapter 20.

Road Testing

Reinstall the tires and torque the lug nuts to specifications using a torque wrench or torque sticks, if available. Before taking the vehicle off the lift, check the parking brake. The parking brake should fully engage with a minimal number of clicks from the apply lever. After all brake service operations are finished, make a road test to ensure that the brakes are operating properly. Use the same procedure that you learned in Chapter 13 to burnish the shoes. The brakes should stop evenly and quickly with no noise or vibration. Check for the same brake conditions as when you originally road tested the vehicle. If any problems are noted, diagnose and repair the problem and recheck brake operation.

Summary

Typical drum brake problems are poor braking, pulling, fading, dragging pulsation, noises, and leaks. If possible, make a road test before disassembling the brakes. If there is any evidence of brake problems during the road test, remove the drums and inspect the brakes.

Brake drums are either integral or non-integral designs. The wheel bearing retainer must be removed to remove integral drums. Most non-integral drums can be pulled from the axle once the wheel and rim are removed.

Once the brake drum is removed, check the brake shoes for wear and damage. Check the linings for glazing, cracks, scoring, and wear. Lining thickness can be checked with calipers if necessary. Check the brake hardware for distortion, signs of overheating and lack of tension. Check the wheel cylinders for leaks. Check the drums for scoring, out-of-round, and excessive diameter.

To remove the brake shoes, remove the holddown springs, retracting springs, and other hardware. Remove the brake shoes and remove any parts that are attached to the shoes. If necessary, remove the wheel cylinder. Some wheel cylinders can be overhauled without removing them from the backing plate.

Reinstall any parts on the new shoes. Then install the shoes on the backing plate, being sure not to mix up any springs or other hardware. Ensure that the primary shoe faces front. Install the self-adjusters and related springs as needed.

Before installing the drum on the brake lathe, make sure the lathe arbor and adapters are clean and in good condition. Remove debris from the drum, and remove all grease from the interior of integral drums. Then install the drum with the proper adapters, making sure the silencer band is installed. Adjust the cutting speed and depth for a rough or fine cut as necessary. Repeat cutting until all scoring is removed and the drum is round. Recheck drum diameter when finished.

Make an initial shoe adjustment before installing the drum. On some brakes, the initial adjustment is the only adjustment provided. Install the drum, installing and adjusting the bearings on an integral drum. After the drum is installed, finish the adjustment if an external adjustment is provided. Bleed the brakes if necessary, then recheck shoe adjustment. Road test the vehicle after all repairs are complete.

Review Questions—Chapter 15

Please do not write in this text. Write your answers on a separate sheet of paper.

1. Sometimes a drum cannot be removed because the drum and shoes are _____.

2. If the star wheel lever is under the star wheel, move the free end of the adjuster tool _____.

3. Explain why you should not step on the brake pedal once a drum is removed.

4. What are some lining conditions that mean that the linings should be replaced?

5. *Yes or No.* If a brake spring shows any of the following conditions, can it be reused?
   (A) Dusty.
   (B) Stretched.
   (C) Paint burned off.
   (D) Oil soaked.
   (E) Brake fluid soaked.

6. If the wheel cylinder cannot be loosened, the wheel cylinder cannot be _____ and should be _____.

7. On most _____ drum brake assemblies, the shoes and star wheel assembly can be removed as a unit.

8. After drum brake disassembly, clean all parts thoroughly without using _____.

9. What part of the self-adjuster should be lightly lubricated before reassembly?

10. On servo brakes, the _____ shoe should face the rear of the vehicle.

11. If you forget to install the silencer band before turning the drum, what will happen?

12. To begin adjusting the cutter, turn the feed dial until the bit _____ the drum surface.

13. After adjusting the clearance, press on the _____ several times to center the anchors, then recheck adjustment.

14. After all adjustment is complete, install a _____ in the adjustment hole.

15. After all service is completed, make a _____ test to ensure that the brakes work properly.

ASE Certification-Type Questions

1. All of the following should be checked before deciding that the drum brakes are defective, EXCEPT:
   (A) disc brakes.
   (B) brake lights.
   (C) power assist.
   (D) tires.

2. Technician A says that brakes that grab when cold, but operate normally when hot are a result of thin drums. Technician B says that fading can result if the drums are improperly turned. Who is right?
   (A) A only.
   (B) B only.
   (C) Both A & B.
   (D) Neither A nor B.

3. All of the following statements about brake pulling are true, EXCEPT:
   (A) pulling to one side is usually caused by a brake problem on the opposite side.
   (B) pulling can be caused by oil or brake fluid on the linings.
   (C) hydraulic system defects will not cause pulling.
   (D) pulling during wet weather only may be caused by a missing adjuster hole plug.

4. When removing a non-integral drum, which of the following procedures does not need to be performed?
   (A) Remove the wheel rim.
   (B) Remove the wheel bearing.
   (C) Back off the shoe adjustment.
   (D) Slide the drum from over the shoes.

5. Once the drum is off, which of the following should not be done?
   (A) Press on the brake pedal.
   (B) Remove the shoes.
   (C) Check drum diameter.
   (D) Turn the star wheel.

6. Technician A says that if brake fluid drips from the wheel cylinder when the dust boot is pried away, the wheel cylinder should be overhauled or replaced. Technician B says that if water drips from the wheel cylinder when the dust boot is pried away, the wheel cylinder should be overhauled or replaced. Who is right?
   (A) A only.
   (B) B only.
   (C) Both A & B.
   (D) Neither A nor B.

7. All of the following drum defects can be repaired by turning, EXCEPT:

(A) scoring.

(B) out-of-round.

(C) excessive diameter.

(D) excessively smooth surface.

8. Technician A says that the threads of the star wheel adjuster should be well lubricated before reassembly. Technician B says that the star wheel socket interior should be lightly lubricated before reassembly. Who is right?

(A) A only.

(B) B only.

(C) Both A & B.

(D) Neither A nor B.

9. Modern brake shoes do not require _____.

(A) grinding

(B) holddown springs

(C) retracting springs

(D) adjusting

10. Technician A says that non-servo brake shoes can be installed as an assembly. Technician B says that most servo brake shoes are installed as an assembly with the self adjuster. Who is right?

(A) A only.

(B) B only.

(C) Both A & B.

(D) Neither A nor B.

11. The primary purpose of using the drum silencer band is to prevent ____.

(A) squealing

(B) lathe damage

(C) a rough drum finish

(D) a shiny drum finish

12. Technician A says that a cut made to remove a great deal of metal a slow speed is a rough cut. Technician B says that a cut made to remove a small amount of metal at fast speed is a finish cut. Who is right?

(A) A only.

(B) B only.

(C) Both A & B.

(D) Neither A nor B.

13. After making a finish cut, recheck the drum _____.

(A) thickness

(B) weight

(C) color

(D) hub

14. Which of the following drum defects cannot be removed by cutter bits?

(A) Scoring.

(B) Hot Spots.

(C) Out-of-round.

(D) Taper.

15. When adjusting the brake shoes, try to have a _____ when the wheel is turned.

(A) light drag

(B) heavy drag

(C) complete wheel lockup

(D) complete lack of drag

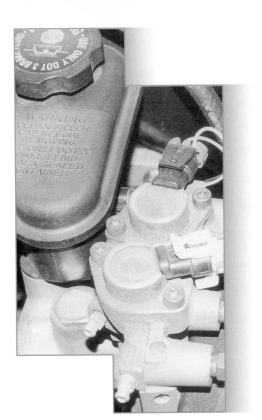

Chapter 16

Wheel Bearings and Oil Seals

After studying this chapter, you will be able to:

❑ Identify types of wheel hubs and axle flanges.
❑ Identify types of wheel bearings.
❑ Identify common locations and usage of each wheel bearing type.
❑ Explain how each type of wheel bearing is installed on spindles and bearing hubs.
❑ Explain how oil seals are constructed.
❑ Identify places where oil seals are installed.
❑ Identify brake related locations using gaskets or O-rings.

Important Terms

Wheel hubs	Rolling element	Tapered roller bearings	Hypoid oil
Axles	Outer race	Nut	Viscosity
Axle flange	Bearing cage	Cotter pin	Oil seals
Wheel studs	Seals	Staking	Lip
Lug bolts	Shields	Gear oil	Garter spring
Lug nuts	Radial load	Grease	Oil film
Wheel bearings	Axial loads	Polymers	Wiper
Friction bearing	Shock loads	Lithium grease	Dust shield
Bushings	Preload	Extreme pressure	Gaskets
Antifriction bearings	Ball bearings	(EP) lithium	O-rings
Inner race	Straight roller bearings	Lithium soap	

In the past, wheel bearings and oil seals were often neglected during brake service. On modern vehicles, most bearings are sealed and do not require service. Since bearings do not have to be removed for service, seals often last the life of the bearing. This chapter covers the basic principles of wheel bearings and oil seals. This chapter also discusses the design and application of gaskets and O-rings where they are used.

Hubs, Axles, and Wheel Bearings

The brake drums or rotors are always attached to *wheel hubs* or *axles*. A wheel hub contains the bearings and forms a mounting surface for the wheel and drum or rotor, **Figure 16-1A.** Wheel hubs on front-wheel drive vehicles are separate from the axles, although they may be pressed onto the axle through the bearings, **Figure 16-1B.** These wheel hub and bearing assemblies are often self-contained units, as shown in **Figure 16-1C.** Some of these assemblies also include the wheel speed sensor tone wheel and in some cases, the sensor. These hubs are bolted to the knuckle or axle. In many cases, especially on older vehicles, the hub and the rotor or drum are an integral unit which uses separate bearings, **Figure 16-1D.**

When solid axles are used, such as on rear-wheel drive vehicles, the axle and drum are connected through the use of an *axle flange,* **Figure 16-2.** The axle flange is welded on or forged as part of the axle. It forms a mounting surface for the wheel and drum or rotor. Axle lengths vary with the design of the vehicle's drivetrain.

In almost all cases, the hub, brake disc or drum, or axle flange is also the mounting surface for the *wheel studs,*

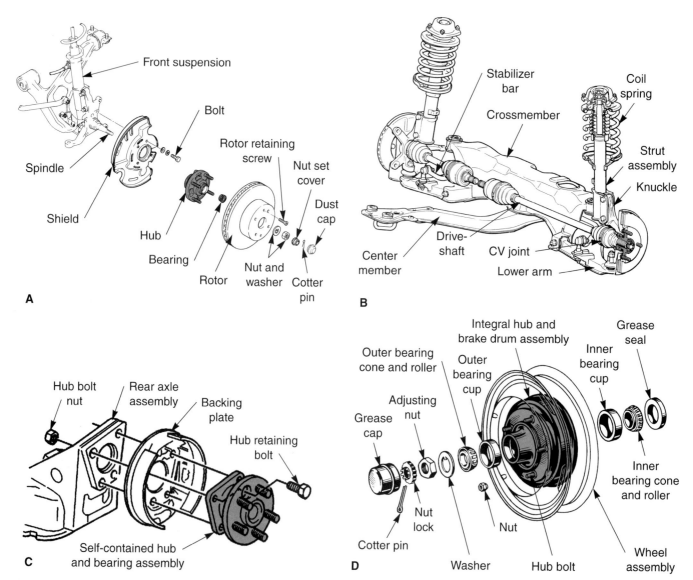

Figure 16-1. A—A wheel hub which houses the bearings and also forms a mounting surface for the brake rotor. B—Wheel bearing assembly used on front-wheel drive axles. C—A self-contained hub and bearing assembly. This unit bolts to the rear axle of a front-wheel drive car. D—An integral hub and brake drum assembly used on the front of a rear-wheel drive car. (Mazda, Hyundai, Pontiac, Ford)

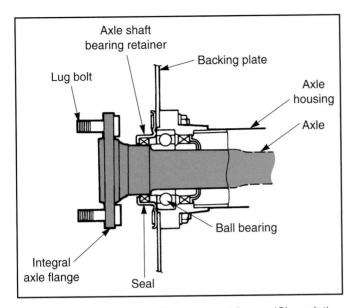

Figure 16-2. *A solid axle with an integral flange. (Chevrolet)*

sometimes called *lug bolts,* **Figure 16-3A.** Most wheel studs are threaded bolts or studs pressed into the hub or flange. A knurled area on the rear section of the stud cuts into the hub or axle metal to keep the stud from becoming loose. The head of the stud is wider, resembling a bolt head. The head keeps the stud from coming completely through the hole.

Lug nuts are installed on the wheel stud and tightened to hold the wheel in place. Some Asian and European built vehicles have threaded holes in the hub or flange. Tapered head lug bolts, **Figure 16-3B,** are installed through the wheel and threaded into the holes.

The purpose of the *wheel bearings* is to form a low friction connection between the rotating wheels and the stationary vehicle. Bearings must also be tight enough to prevent unwanted movement and vibration as they operate, but loose enough to keep friction at a minimum, and to allow lubricant to enter the moving parts and compensate for expansion as the bearings heat up.

Friction and Antifriction Bearings

The simplest type of bearing is the friction bearing shown in **Figure 16-4.** A *friction bearing* is a metal sleeve that surrounds a rotating shaft. Friction bearings rely on a sliding action between the stationary bearing surface and the rotating shaft. Friction bearings are often used where space is limited, such as in engines and transmissions. They are also used in the pivot points of the vehicle suspension and steering and are called *bushings.* However, friction bearings are impractical for use in wheel bearings which must turn at high speed under severe heat and loads, with no provision for pressure lubrication.

All wheel bearings are *antifriction bearings.* Antifriction bearings consist of three basic parts, the *inner race, rolling element,* and *outer race.* These parts are shown in **Figure 16-5.** The rolling elements are separate

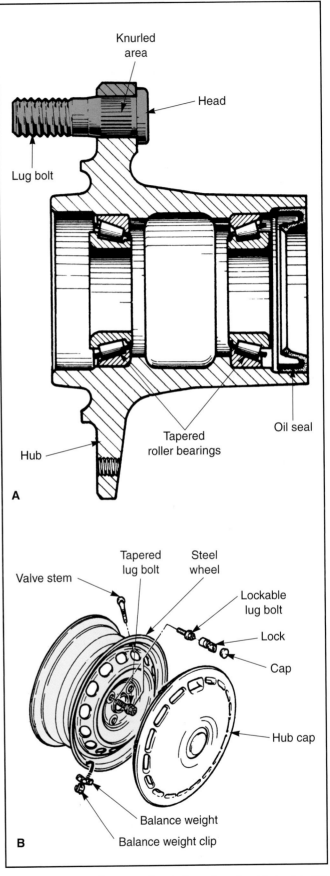

Figure 16-3. *A—Cross-sectional view of a wheel hub showing a lug bolt. Note the use of tapered roller bearings. B—A steel wheel which is fastened to the hub or flange (not shown), with tapered head lug bolts. (Saab and Audi)*

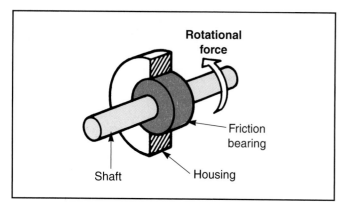

Figure 16-4. *A simple friction bearing supporting a rotating shaft. (Federal Mogul)*

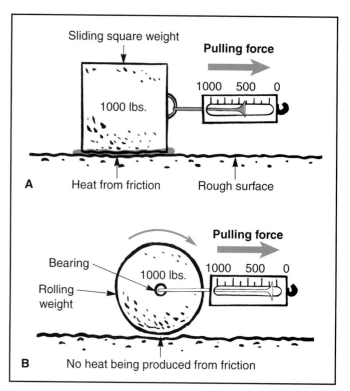

Figure 16-6. *A—Heat being generated as a square block is quickly moved over a surface. Note the high (large) pulling force required to skid the block. B—If the weight is shaped into a ball, little or no friction is involved as the ball rolls across the surface. Note the low pulling force needed to roll the ball. (Bower/BCA)*

and roll between the two races, allowing them to move in relation to each other. This creates much less friction than the sliding action of friction bearings. Since their rolling action draws lubricant between the moving parts, antifriction bearings can operate for long periods with an initial unpressurized supply of lubricant. For maximum durability, the rolling elements and races are made of heat treated steel.

Other parts used with antifriction bearings are the **bearing cage** which keeps the individual rolling elements separated as they turn, and **seals** or **shields** which keep lubricant in and dirt and water out.

Antifriction Bearing Operation

The rolling action of antifriction bearings reduces friction between two moving parts. Note how heat is generated by briskly moving the square block over a surface, **Figure 16-6.** As a more practical example of this, rub your hands together briskly. Note that heat is quickly produced by friction.

Now place a pencil between your hands and rub them together. You will feel almost no heat, since the rolling friction created by the pencil is much less than the sliding friction created as you rubbed your hands together without the pencil. If the square were made into a ball, it would easily roll over the surface.

Wheel bearings create this same rolling friction by placing a series of balls or rollers between the two races, **Figure 16-7.** As the races move in relation to each other, the balls or rollers roll between them, greatly reducing friction.

Wheel Bearing Design Factors

As with most automotive parts, wheel bearing design is determined according to vehicle size, drivetrain layout, and operating conditions. The following sections cover design factors affecting wheel bearing use.

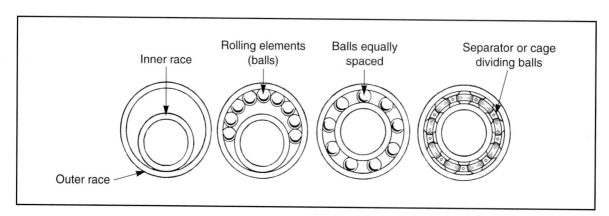

Figure 16-5. *The basic parts of a ball bearing assembly. (New Departure)*

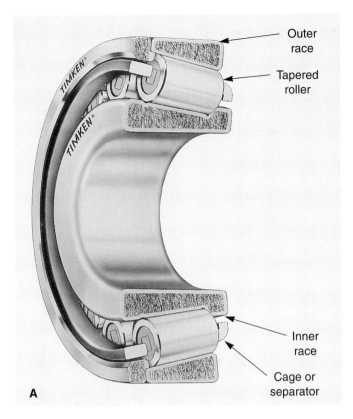

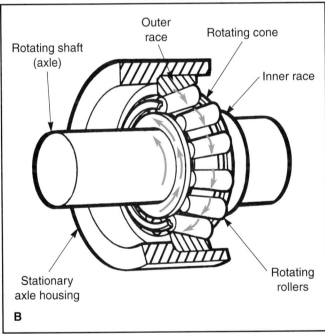

Figure 16-7. *A—Cutaway of a tapered roller wheel bearing. B—This illustrates the rollers direction of travel as opposed to the inner race (cone) and axle shaft. (Federal Mogul & Timken)*

Wheel Bearing Load

As with any vehicle part, the wheel bearing is subjected to loads. Any load on the bearing tends to press the bearing parts together. The resulting pressure increases friction and heat. There are two main types of loads, radial and axial.

Radial Loads

An obvious wheel bearing load would be the weight of the vehicle. Vehicle weight passes through the spindle or axle, bearing, and hub, to the rim and tire. This weight varies as the vehicle accelerates, decelerates, and moves over bumps. In some cases, the weight change is very abrupt and severe. Also, when the vehicle is moving, bearing rotation causes centrifugal force, which tends to make the bearing rollers move outward. If the outer race was not in place, the bearing would fly apart. The combination of weight and centrifugal force is called *radial load.* Radial loading occurs at a right angle to the bearing and shaft. This is shown in **Figure 16-8.**

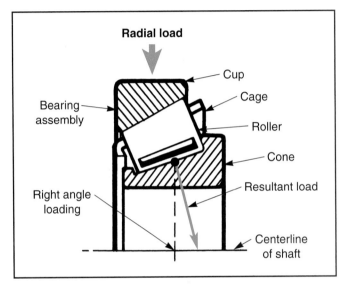

Figure 16-8. *A cutaway of a tapered roller bearing with a radial load being applied. (Federal Mogul)*

Axial Loads

Another force occurs when the vehicle is turned. The front wheels are turned by the steering linkage, while the body wants to keep moving forward. This places a sideways load on the front bearings. When the vehicle is being turned, the rear tires tend to keep moving in the same direction. This places sideways loads on the rear bearings. These sideways loads are called *axial loads,* **Figure 16-9.** Axial loads are sometimes called *thrust loads.* Axial loads are always parallel to the shaft. All of these loads, as well as maximum bearing speed, where the bearing will be used on the vehicle, and the ultimate use of the vehicle itself, are calculated when the manufacturer decides what kind of wheel bearing to use.

Bearing Size

Since most of the load and turning effort are placed on the front wheels of cars and light trucks, the front bearings are usually larger than the rear bearings. However, on larger trucks, the rear bearings may be larger in order to handle the heavy loads placed in the truck bed.

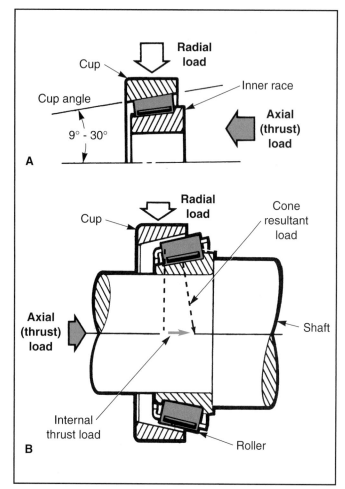

Figure 16-9. *A—Cutaway of a roller bearing showing axial load forcing the rollers into contact with the bearing cup and race. B—Axial load pushing on the bearing from the opposite side. Note how the rollers would tend to move away from the cup. (Federal Mogul)*

Rolling Element Size

The size and type of rolling element must also be considered. Smaller rolling elements can distribute the load more evenly through the races, but may shatter if subjected to shock loads. *Shock loads* are rapid, severe increases in load, usually caused when the vehicle drives over bumps or potholes. Large rolling elements are more durable, but could damage the races when loads are applied. The type of rolling element affects the bearing's ability to handle various combinations of loads. Most bearing selection is a compromise between the largest and smallest rolling element sizes.

Bearing Preload

Preload is the amount of pressure placed on the bearing before it is put into service. Too little preload will cause the rolling elements to rock and vibrate, damaging the bearing. Insufficient preload also intensifies the effect of shock loads. Too much preload will press the rolling elements too tightly against the races, creating unnecessary friction and heat.

Wheel Bearing Types

All modern vehicles use antifriction wheel bearings. However, the type of bearing varies with the weight placed on the bearing, whether it is used on a steering or non-steering axle, and whether or not it is used on a driving axle. To meet these varying requirements, three types of antifriction bearings are used on modern vehicles.

Ball Bearings

Ball bearings are used on the front axles of some front-wheel drive vehicles. Most applications have two ball bearing assemblies. A dual ball bearing assembly is shown in **Figure 16-10.** In addition to splitting the weight, each bearing assembly can take axial loads in one direction. Pairing the ball bearings handles axial loads in both directions. Note in **Figure 16-11** that the inner and outer races are designed to accept thrust loads in opposite directions. Modern ball bearings have the balls and races in a single sealed unit. Preload is factory set and cannot be adjusted.

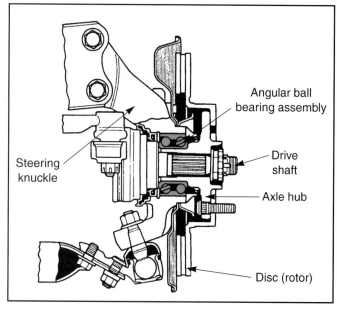

Figure 16-10. *A front-wheel drive axle and hub which incorporates a dual ball bearing assembly. (Toyota)*

Straight Roller Bearings

Straight roller bearings, **Figure 16-12,** are used on the rear axles of many rear-wheel drive vehicles. Flat roller bearings can absorb radial loads but cannot handle any axial loads. Straight roller bearings are self-contained units with all bearing components in a single unit. Preload is not adjustable. On most cars, the axle shaft is the inner race.

Tapered Roller Bearings

Tapered roller bearings are used in front and rear axles. They can be found on driving and non-driving axles. The tapered design allows them to handle any combination of

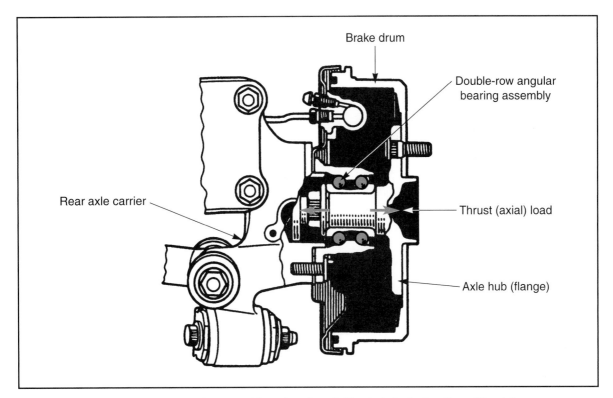

Figure 16-11. *By pairing the ball bearings, they are able to handle axial loads in both directions. (Toyota)*

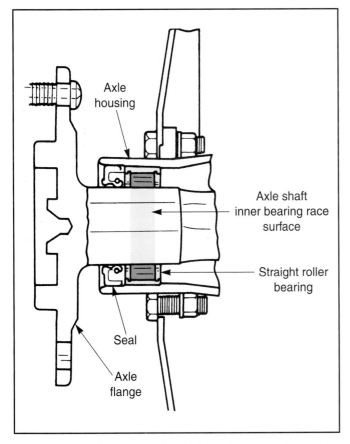

Figure 16-12. *A cross-sectional view of a rear axle housing which supports the axle with a straight roller bearing assembly. The outer surface area of this axle has been hardened and polished. This allows it to act as the bearing inner race for the rollers. (General Motors)*

axial and radial loads. As with ball bearings, tapered roller bearing are installed in pairs to absorb sideways loads. See **Figure 16-13.** On most tapered roller bearing designs, the inner race, rollers, and cage are a single unit. The outer race is pressed into the hub.

The preload of tapered roller bearings is always adjustable. The usual adjustment method is a **nut** which is installed on the spindle, **Figure 16-14.** Tightening the nut adjusts the bearing preload. After the preload is set, the nut is held in position by a nut retainer and/or **cotter pin,** **Figure 16-15,** or by **staking** (bending) the outer part of the nut into a slot on the spindle, **Figure 16-16.**

Wheel Bearing Lubrication

Although they are designed to operate with a minimum of resistance, wheel bearings develop some friction, and must be lubricated. Bearing lubricant must maintain a film of lubricant between the rotating and stationary elements of the bearing to reduce friction and wear. It must also coat the surfaces of the bearing to prevent corrosion.

Lubrication Methods

Wheel bearings are lubricated in one of four ways. Most tapered wheel bearings are packed (filled) with **wheel bearing grease.** The grease in these bearings must be periodically renewed. This is done by removing and cleaning the wheel bearings and repacking them with grease.

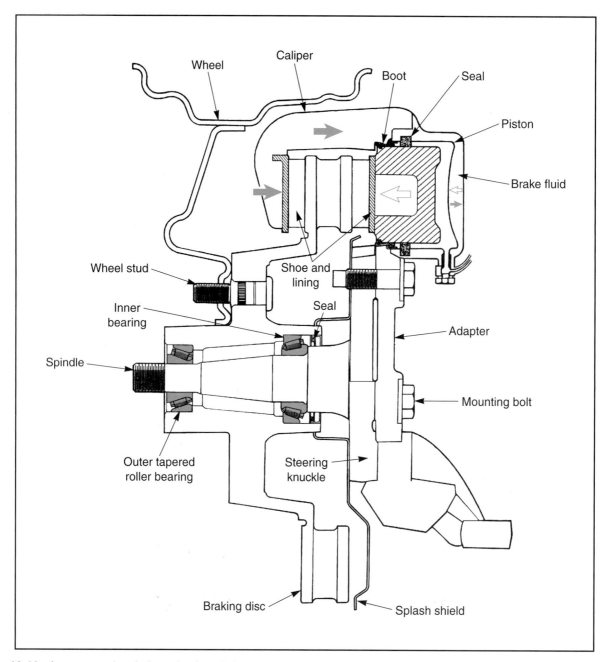

Figure 16-13. *A cross-sectional view of a front hub, disc, and wheel. This assembly uses two tapered roller bearings to handle radial and axial bearing loads. (Chrysler)*

Some bearings are packed with grease and permanently sealed. Most front wheel bearings on front-wheel drive vehicles are lubricated this way. Some solid rear axle bearings are also greased for life. If the seals fail, the bearings must be replaced.

A common method used on solid rear axles of rear-wheel drive vehicles is to splash the bearings with **gear oil** as the bearings rotate. Oil in the rear axle housing is thrown upward by the action of the ring gear and other parts, and makes its way to the outer bearings as the vehicle is turned. On some solid rear axles in large trucks, the bearing operates partially or completely submerged in oil. This lubrication method is uncommon on light trucks and cars.

Types of Lubricant

The two main classes of wheel bearing lubricants are wheel bearing grease and gear oil. Most wheel bearings are lubricated by grease. Some solid rear axle wheel bearings are lubricated with gear oil.

Grease is a semisolid lubricant, thick enough to stay in place and provide lubrication, but thin enough to flow between parts. Quality greases can resist water entry and corrosion, help the bearing absorb shock loads, and do not easily melt when hot.

Modern wheel bearing grease is a combination of various oils and **polymers,** which are molecules arranged into long chains. The long chain polymers allow the grease to stay in place under heavy loads, while flowing easily

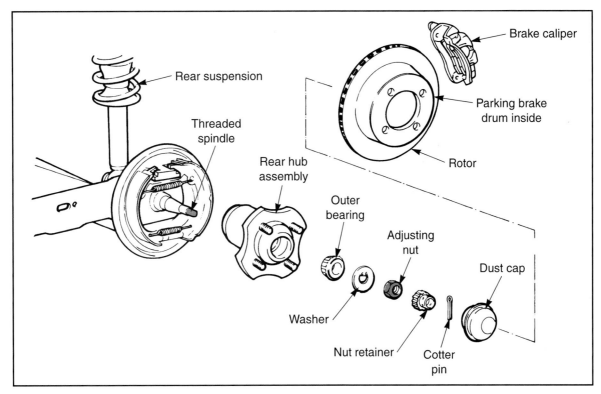

Figure 16-14. *An exploded view of a rear hub and bearing setup. A bearing adjusting nut is used to obtain the correct amount of preload (pressure on bearing). The nut retainer and cotter pin will hold the adjusting nut in the proper position. (Hyundai)*

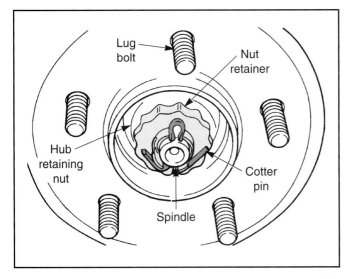

Figure 16-15. *A nut retainer and cotter pin securing the hub retaining nut in its correct position. (Chrysler)*

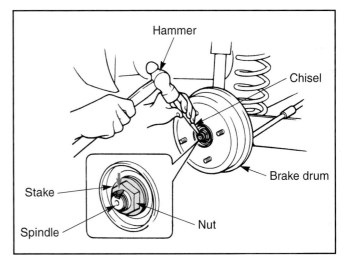

Figure 16-16. *Using a hammer and chisel to bend (deform) a portion of the nut flange into a groove cut in the spindle. This will prevent the nut from becoming tighter or coming off. (General Motors)*

under high pressures and at low temperatures. The wheel bearings on most modern vehicles call for **lithium grease,** often called **extreme pressure (EP) lithium,** or **lithium soap.** Lithium is a metallic element which provides additional lubrication and corrosion protection.

Some older vehicles used what is known as long fiber grease. This grease contained additives which kept it from flying out of the bearing at high speeds and temperatures. In most cases, it is acceptable to use lithium grease in older vehicles. Lithium grease should always be used when it is specified by the manufacturer.

Wheel bearings used in solid rear axles are sometimes lubricated with the same gear oil used to lubricate other rear axle parts. This gear oil is commonly known as **hypoid oil.** Gear oil is measured in terms of **viscosity** (weight) like motor oil. However, gear oil contains high pressure additives that resist being squeezed from between the gears of the ring and pinion. Other additives provide lubrication for the differential assembly and bearings. Gear oils must stay liquid at low temperatures, while providing good lubrication when hot.

Gear oils are classified according to weight (which does not correspond to motor oil weight), and by their GL number. The GL number is set by the American Petroleum Institute (API), with GL 1 being the lowest. Most cars and light trucks require a GL 4 oil as a minimum, with GL 5 and 6 for more severe conditions.

Oil Seals

Oil seals have two purposes, keeping lubricant in and keeping dirt and water out. Most oil seals are constructed as shown in **Figure 16-17.** The flexible rubber part of the seal is called the *lip.* The lip contacts the rotating shaft to seal it. The lip is crimped or bonded to a steel outer shell. A **garter spring** is installed behind the seal lip to hold it tightly to the shaft. A small amount of lubricant enters the lip area and forms an *oil film* between the lip and shaft. This oil film reduces seal and shaft wear, and provides additional sealing. If the seal is in good condition, the oil film will not leak past the seal lip.

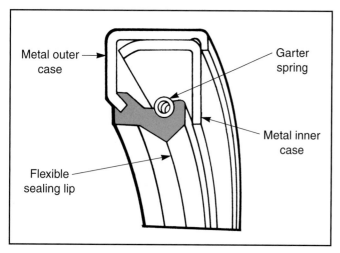

Figure 16-17. *A cutaway view of a typical single lip oil seal. (Federal Mogul)*

The area behind the oil seal lip is designed so any pressure behind the seal tends to press the lip tightly against the shaft. This helps to prevent leaks when pressures are high, such as when the grease in a bearing assembly becomes hot and expands. When pressure is low and leaks are unlikely, the lower tension on the lip reduces seal and shaft wear.

Most oil seals are pressed into the hub or axle as shown in **Figure 16-18.** Some seals have a felt **wiper** to absorb any slight amounts of lubricant that get past the lip. Seals used on front wheel drive vehicles often have a **dust shield** to protect the seal against dirt and water. A dust shield is shown in **Figure 16-19.**

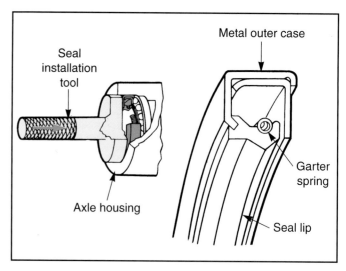

Figure 16-18. *An oil seal being pressed into its recess in the end of the axle housing. Use the proper seal tool to prevent seal damage. Be sure the seal is facing in the correct direction. (Federal Mogul)*

Gaskets and O-Rings

Most sealing jobs are done by the oil seals. However, in a few places on the vehicle axles, it would be impractical to install a seal. In these instances, gaskets and O-rings are used. The placement and purposes of gaskets and seals are discussed in the following paragraphs.

Gaskets

Gaskets are used to prevent leaks between two stationary parts. Gasket material can be made of cork, rubber, or various kinds of treated paper. Some gaskets are made of combinations of materials. A few gaskets are made of thin metal with a central ridge. The ridge is compressed when the parts are tightened, creating a seal.

Typical gasket locations are the rear axle bearing retainer plates of some rear-wheel drive vehicles, **Figure 16-20,** and the front axle housings of many four-wheel drive vehicles.

O-Rings

The O-rings used with axle and brake related parts are stationary seals. Most **O-rings** serve the same function as gaskets, but in a smaller space. O-rings are sometimes used as an inner or outer seal on some bearings, **Figure 16-21.** O-rings are used to seal brake hydraulic components such as the ABS hydraulic modulator.

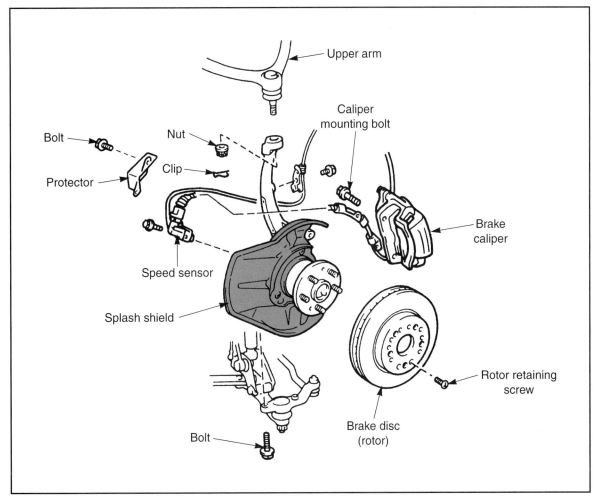

Figure 16-19. *A splash shield, which also acts as a dust shield on one type of front suspension and hub arrangement. (Lexus)*

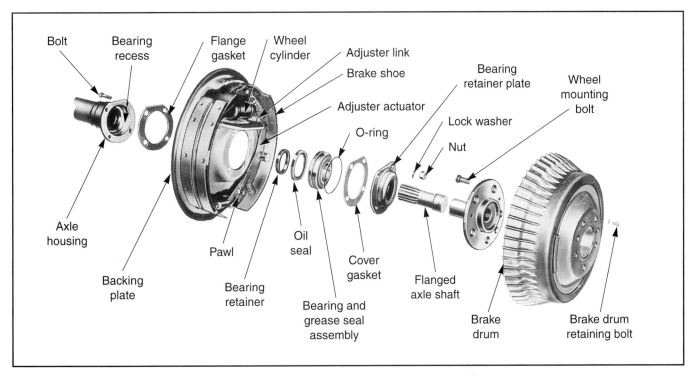

Figure 16-20. *Gasket locations used with one particular rear-wheel drive vehicle. (Cadillac)*

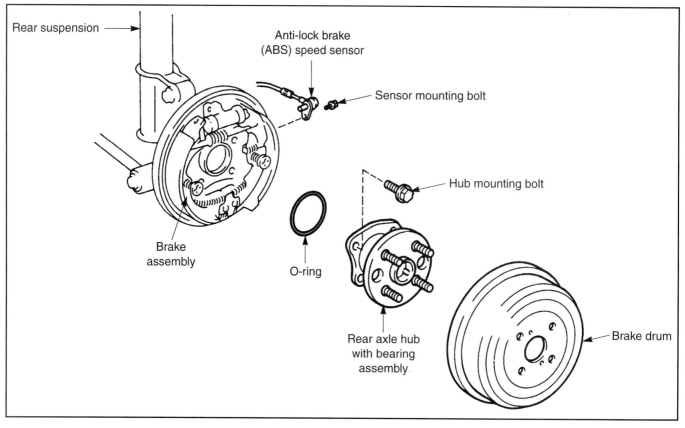

Figure 16-21. *An O-ring used as a bearing seal on a rear axle hub assembly. (Toyota)*

Summary

The brake rotors or drums are always attached to the wheel hubs or axles. Wheel hubs rotate with the axle shaft. The drum or rotor is attached to the axle by a flange. The axle and flange usually contain the wheel lugs. Wheel studs are pressed into the hub or flange, although a few cars have threaded holes and use tapered nuts to attach the wheel rim.

The purpose of the wheel bearings is to form a low friction connection between the rotating wheels and the stationary vehicle. The two major types of bearings are friction and antifriction bearings. Friction bearings are simple and will fit in small areas, but are unsuitable for wheel bearings. All wheel bearings are antifriction bearings. Antifriction bearings consist of an inner race, rolling elements, and an outer race. The rolling elements rotate between the races as they move in relation to each other. Antifriction bearings also use a bearing cage to keep the rolling elements separated, and seals to keep lubricant in and dirt and water out.

Wheel bearings are subjected to two types of loads, radial and axial. Radial loading occurs at right angles to the rotating shaft, while axial loading occurs parallel to the shaft.

Manufacturers must select the proper wheel bearing based on several factors, including weight, speed, and which axle the bearing will be used on. After determining these factors, the manufacturer selects the proper bearing size, rolling element size, and preload.

The main types of bearings are the ball, straight roller, and the tapered roller bearing. Ball bearings are used on the front axle of most front wheel drive vehicles. Straight roller bearings are used on solid rear axles. Tapered roller bearings are used on front and rear axles. Tapered roller bearings are the only wheel bearings which can be adjusted.

Wheel bearings can be lubricated by grease which must be renewed occasionally. Other bearings are greased for life and must be replaced if the grease leaks out. Rear wheel bearings may be lubricated by oil splash or by partial submergence in oil. Greases used in bearings must be thick enough to stay in place, but thin enough to flow between moving parts. Gear oil must conform to API standards.

Oil seals are designed to keep lubricant in, and dirt and water out. The lip of the oil seal fits against the rotating shaft, and is further sealed by a garter spring and a thin film of lubricant between the lip and shaft. The seal is designed so that the lip is pressed tightly by any pressure behind the seal. Most seals are pressed into the hub or axle. Gaskets and O-rings are used where using a seal would be impractical. Gaskets and O-rings always seal stationary parts.

Review Questions—Chapter 16

Please do not write in this text. Write your answers on a separate sheet of paper.

1. Most wheel studs are _____ into the hub or axle flange.

2. Are friction bearings ever used as wheel bearings?

3. What are three main parts of an antifriction bearing?

4. Rolling friction produces _____ heat than sliding friction.

5. Match the type of load to its description.

_____ Occurs at right angles   (A) Radial load
       to the bearing   (B) Axial load
_____ Weight of the vehicle
       affects this load
_____ Turns affect this load
_____ Bumps affect this load
_____ Centrifugal force affects this load
_____ A sideways load
_____ This load occurs parallel to the shaft

6. What are the advantages and disadvantages of using small rolling elements?

7. Too much bearing preload will create unnecessary _____ and _____.

8. Ball bearing _____ is preset.

9. _____ are used on the rear axles of many rear-wheel drive vehicles

10. If the seal fails on a _____, the entire bearing must be replaced.

11. Most modern vehicles use _____ grease in their wheel bearings.

12. Older vehicles used _____ fiber grease.

13. Gear oil is similar to motor oil, but contains _____ additives.

14. What are the two jobs of an oil seal?

15. Gaskets prevent leaks between _____ parts.

ASE Certification-Type Questions

1. All of the following statements about wheel hubs are true, EXCEPT:
   (A) the wheel hub contains the bearings.
   (B) the wheel hub is welded or forged to the axle shaft.
   (C) the wheel hub usually contains the wheel studs.
   (D) the wheel hub is a mounting surface for the drum or rotor.

2. The wheel bearings form a low friction connection between the wheels and _____.
   (A) rim
   (B) hub
   (C) axle
   (D) vehicle

3. Friction bearings are used in all of the following places of the vehicle, EXCEPT:
   (A) axles.
   (B) engines.
   (C) transmissions.
   (D) steering pivots.

4. Technician A says that all wheel bearings are antifriction bearings. Technician B says that all antifriction bearings are ball bearings. Who is right?
   (A) A only.
   (B) B only.
   (C) Both A & B.
   (D) Neither A nor B.

5. The bearing cage keeps the _____ apart.
   (A) rolling elements
   (B) rollers and races
   (C) inner and outer races
   (D) hub and outer race

6. Thrust load is another term for _____ load.
   (A) radial
   (B) axial
   (C) weight
   (D) shock

7. Technician A says that preload is factory set on tapered roller bearings. Technician B says that the most common type of preload adjuster is a nut on a threaded spindle. Who is right?
   (A) A only.
   (B) B only.
   (C) Both A & B.
   (D) Neither A nor B.

8. All of the following wheel bearings are lubricated with wheel bearing grease, EXCEPT:
   (A) ball bearings.
   (B) straight roller bearings.
   (C) tapered roller bearings.
   (D) bearings installed in sets of two.

9. A good wheel bearing grease should have all of the following properties, EXCEPT:
   (A) be thick enough to stay in place.
   (B) be thin enough to flow between parts.
   (C) melt when hot.
   (D) resist water entry.

10. Technician A says that an oil seal should keep lubricant in the bearing. Technician B says that an oil seal should keep dirt out of the bearing. Who is right?
    (A) A only.
    (B) B only.
    (C) Both A & B.
    (D) Neither A nor B.

A self-contained hub and bearing assembly. The hub portion containing the wheel studs turns. The stationary section contains the bearings and is bolted to the wheel spindle.

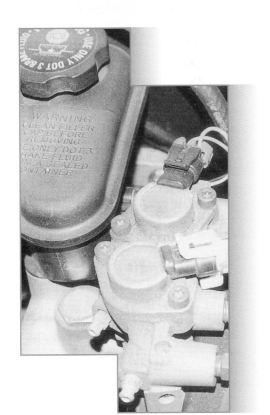

Chapter 17

Wheel Bearing and Oil Seal Service

After studying this chapter, you will be able to:

❑ Identify common wheel bearing defects.
❑ Diagnose wheel bearing problems.
❑ Remove and clean tapered wheel bearings.
❑ Inspect tapered roller wheel bearings and identify defects.
❑ Remove and replace pressed in wheel bearing races.
❑ Lubricate, install, and adjust tapered roller wheel bearings.
❑ Remove and replace ball and flat roller wheel bearings.
❑ Identify leaking oil seals, gaskets, and O-rings.
❑ Remove and replace oil seals.
❑ Remove and replace gaskets and O-rings.
❑ Remove and replace wheel studs.

Important Terms

Pitting	Fretting	Knuckle	Differential cover
Spalling	Brinelling	Bearing collar	Rear axle cover
Smearing	Wheel bearing packer	C-lock	Inspection cover

This chapter is designed to familiarize you with the principles of wheel bearing service. The main emphasis of this chapter is the service of tapered wheel bearings, as these are the only bearings that can be lubricated and adjusted. Removal and replacement procedures for non-serviceable straight roller and ball bearings are also covered. This chapter also contains instructions on the removal and replacement of seals, gaskets, O-rings, and wheel studs.

Common Wheel Bearing Defects

Wheel bearings are generally trouble free. Most tapered wheel bearing problems are caused by lack of maintenance. Ball and straight roller bearings require no maintenance, and will usually last the life of the vehicle if the bearing seals are not damaged. Occasionally, when the vehicle is driven through deep water, some water may be forced past the seals and enter the bearings. This water will dilute the lubricant and cause corrosion that can eventually ruin the bearings.

Using the wrong lubricant, allowing new lubricant to be contaminated during service, careless bearing or race installation, or incorrect adjustment can cause rapid bearing failure. Bearing looseness can be caused by wear, or in the case of tapered roller bearings, improper adjustment. In some cases, a loose tapered roller bearing may still be serviceable, however, it must be carefully inspected before it is reused. Any looseness in a ball or straight roller bearing is cause for replacement.

Excessive wear is a common wheel bearing problem. Wear is usually caused by worn out lubricant or dirt entering the bearing. When wear results in extreme bearing looseness, the seal lip cannot maintain tight contact with the shaft, and will leak lubricant.

Damage to the bearing surfaces is caused by lubricant failure, overloading, or foreign material in the lubricant. Common roller and race damage includes:

- ❑ *Pitting,* which consists of small holes, or pits, in the rollers and races. Severe pitting or cratering is called *spalling.*
- ❑ Depressions or dents in the races and rollers.
- ❑ *Smearing,* which is metal transfer between the rolling elements and races.
- ❑ *Fretting,* also a type of metal transfer.
- ❑ *Brinelling,* a series of lines on the races of roller bearings.
- ❑ Bearing cage damage.
- ❑ Cracks in the races.

Bearings with any significant defects should be replaced. A loose tapered wheel bearing is still serviceable, however, it must be carefully inspected before it is reused. Any damage present on a bearing is cause for replacement.

Wheel Bearing Problem Diagnosis

The vehicle driver will usually notice a bearing defect as noise. Most wheel bearing noises are whines, growls, or roaring noises that rise and fall with road speed. Very dry bearings may squeak. A loose bearing may create a clunking noise as the vehicle begins to move, is braked, or turned. Loose bearings on a drive axle may cause vibration. Keep in mind that more than one problem may be present.

Drive Test

To diagnose whining, growling, or roaring noises, drive the vehicle on a quiet, dry street with light traffic. While maintaining a constant speed, lightly swerve the vehicle from side to side. Swerving changes the axial (thrust) load on the bearings. If the sound changes pitch or disappears as the vehicle is swerved, the wheel bearings are loose or defective. If the sound does not change, the problem is probably in the tires or another part of the vehicle.

 Note: If the bearing noise changes pitch or goes away during the swerve test, the defective bearing is usually on the side opposite the direction that caused the change in pitch.

To diagnose clunking or knocking noises drive the vehicle and firmly apply the brakes, then accelerate. Clunking or knocking noises should occur as the vehicle is accelerated and braked.

Shop Tests

If the road test indicates a wheel bearing problem, raise the vehicle so the wheels are off the ground. Then shake the suspected wheel(s) vigorously, **Figure 17-1.** There should be almost no play in the wheel. If any play is observed, check the steering and suspension components carefully to ensure that any looseness is actually in the

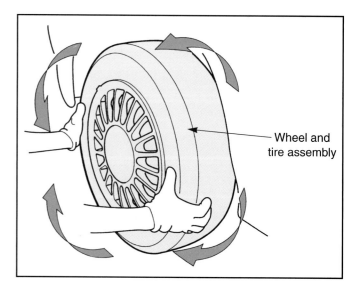

Wheel and tire assembly

Figure 17-1. *With the vehicle off the ground, shake the tire and wheel assembly vigorously. If there are signs of excessive play, find and correct the problem. (Honda)*

wheel bearings. It is best to have an assistant shake the wheel while you observe the steering and suspension components.

 Note: On vehicles equipped with solid front axles, the results of this test may be inconclusive due to the firm mounting of the bearings and axle. The axle should be removed and the bearing visually inspected.

If these tests indicate any wheel bearing problems, the bearings should be removed for visual inspection.

Tapered Roller Bearing Service

Tapered roller bearings are usually removed, cleaned, inspected, and repacked as part of normal brake service. Tapered roller bearings are the only type of wheel bearing that can be disassembled for cleaning, inspection, and lubrication. Ball and straight roller bearings are usually not serviced and are replaced when defective. The following sections explain how to remove, clean, inspect, reinstall, and adjust tapered roller bearings. Ball and straight roller bearing service is discussed later in this chapter.

 Caution: When servicing both sets of bearings on an axle, do not mix right and left side bearings. If a used bearing is not installed with its original race, it will wear quickly. If possible, remove and service the bearings one side at a time.

Removing Tapered Roller Wheel Bearings

Tapered wheel bearing and hub removal will vary with the type of brake and whether the rotor or drum is integral with the hub. On a disc brake system, begin by removing the caliper, **Figure 17-2.** On a drum brake system, back off the shoes if necessary, **Figure 17-3.** If the drum or rotor is a non-integral design, remove it at this time.

When each step has been performed as necessary, remove the grease cap covering the adjusting nut assembly. This is shown in **Figure 17-4.** Using diagonal pliers, remove the cotter pin and discard, **Figure 17-5.**

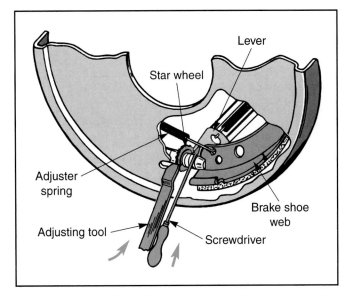

Figure 17-3. *Backing off the star wheel adjuster allowing the brake shoes to retract away from the drum. This makes drum removal easier. (Chrysler)*

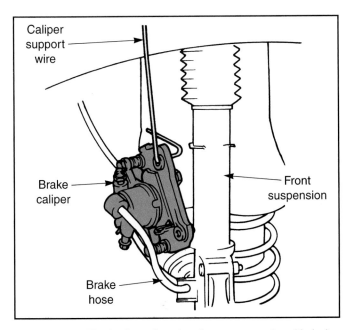

Figure 17-2. *The brake caliper has been removed and is being correctly supported before working on the wheel bearings. Be careful also with anti-lock brake system (ABS) sensors and wires. (Chevrolet)*

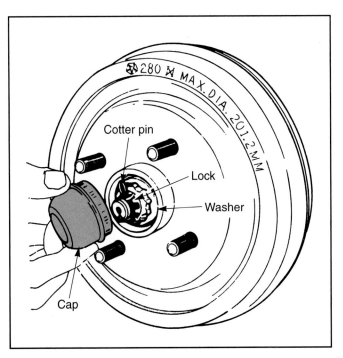

Figure 17-4. *Removing the grease cap to allow access to the bearing adjusting nut assembly. (Plymouth)*

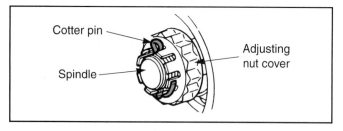

Figure 17-5. *Remove the adjusting nut cotter pin with pliers or a special cotter pin tool. (British-Leyland)*

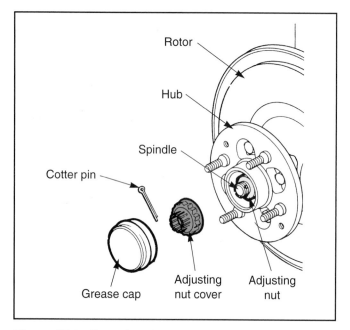

Figure 17-6. *Removing the adjusting nut cover so that the spindle nut may be unscrewed. (British-Leyland)*

Then remove the adjusting nut cover, (if used) nut, and thrust washer from the spindle, **Figure 17-6.** Lightly shake the hub to unseat the outer bearing, then slide the bearing from the spindle. Once the outer bearing is removed, remove the hub from the spindle. In some cases, a puller may be needed to separate the hub from the spindle.

Note: Some wheel bearing retainers used on four-wheel drive vehicles are held by a threaded ring or other holding device, Figure 17-7. Some nuts are retained by bending the outer edge into a slot on the hub. Other nuts are jam nuts with a deformed section in the internal threads. Do not drag the inner bearing and seal against the spindle threads.

Place the hub on a clean bench with the inner bearing side up. If possible, service bearings on a metal bench. Use a special tool or pry bar to remove the seal. Seal removal is shown in **Figure 17-8.** Do not attempt to save the seal for reuse. Once the seal has been removed, lift the inner bearing from the hub.

Note: The races are pressed into the hub, and should not be removed unless they or the bearing(s) are defective.

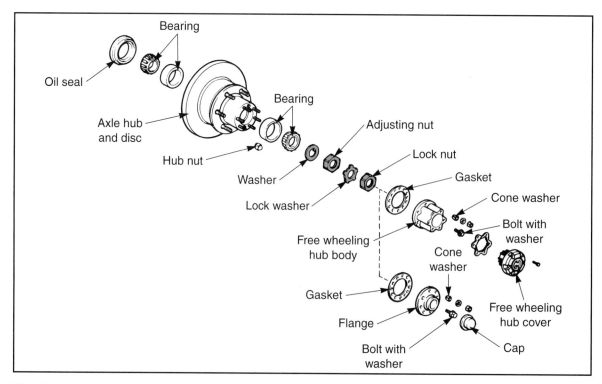

Figure 17-7. *An exploded view of a four-wheel drive front hub and wheel bearing assembly. (Toyota)*

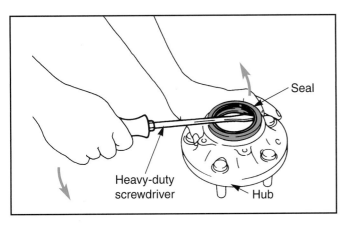

Figure 17-8. *Removing a grease seal from the hub. Be careful not to damage the seal recess in the hub. This can allow leakage. (Mazda)*

Cleaning Tapered Roller Wheel Bearings

Clean the bearings, grease cap, thrust washer, and adjusting nut using clean solvent. If the grease has hardened, scrub the bearings with a brush. Dry the bearings with compressed air.

Clean the inside of the hub, being sure to remove all old grease from the races and hub center cavity. Cleaning solvent can be used, but must be removed from any braking surfaces before the hub is reinstalled. Wipe the spindle clean, and remove any grease from the cotter pin hole.

 Caution: Do not spin a bearing with compressed air. The bearing rollers will fly out of the cage.

Inspecting Tapered Roller Wheel Bearings

After cleaning, inspect the bearings. Check the surface of each bearing roller, and check each of the races. Refer to **Figure 17-9** and look for the following problems:

❑ Discoloration, if not severe, is not cause for bearing replacement. A black or brown color is the result of staining agents in the lubricant. However, if the bearing shows a blue color, it has overheated and should be replaced.

❑ Some wear is normal, but an excessively worn bearing should be replaced. Excessive wear is usually due to abrasive particles in the grease. Severe abrasive wear may appear as scratches on the rollers.

❑ Pitting and depressions on the races and rollers are caused by dirt in the bearing, or by surface metal flaking off the rollers. Sometimes pitting is caused by a defective vehicle ground that causes metal transfer between the rollers and bearings.

❑ Smearing or fretting are caused when the rollers slide across the race instead of rolling. This can be caused by overtightening (too much preload, discussed in Chapter 16), lack of lubricant, failed lubricant, or overheating. Overheating is usually the result of excessive preload.

❑ Brinelling can be seen as a series of lines on the races. It occurs when the bearing surface is broken down or indented due to shock loads or loose adjustment (too little preload).

If the visual inspection does not turn up any problems, make a final check by pressing the bearing tightly into the outer race while turning it, **Figure 17-10.** If there is any roughness when the bearing is turned, reclean the bearing and try again. If the bearing is still rough, replace it.

If the rollers or races show any signs of defect, the bearing must be replaced. Replacement bearings always consist of both the roller and inner race assembly, and the outer race. Always compare the old parts with the new ones to be sure they are correct.

 Note: If a bearing race turns inside the hub, replace the hub. This is an indication the hub has excessive clearance. Installing a new race and bearing will not correct the problem.

Remove and Replace Wheel Bearing Races

When it is necessary to remove a bearing race, place the hub on a clean workbench with the side having the defective bearing facing down. Place a long punch behind the race (there is usually an indented area in the drum to permit access to the rear of the race). Strike the punch with a hammer until the race begins to move. Then move the punch to another spot on the rear of the race, directly across from the first spot. Continue hammering until the other side of the race begins to move. Repeat this procedure, alternating sides, until the race falls out of the hub. See **Figure 17-11.**

 Caution: Wear eye protection when driving bearing races in or out. Bearing metal is brittle, and the parts can shatter. If the race will be reused (such as when exchanging bearings between old and new hubs), use a brass drift to remove it, since a hardened steel punch may damage the race.

If possible, use a bearing driver to install the new race in the hub, **Figure 17-12.** Begin by setting the new race in the hub as squarely as possible. Drive the race into the hub by hammering on the end of the bearing driver. Remove the driver every few blows and ensure the race is squarely entering the hub. If a bearing race driver is not available, use a brass drift to carefully install the bearing. If using a drift, check carefully to ensure the race is squarely entering the hub. If the bearing is entering unevenly, carefully hammer on the highest side until the race is square with the bore. The race must seat completely against the bottom of the hub.

Once the race is fully seated, thoroughly clean the hub to remove any metal chips that may have been produced during the installation process.

ABRASIVE ROLLER WEAR

Pattern on races and rollers caused by fine abrasives.

Clean all parts and housings. Check seals and bearings and replace if leaking, rough or noisy.

GALLING

Metal smears on roller ends due to overheat, lubricant failure, or overload (wagon's).

Replace bearing—check seals and check for proper lubrication.

BENT CAGE

Cage damage due to improper handling or tool usage.

Replace bearing.

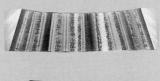

ABRASIVE STEP WEAR

Pattern on roller ends caused by fine abrasives.

Clean all parts and housings. Check seals and bearings and replace if leaking, rough, or noisy.

ETCHING

Bearing surfaces appear gray or grayish black in color with related etching away of material usually at roller spacing.

Replace bearings—check seals and check for proper lubrication.

BENT CAGE

Cage damage due to improper handling or tool usage.

Replace bearing.

INDENTATIONS

Surface depressions on race and rollers caused by hard particles of foreign material.

Clean all parts and housing. Check seals and replace bearings if rough, or noisy.

CAGE WEAR

Wear around outside diameter of cage and roller pockets caused by abrasive material and inefficient lubrication.

Clean related parts and housings. Check seals and replace bearings.

MISALIGNMENT

Outer race misalignment due to foreign object.

Clean related parts and replace bearing. Make sure races are properly seated.

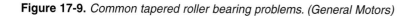

Figure 17-9. *Common tapered roller bearing problems. (General Motors)*

Figure 17-10. *To test a wheel bearing, press down with your hand and rotate the bearing. Any pits, roughness, or broken parts will be felt. (Deere & Co.)*

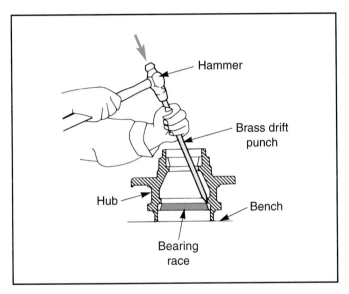

Figure 17-11. *Removing a bearing race from the hub with a brass drift punch and hammer. Wear safety glasses. (Nissan)*

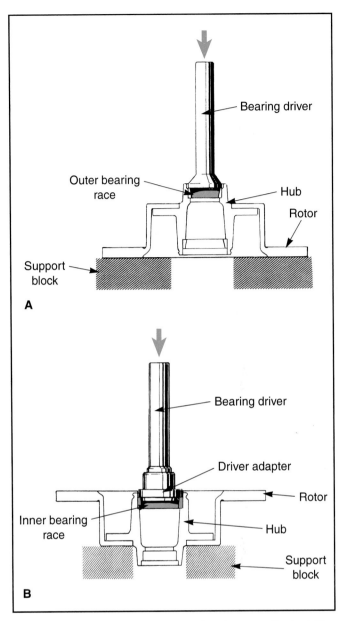

Figure 17-12. *A—Installing an outer bearing race. B—Installing an inner bearing race. Note the hub and rotor are properly supported. (Volvo)*

Lubricate Tapered Roller Wheel Bearings

Always follow the bearing manufacturer's recommendations for lubricant type, amount, and service intervals. Bearing damage can occur when either too much or too little lubricant is used. Excessive grease can be churned, allowing air to become mixed with the grease, which can produce friction and heat. Not enough grease can cause bearing failure from lack of lubrication.

Most vehicles made in the last 30 years use lithium-based extreme pressure (EP) grease. Check the proper service manual if there is any doubt as to the proper grease. It is especially important to use high temperature wheel grease in disc brake hubs. The higher temperatures produced by disc brakes can cause substandard greases to fail.

Always use quality wheel bearing grease. There are two methods of packing bearings: by hand or with a wheel bearing packer.

 Note: Always close wheel bearing grease containers when not in use. Dust and metal particles can fall into open containers. This debris will cause bearing damage.

Hand Packing Wheel Bearings

Hand packing is the oldest, and perhaps, the best method of lubricating wheel bearings. To hand pack wheel bearings, place a small amount of wheel bearing grease in your palm. Use your other hand to push the largest side of the bearing into the grease. Press down until grease begins to exit from the roller and cage assembly. See **Figure 17-13.** Turn the bearing slightly and repeat this procedure until the entire bearing is packed. Ensure that grease is present at each roller. Turn the bearing over and repeat. After the bearings have been thoroughly lubricated, wipe off excess grease.

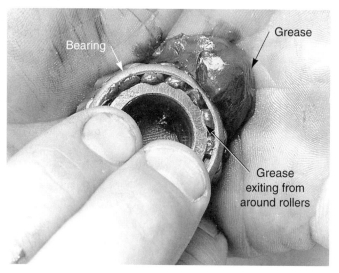

Figure 17-13. *Hand packing a tapered roller bearing with new grease. (Timken)*

Using a Wheel Bearing Packer

Use of a **wheel bearing packer** depends on its design. Place the bearing and holding cone over the spindle of the bearing packer. Push on the cone to tightly install the bearing, then pump grease from the reservoir into the bearing. Some bearing packers require you to press down on the cone to pack the bearing. When grease exits from between the rollers and cage of the bearing, it is packed. Remove the cone and any excess grease from the bearing. A typical bearing packer is shown in **Figure 17-14.**

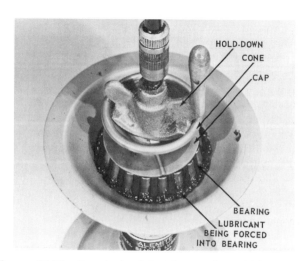

Figure 17-14. *One type of bearing packer which uses air pressure to force the grease into the bearing. (Ammco)*

Install and Adjust Tapered Roller Wheel Bearings

Begin by installing the inner bearing first. The inner side of the hub should be on the bench, facing up. Ensure the hub cavity is clean, then place a small amount of wheel bearing grease in the cavity. Use the same grease as used in the bearings. Place approximately two tablespoons of grease in the cavity. Excessive grease will increase the possibility that grease will get past the seal and contaminate the brakes. Lightly coat the inner and outer races with wheel bearing grease.

Carefully insert the inner wheel bearing into the hub. The smallest side of the bearing always faces the center of the hub. After the bearing is in place, install a new oil seal. Before installation, compare the old and new seals to be sure the new seal is correct.

 Note: A new seal must be installed whenever the bearings are repacked. An old seal is usually hardened and worn and cannot seal properly.

Lightly coat the sealing lip with wheel bearing grease. The outer metal housing is usually coated with sealant and does not require either grease or additional sealant. Next, place the seal in position in the hub, making sure the lip faces the bearing. Place the proper size driver over the seal and gently tap it into position using a hammer. See **Figure 17-15.**

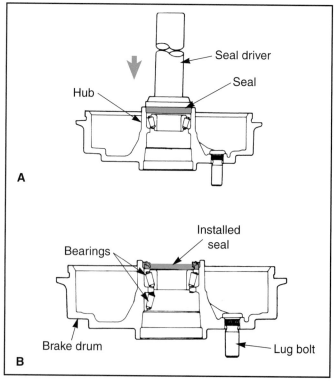

Figure 17-15. *A—Using a proper size seal driver to install a new seal into position. B—Seal is now correctly installed. (Hyundai)*

If the proper driver is not available, the seal can be driven into place with a hammer and drift, if extreme care is taken.

After the inner bearing and seal are installed, ensure the wheel spindle is clean. Then carefully install the hub over the spindle. Be careful not to allow the new seal to contact the spindle threads. Push the hub on far enough so that the seal contacts its riding surface on the spindle. Then install the outer bearing, smallest side facing inward. Install the thrust washer and adjusting nut. The nut should be tightened snugly, but not excessively tight. If the hub is integral with the brake drum, make sure the shoes are not dragging.

Wheel Bearing Adjustment

Wheel bearing adjustment must be performed accurately. Excessively tight bearings will overheat and fail; loose bearings will be damaged by vibration and shock loads. Always check the appropriate service manual for the correct adjustment procedure and specifications. A few four-wheel drive vehicles use a threaded ring for adjustment instead of a nut. Some manufacturers recommend the use of a torque wrench to make the initial tightening, and for final adjustment.

Begin by tightening the adjusting nut to about 100 ft-lbs. (135.5 N•m) or the recommended amount of torque while turning the hub. This fully seats the wheel bearing components so an accurate adjustment can be made. Then back off the nut until there is no load on the bearings.

Hand tighten the nut, then loosen it until a hole in the spindle lines up with a slot in the nut, **Figure 17-16.** If the nut has a separate lock cover, place it over the nut and then back off until one of the cover slots lines up with the spindle hole.

After the nut is in the proper position, install a new cotter pin, **Figure 17-17.** Bend the ends of the cotter pin against the nut and spindle, and cut off any excessive length to prevent contact with the dust cap. If the nut is retained by bending its outer edge into a slot in the spindle, use a small punch and hammer to bend it. If the nut edge has been deformed several times, it should be replaced. If a jam nut is used, be sure to obtain and install a new nut.

As a final check, measure the looseness in the hub assembly. There should be from .01-.05" (.025-.127 mm) endplay when properly adjusted. When the adjustment is correct, install the grease cap. Then install the caliper and rotor or drum as applicable and replace the wheel and tire.

Ball and Straight Roller Wheel Bearings Replacement

Most ball and straight roller bearings are pressed into the hub or axle. They cannot be adjusted or lubricated. These bearings are replaced by pressing or pulling the hub and bearing from the housing. Procedures for bearing removal are discussed in the following paragraphs.

CV Axle Wheel Bearings

Ball bearings are used as wheel bearings on front-wheel drive vehicles and rear-wheel drive vehicles with

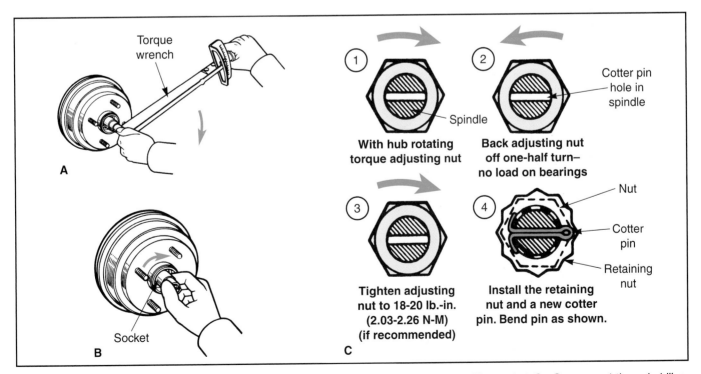

Figure 17-16. A—Torquing the adjusting nut to seat the bearings. B—Hand tightening nut with a socket. C—Sequence 1 through 4 illustrate another bearing adjustment technique. Follow the vehicle manufacturers wheel bearing adjustment procedures. (Toyota and Ford)

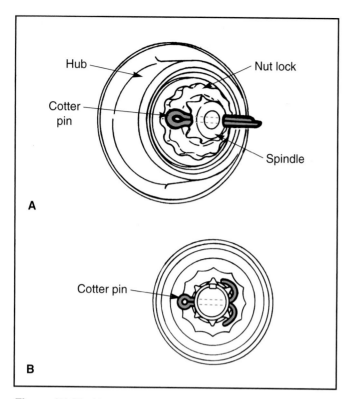

Figure 17-17. *Always use a new cotter pin after bearing adjustment. A—This shows cotter pin in position. B—The cotter pin ends have been properly bent. (Hyundai)*

independent suspensions. These are pressed into the hub, as in **Figure 17-18.** The hub is attached to, or is part of the spindle or *knuckle.* These bearings are generally sealed and must be replaced when they develop problems. To remove and replace most front wheel bearings, the hub must be removed. On a few vehicles, the bearings can be removed without removing the hub assembly, if a special puller is available.

Note: Many bearing and hub assemblies are sealed units. If the bearing is defective, the hub assembly must be replaced. Check the service manual for exact service procedures. Some bearing and hub assemblies contain wheel speed sensors used for the anti-lock brake system.

Hub Removal

Start by removing the wheel and tire. Remove and support the disc brake caliper. Remove the rotor if it is not an integral unit. Then remove the CV axle shaft retaining nut, **Figure 17-19.** The nut may be held in place with a cotter pin, a deformed spot on the outer surface of the nut, or may have no retainer. After the retainer has been removed, loosen and remove the nut. After the nut has been removed, remove the washer, if used.

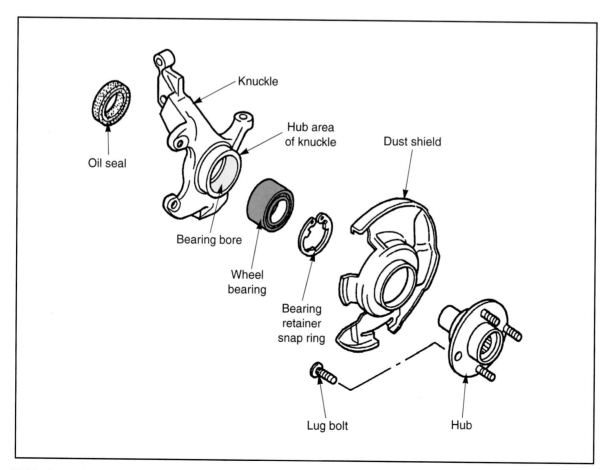

Figure 17-18. *An exploded view of a steering knuckle and wheel bearing assembly. Bearing fits into the bore and is held in position with a bearing retainer snap ring. (Kia)*

Note: Some nuts, especially those used on rear axles, have left hand threads. Check the service manual before attempting to remove the nut.

Caution: Exercise care when handling wheel speed sensor.

If the hub is part of the knuckle assembly, the entire assembly must be removed so the bearings can be pressed out. This involves removing the lower ball joint, and if used, the upper ball joint. If the vehicle has a MacPherson strut assembly, the knuckle must be removed from the strut.

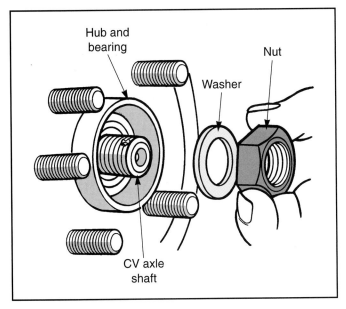

Figure 17-19. *Removing the CV (constant velocity) axle shaft nut and washer. (Chrysler)*

Next, the CV axle shaft must be separated from the bearing. Some axle shaft bearings are slip fit, and can be removed simply by pushing them inward. Other CV axle shaft bearings are pressed, and must be removed by a special tool that exerts pressure on the shaft. A shaft removal tool is being used in **Figure 17-20.**

Once the shaft is loose from the bearing, the hub can be removed from the vehicle. If the hub is bolted to the knuckle assembly it can be removed by removing the fasteners. A bolted hub is shown in **Figure 17-21.**

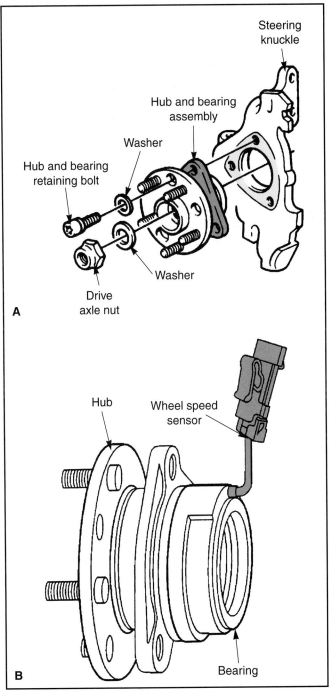

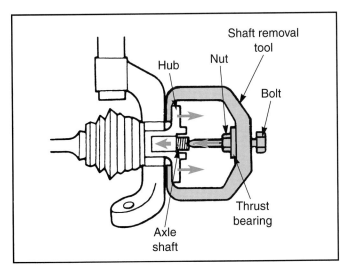

Figure 17-20. *An axle shaft removal tool being used to separate the hub and axle shaft. (Pontiac)*

Figure 17-21. *A—A bolted-on hub assembly being removed from the steering knuckle. B—Use caution not to damage any anti-lock brake wires, sensors, reluctor rings, etc., which in some cases, is part of the wheel bearing assembly. (Pontiac)*

Warning: Many suspension parts are under spring tension. You could be injured if parts under tension are carelessly removed. Always consult the proper service manual for suspension disassembly procedures.

To remove the knuckle, begin by removing the outer tie rod. Skip this step if the knuckle is installed on the rear axle. Remove the cotter pin and nut, then use the proper puller to remove the tie rod, **Figure 17-22**.

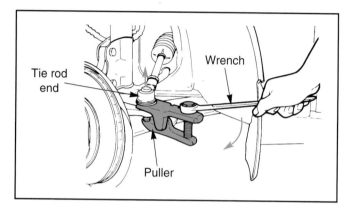

Figure 17-22. *Removing the outer tie rod end with a special puller tool. (Hyundai)*

Next, remove all tension from the lower ball joint. On MacPherson strut systems, tension will be removed from the lower ball joint when the wheel is raised from the ground. On most conventional suspensions, the lower control arm must be supported to remove tension from the ball joint. Check carefully before proceeding.

 Caution: If the brake hose bracket is connected to the knuckle, remove it now.

Once tension is removed, locate the ball joint stud. If the stud is held by a nut, remove the cotter pin and nut. If the ball joint stud is held by a through bolt, remove it. Then separate the lower ball joint from the knuckle. You may need to use a ball joint puller. See **Figure 17-23**.

After the lower ball joint is removed, remove the upper ball joint or MacPherson strut assembly as necessary. The upper ball joint can be removed in the same manner as the lower joint. Most knuckles are attached to the MacPherson strut by one or two bolts and nut assemblies. Remove the bolts and pull the knuckles from the strut assembly.

Bearing Removal and Installation

On a few vehicles, the bearings are a slip fit into the hub, and can be removed without completely removing the spindle from the vehicle. These bearings are usually

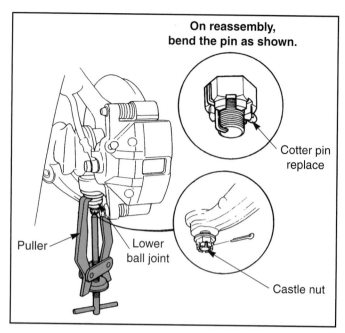

Figure 17-23. *A ball joint being separated with one particular type of puller. (Honda)*

held in place by a large snap ring or a bolted retainer. However, most bearings are pressed in and require the use of a hydraulic press for removal. A few bearings are non-removable, and the entire hub and spindle assembly must be replaced if they are defective.

Note: Wheel speed sensors are part of some wheel bearing assemblies. The sensors in these bearings are not serviceable.

To remove a pressed in bearing from the hub, you will need a hydraulic press. Place the hub and knuckle in the press, using the correct adapters to exert pressure on the outer bearing race. Remove any snap rings or dust shields before beginning the pressing operation.

The press operation should be performed carefully. Refer to **Figure 17-24** for general bearing pressing information. Note especially the precautions against pressing the inner bearing race. As you exert pressure on the hub and knuckle assembly, be sure the pressure is actually removing the bearing. If more than 2 tons (17 792 N) of pressure does not result in some bearing movement, make sure the adapters are positioned correctly.

After the old bearings are removed, check them against the new bearings to ensure they are correct. To install the new bearing, lubricate and position the bearing and adapters to press the new bearing into the hub, **Figure 17-24**. If any spacers are installed under the bearing, do not forget to place them in the hub before installing the bearing.

 Warning: Position the adapters carefully to ensure the seals are not damaged. Press on the outer race only.

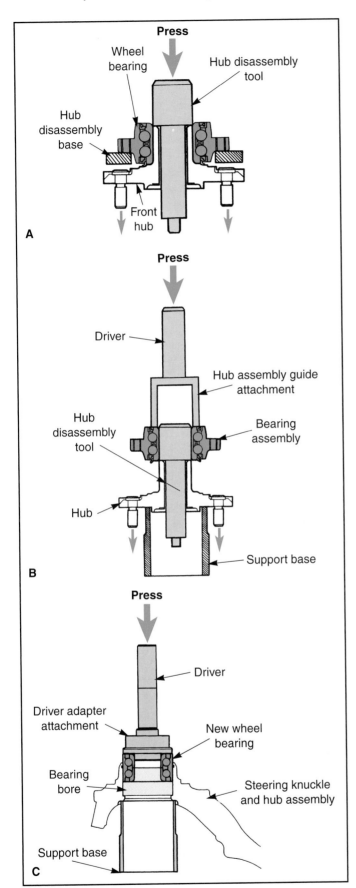

Figure 17-24. *A—Wheel bearing being pressed from the hub. B—A new bearing that is being pressed into position. C—Pressing a new wheel bearing into the steering knuckle hub. Wear safety glasses. (Honda)*

Carefully observe the hub and bearing during the pressing operation to ensure the bearing is fully seated. Install the next bearing and any spacers between the two bearings. Then, press in the second bearing. After the pressing operation is complete, reinstall any snap rings or dust shields, **Figure 17-25.**

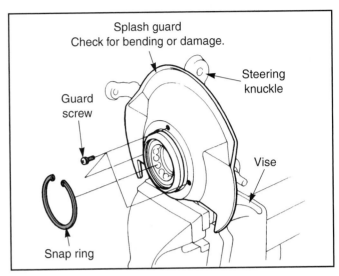

Figure 17-25. *A bearing retaining snap ring ready to be installed. If these show any signs of damage, replace them. (Honda)*

Hub Installation

Reinstall the hub and knuckle assembly on the vehicle, being sure the CV axle shaft meshes with any splines in the hub assembly. See **Figure 17-26.** Then attach the upper ball joint (when used), MacPherson strut, lower ball joint, and tie rod end. Tighten all fasteners to specifications, and install new cotter pins when necessary.

Ensure the CV axle is correctly aligned with the hub and bearings, then install the lock nut and washer. Tighten the nut to specifications, and install a new cotter pin. Reinstall the rotor and caliper, and the hose brackets if applicable. Reinstall the wheel speed sensor, if equipped, and check the air gap if needed. Reinstall the wheel and tire.

Note: Some suspensions must be aligned after the knuckle has been removed. Consult the proper service manual.

Solid Axle Wheel Bearings

Rear wheel bearings used on solid rear axles are either installed on the axle shaft or pressed into the axle housing, **Figure 17-27.** Some shafts are held by a retainer plate, while others are held by a C-lock in the differential assembly. The following sections explain how to remove and replace the various types of rear wheel bearings. Consult the correct service manual to determine which type of axle is being used on a particular vehicle.

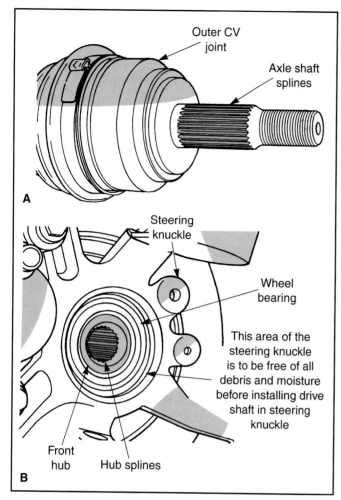

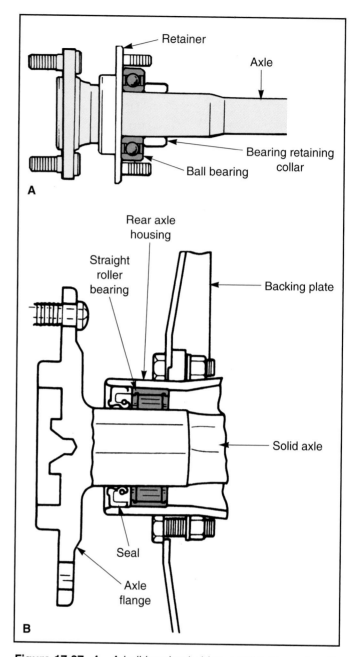

Figure 17-26. *A—The axle shaft splines must correctly mesh with the hub splines. B—Make sure the bearing splines are free of dirt and debris before installing the shaft. (Chrysler)*

Solid Axle—Bearing on Shaft

If the axle bearing is pressed on the shaft, the bearing and shaft are removed as a unit. The majority of these rear axle bearings are held in the axle housing by a retainer plate on the brake backing plate. The bearing is held on the shaft by a ***bearing collar.*** The bolts holding the retainer also hold the backing plate. Most bearings used with this design are greased and sealed. There is a seal between the bearing and the interior of the axle housing.

Axle Shaft Removal

To begin removal, remove the wheel and tire, and the brake drum if necessary. Then remove the retainer plate bolts. The bolts can be reached through a hole in the axle flange, **Figure 17-28.** After the bolts are removed, the shaft and bearing can be pulled from the axle. If the axle sticks, it can be loosened with a slide hammer, **Figure 17-29.**

 Note: In some cases, the axle can be removed by bolting the brake drum backward on the axle and using it to pull the axle with your hands.

Figure 17-27. *A—A ball bearing held to a solid drive axle with a bearing retaining collar. B—A solid rear axle which rides in a straight roller bearing, which has been pressed into the rear axle housing. (General Motors)*

Bearing Removal and Installation

The bearing can be removed after the retaining collar is split with a chisel, as shown in **Figure 17-30.** With the collar removed, the bearing and retainer plate will slide off the shaft. Check the retainer plate for the presence of a seal, and install a new seal if necessary. Check the shaft for damage, and replace if necessary. Slide the new bearing onto the shaft.

 Caution: Do not forget to install the retainer plate before the bearing.

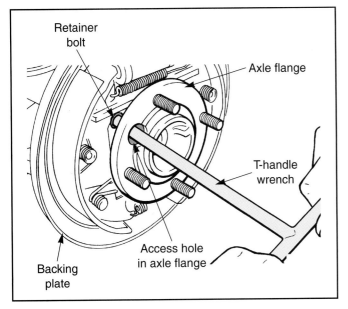

Figure 17-28. *Axle shaft retainer bolts being removed with a T-handle wrench. Note the access hole in the axle flange. (Chrysler)*

Install a new collar using a hydraulic press, **Figure 17-31.** If the collar does not take at least 2 tons (17 792 N) of pressure to press on, the shaft may be worn, and should be replaced as the collar might come loose during use. Some collars must be heated before installation.

Axle Installation

Before reinstalling the shaft, install a new seal in the housing, **Figure 17-32.** If a gasket is used between the backing plate and retainer plate, install it now. If the bearing has an external O-ring, ensure that it is in place and lightly lubricate it with bearing grease.

Slide the shaft into the axle housing, being careful not to damage the new seal. Position the retainer plate and install the mounting bolts. Tighten the bolts to the proper torque.

Solid Axle—Bearing in Axle Housing

If the bearing is installed in the axle housing, the shaft must be removed to gain access to the bearing. A **C-lock** axle will always have a **differential cover,** sometimes

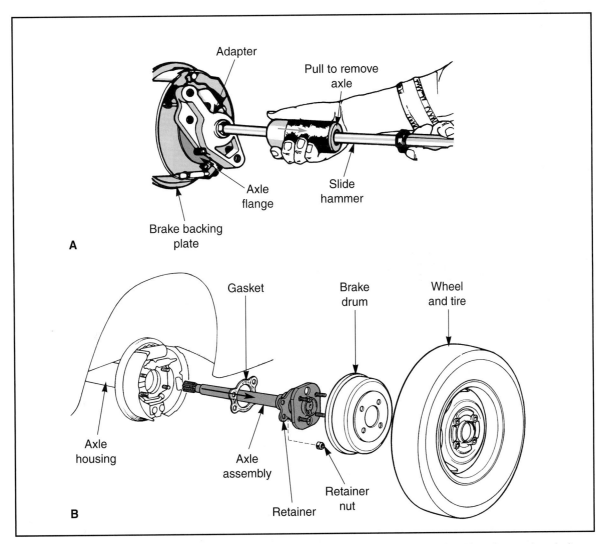

Figure 17-29. *A—A slide hammer may be needed to remove the axle from the housing. B—After the retainer bolts or nuts are removed, the axle and bearing may be pulled from the axle housing. (Toyota)*

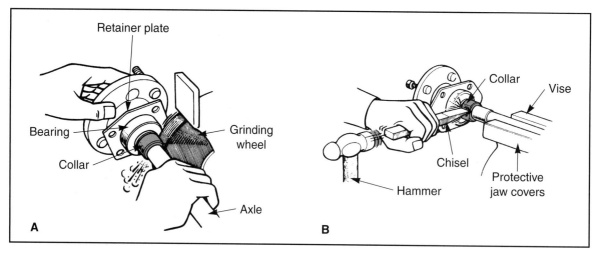

Figure 17-30. *Removing an axle bearing retaining collar. A—Grind the collar partway through, being very careful not to nick the axle shaft. B—Secure the axle shaft in a vise with protective jaws. Using a sharp chisel and hammer, strike the collar on the ground area. The collar should crack open and break apart, making removal easy. The bearing is now free for replacement. (Toyota)*

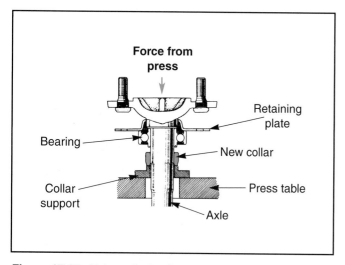

Figure 17-31. *Using a hydraulic press to install a new bearing retaining collar. Use the correct size collar support. Keep your hands and fingers away. If something snaps, they can easily be injured. (Toyota)*

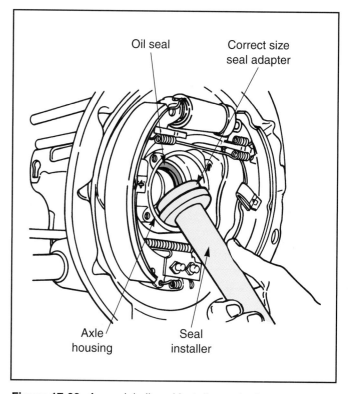

Figure 17-32. *A special oil seal installer and adapter was used to properly seat the new seal in the axle housing. Be sure to install the seal facing the correct direction. (Chevrolet)*

called a ***rear axle cover*** or an ***inspection cover.*** The differential cover is located at the rear of the rear axle assembly, directly behind the pinion yoke, **Figure 17-33.** If the vehicle does not have a differential cover, it does not use a C-lock retainer. Most of these bearings are lubricated by the oil in the rear axle. A seal is installed on the outside of the bearing.

Axle Shaft Removal

To remove an axle shaft held by a retainer plate, use the same procedure as for shaft mounted bearings. To remove a shaft held by a C-lock, begin by removing the wheel and tire and the brake drum if used. If the axle has disc brakes, remove the caliper and rotor. Then place a drain pan under the cover and remove the bolts. If necessary, pry lightly on the cover to loosen it. Allow the oil to drain into the pan. See **Figure 17-34.**

 Note: Do not attempt to save the fluid for reuse. Even slight impurities will cause severe rear axle damage.

With the cover removed, the differential carrier will be visible. The C-lock holding the axle to the side gear is located inside the differential carrier and the shaft holding the differential spider gears must be removed to gain access. Refer to **Figure 17-35** for an illustration of a typical

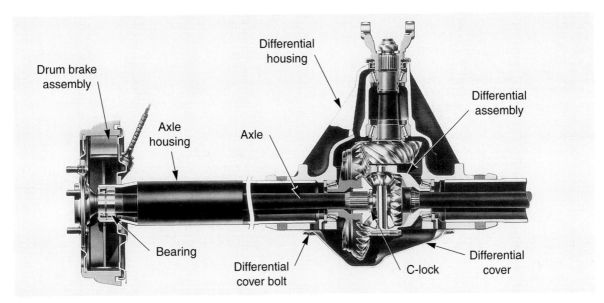

Figure 17-33. *A cutaway view of a rear axle housing and differential assembly. This axle is retained in position with C-locks. (Chevrolet)*

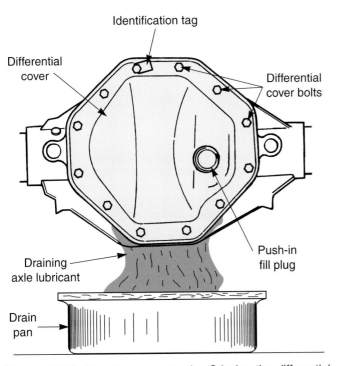

Figure 17-34. *To gain access to the C-locks, the differential cover must be removed. Place a drain pan under the cover and loosen all the cover bolts. Leave several loose bolts at the top of the cover. Tap or pry the cover away from the differential. At this point, the gear lubricant is free to drain out. The cover can now be completely removed. (Chrysler)*

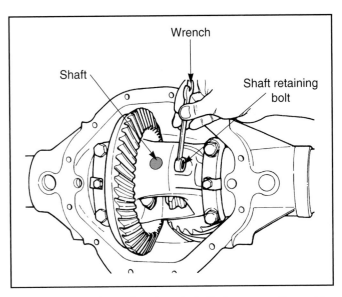

Figure 17-35. *Removing the shaft that secures the spider gears. Once the bolt is removed, the shaft can be pulled out. This will allow the axle shafts to be pushed inward freeing the C-locks for removal. (Dodge)*

Caution: Do not turn the differential carrier or the shaft. This will cause the differential spider gears to rotate out of position. Some technicians prefer to loosely install the differential spider gear shaft to prevent problems.

Bearing Removal and Installation

After the axle shaft has been removed, remove the seal and bearing from the housing. The seal is pressed into the housing, and can be pried out. Some technicians use the axle shaft as a lever to pull some seals. Do not attempt to reuse this seal. Most bearings are loosely installed in the

differential containing a C-lock. Turn the pinion gear until the bolt holding the differential shaft is visible. Remove this bolt and slide the shaft out of the differential carrier. Then go to the wheel and push the axle flange inward. This will move the axle inward, and the C-lock can be removed, **Figure 17-36.** Once the C-lock is removed, pull the axle from the housing.

housing. However, oil deposits may cause the bearing to stick, and it may be necessary to use a slide hammer or bearing puller for removal, **Figure 17-37.**

 Note: If the brake shoes are coated with axle lubricant, replace the shoes.

Inspect the shaft where the bearing rollers ride for damage. This damage will usually consist of excessive wear or pitting. Also check the surface where the seal rides for excessive grooving. If any defects are found, replace the shaft.

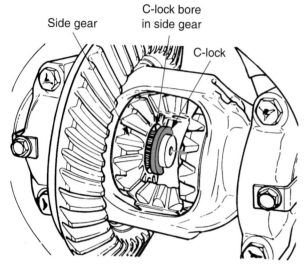

Figure 17-36. *The axle shaft has been pushed in. The C-lock is free of its bore in the side gear, and can now be removed to free axle. (Dodge)*

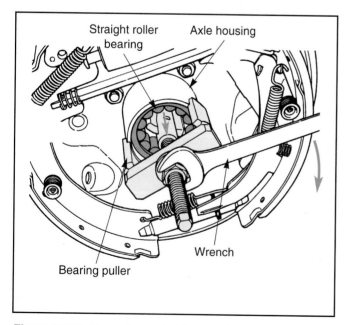

Figure 17-37. *Removing a straight roller bearing from the axle housing with a special bearing puller. (Dodge)*

The new bearing should be carefully tapped into place. After the bearing is installed, press a new seal into place using the proper adapter, **Figure 17-38.** Slide the shaft into place, being sure it engages the splines on the differential side gear. Ensure the side and spider gears are in their proper positions, then reinstall the spider gear shaft and bolt. Tighten the bolt to the proper torque, then turn the differential through several revolutions to ensure that all parts are in place and not binding.

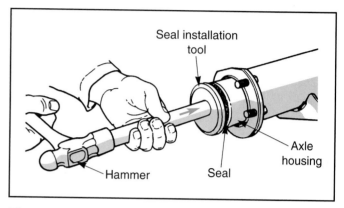

Figure 17-38. *The new bearing is in place, and a new seal is being carefully tapped into position. (Chrysler)*

Install the differential cover using a new gasket. Refill the rear axle assembly with fresh fluid. Reinstall the brake components, wheel, and tire.

Caution: If the differential is a limited-slip or Positraction type, you may need to use a special differential oil or add a limited-slip additive. Ordinary differential oil will cause a limited-slip differential to become noisy and vibrate.

Non-Driving Axle Wheel Bearings

The ball or straight roller bearing assemblies used on non-driving axles are usually self-contained units. To remove the hub and bearing assembly, remove the wheel and tire. Next remove the brake caliper and rotor. The hub can then be unbolted from the knuckle assembly, **Figure 17-39.** In a few cases, the bearings are pressed into the knuckle assembly and the knuckle must be removed from the vehicle to press out the bearings. Bearing removal using a hydraulic press was covered in the CV axle bearing section.

Seal, Gasket, and O-Ring Service

As you learned in Chapter 16, the job of oil seals, gaskets, and O-rings is to keep lubricant in, and dirt and water out. When they fail, the bearing can be ruined by loss of lubricant or the entry of foreign material. A failed seal can

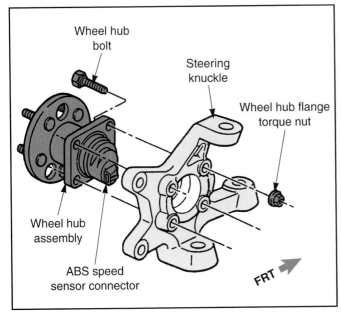

Figure 17-39. *A non-driving axle which has been unbolted and removed from the steering knuckle. Note the anti-lock brake system (ABS) speed sensor connector. Handle all ABS parts with care. (General Motors)*

also allow lubricant to contact the brake linings, causing brake grab and chatter, and ruining the linings. The following sections explain how to service seals, gaskets, and O-rings.

Identify Leaking Seals, Gaskets, and O-Rings

The most obvious sign of leaks is lubricant where it should not be found. If the brake linings, backing plate, or hub exterior show signs of lubricant, you can assume the related seal is leaking. If a bearing shows signs of extreme wear or corrosion from water entry, the seal or gasket has probably failed. Often the seal or gasket will obviously be damaged. In many cases, however, the seal may appear to be in good condition.

Oil seals are usually replaced whenever the related bearing is replaced. However, the seal can be checked for defects as an aid to diagnosis. To check a seal, first observe the condition of the seal's rubber portion. If the rubber is hard, or if the lip appears to be worn, the seal is defective. If the metal housing is bent or cracked, the seal was probably leaking. As a final check, place the seal over the axle shaft. It should fit snugly on the axle at its normal riding position. If the seal feels loose, it will not seal properly. Gaskets are also normally replaced, but can be inspected for cracks, thin spots, or deformation. Any of these can cause leaks, and the gasket should be replaced.

O-rings are not commonly used with wheel bearings. They should be replaced whenever the bearing is serviced. O-rings usually become hard and shrunken from brake and bearing heat. A hard O-ring will usually break when it is removed from its groove.

 Note: Replace brake linings that are saturated with axle lubricant.

Replacing Oil Seals

Seal design varies with the seal's intended function. Some seals used with tapered roller bearings on front and rear axles are designed to retain grease. Other seals hold in gear oil when used with straight roller bearings in solid rear axles. Installation varies with the seal design and location. The following procedures are general, since the number and placement of seals varies widely.

Since the oil seals are usually changed anytime they are removed, damage to the seal during removal is acceptable. Seals can be removed by prying with a screwdriver or pry bar, driving with a punch and hammer, or by using a puller. See **Figure 17-40.** However, care must be taken that the bore into which the seal fits is not damaged during the removal process.

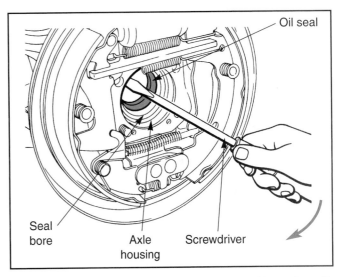

Figure 17-40. *A rear axle oil seal being removed by prying with a special heavy-duty screwdriver. (Chrysler)*

The hub or other part containing the seal bore should be carefully inspected to ensure that it is not damaged. Minor nicks and sharp edges can be sanded off. If the bore is out of round, dented, or cracked, the part should be replaced.

Compare the old and replacement seals. As a further check, place the new seal over the shaft, and ensure that it fits snugly. To install the new seal, obtain the proper driver. If required, lightly coat the metal housing of the seal with gasket sealer.

 Note: Many seals have sealing compound applied to the outer edge of the metal housing, and no further sealant should be applied.

Before installing the new seal, ensure the bearing and any other parts that fit under the seal are installed. Position the seal squarely in the bore and place the driver in position over the seal. Make sure the seal is installed in the proper direction as a seal installed backward will leak. On a tapered roller bearing hub, be sure the lip faces toward the bearing. On a solid rear axle or wheel hub, make sure the seal lip faces inward. Lightly tap the seal into position using light even hammer blows, **Figure 17-41**. Ensure the seal is installed squarely and is fully seated.

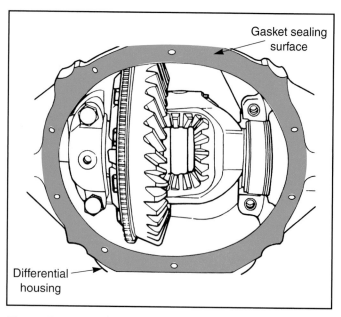

Figure 17-42. *A gasket sealing surface on one particular rear axle housing. Check sealing surfaces for rust, nicks, dents, cracks, etc., that may prevent the new gasket from sealing properly. (Chrysler)*

Note: If necessary, the gasket can be held in position by a light coating of grease.

Install and tighten the attaching bolts. Be careful not to overtighten the fasteners, since the gasket can be crushed by excessive bolt torque.

To replace an O-ring, use a pointed tool such as a pick or awl, if necessary, to pull the O-ring from its groove, **Figure 17-43**. Be careful not to damage the groove. Check for damage, deposits, or debris in the groove. If necessary, carefully clean the groove using a brass or plastic tool. Lightly lubricate the new O-ring and install it in the groove, being sure that it is not twisted. When installing the part with the O-ring, lubricate the outside of the ring to ease installation.

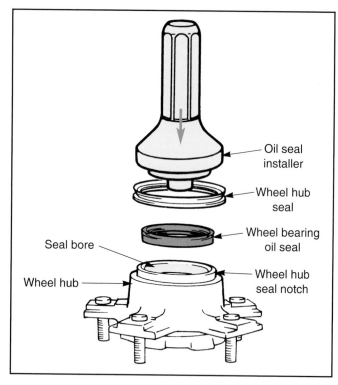

Figure 17-41. *A wheel hub seal that will be lightly hammered into position, using the correct size oil seal installer and a hammer. The outer oil seal will be placed by hand, onto its notched area of the hub, after the oil seal is installed. (Chevrolet)*

Remove and Replace Gaskets and O-Rings

Gaskets and O-rings are simply replaced, but some precautions must be taken. To replace a gasket, scrape the old gasket from the surface to which it is attached. In many cases, the gasket can be lifted from the part. After the gasket is removed, scrape any old gasket material and sealant from the sealing surfaces. After scraping and cleaning, carefully inspect the sealing surfaces for deep scratches, dents, and foreign material. Refer to **Figure 17-42**. Check sheet metal parts (such as retainer plates) for flatness. If a sealing surface shows any signs of defects, replace the part.

Before installing the new gasket, compare it with the old gasket. Place the new gasket in position, using sealant if specified. Make sure all bolt and alignment holes line up with the mating parts. Check that the gasket covers the sealing surfaces and is facing in the proper direction.

Replacing Wheel Studs

Sooner or later, you will encounter a stripped or broken wheel stud. Some manufacturers recommend removing the hub or axle flange and pressing in the new stud as shown in **Figure 17-44**. Other manufacturers recommend replacing a damaged stud without removing the hub or axle from the vehicle. This procedure is explained below.

Begin by removing the wheel and tire. Ensure there is access to the rear of the stud. If access is limited, remove the caliper, drum, or other parts necessary to gain access. Then use a hammer or special stud press to remove the old stud. See **Figure 17-45**. Make sure the broken stud does not fall into the hub where it could be caught between moving

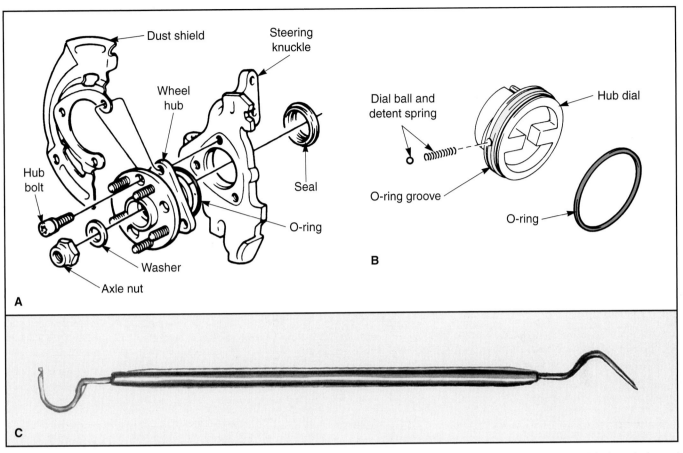

Figure 17-43. *A—Wheel hub which uses an O-ring seal. B—A hub engagement dial from a four-wheel drive vehicle front hub, and its O-ring seal. C—One type of O-ring removal tool. (OTC, General Motors & Toyota)*

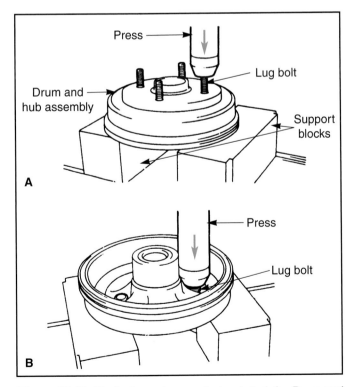

Figure 17-44. *Replacing a damaged wheel stud. A—Drum and hub are supported as the hydraulic press forces the old stud out. B—The hub and drum have been turned over, and the new lug bolt is being pressed into position. (Toyota)*

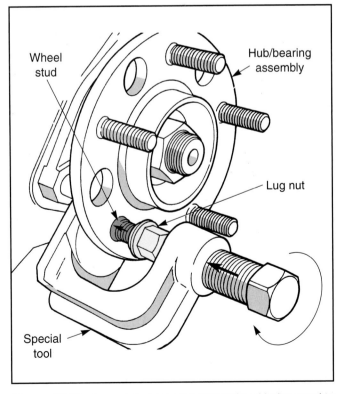

Figure 17-45. *A special wheel stud removal tool being used to press out a broken stud. The hub and bearing assembly are still attached to the vehicle. (Chrysler)*

parts. Compare the replacement stud with the broken stud to ensure the diameter of the knurled mounting section, as well as thread size and length, are correct.

Place the new stud through the hole in the hub or flange. Install several washers over the stud and then install one of the lug nuts. The assembly is shown in **Figure 17-46.** Thoroughly lubricate the stud threads and tighten the nut against the washers. As the nut is tightened, the stud will be drawn into the hole. The knurled area will cut into the sides of the hole, and the stud will be held tightly.

> **Caution: If the knurled area of the replacement stud does not enter the hub or flange, its diameter may be too large, Recheck to ensure that you have the right stud. Do not mix standard and metric studs. Do not use an impact wrench to install the replacement stud.**

As you continue to tighten the nut, the knurled area will be drawn through the hole. When the shoulder of the stud is flush with the rear of the hub or flange, the stud is fully installed. Remove the nut and washers and check that the stud is tight. Then reinstall the wheel and tire.

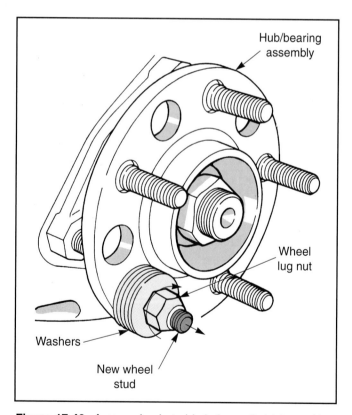

Figure 17-46. *A new wheel stud is being pulled into position. Install several washers over the stud, then place the lug nut, flat side against the washers, onto the stud. Continue to tighten the lug nut until the stud is fully seated. Add or subtract washers as needed for seating. Remove the lug nut and washers. Wheel and tire can now be reinstalled. (Chrysler)*

Summary

Wheel bearings are generally trouble free. Tapered roller bearings should be periodically cleaned and lubricated for maximum life. Ball and straight roller bearings require no maintenance, and will usually last the life of the vehicle if the bearing seals are not damaged.

Typical bearing problems include pitting, spalling, depressions and dents, smearing, fretting, brinelling, cage damage, and cracks. Bearing defects are caused by lack of lubricant, overloading, incorrect preload, and seal failure. Any damage to a bearing is cause for replacement.

Tapered roller wheel bearings are often cleaned and repacked as part of brake service. If possible, remove and service the bearings on one side at a time. To service tapered bearings, remove the adjusting nut and remove the hub. After thoroughly cleaning the bearings, check them for visible damage, roughness, and wear.

If the bearings can be reused, repack them by hand or with a bearing repacker. Always use quality wheel bearing grease. Always close wheel bearing grease containers when not in use. Defective races can be removed from the hub with a punch and hammer. New races should be installed with a driver. Wear eye protection when driving bearing races.

Install the inner wheel bearing and seal, then place the hub on the spindle and install the outer bearing, washer, and adjusting nut. Adjust preload and install a new cotter pin when needed.

Most ball and straight roller bearings are pressed into the hub or axle. Ball bearings are used for wheel bearings on front-wheel drive vehicles, and rear-wheel drive vehicles with independent suspensions. To remove and replace most front wheel bearings, the hub must be removed and the bearings pressed from the hub. Some bearings can be removed without removing the hub assembly. Other bearing and hub assemblies are sealed and must be replaced as a unit.

To remove the bearings, the CV axle shaft must be separated from the bearing and the knuckle removed from the vehicle. After the knuckle is removed, the bearings are removed, and new bearings installed, using a hydraulic press.

Rear wheel bearings used on solid rear axles are either attached to the axle shaft or pressed into the axle housing. If the axle bearing is pressed on the shaft, the bearing and shaft can be removed as a unit after the bearing retainer is unbolted. A new bearing can be installed after the collar is removed. The seal between the bearing and the interior of the axle housing should also be changed.

If the bearing is installed in the axle housing, the shaft must be removed to gain access to the bearing. Some shafts are held by a retainer plate, but most are installed with C-locks. To remove the C-lock, the differential inspection cover must be removed and the differential partially disassembled. Non-driving axle wheel bearings are removed in the same way as CV axle bearings.

Oil seals, gaskets, and O-rings should be checked for leakage and replaced when necessary. These parts are usually replaced when the related part is serviced. Removal and replacement procedures are similar for all types of seals. Seals are usually damaged during removal. Check the seal bore for damage after the seal is removed. The new seal should be installed facing the proper direction. Gaskets and O-rings can be replaced by simple procedures. Always ensure that you have the right part.

Most wheel studs can be replaced without removing the hub or axle flange from the vehicle. Be sure to select the right replacement stud. Use a lug bolt and washers to install the new stud.

Review Questions—Chapter 17

Please do not write in this text. Write your answers on a separate sheet of paper.

1. Of the three major classes of bearings, which can be ruined by improper adjustment?

2. If any bearing defects are found, what should you do to repair the damage?

3. The vehicle driver will usually become aware of a defective bearing when it becomes _____.

4. Swerving a vehicle from side to side changes the _____ load on a bearing.

5. Place the following bearings removal steps in order:

   _____ 1.
   _____ 2.
   _____ 3.
   _____ 4.
   _____ 5.
   _____ 6.

   (A) Remove the inner bearing and hub assembly.
   (B) Remove the dust cap.
   (C) Remove the adjusting nut.
   (D) Remove the outer bearing.
   (E) Remove the cotter pin.
   (F) Lightly shake the hub assembly.

6. Two parts that should always be replaced when tapered roller bearings are serviced are the _____ and _____.

7. The bearings should be _____ with compressed air, but under no circumstances should they be _____.

8. If a bearing race will be reused, always remove it using a drift made of _____.

9. The lip of a replacement oil seal should face the _____.

10. What is the purpose of tightening the adjusting nut to 100 ft-lbs (135.5 N•m) before making the final wheel bearing adjustment?

11. A special tool must be used to remove the CV axle shafts from the bearing when the shaft is _____ into the bearing.

12. If a rear axle bearing is pressed on the shaft, it is usually lubricated by _____.

13. If a rear axle bearing is installed in the axle housing, the shaft is held in the housing by a _____.

14. If a bearing has been ruined by water, the related _____ has failed.

15. How do you remove a broken stud from a hub?

ASE Certification-Type Questions

1. Technician A says that ball bearings usually fail due to lack of maintenance. Technician B says that straight roller bearings usually fail due to lack of maintenance. Who is right?
   (A) A only.
   (B) B only.
   (C) Both A & B.
   (D) Neither A nor B.

2. A vehicle has a whining noise at cruising speeds. If the vehicle is swerved as it is driven, the sound changes. Which of these is the *least likely* cause of the whining noise?
   (A) Defective front ball bearing.
   (B) Tight grease seal.
   (C) Loose tapered wheel bearing.
   (D) Worn rear straight roller bearing.

3. Which of the following is the most common method of holding the adjusting nut in place on a tapered roller bearing assembly?
   (A) Cotter pin.
   (B) Jam nut.
   (C) Deformed nut.
   (D) Locking tab.

4. Technician A says that bearing overheating can be caused by excessive preload. Technician B says that bearing brinelling can be caused by excessive preload. Who is right?
   (A) A only.
   (B) B only.
   (C) Both A & B.
   (D) Neither A nor B.

5. If a bearing is damaged but the outer race appears to be in good condition, which of the following steps should the technician take?
   (A) Replace the bearing but reuse the outer race.
   (B) Replace the bearing and race.
   (C) Swap the bearings from side to side.
   (D) Reuse the old bearing and race.

6. Bearing races are always _____ into the hub.

    (A) bolted

    (B) welded

    (C) staked

    (D) pressed

7. Modern wheel bearing grease include all of the following, EXCEPT:

    (A) extreme pressure.

    (B) lithium based.

    (C) long fiber.

    (D) from a covered container.

8. When adjusting tapered wheel bearings, which of the following should be done *first*?

    (A) Tighten the adjusting nut to 100 ft-lbs (135.5 N•m).

    (B) Install the cotter pin.

    (C) Adjust the nut hand tight.

    (D) Loosen the nut until a slot lines up with the spindle hole.

9. Technician A says that bearing endplay of .02" (.025 mm) indicates a loose bearing. Technician B says that no bearing endplay indicates a tight bearing. Who is right?

    (A) A only.

    (B) B only.

    (C) Both A & B.

    (D) Neither A nor B.

10. All of the following parts must be removed from the steering knuckle before it can be removed from the vehicle, EXCEPT:

    (A) CV axle nut.

    (B) tie rod end.

    (C) bearing seal.

    (D) brake caliper.

11. Which of the following types of bearings usually require that the technician have access to a hydraulic press?

    (A) Ball.

    (B) Tapered roller.

    (C) Straight roller.

    (D) None of the above.

12. Technician A says that some rear axle bearings are held to the shaft by a collar. Technician B says that some axle bearings are held in the axle housing by a retainer plate. Who is right?

    (A) A only.

    (B) B only.

    (C) Both A & B.

    (D) Neither A nor B.

13. If a vehicle rear axle does not have _____, it does not have a C-lock axle.

    (A) drum brakes

    (B) disc brakes

    (C) an inspection cover

    (D) solid axle shafts

14. Which of the following is *not* an oil seal defect?

    (A) Bent metal housing.

    (B) Hardened lip.

    (C) Torn lip.

    (D) Snug fit on shaft.

15. Technician A says that some wheel studs can be replaced without removing the hub or flange. Technician B says that a stud that will not enter the hub indicates a defective hub. Who is right?

    (A) A only.

    (B) B only.

    (C) Both A & B.

    (D) Neither A nor B.

Chapter 18

Parking Brakes

After studying this chapter, you will be able to:

❏ Explain the function of the parking brake.
❏ Identify the components and explain the operation of driver applied parking brake levers.
❏ Describe the construction and explain the operation of automatic parking brake release mechanisms.
❏ Identify the components and operation of parking brake warning lights.
❏ Identify the components of parking brake linkage.
❏ Explain the operation of parking brake levers and linkage.
❏ Describe the construction and operation of drum brake parking brakes.
❏ Describe the construction and operation of disc brake parking brakes.

Important Terms

Parking brake
Foot-operated parking
  brake
Pedal arm
Ratchet
Brake release handle
Multi-stroke foot parking
  brake
Vacuum release
Vacuum canister
Vacuum motor
Hand-operated parking
  brake

Push-pull hand brake
T-handle
Floor-mounted lever
Parking brake
  warning light
Parking brake
  cable
Cable fittings
Cable sheath
Cable pulleys
Linkage rods
Brake multiplier

Equalizer
Linkage adjustment
  devices
Service brakes
Integral drum parking
  brake
Retaining prong
Parking brake apply
  lever
Lever strut
Self-contained drum
  parking brakes

Screw disc parking
  brake
Nut and cone
Actuator screw
Ball-and-ramp disc
  parking brake
Operating shaft
Thrust screw
Cam disc parking
  brake
Cam
Actuator rod

This chapter is designed to explain the purpose, design, and operation of parking brakes, sometimes called emergency brakes. The parking brake is often ignored since most drivers use the parking pawl in the automatic transmission to hold the vehicle. Some technicians also overlook the parking brake during routine brake service, since it is seldom the source of customer complaints. This chapter covers the design and operation of both drum and disc parking brakes. It also covers the use of manual and automatic parking brake release mechanisms.

Parking Brake Function

The primary purpose of the **parking brake** is to keep the vehicle from being moved or moving while it is not being driven. Although the transmission can be used to keep the vehicle from moving, the parking brake provides much better holding power, which is especially important when the vehicle is parked on a steep grade. On vehicles with manual transmissions, it can be used by the driver to hold the vehicle on an incline while releasing the clutch and moving his or her foot from the brake to the accelerator. This allows the driver to start without rolling backwards. If the brake hydraulic system fails, the vehicle can be slowed and eventually stopped by applying the parking brake, although the braking effect will be much lower.

Parking Brake Linkage

Before discussing the wheel mounted parking brake components, we will cover the parking brake actuating linkage. The arrangement of levers, cables, and linkages that make up parking brakes is similar on all vehicles. The major difference is how the driver operates the linkage, and whether the release mechanism is manual or automatic. All parking brake levers give the driver mechanical advantage by increasing mechanical leverage. Force can be multiplied by arranging the lever inputs and outputs in relation to their pivot points. The development and use of mechanical leverage was discussed in Chapter 4.

Driver-Operated Levers

The driver operates the linkage through one of two types of hand lever or a foot lever with a pedal. Most modern vehicles have a foot-operated lever, or a hand-operated lever installed beside or between the front seats. A few vehicles have hand operated push-pull levers located under the dashboard. Some foot pedals are released by an automatic device.

Foot-Operated Parking Brake

The basic component of a **foot-operated parking brake** is a lever that is operated by the driver's foot. The lever is usually called the **pedal arm.** Pushing down on

the pedal arm causes it to pivot and pull up on the parking brake cable. The pedal arm is held in the applied position by a pawl which engages teeth on the lever assembly. This holding design is known as a **ratchet.** The pedal arm is released by pulling on the **brake release handle.** Pulling on the handle moves a linkage rod which disengages the ratchet from the teeth on the lever assembly. A spring returns the pedal arm to the released position. A manually released foot actuated parking brake can be seen in **Figure 18-1.**

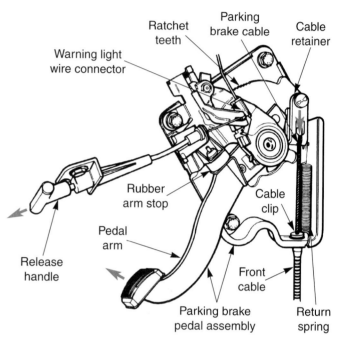

Figure 18-1. *An overall view of a typical manually released foot actuated parking brake mechanism. (Jeep)*

When the driver presses down on the foot brake, the pedal arm pivots so that the point where it is attached to the cable moves up. This causes the parking brake cable to tighten, actuating the parking brakes. As the pedal arm swings down, the pedal arm teeth skip over the pawl. Once the desired parking brake tension is reached and the driver releases foot pressure, the pawl engages the nearest tooth, holding the brake in the applied position. A switch is connected to the red Brake warning light, which tells the driver the parking brake is applied when the ignition is on.

To release the brake, the driver pulls on the release handle, moving the rod. The rod disengages the pawl from the pedal arm teeth, allowing the brake to be pulled into the released position by the return spring.

Multi-Stroke Foot Parking Brake

The **multi-stroke foot parking brake** requires several applications of the pedal arm to set the brake. In other words, the pedal must be pumped to apply the parking brake. When the driver applies the multi-stroke foot lever pedal, it returns to its normal released position, even though the parking brakes are applied. Additional applications of the pedal increase the tension on the parking brake cable. By allowing more than one pedal application, this parking brake design can compensate for play in the cables and connecting linkage caused by lining wear. This is important when rear disc brakes are used, since pad wear can greatly increase the amount of distance required to apply the brakes. See **Figure 18-2.**

Foot Brake Vacuum Release

Some foot brakes are released by an automatic vacuum release. The *vacuum release* is always used on vehicles with automatic transmissions. This release consists of a small diaphragm assembly attached to the brake release lever. The chamber can be called a *vacuum canister* or *vacuum motor.* Inside, a flexible rubber diaphragm divides the chamber into two sections. One side is attached to a rod and return spring, and the other side is sealed, except for a vacuum port. The vacuum port is connected to engine vacuum by hoses and a control valve. Some vehicles use a separate pump to create the vacuum.

The vacuum control valve is attached to the shift selector. When the shift lever is in the park position, no vacuum can reach the vacuum chamber. The return spring keeps the rod in the applied position. When the shift lever is moved to put the transmission into any other gear position, the control valve is moved by the shifter. The valve allows vacuum to reach the canister, which draws the diaphragm in, moving the rod and compressing the return spring. This releases the parking brake.

Most automatic vacuum releases have an emergency manual release lever. If vacuum fails to reach the chamber, the brake may still be released by pulling the manual release lever. On some vehicles, a small hand operated vacuum pump is installed next to the vacuum chamber. Operating this pump creates enough vacuum to release the brake when the shift lever is moved. One type of vacuum operated brake assembly is illustrated in **Figure 18-3.**

Hand-Operated Parking Brake

The two major types of *hand-operated parking brake* levers differ only in where they are located. The dashboard operated push-pull hand brake lever is used on older cars and some current imported trucks. The floor-mounted hand lever is common on vehicles with bucket seats. Each type is shown in **Figure 18-4.**

Push-Pull Hand Brake

The *push-pull hand brake* is a metal rod which travels inside a support tube. The push-pull brake lever is located directly under the dashboard near the steering wheel. One side of the rod is flat, and contains a series of teeth. The rod is attached to a *T-handle.* When the T-handle is pulled outward, the end of the rod pulls on the cable, engaging the parking brakes. A pawl assembly located near the end of the support tube engages the

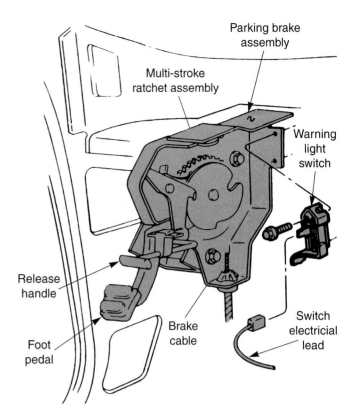

Figure 18-2. *A multi-stroke foot operated parking brake unit. (Pontiac)*

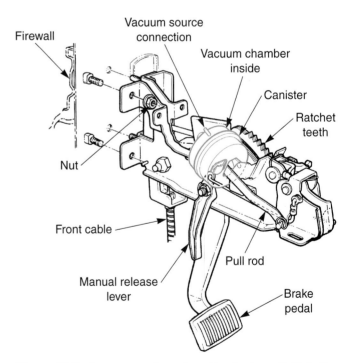

Figure 18-3. *A vacuum released parking brake unit. Note the manual release lever which is used to release the brake in the event of a vacuum or vacuum operated part failure. (Chrysler)*

teeth on the rod, holding it in the applied position. On most modern vehicles, a warning light will be illuminated when the parking brake is applied and the ignition switch is in the on position.

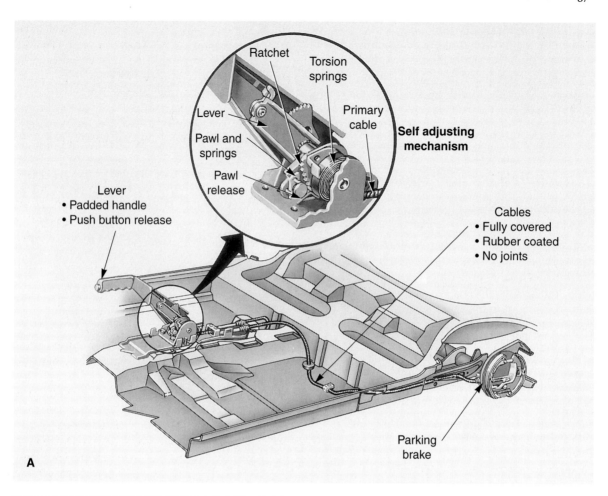

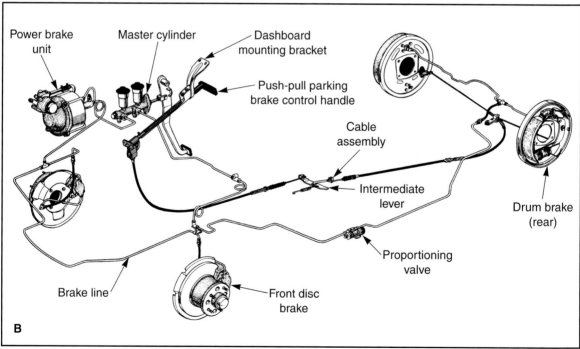

Figure 18-4. *A—A floor mounted hand brake lever and cable assembly. B—A dashboard, hand operated, push-pull hand brake assembly with related parts. (Chrysler, Toyota)*

To release the parking brake, the T-handle is turned to the left or right. This twists the rod teeth away from the pawl, disengaging them. The driver can then push the T-handle back in, releasing the parking brake. On some push-pull levers, the pawl is disengaged by pushing a button located on the T-handle. See **Figure 18-5.**

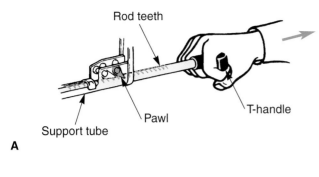

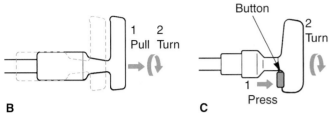

Figure 18-5. *A T-handle push-pull hand brake. A—T-handle being pulled to set parking brakes. The pawl is engaging the rod teeth. B—To disengage the pawl from the rod teeth, pull outward slightly, turn the T-handle, and push in. C—Some T-handles use a push button to disengage the pawl and teeth. (Hyundai, Toyota)*

Floor-Mounted Lever

The operation of the **floor-mounted lever** is similar to the foot-operated parking brake. The hand lever is a hollow steel tube or flat steel covered by a plastic sleeve and/or leather boot attached to the floorboard through a support plate. The support plate contains teeth which engage a pawl in the bottom of the lever assembly. A spring-loaded push rod inside the tube connects the pawl with a push button at the top of the lever. See **Figure 18-6.**

When the lever is pulled up, it pivots on the support plate, tightening the brake cable and applying the parking brake. As the lever moves up, the pawl will slide over the teeth on the support bracket. The pawl holds the lever in place when it is applied. The red Brake warning light is illuminated when the ignition switch is on and the parking brake is applied.

To release the brake, the driver pulls up slightly on the lever and depresses the push button. Depressing the push button releases the pawl from the support bracket teeth. The driver can then push the lever downward, removing cable tension and releasing the parking brake. A few floor-mounted brake levers have an automatic adjustment device. This device will take up any slack in the parking brake cables when the lever is operated.

Parking Brake Warning Light

A **parking brake warning light** is used on almost all modern vehicles to inform the driver when the parking brake is applied. Some systems also have a buzzer that works in conjunction with the warning light. The audible tone alerts the driver in case he or she fails to notice the warning light or if the lightbulb burns out.

The parking brake warning light is the red Brake warning light on the instrument panel, **Figure 18-7.** The switch is installed on the parking brake support bracket in a position where it can contact the movable parking brake linkage. Switch operation is the same for hand and foot parking brakes. Switch operation and other electrical fundamentals will be discussed in more detail in Chapter 20.

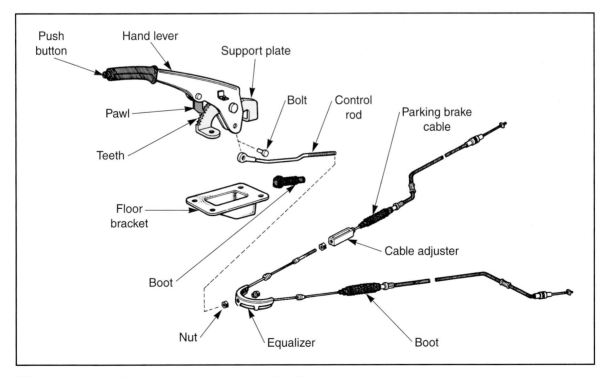

Figure 18-6. *A floor mounted, hand operated, lever assembly. (Toyota)*

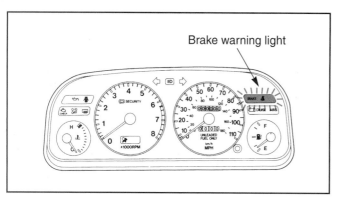

Figure 18-7. *A brake warning light and its location on the instrument cluster. (Toyota)*

Most warning light switches have a single positive terminal. Battery voltage is available at the terminal whenever the ignition switch is on. The circuit supplying this voltage is wired through the ignition switch, a fuse, and the warning lightbulb, **Figure 18-8.**

When the parking brake is actuated, some part of the linkage will contact the switch, depressing the plunger. Depressing the plunger causes contacts on the switch to close. This completes the circuit by grounding the positive terminal to the switch body. The switch body is grounded to the battery negative terminal through the support bracket and frame. Current flows through the ignition switch, fuse, and red Brake warning lamp bulb, causing it to illuminate. A parking brake assembly showing one type of switch and its location is illustrated in **Figure 18-9.**

Parking Brake Apply Linkage

The following sections cover the construction and operation of the linkage which connect the driver operated parking brake lever with the wheel brakes. This linkage consists of cables, rods, and levers. These and other related parts are covered in this section.

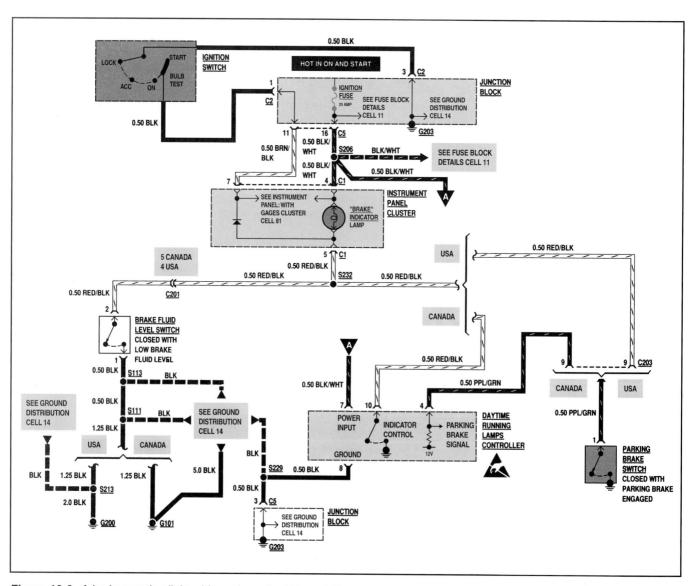

Figure 18-8. *A brake warning light wiring schematic. (Chevrolet)*

Parking Brake Cables

Parking brake cables are usually made from stranded steel wire. Cable thickness is usually around 3/16" (4.76 mm). Because cable is flexible, it can be bent around corners and through openings under the vehicle body. **Figure 18-10** shows a typical parking brake cable setup. Most vehicles have at least three cables, one from the apply lever to the equalizer, and two from the equalizer to the wheel brakes.

Cable Fittings

Formed **cable fittings** are pressed on or bolted to the ends of the cables. A common fitting is a round ball, cylinder, or flat end, which installs into a slot in a bracket or lever, **Figure 18-11.** These fittings allow the cable to be attached to the parking brake apply lever and other levers and components.

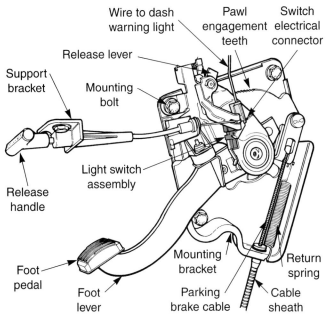

Figure 18-9. *A foot operated brake showing the brake warning light switch and its location. (Jeep)*

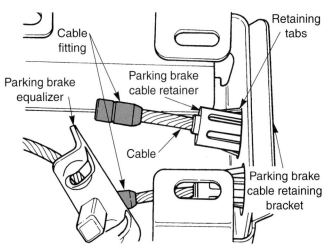

Figure 18-11. *A cylinder-type cable fitting that has been swaged on (pressed) under pressure. This particular one is used to connect the cable to the equalizer. (Chrysler)*

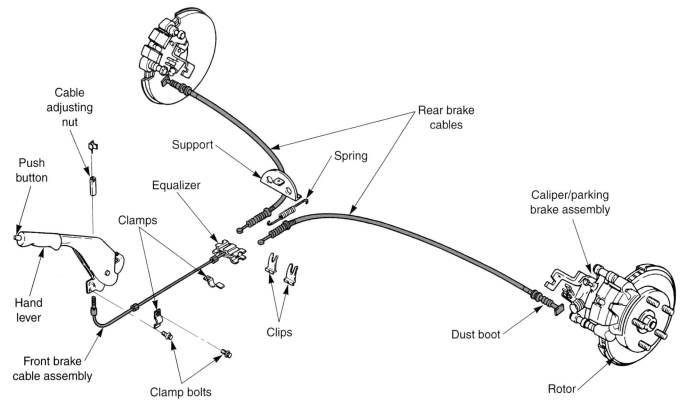

Figure 18-10. *A floor-mounted hand lever and cable assembly used with rear disc brakes. Note that three separate cables are used. (Mazda)*

Cable Sheaths

Parking brake cables are furnished as complete units. For much of its length, the cable is enclosed in a **cable sheath.** These sheaths, sometimes called *conduits,* are made of steel wire wrapped tightly together. The sheath guides and protects the cable. Some sheaths are covered with rubber to help prevent the entry of dirt and water. The sheath is flexible, allowing the cable to move inside. A typical cable sheath is shown in **Figure 18-12.**

 Note: The entire cable assembly is simply referred to as a cable.

Cable Pulleys

Sometimes **cable pulleys, Figure 18-13,** are used to carry the cable around sharp bends. The pulleys are often used at the apply lever to reduce wear on the cable. Pulleys are sometimes used under the vehicle. Occasionally a piece of solid conduit is bent into position and used to carry the cable around a sharp point.

Linkage Rods

On a few vehicles, **linkage rods** are used as part of the parking brake linkage. The entire linkage system or just a portion can make use of rods. One brake system utilizing a control rod is shown in **Figure 18-14.**

Brake Multiplier

The parking brake lever is designed so that driver force is all that is needed to apply the wheel brakes. In

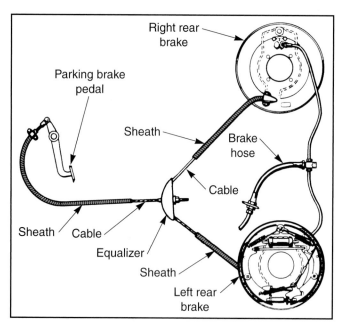

Figure 18-12. *Brake cable sheaths as used on one parking brake assembly. (Bendix)*

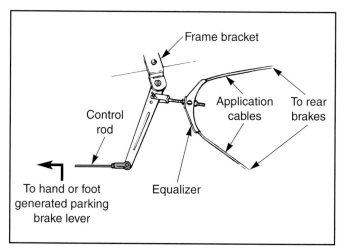

Figure 18-14. *A brake cable apply system that incorporates a control rod. (FMC)*

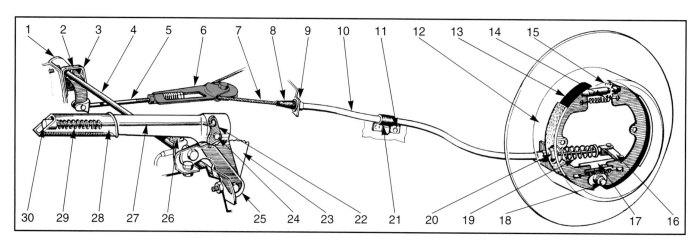

Figure 18-13. *A hand operated parking brake which uses a cable pulley assembly (6) on the equalizer. The pulley provides smoother operation, longer cable wear, etc. 1—Inside support attachment. 2—Rubber cover. 3—Lever. 4—Shaft. 5—Pull rod. 6—Pulley. 7—Cable. 8—Rubber cover. 9—Front attachment. 10—Cable sleeve. 11—Attachment. 12—Brake drum. 13—Secondary brake shoe. 14—Return spring. 15—Star wheel. 16—Lever. 17—Movable rod. 18—Anchor bolt. 19—Return spring. 20—Rear attachment. 21—Rubber cable guide. 22—Pawl. 23—Ratchet segment. 24—Rivet. 25—Outside support attachment. 26—Warning valve switch. 27—Push rod. 28—Parking brake. 29—Spring. 30—Pushbutton. (Chevrolet)*

some applications, however, more leverage is needed. The **brake multiplier** (also called an intermediate lever) is incorporated into some parking brake systems to increase cable pull leverage. The multiplier creates extra mechanical advantage by careful placement of the input, output, and pivot points. This creates more leverage, but at the expense of distance traveled. If the brakes are properly adjusted, this decrease in distance will not be a factor. A brake multiplier is illustrated in **Figure 18-15.**

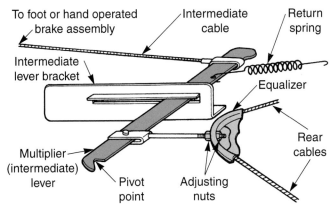

Figure 18-15. *The brake multiplier as used on one parking brake control setup. The multiplier acts somewhat like a pry bar (fulcrum). The longer the leverage, the more pulling force it will generate. The equalizer takes up slack in the cables. (General Motors)*

Brake Equalizer

A brake **equalizer** is installed in the parking brake system to provide a balanced, or equal pull on the brake levers. The equalizer does this by allowing the application cable to center itself. The equalizer is a steel bar with attachments for the brake cables or rods. The input is from the application cable or control rod. There are one or two output cables. Note that the cable adjuster is usually installed at the equalizer, **Figure 18-15.**

In operation, the input cable pulls on the equalizer, which then pulls the wheel brake cables. If one cable tightens the wheel brake before the other, the equalizer will pivot on that cable, **Figure 18-16.** This allows the cable on the other side to continue to move until the tension on that wheel brake matches the tension on the first wheel.

Support Brackets, Guides, and Heat Shields

The parking brake cables and other parts must be secured and protected. As mentioned earlier, many cables are routed through rigid or flexible conduit and some are covered with rubber. The cables, control rods, multiplier, and equalizer are attached to the vehicle body by various brackets and clips. To protect the parking brake components from road splash and exhaust heat, shields are often installed. Some typical brackets, guide, supports, and heat shields are depicted in **Figure 18-17.**

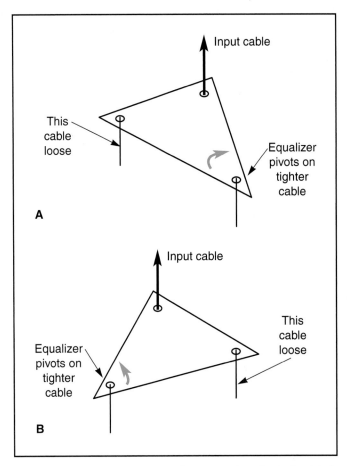

Figure 18-16. *Brake cable equalizer operation. A—Left cable is loose, equalizer pivots (travels) to the right to take up the slack. B—Right cable is loose, equalizer travels to the left to take up the slack.*

Linkage Adjusting Devices

As the brake linings and linkage parts wear, it is sometimes necessary to adjust the parking brake to ensure that it operates. Many times, the adjustment is made at the wheel brakes. This is common on drum brake systems. Some disc brake designs are self adjusting. However, many parking brake systems have **linkage adjustment devices.** Most of these are located on the equalizer and consist of a threaded rod held in position by locknuts. Other vehicles may have turnbuckles or adjustable levers.

Two adjusters are shown in **Figure 18-18.** Most of them are operated by loosening the holding fasteners and tightening the adjuster device to remove any slack in the cable.

Parking Brake Types

Previous sections discussed the linkage that transferred the driver's hand or foot application to the wheels. This section covers the parking brake components installed at the wheels. The wheel parking brakes operate by using mechanical linkage to apply brake friction linings. The brake linings normally operated by the brake hydraulic

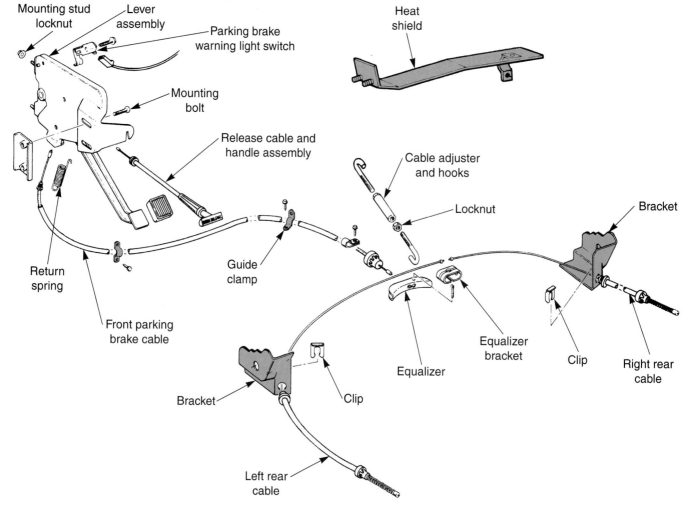

Figure 18-17. *A foot operated parking brake system illustrating the support brackets, guide clamps, and a heat shield. (Chrysler)*

system are called the ***service brakes.*** Most parking brake mechanisms apply the service brakes mechanically. A few parking brakes operate separate friction linings that are not connected to the hydraulic system. Modern vehicles are equipped with three main types of parking brakes:

❑ Integral (built-in) parking brake which applies the brake shoes on a rear drum brake system.

❑ An independent and self-contained drum brake used with some rear disc brakes.

❑ Integral parking brake which applies the brake pads on a rear disc brake system.

Use of these parking brakes depends on the type of service brake used on the wheel. Most parking brakes are installed on the rear wheels. A few vehicles have the parking brake installed on the front wheels.

Note: A few very old cars and large trucks use a parking brake mounted on the rear of the transmission. This type is known as a driveshaft parking brake, and is generally used on heavy-duty trucks.

Integral Drum Parking Brake

The ***integral drum parking brake*** is part of the vehicle's drum brake assembly. It applies the same brake shoes used by the hydraulic system. The parking brake cable sheath is solidly attached to the backing plate. Most are attached by a retaining prong. The ***retaining prong*** is a set of metal fingers, **Figure 18-19,** which can be pressed through the backing plate. After the fingers enter the backing plate opening, they expand and cannot be removed. Other cables are held by clamps, clips, etc.

Inside the drum, the cable is attached to the lower end of the ***parking brake apply lever.*** The lever is attached to the web of the secondary shoe. The secondary and primary shoes are linked through the parking brake ***lever strut.*** The apply lever pivots on the lever strut.

Operation—Servo Drum Brakes

When the parking brake is applied on servo drum brakes, the cable pulls on the lower end of the parking brake lever. The lever moves forward and pushes the lever strut, which moves the secondary shoe into contact with the drum. The secondary shoe contacts the drum, and also directs apply pressure through the star wheel assembly to

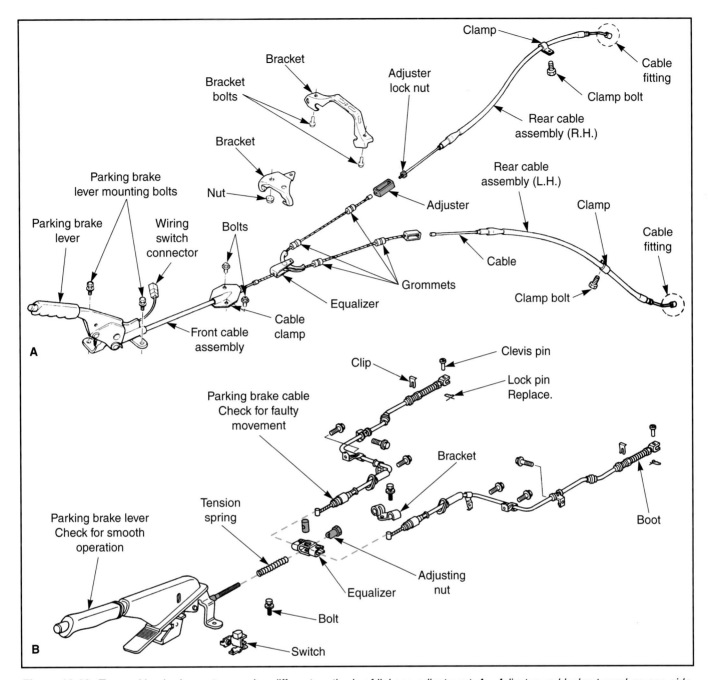

Figure 18-18. *Two parking brake systems using different methods of linkage adjustment. A—Adjuster and locknut used on one side. B—When both cables are attached to the equalizer, correct adjustment can be obtained by turning the adjusting nut. (Honda, Nissan)*

the primary shoe. After the lever strut has applied the secondary shoe, it cannot move any more. The parking brake lever then pivots on the lever strut and applies additional pressure to the primary shoe. This action is shown in **Figure 18-20.**

When the parking brake is released by the driver, the lever returns to its released position and the brake shoe return springs release the shoes.

Operation—Non-Servo Drum Brakes

Parking brake operation on a non-servo brake drum is similar to that on a servo brake. However, the action of the lever strut is taken by the adjuster assembly. Since the

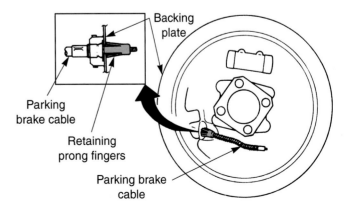

Figure 18-19. *Parking brake cable assembly held in the backing plate with a retaining prong unit. (Chevrolet)*

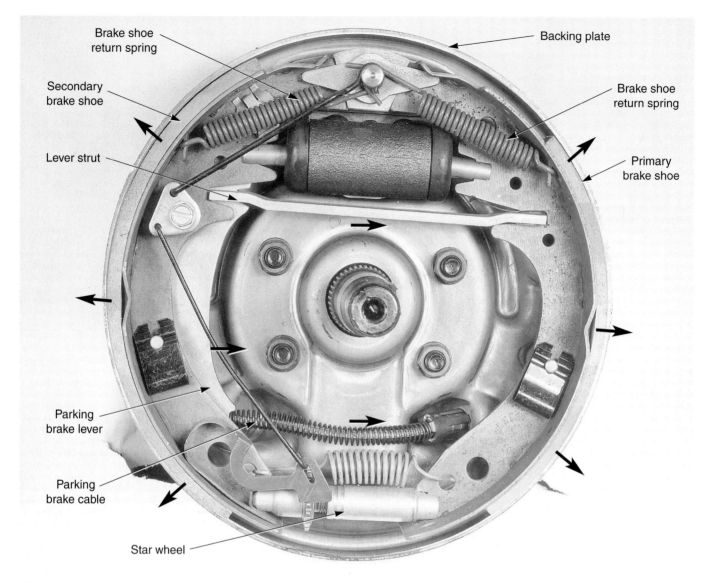

Figure 18-20. *Operation of one particular servo brake parking brake assembly. There are a number of different servo brake styles. (Chrysler)*

bottom anchor is fixed, no pressure is transferred between the primary and secondary shoes. The brake lever pivoting action through the adjuster assembly applies both shoes. **Figure 18-21** shows the parking brake operation on a typical non-servo drum brake.

Self-Contained Drum Parking Brake

Self-contained drum parking brakes are used on some vehicles with rear disc brakes. The parking brake assembly consists of two shoes and a brake drum. The shoes are smaller than conventional brakes since they act only as parking brakes. The drum is an integral part of the rotor assembly.

A backing plate is fastened to the axle housing as a place for brake shoe attachment and a splash shield. A fixed anchor or star wheel is installed at the bottom of the backing plate. The shoes are attached to the backing plate and anchor with conventional hold-down and retractor

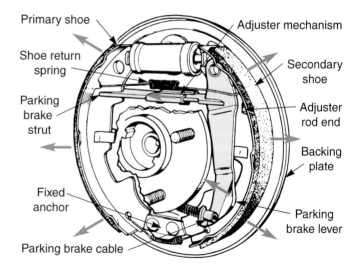

Figure 18-21. *Operation of one type of non-servo parking brake assembly. This assembly uses a fixed anchor at the bottom of the backing plate. This prevents a transfer of pressure between the primary and secondary brake shoes. (Delco)*

springs. Brake shoe operation is similar to the operation of conventional drum brakes. There is a lever and strut, which moves the tops of the shoes into contact with the drum, **Figure 18-22.** However, there is no hydraulic wheel cylinder or automatic adjuster.

Pulling on the brake cable causes the lever to move, forcing the shoes apart. This presses the shoes into the drum. When the cable tension is released the return springs draw the shoes away from the drum, releasing the brakes. This action is shown in **Figure 18-23.**

Integral Disc Parking Brake

The use of rear disc brakes is increasing. On vehicles with no drum brakes, there must be a method of using the disc brakes as parking brakes. Several methods are used to apply disc brakes with mechanical linkage. They include screw, ball-and-ramp, and cam types.

 Note: A few older cars and large trucks use a lever parking brake which clamps an extra set of pads to the rotor. This type of parking brake is independent of the hydraulic disc brake caliper.

Screw Disc Parking Brake

The *screw disc parking brake* is the most common type of caliper parking brake actuator. All parts except for the apply lever are located inside the caliper housing.

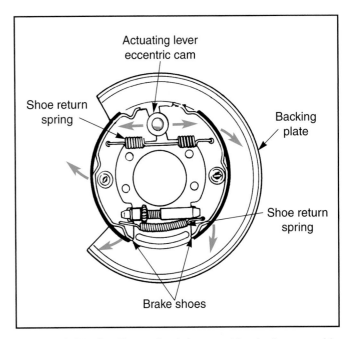

Figure 18-23. *A self-contained drum parking brake assembly. The actuating lever turns the eccentric cam which moves the shoes into drum contact. (Parker)*

Hydraulic system operation is similar to that of a conventional floating caliper disc brake.

In the simplest version of this design, the piston assembly contains a *nut and cone* assembly. The nut is threaded into a lever operated *actuator screw.* The lever is bolted to the actuator screw and has a return spring. The

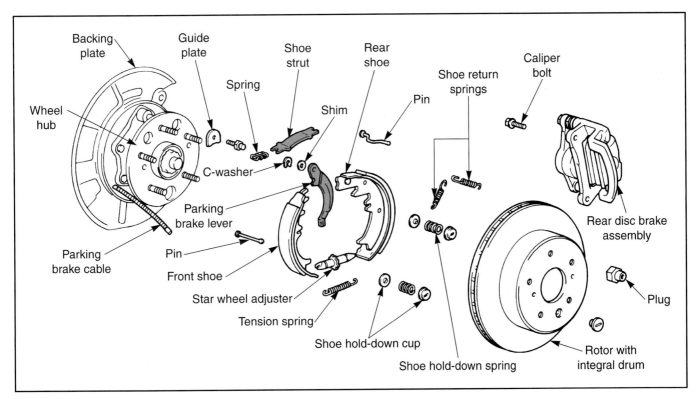

Figure 18-22. *An exploded view of a self-contained drum parking brake setup. The parking brake lever and strut moves the shoes into drum contact. (Lexus)*

actuator screw and lever can rotate, but cannot slide in or out. The nut is splined to the cone allowing it to slide inside the cone and piston. This allows the cone and piston to move independently of the nut when the piston is applied with hydraulic pressure. This allows the screw to remain stationary when the service brakes are applied.

When the lever is rotated by the cable, the screw turns. Since the screw is fixed in place and cannot travel in or out, turning it moves the nut outward. Movement of the nut forces the cone and piston outward. The piston presses the brake pads against the rotor, applying the parking brake.

When the parking brake is released, the return spring pulls the lever back to its released position. The actuator screw will then draw the nut back, allowing the piston seal and boot to retract the piston slightly, releasing the brakes. A cutaway of the brakes being applied is shown in **Figure 18-24.** The screw type brake adjusts itself automatically each time it is applied, if any pad wear has taken place.

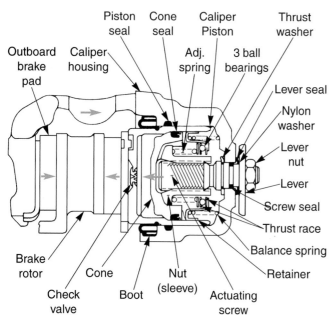

Figure 18-24. *A cutaway view of a screw type disc parking brake assembly showing the brakes being applied. The actuating screw forces the piston pads in contact with the rotor. The caliper is a sliding type, which applies the outboard brake pad to the rotor. (EIS)*

Ball-and-Ramp Disc Parking Brake

The *ball-and-ramp disc parking brake* is located inside the caliper housing. Ball-and-ramp brake units are always used with single piston, sliding brake calipers. The ball-and-ramp assembly consists of three steel balls between an **operating shaft** and a **thrust screw, Figure 18-25.** The operating shaft is attached to the apply lever and has three slots which hold the steel balls in position. The operating shaft and lever can turn but cannot move in or out. The thrust screw has ramps, or tapered areas, where it contacts the balls.

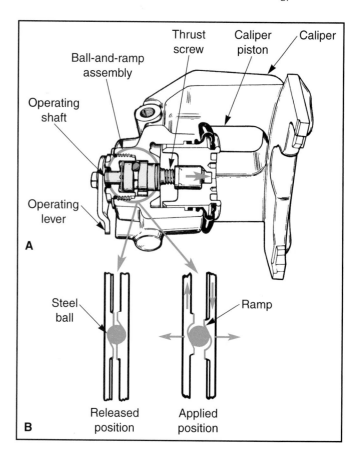

Figure 18-25. *A—A cross-sectional view showing an assembled ball-and-ramp unit. B—Ball-and-ramp assembly in the released and the applied positions. (Bendix)*

When the parking brake is applied, the cable rotates the lever and operating shaft. The three steel balls turn with the operating shaft, and ride up on the thrust screw ramps. This increases the space between them, forcing the thrust screw away from the operating shaft. Since the operating shaft cannot move in or out, the effect of this increase in space is to move the thrust screw outward. When the thrust screw is attached to the piston, outward movement of the thrust screw moves the piston outward, forcing the pads against the rotor.

When the parking brake is released, the balls roll back to their lowest point, allowing the seal to retract the piston. If pad wear has caused excessive play in the apply unit, the threaded shaft on the thrust screw turns slightly during release to maintain minimum clearance.

Cam Disc Parking Brake

The *cam disc parking brake* is used on some Asian-built vehicles. It consists of a lever operated *cam,* sometimes called an *eccentric,* which pushes on a pin or **actuator rod.** The actuator rod is attached to the caliper piston. **Figure 18-26** shows a typical cam type parking brake.

When the parking brake is applied, the lever rotates, turning the cam. As the cam rotates to its high point, it moves the actuator rod outward. As the rod moves outward, it pushes the piston outward, causing the pads to contact the rotor.

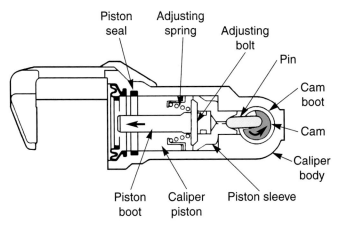

Figure 18-26. *A cutaway view of a cam disc parking brake. As the cam rotates (brakes being applied), the pin will push on the piston sleeve, adjusting bolt and caliper piston. This action forces the caliper piston outward, thus applying the brakes by pinching the rotor between the pads (not shown). (Honda)*

When the parking brake is released, the cam moves to its lowest position, and the piston seal moves the piston and actuator rod inward. To adjust for pad wear, the cam is tapered so as it turns, the actuator rod moves outward. If clearance is at the minimum, the cam turns the rod in the opposite direction when the brake is released. If there is excessive play in the apply unit, the cam loses contact with the apply rod and the threaded shaft does not turn back. This maintains minimum clearance.

Summary

The parking brake is used to keep the vehicle from moving while it is not being used. It can also be used as a backup brake system in an emergency. The brake linkage consists of a series of cables, rods, and levers. A hand or foot lever is used to move the linkage and apply the parking brake.

Foot brakes are held in the applied position by a ratchet mechanism. Some foot brakes used with disc brakes require several applications to apply the parking brake. Most foot brakes are released by hand, but a few are automatically released by a vacuum operated diaphragm and control system. Hand brakes can be the common floor lever, used on bucket seat vehicles, or the push-pull type with a T-handle mounted under the dashboard. A warning light illuminates when the ignition is on, telling the driver the parking brake is applied.

Brake cables are made of stranded steel with fittings on each end for attachment to other parking brake components. Cables are partially or completely installed inside of a cable sheath to protect them from road splash and damage. A parking brake multiplier is used on some parking brake linkages to increase brake pedal force. The equalizer is used to provide a balanced pull between the brake cables at each wheel. These parts are held and pro-

tected by support brackets and guides. Some parking brake linkages are protected from the exhaust system and road splash by heat and splash shields.

The integral drum parking brake is part of the service drum brake system. It applies the service brakes through an apply lever installed inside the drum. A lever strut helps to transmit braking force between the primary and secondary shoes. The brake linkage can be adjusted to compensate for wear.

One type of parking brake used with rear disc brakes is the self-contained drum brake, which is mechanically operated and has no connection with the brake hydraulic system. Most rear disc systems have an integral parking brake which mechanically applies the brake pads. The three main kinds of integral disc parking brakes are the screw type, ball-and-ramp type, and cam type. All rely on the movement of a lever that is converted into pressure on the disc brake pads.

Review Questions—Chapter 18

Please do not write in this text. Write your answers on a separate sheet of paper.

1. What is the primary purpose of the parking brake?

2. What is another use of the parking brake?

3. The design of the parking brake lever increases mechanical _____.

4. A multi-stroke foot parking brake is usually used when the rear brakes are _____ brakes.

5. Automatic foot brake releases are operated by engine _____.

6. The floor mounted hand lever is used on vehicles with _____.

7. What are parking brake cables usually made from?

8. A parking brake multiplier increases parking brake _____ at the expense of _____.

9. The equalizer ensures that the same amount of force is applied to each _____ brake.

10. Match the brake with its description.

    _____ Integral drum parking brake

    _____ Self-contained drum parking brake

    _____ Integral disc parking brake

    (A) Applies the service disc brakes.

    (B) Applies separate drum brakes.

    (C) Applies a driveshaft brake.

    (D) Applies the service drum brakes.

11. On a drum type parking brake, the parking brake cable is attached to the _____ brake shoe through a lever.

12. The most common type of disc brake parking brake is the _____ type.

13. Moving the cable causes the screw to turn, moving the _____ forward. This causes the brake pads to press against the rotor, setting the brake.

14. The cam type disc parking brake is found on _____ vehicles.

15. The cam, sometimes called an eccentric, pushes on an _____ attached to the hydraulic piston.

ASE Certification-Type Questions

1. Technician A says that the foot brake is held in the applied position by a ratchet. Technician B says that the vacuum release brake will release the brake when vacuum is removed. Who is right?
   (A) A only.
   (B) B only.
   (C) Both A & B.
   (D) Neither A nor B.

2. The multi-stroke parking brake is usually used with vehicles having _____.
   (A) rear disc brakes
   (B) rear drum brakes
   (C) front disc brakes
   (D) manual transmissions

3. Technician A says that most hand operated parking brake levers are dash mounted. Technician B says that the brake warning lights are illuminated only when the ignition is on with the parking brake applied. Who is right?
   (A) A only.
   (B) B only.
   (C) Both A & B.
   (D) Neither A nor B.

4. All of the following statements about parking brake cables are true, EXCEPT:
   (A) brake cables have fittings on their ends.
   (B) brake cables are made to be very rigid.
   (C) brake cables are often enclosed.
   (D) many vehicles have three brake cables.

5. The multiplier is a kind of _____.
   (A) adjuster
   (B) strain relief device
   (C) lever
   (D) pulley

6. Technician A says that the integral drum parking brake operates the service brakes. Technician B says that the integral drum parking brake operates a separate set of brake shoes. Who is right?
   (A) A only.
   (B) B only.
   (C) Both A & B.
   (D) Neither A nor B.

7. The parking brake _____ is attached to the secondary shoe.
   (A) apply lever
   (B) equalizer
   (C) cable
   (D) sheath

8. When a drum parking brake is applied, the brake _____ transfers parking brake force between the shoes.
   (A) multiplier
   (B) apply lever
   (C) front cable
   (D) lever strut

9. All of the following disc parking brake mechanisms are located inside the caliper, EXCEPT:
   (A) ball-and-ramp.
   (B) screw.
   (C) lever.
   (D) cam.

10. Technician A says that a screw type disc parking brake adjusts itself automatically each time it is applied. Technician B says that a ball-and-ramp disc parking brake adjusts itself automatically each time it is applied. Who is right?
    (A) A only.
    (B) B only.
    (C) Both A & B.
    (D) Neither A nor B.

Chapter 19

Parking Brake Service

After studying this chapter, you will be able to:
- ❏ Identify common parking brake problems.
- ❏ Diagnose parking brake problems.
- ❏ Adjust parking brakes.
- ❏ Service parking brake cables and related parts.
- ❏ Service automatic vacuum brake release mechanisms.
- ❏ Service warning lights.
- ❏ Adjust drum parking brake wheel assemblies.
- ❏ Adjust disc parking brake wheel assemblies.

Important Terms

Parking brake
   performance test

Click test

Vacuum pump

Vacuum gauge

Ohmmeter

The parking brake is a relatively simple system that seldom gives trouble. However, problems in the parking brake can be difficult to isolate from other brake complaints since they do not occur frequently. This chapter builds on the information in Chapter 18 to identify common parking brake problems and service procedures.

Parking Brake Problems

The most common parking brake problem is failure to hold the vehicle stationary. Poor holding power is most often caused by misadjusted or worn service brakes. However, stretched parking brake cables or bent levers can cause poor holding even when the service brakes are in good condition. If the parking brake suddenly becomes completely inoperative, and the service brakes work properly, a linkage part has broken or has become disconnected.

Another common problem is parking brake sticking. Many drivers simply use the automatic transmission park position or place a manual transmission in gear to hold the vehicle. The parking brake is not always or seldom used. Over time, the cables become dirty and corroded. If this happens, the cables may not apply the parking brake or will stick in the applied position. When this happens, the vehicle cannot move, and the driver may blame the service brakes or possibly the transmission.

Another cause of sticking parking brakes is a broken release mechanism. If the release lever is pulled and the parking brake does not release, the linkage has probably come loose. Some cable operated releases can stick or the cable can break.

Warning light problems can result in the light not coming on when the parking brake is applied, or staying on at all times. Warning light problems are caused by a misadjusted or defective switch, a blown fuse, or a burned out bulb.

Diagnosing Parking Brake Problems

To diagnose parking brake problems, you need to determine whether the problem is in the parking brake linkage or the service brakes. Once you have done this, then you can proceed to check individual components.

Parking Brake Performance Test

The simplest way to diagnose parking brake problems is to make a *parking brake performance test.* Start by driving the vehicle to a spot with enough room to allow it to move forward. Then perform the following tests depending on whether the vehicle has an automatic or manual transmission.

Click Test

The first test can be used on vehicles with either a manual or automatic transmission. This test is the so-called *click test.* Simply put, you either pull or push the parking brake pedal or lever, apply and count the number of clicks from the mechanism until it stops. If you count more than 8-10 clicks, the parking brake is in need of adjustment.

 Note: This test is not accurate on vehicles with ratcheting foot levers.

Automatic Transmission Vehicles

Stop the vehicle, place the transmission in Park, and leave the engine running. Apply the parking brake and note whether the warning light comes on. See **Figure 19-1.** Then place the transmission in Drive. If the vehicle has an automatic parking brake release, use your foot to hold the parking brake in the applied position, do not apply the service brakes. If the vehicle moves when the transmission is placed in Drive, the parking brake is not holding. Then release the parking brake. The warning light should go out and the vehicle should idle forward. If the vehicle does not move easily, the parking brake is sticking.

If the release mechanism is a manual type, ensure the ratchet and pawl assembly holds the lever in the applied position when you remove your hand. Check that the pawl disengages easily when you operate the release handle or button. If the vehicle has a vacuum release parking brake, ensure that it holds in Park with the engine running and releases when the shifter is moved to the Drive position.

 Note: In some vehicles with rear disc brakes, movement backward (with noticeable resistance) may occur if the accelerator is pressed. This is considered normal.

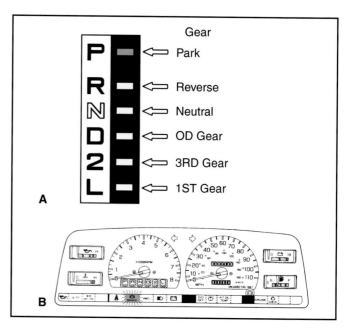

Figure 19-1. *A—The vehicle's automatic transmission is in park. B—The parking brake has been applied and the brake on warning light is now turned on. (Toyota)*

Manual Transmission Vehicles

Stop the vehicle and leave the engine running. Depress the clutch pedal and place the transmission in first gear. Apply the parking brake and observe the warning light, it should be on. Then slowly release the clutch pedal. The vehicle should not move and the engine should stall when the clutch is fully released. If the vehicle moves forward, the parking brake is not holding. Keep in mind that some forward movement with resistance is considered normal in most vehicles with rear disc brakes. Then release the parking brake and slowly apply the clutch. The warning light should be out and the vehicle should idle forward. If the vehicle does not move easily, the parking brake is sticking. Also check that the ratchet and pawl assembly holds and releases correctly.

Checking for Damage

If the performance test indicates the parking brake is not operating properly, check for damaged or misadjusted parts. Improper adjustment due to wear or incorrect adjusting procedures is common. Almost any part of the parking brake system can become damaged. Start checking for problems in the passenger compartment of the vehicle.

Driver-Operated Levers

Start by checking the driver-operated levers for damage. This can be done inside the vehicle. However, it is often necessary to remove some interior trim components to gain access to the lever mechanism. Be careful not to get any dirt or grease on the seats or trim parts. Also keep in mind that trim pieces that are more than a few years old may be brittle. Handle them carefully to avoid damage.

Hand-Operated Levers

Hand-operated parking levers should be inspected for cracks in the lever and at the mounting fasteners. See **Figure 19-2.** Also check for loose fasteners, broken parts, damaged ratchet teeth, lack of ratchet and cable lubrication, lack of cable release, and damaged or misadjusted warning switches.

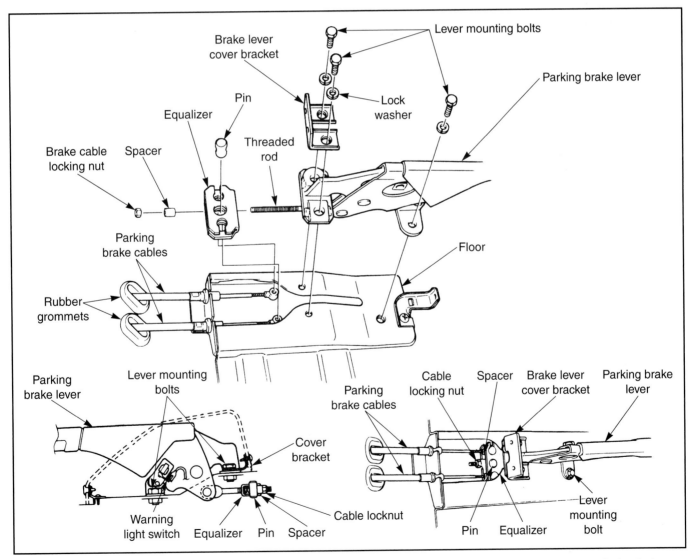

Figure 19-2. *One particular type of hand-operated parking brake lever assembly. Check for metal cracks, loose or missing fasteners, damaged cables. (Chevrolet)*

Foot-Operated Levers

The foot-operated lever unit should also be checked for loose fasteners, cracked or broken parts, damaged ratchet teeth, proper release mechanism function, warning switch damage or misadjustment, and release operation. If the parking brake has a vacuum release device, check for loose or missing hoses and vacuum leaks at the diaphragm. A *vacuum pump* and/or *vacuum gauge,* can be used to check the vacuum components. Vacuum pumps were discussed in Chapter 3.

Warning Lights

Warning lights develop one of two problems: they do not come on or they do not turn off. The following sections address each problem.

Warning Light Does Not Illuminate

Begin by checking the fuse that controls the warning light circuit, **Figure 19-3.** If the warning lightbulb is easily accessible, make sure the bulb is not burned out. If the bulb cannot be easily reached, skip this step until you are sure that the switch and fuse are good.

There are two ways to check warning light switches: checking the switch itself, or bypassing it. To check the switch, obtain an **ohmmeter.** Disconnect the switch connector and attach the ohmmeter leads between the parking brake switch terminal and a mounting bolt. If there is continuity (very low resistance) when the lever is pulled (hand brakes) or depressed (foot brakes), and there is no continuity when the lever is returned to the released position, the switch is functioning properly. A switch being tested with an ohmmeter may be viewed in **Figure 19-4.** A self-powered test light can also be used to make this test.

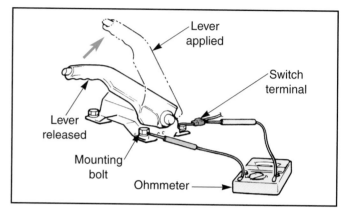

Figure 19-4. *Checking one type of parking brake warning light switch with an ohmmeter. Test with the brake lever in both the applied and released positions. This will turn the switch on and off. (Chrysler)*

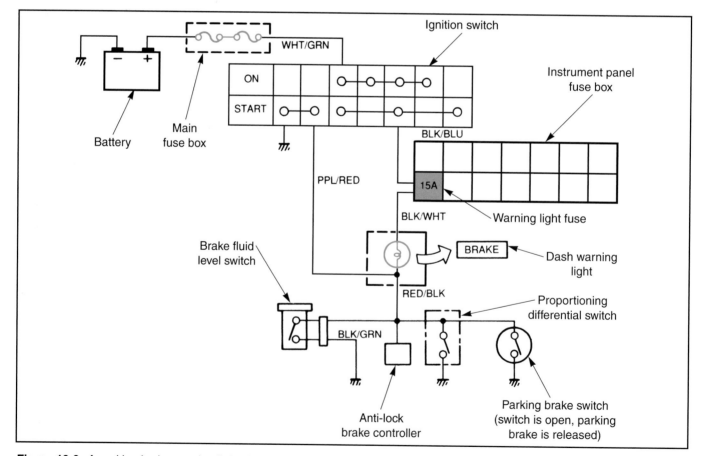

Figure 19-3. *A parking brake warning light electrical circuit schematic. The circuit is open, the parking brake switch contacts have not made contact. Note the brake fluid level switch in the master cylinder. This switch may also turn on the warning light when the sensor detects a low fluid level. (Chevrolet)*

To bypass the switch, disconnect the switch lead and connect a jumper wire to the lead. Ground the other end of the jumper wire and turn the ignition switch to the *on* position. If the light comes on, the switch is defective or misadjusted. If the light does not come on, the problem is in the bulb, fuse, or wiring.

If the proper readings are not obtained, adjust or replace the switch as necessary. If a switch is defective, it must be replaced.

Warning Light Always On

If the brake warning light is always on, make sure the problem is in the parking brake switch. Many vehicles use the same light to warn of lining wear or problems with the hydraulic system via the pressure differential switch. If these check ok, disconnect the switch at the parking brake. If the light goes out, the switch is stuck or misadjusted. Adjust or replace it as necessary. If the light stays on, there is a short in the wiring leading to the switch.

Cables and Linkage

Operating cables and linkage should be inspected for damage and corrosion. Check closely for frayed cables. Any loose cable strands are cause for replacement. Look for damaged sheaths and missing guides. Check for damage at the rubber boots where the cables enter the vehicle's interior. Also check the cable ends, equalizer, and multiplier for bends and wear. If any cable pulleys are used, make sure they turn freely and do not bind.

Adjusting Parking Brakes

If the foot or hand lever can be moved out to the limit of its travel without effectively applying the parking brake, the system requires adjustment. To adjust parking brakes, begin by checking the condition of the service brakes. Often, a loose parking brake adjustment is an indication the service brakes are worn. If the service brakes are in good condition, adjust them as necessary. Then adjust the parking brake linkage.

 Note: Self-contained drum brakes used on some rear disc brake vehicles are adjusted in the same way as conventional service drum brakes.

Adjusting Wheel Brake Components

This section is a review of wheel brake adjustment. Parking brake component service on the wheel brake assemblies is covered in Chapters 13 and 15. When working on any wheel brake components be sure to use an approved vacuum or washer to clean parts.

Drum Brakes

Service of integral rear drum brake system components was covered in Chapter 15. However, drum brakes do require periodic adjustment as part of normal maintenance and whenever the front pads are replaced.

To adjust integral drum brakes, begin by placing the vehicle on a lift. If the vehicle is a rear-wheel drive, shift the transmission to Neutral. Make sure the parking brake lever is not applied. Raise the vehicle and remove the rear wheels. Pull the drum, clean any dust, and check the linings for wear. There should be a minimum of .059" (1.5 mm) of lining on the shoes. Replace the linings if they are worn.

Reinstall the drums and install 2-3 lug nuts to hold the drum in place. Most drum brake assemblies have an access hole, either in the backing place or on the drum face. The access hole is in line with the drum brake star wheel. Most are sealed by a punched metal cover. Remove this cover and discard. Insert a brake adjustment tool and/or standard screwdriver, **Figure 19-5.** Using a lever action, rotate the star wheel. Direction for adjustment varies by vehicle, so check the service manual. Adjust the wheels until you feel a light drag. Check parking brake application and adjust the cables as needed. Make sure you install a rubber plug in the access hole when you are completed.

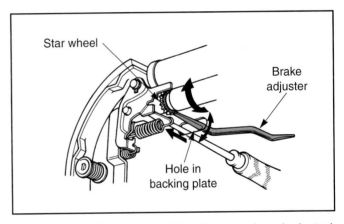

Figure 19-5. *Adjusting integral drum brakes using a brake tool. In some cases, you may need to use a screwdriver to hold the adjuster lever away from the star wheel. (Toyota)*

Self-contained drum brakes used on rear-wheel disc brake systems are similar to those used in integral drum systems. The drum is part of the rotor assembly. The parking brake setup shown in **Figure 19-6,** does not require periodic service.

If the caliper is removed for pad replacement or other service, it is a good idea to remove the rotor and check the linings for wear. Shoe replacement procedures are similar to those for integral drum brakes. The self-contained drum brake adjuster can usually be reached though the front of the drum and rotor assembly, **Figure 19-7.** Obtain the correct manufacturers' service manual before attempting to adjust the brakes.

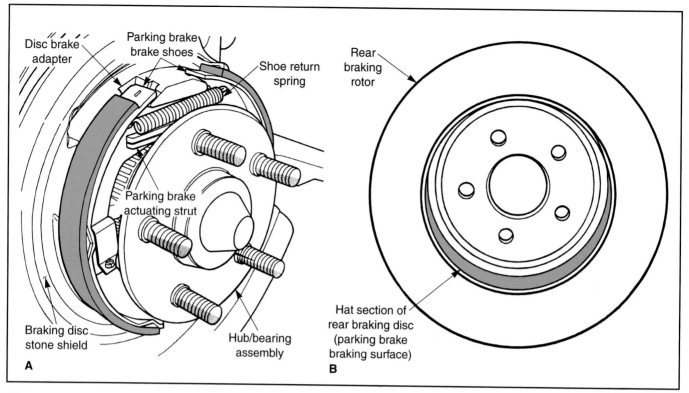

Figure 19-6. *An integral brake drum and rotor assembly. A—Rotor and drum has been removed to show the parking brake shoes. B—Disc brake rotor illustrating the parking brake section. (Chrysler)*

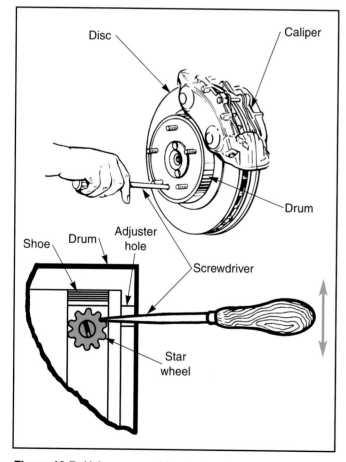

Figure 19-7. *Using a screwdriver to turn the star wheel, setting the correct shoe-to-drum operating clearance. Be sure to replace the adjuster hole plug. (FMC & ATE)*

Disc Brakes

Three common parking brakes are used with rear-wheel disc brakes. They are the screw, ball-and-ramp, and cam. Service procedures for each type are discussed in Chapter 13. You may need to adjust the rear caliper pistons to the rotors after brake service and occasionally due to lack of parking brake use. In some cases, the brake pedal only needs to be pumped several times to bring the pads in adjustment. However, a special adjustment procedure is sometimes needed to bring the pads into position and to obtain a good pedal.

Using a flat-head screwdriver, carefully position the tip against the top lip of the caliper piston, **Figure 19-8.** Clamp a pair of vice grips on the caliper's parking brake actuating mechanism.

 Note: In some cases, this is easier to perform using a box-end wrench.

While carefully prying against the piston, use the vice grips to apply and release the parking brake. The caliper piston should adjust the pads until they just rest against the rotor. If the caliper piston fails to move, try tapping the caliper piston area with a ball peen hammer. Be careful not to damage the bleed screw. If the piston continues to stick, remove the caliper and perform an overhaul.

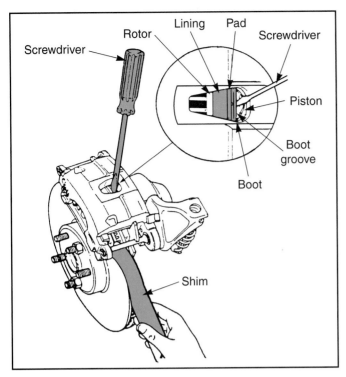

Figure 19-8. *Adjustment for a rear disc brake caliper. Not all calipers need this type of adjustment. (General Motors)*

Linkage Adjustment

Normal usage will stretch the parking brake cable and cause some wear in the related guides or levers. This creates looseness and prevents full application of the parking brake. Linkage adjustment consists of reducing the length of the front cable to compensate for looseness. Most linkage adjustment devices are located at the equalizer, **Figure 19-9.**

To adjust the tension at the equalizer, slightly engage the hand or foot lever about two or three notches of the ratchet mechanism. This will take up a small amount of slack and prevent overtightening. Then loosen the locknut holding the equalizer adjuster. If necessary, spray the locknut and adjuster shaft with penetrating oil before trying to turn the nuts.

Turn the adjusting nut until you can feel a drag when turning the rear wheels. Then back off the nut until all brake drag is gone. Tighten the locknut and firmly apply the parking brake. Then recheck to ensure the brakes are not dragging when the parking brake is fully released. Perform the parking brake performance test as discussed earlier in this chapter to ensure the parking brakes are properly adjusted.

Freeing Brake Cables

Ideally, a sticking or seized brake cable should be replaced. However, if a replacement cable is unavailable, you may be able to free the stuck cable. Begin by removing the cable ends. Then, using penetrating oil, spray the cable at the point where it enters the sheath. Allow the penetrating oil to soak in, then repeat. Work the cable back and forth, applying penetrating oil as needed, until the cable loosens. Clean as much corrosion and contaminants from the inner cable as

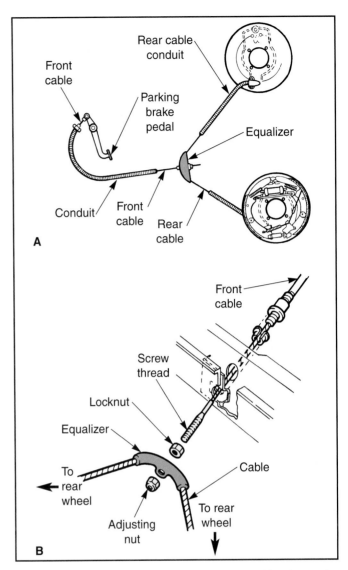

Figure 19-9. *Parking cable adjustment. A—This shows the overall layout. B—An exploded view of the equalizer and related parts. (TRW Inc.)*

possible. Once the cable is free, apply liberal amounts of lithium grease to the inner and outer cable sections, then reinstall the cable and adjust the brakes.

Replacing Parking Brake Linkage and Cables

Some parts of the parking brake assembly must be replaced if they become damaged or worn. Removal and installation procedures for these parts are given in the following sections.

 Caution: Before removing any part of the parking brake assembly, be sure that the vehicle is in Park, or otherwise secured to keep it from rolling.

Hand-Operated Levers

Hand-operated levers are usually replaced as an assembly. Begin by carefully removing any trim panels that block access to the lever fasteners. You may need to remove one or more of the front seats and partially pull the

carpet back to reach the lever fasteners. Remove the fasteners holding the unit to the mounting bracket or body panel. Unhook the cable end from the lever slot and remove the warning switch electrical connector if necessary, then remove the lever from the vehicle.

 Note: You may need to disconnect the lever apply cable from the brake cable(s) before removal from the lever.

Check the old and new levers to ensure the replacement is correct. Place the new lever assembly in position and attach the cable end and switch electrical connector. Loosely install all fasteners, then tighten. Recheck parking brake and warning light operation, and adjust as needed, **Figure 19-10.** This hand brake assembly is located between the seats. A dashboard mounted hand brake is shown in **Figure 19-11.**

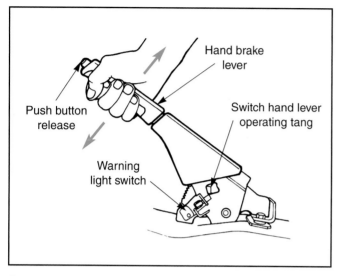

Figure 19-10. *Checking the parking brake switch operation by applying and releasing the lever. (Chevrolet)*

Foot-Operated Levers

Foot-operated levers are also replaced as an assembly. Start by removing any trim panels and repositioning the carpet and floor mat as necessary to access the foot lever assembly. Remove the warning switch electrical

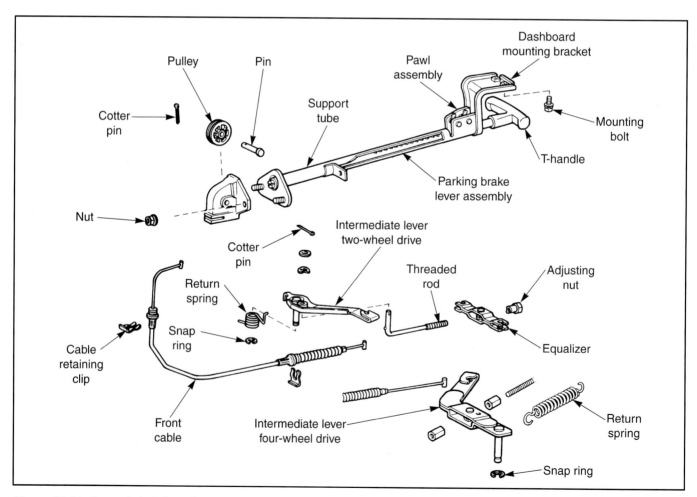

Figure 19-11. *An exploded view of one particular dashboard mounted hand brake assembly. (Toyota)*

connector if the switch is mounted on the lever, and remove the vacuum release hose if used. Remove the fasteners holding the lever to the kick panel bracket and remove the cable end from the lever mounting slot. Then remove the lever from the vehicle.

Check the old and new levers to ensure the replacement is correct. Place the new lever assembly in position and attach the cable end and switch electrical connector. Install and tighten all fasteners. Recheck parking brake and warning light operation, and adjust as necessary. Reinstall any trim panels that were removed as well as the carpeting and/or floor mat. **Figure 19-12** illustrates one type of foot operated manually released parking brake mechanism.

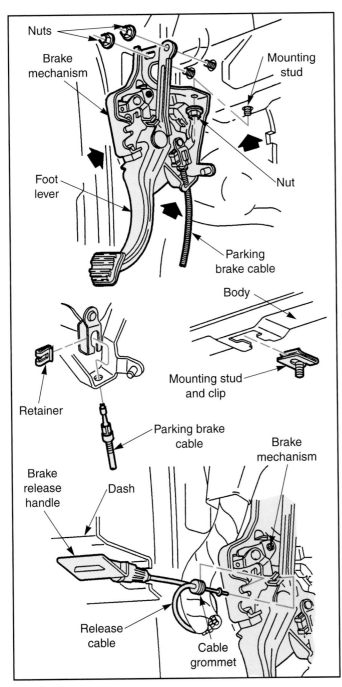

Figure 19-12. *A foot-operated, manually released parking brake. (Pontiac)*

Vacuum Release Vacuum Unit and Switch

The vacuum release vacuum unit is located at the parking brake assembly. Remove the vacuum hose and disconnect the linkage from the parking brake. Remove the fasteners and remove the vacuum unit from the vehicle. Install the new unit and connect the linkage and vacuum hose. After installation is complete, recheck parking brake release operation.

The vacuum switch is located on the steering column or floor shift linkage, often near the neutral safety, **Figure 19-13.** Remove the vacuum lines and fasteners and remove the switch. Place the new switch in position, being sure the operating tang is lined up with the proper slot on the shift linkage. Then install and tighten the fasteners. Reconnect the vacuum lines and recheck parking brake release operation. Be sure the release operates only when the transmission is placed in Drive.

Warning Lights and Switches

Warning lightbulbs are located in the dashboard. Replacement may be easy or very difficult, depending on the interior dash and instrument cluster design. Check the manufacturers' service manual for the location and service procedures for dashboard mounted bulbs.

Warning switches are usually attached to and operated by the lever. See **Figure 19-14.** To remove the switch, remove any trim pieces and remove the fasteners holding the switch to the body. Install the new switch and check warning light operation. Most warning light switches are adjustable. If the light does not come on when the lever is applied and go off when the lever is released, adjust as necessary. Refer to the warning light checking procedures earlier in this chapter.

Cables and Levers

Cables and levers usually last the life of the vehicle, unless they are damaged by road debris or become corroded and stick. Replacement is usually straightforward, but new cables must be routed and secured exactly as specified. Improper cable installation will lead to sticking and damage. The following general procedure can be used to replace cables and linkage on most vehicles. Obtain the correct service manual before starting any replacement procedures.

Front Cable Replacement

Raise and properly support the vehicle, then loosen the equalizer enough to provide cable slack. Loosen the locknut and unthread the adjuster nut from the threaded shaft on the end of the cable. After the nut is removed, slide the threaded shaft out of the equalizer. Remove any cable guides under the vehicle, and remove any boots that seal the cable opening into the passenger compartment. Remove the cable end from the hand or foot apply lever, and slide the cable down through the boot, removing it from under the vehicle.

Compare the old and new cables. Ensure the new cable has the same fittings and is the same length as the

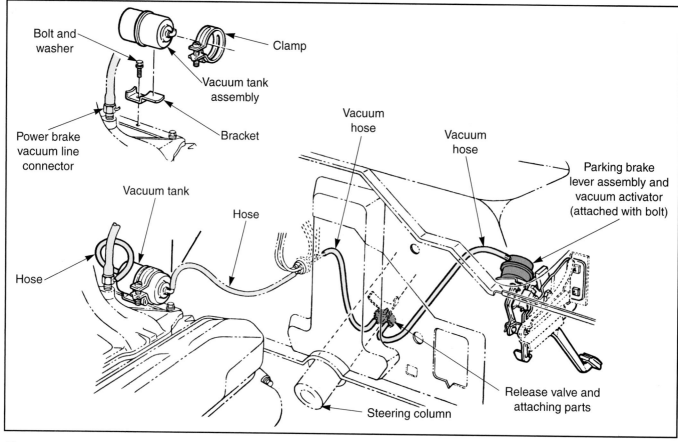

Figure 19-13. *Layout of a vacuum powered parking brake release system. (Chrysler)*

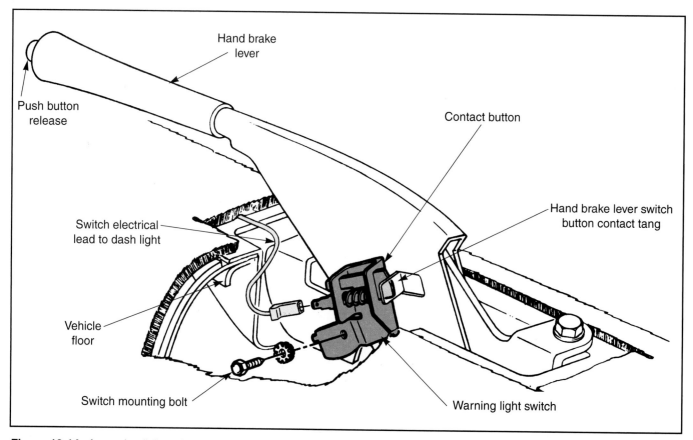

Figure 19-14. *A warning light switch and its mounting location on a floor-mounted hand brake. (Chevrolet)*

old cable. Install the new cable, being careful not to damage the cable sheath or the boot at the entrance to the passenger compartment. Attach the cable end at the apply lever, and install the cable into any guides used. Then slide the cable end threaded shaft through the equalizer and install the adjusting nut. Adjust the cable as needed and install and tighten the locknut. **Figure 19-15** shows a typical front cable layout.

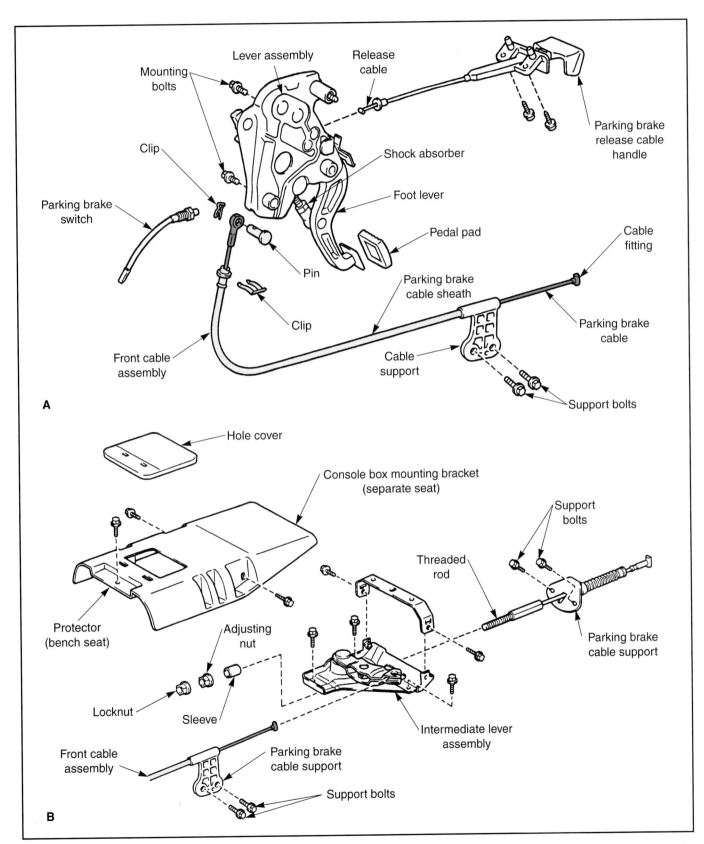

Figure 19-15. *A typical foot operated, manually released parking brake. A—This shows the lever assembly and the front cable. B—Front cable to intermediate lever assembly. (Toyota)*

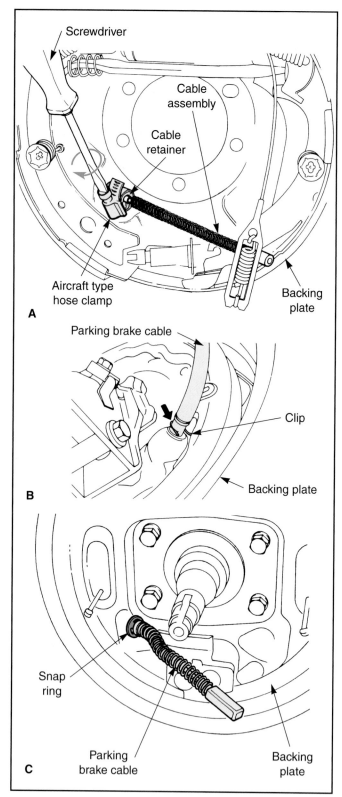

A—Removing the parking brake cable retainer

Intermediate Cable Replacement

If the vehicle has an intermediate cable, removal procedures are similar to those for the front brake cable. However, the intermediate cable can be removed from under the vehicle. Start by raising and supporting the vehicle. After removing cable tension, remove the cable ends at the multiplier and equalizer. Check that the new cable is correct, then install it and readjust the parking brake.

Rear Cable Replacement

To remove a rear cable, raise and support the vehicle. Remove the tire and wheel. Loosen the equalizer adjuster enough to provide some cable slack. Remove the cable at the equalizer and remove any guides between the equalizer and wheel. Then follow the procedures necessary for a drum or disc brake.

Drum Brake

Remove the brake drum and remove any brake dust with an approved vacuum or washer. Remove the primary shoes and the brake strut. Remove the cable lever from the secondary shoe. If a finger type retainer is used to hold the cable sheath to the backing plate, compress the fingers with a hose screw clamp, **Figure 19-16.**

> **Note: A proper size box-end wrench can be used to compress the finger retainer on some parking brakes.**

If a clip holds the sheath, remove the clip, **Figure 19-16.** Remove the sheath from the backing plate and pull out or unbolt the cable retainer. Then disconnect the end of the cable from the parking brake lever. See **Figure 19-17.**

Install the cable assembly into the backing plate and reinstall the fastener. Be sure a clip retainer is fully installed, and that a finger type is fully inserted until the fingers expand outward and grip the backing plate. Install the parking brake lever on the end of the cable. Then reassemble the brake shoe assembly and strut. Reinstall the drum, and adjust the brakes as necessary. Reinstall the wheel and tire.

Figure 19-16. *A—Removing the parking brake cable retainer (finger or prong type) with a screw-type hose clamp. The clamp is placed around the retainer fingers, and tightened just enough to allow the cable and retainer to be removed from the backing (support) plate. Some may also be removed by placing a box end wrench of the correct size over the fingers. B—Removing the parking brake cable sheath from the backing plate. Brake cable being held with a clip. C—Cable being retained by a snap ring. Clamps and other methods are also used. (Chrysler & Hyundai)*

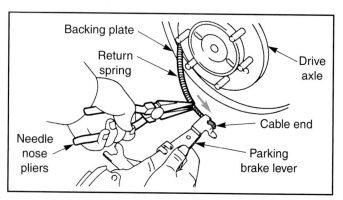

Figure 19-17. *Disconnecting the parking brake cable from the parking brake lever. Release the return spring tension with pliers. Then push the cable fitting end out of its retaining slot in the parking brake lever. (Toyota)*

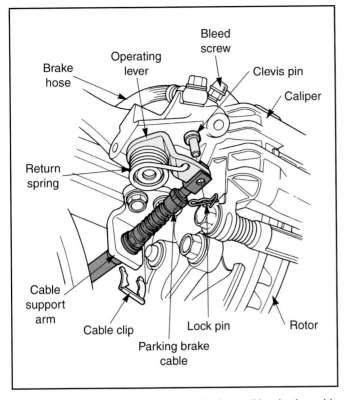

Figure 19-18. *Disconnecting a disc brake parking brake cable from the operating lever and cable support. (Honda)*

Install the cable into the guides. Install the cable end into the equalizer and readjust the cable as necessary. Be sure to recheck service brake adjustment after applying the parking brake.

Disc Brake

Disconnect the cable at the lever and the sheath at the lever bracket, **Figure 19-18**. Remove any return springs, being careful not to interchange the springs between right and left. Install the new cable and guides. Replace the return springs. Install the cable end into the equalizer and adjust the cable as necessary. In some cases, a plastic tie strap or clip may be used to help retain the cable to the lever bracket. Make sure the clip or a replacement strap is in place before installing the wheel on the vehicle.

Levers

Common levers used in parking brake linkage are the equalizer and multiplier. Replacement procedures are similar.

Equalizer

The equalizer is held in place by its cable connections and is usually not mounted to the vehicle body. See **Figure 19-19**. To remove the equalizer, raise and support the vehicle. Then loosen the cable adjuster and remove

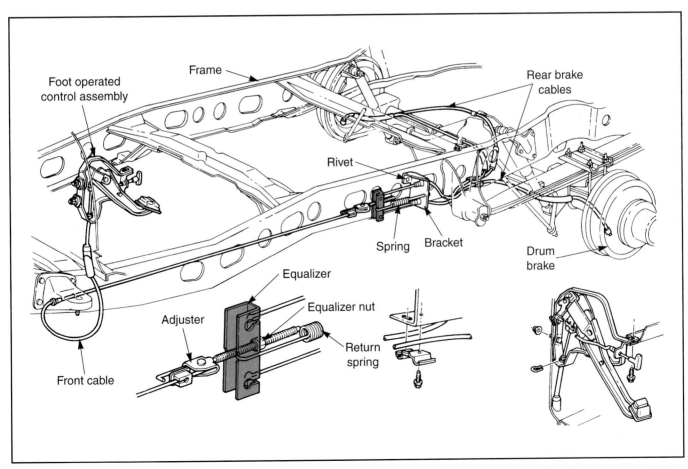

Figure 19-19. *An equalizer unit held in place where the cable connects. This particular foot operated assembly is used on a 3/4 ton truck. (Ford)*

the front and rear cables from the equalizer. Install the cables onto the new equalizer and adjust for proper parking brake operation.

Multiplier

The multiplier depends on a body connection for its leverage action, **Figure 19-20.** To replace a multiplier, raise and support the vehicle. Loosen the cable adjuster and remove the front and intermediate cables from the multiplier. Remove any return springs and remove the multiplier. Install the cables onto the new multiplier and adjust the equalizer for proper parking brake operation.

Summary

The most common parking brake problem is failure to hold the vehicle stationary. Parking brakes also stick in the applied position. Warning lights can fail to come on or fail to turn off. The easiest way to check parking brake operation is the performance test.

A visual check often turns up bent, corroded, or stretched parts. Checking the vacuum release and warning light system requires the use of vacuum pump and gauge. The warning light system can be checked with an ohmmeter, self-powered test light, or jumper wires.

In many cases, parking brake problems can be solved by adjustment. Always adjust the service brakes before adjusting the cables. Always raise and secure the vehicle properly before going under it. Cable adjustment is usually made at the equalizer.

When replacing cables, be sure to check the old and new parts. Cable installation varies depending on the type of cable replaced. Route new cables exactly as the old cable was routed. Driver operated levers, warning light switches, and vacuum release parts can be changed inside of the vehicle passenger compartment. Equalizers and multipliers can be changed from under the vehicle.

Drum brake component service is similar to that for conventional drum brakes. Disc brake part replacement requires caliper disassembly. Inspect all parts for damage, and replace as necessary. Adjust clearances and free-travel to the manufacturers specifications. Always check the parking brake function and holding ability after completing repairs.

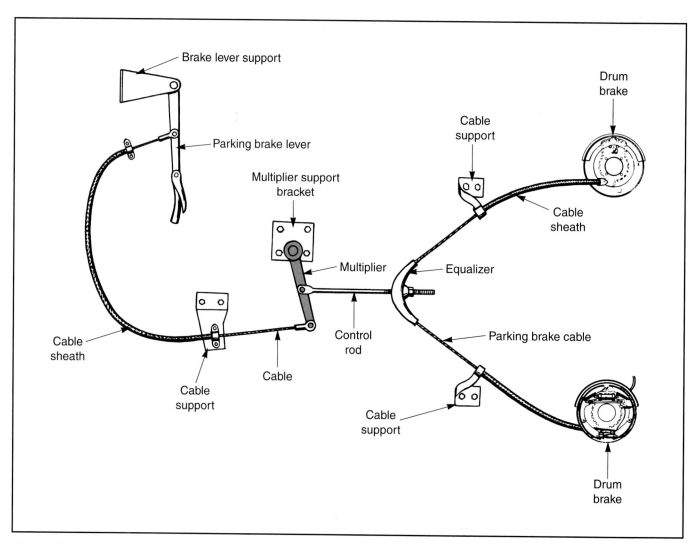

Figure 19-20. *A body or frame mounted multiplier assembly. (Bendix)*

Review Questions—Chapter 19

Please do not write in this text. Write your answers on a separate sheet of paper.

1. What is the most common parking brake complaint?

2. If the parking brakes are not holding, what should you check *first*?

3. If the vehicle moves forward from rest with the parking brake applied, is the parking brake working properly?

4. When removing trim pieces to gain access to the parking brake lever mechanism, treat them gently because they may be _____.

5. What are the two main types of parking brake warning light problems?

6. The parking brake _____ should be checked for loose strands.

7. After adjusting the parking brake, make sure that the brakes do not _____ with the lever fully released.

8. Before a cable can be replaced, all _____ must be removed from the cable.

9. *True or False?* To remove play from the rear caliper pistons after brake service, the pedal may have to be pumped.

10. *True or False?* There are no disc parking brakes that have provisions for manual adjustment.

ASE Certification-Type Questions

1. Technician A says that worn out service brakes are a common cause of parking brake problems. Technician B says that parking brakes stick when they are not used frequently. Who is right?
   (A) A only.
   (B) B only.
   (C) Both A & B.
   (D) Neither A nor B.

2. A hand parking brake is fully applied on a car with an automatic transmission. When the shifter is moved to the drive position, the car moves forward. This means that the parking brake is _____.
   (A) sticking
   (B) binding
   (C) not holding
   (D) not releasing

3. All of the following statements about warning lights are true, EXCEPT:
   (A) if the light does not come on, the bulb may be burned out.
   (B) if the light does not go off, the fuse may be blown.
   (C) if the switch is out of adjustment, the light may not go off.
   (D) if the switch is out of adjustment, the light may not come on.

4. Technician A says that disc type parking brake units cannot be adjusted. Technician B says that self-contained drum brakes cannot be adjusted. Who is right?
   (A) A only.
   (B) B only.
   (C) Both A & B.
   (D) Neither A nor B.

5. When adjusting drum brakes, there should be ____ of lining on the shoes.
   (A) .059″ (1.5 mm)
   (B) 1″ (25.4 mm)
   (C) .001″ (.0015 mm)
   (D) .030″ (.076 mm)

6. Technician A says that you may need to adjust the rear disc brake calipers after brake pad service. Technician B says that this is never needed. Who is right?
   (A) A only.
   (B) B only.
   (C) Both A & B.
   (D) Neither A nor B.

7. If a rear caliper piston fails to move during adjustment, what can be done?
   (A) Rebuild the caliper.
   (B) Tap on the caliper with a ball-peen hammer.
   (C) Bleed the brakes.
   (D) Both A & B.

8. Most linkage adjustment devices are located at the _____.

(A) hand or foot lever
(B) multiplier
(C) equalizer
(D) wheel brakes

9. When removing a parking brake cable stopper ring (retainer) from the backing plate, use a _____ to compress the fingers.

(A) needle nose pliers
(B) vise grip pliers
(C) C-clamp
(D) hose screw clamp

10. The _____ is usually not directly connected to the vehicle body.

(A) equalizer
(B) multiplier
(C) foot lever
(D) hand lever

Chapter 20

Brake System Electrical and Electronic Components

After studying this chapter, you will be able to:

❑ Explain theories of electricity.
❑ Identify basic electrical measurements.
❑ Identify basic vehicle electrical circuits.
❑ Identify and explain the purpose of vehicle wiring and connectors.
❑ Explain how to diagnose brake and warning light problems.
❑ Explain the construction and operation of the automotive computer.
❑ Identify the major parts of vehicle computers.
❑ Explain the operation of control loops.

Important Terms

Conductors	Parallel circuit	Fusible link	Electromagnetic
Insulators	Series-parallel	Circuit breakers	interference (EMI)
Flow	circuit	Filament	Semiconductors
Circuit	Short circuit	Magnetic	Diodes
Current	Grounded circuit	Polarity	Rectified
Amperes	Open circuit	Electromagnetism	Integrated circuit (IC)
Voltage	Harnesses	Electromagnet	Control loop
Volts	Color coded	Induction	Input sensors
Resistance	Ground	Relay	Output devices
Ohms	wire	Solenoid	Central processing unit (CPU)
Direct current (dc)	Wire gage	Motors	Memory
Alternating current (ac)	Plug-in connector	Armature	Read only memory (ROM)
Waveform	Schematic	Field coils	Random access memory (RAM)
Ohm's law	Circuit protection devices	Static electricity	Trouble code
Complete circuit	Fuses	Radio frequency	Scan tool
Series circuit	Fuse block	interference (RFI)	Self-diagnosis

You cannot diagnose problems on modern vehicles without a thorough knowledge of electronic theory. In addition to discussing the electrical and electronic components related to the brake system, this chapter also contains a brief review of the most important principles of electricity, electronics, and computer operation. This chapter will also prepare you for Chapters 21 and 22, which discuss anti-lock brakes and traction controls.

Electrical Basics

The following sections outline the fundamental operating principles of all electrical equipment. All modern vehicles use 12-volt electrical systems. This information applies to any make of vehicle.

Conductors and Insulators

Everything is made of atoms. Every atom has a center of protons and neutrons. The neutrons have no charge and the protons have a positive charge, making the center of the atom positively charged. Revolving around this center are negatively charged electrons, **Figure 20-1**.

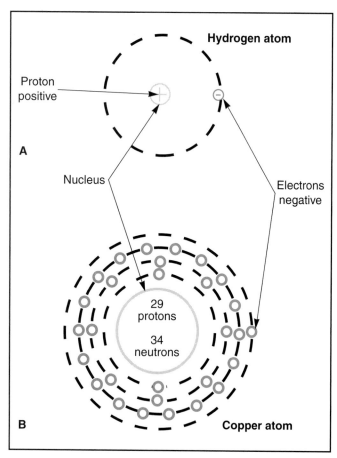

Figure 20-1. *Two different types of atoms. A—Hydrogen is one of the simplest atoms containing one proton and one electron. B—A copper atom. Copper is widely used in electrical systems. It is made up of 29 electrons, revolving around a nucleus of 29 protons and 34 neutrons. (Prestolite)*

Some atoms easily give up or receive electrons. These atoms make up elements that are good electrical **conductors.** Examples of good conductors are copper and aluminum. Materials whose atoms resist giving up or accepting electrons are **insulators.** Glass and plastic are examples of good insulators. Some materials can alternate between conducting and insulating. These materials are discussed in the computer section later in this chapter.

Electrical Measurements

Electricity is the movement, or **flow,** of electrons. The path through which the electrons move is called a **circuit.** The flow of electricity through a circuit depends on three electrical properties, all of which can be measured.

Current

Current is the number of electrons flowing past any point in the circuit. Current is measured in **amperes,** usually shortened to *amps.* The higher the amp reading, the more electrons are moving in the circuit.

Voltage

Voltage is electrical pressure, created by the difference in the number of electrons between two terminals. It provides the push that makes electrons flow and is measured in **volts.**

Resistance

Resistance is the opposition of atoms in a conductor to the flow of electrons. All conductors, even copper and aluminum, have some resistance to giving up their electrons. Resistance is measured in **ohms.**

Direct and Alternating Current

The vehicle battery and alternator always have two terminals: positive and negative. One of these terminals has more electrons than the other. There is always a shortage of electrons at the positive terminal, and an excess of electrons at the negative terminal. This type of current flows in only one direction, from negative to positive and is called a **direct current (dc)** system.

In the electrical systems used in homes, schools, and offices, the flow of electrons changes direction many times every second. These are **alternating current (ac)** systems. Since the vehicle battery cannot be charged by alternating current, automotive electrical systems are always direct current systems.

However, alternating current is used in the vehicle as a means of signal transmission. For instance, the wheel speed sensors used with an ABS system produce an ac current, which is used to calculate wheel speed.

Waveforms

Every voltage has a particular shape or **waveform.** The waveform is always read across as the time and up as the amount of voltage. The waveform shape is a clue to the operation of the particular circuit. If the waveform does not match the standard pattern, a problem is most likely present. The ac signal produced by a wheel speed sensor is shown in **Figure 20-2.**

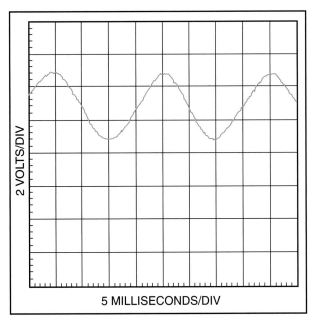

Figure 20-2. *The vehicle speed waveform. The frequency of this alternating current (ac) voltage is in proportion to the revolving speed of the rotor. (General Motors)*

Ohm's Law

Sometimes one of the electrical properties of a circuit is not known. However, if you know two of the properties, you can calculate the third using **Ohm's law. Figure 20-3** is a graphic representation of Ohm's law. It is sometimes called the Ohm's law triangle. Using Ohm's law requires no more than simple multiplication or division.

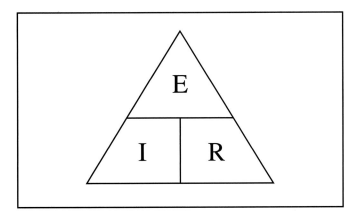

Figure 20-3. *Ohm's Law is used to calculate the current (I), voltage (E), and resistance (R), in an electrical circuit.*

For example, you want to find the total amperage draw of a set of aftermarket (add-on) brake lights. If the lights have a resistance of 3 ohms and the vehicle has a 12-volt electrical system, you can use Ohm's law to calculate the amperage draw:

$$\frac{12 \text{ volts}}{3 \text{ ohms}} = 4 \text{ amps}$$

If you know the amperage and voltage, and want to calculate the resistance in the above illustration, use the following calculation:

$$\frac{12 \text{ volts}}{4 \text{ amps}} = 3 \text{ ohms}$$

If you know the amperage and resistance, calculate the voltage by the following formula:

$$4 \text{ amps} \times 3 \text{ ohms} = 12 \text{ volts}$$

These formulas make up Ohm's law. When applying Ohm's law to find an unknown electrical property, remember the following:

❑ When amperage or resistance is unknown, divide voltage by the other known value to obtain the unknown value.

❑ When voltage is unknown, multiply amperage and resistance to get voltage.

Types of Electrical Circuits

A typical automotive electrical circuit is the brake light circuit as shown in **Figure 20-4.** In the brake light circuit, the electrical path is from the battery through the negative battery cable, frame, brake lightbulbs, brake light switch, fuse, and positive cable, returning to the battery. This is called a **complete circuit,** since the electricity makes a loop through the battery, cables, fuse, switch, bulbs, frame, and back to the battery.

The path for the electrons to return to the battery is as important as the path from the battery to the electrical device. There are three types of automotive circuits. Every wire in a car or truck is part of one of these circuits. The three types of circuits are series, parallel, and series-parallel.

The circuit in **Figure 20-5** is the simplest type of automotive circuit, the **series circuit.** This series circuit consists of the vehicle battery, switch, lightbulb, and connecting wiring. Electrons flow through the wiring from the battery, through the switch and bulb, and back to the battery. The same amount of current flows through every part of the series circuit. A common example of the series circuit is the parking brake warning light.

The circuit in **Figure 20-6** is a **parallel circuit.** In a parallel circuit, current flow is split so each electrical component has its own path. Varying amounts of current will flow in each path, depending on the resistance of each part of the circuit.

The **series-parallel circuit** has some components that are wired in series, and some that are wired in parallel, as shown in **Figure 20-7.** All current flows through

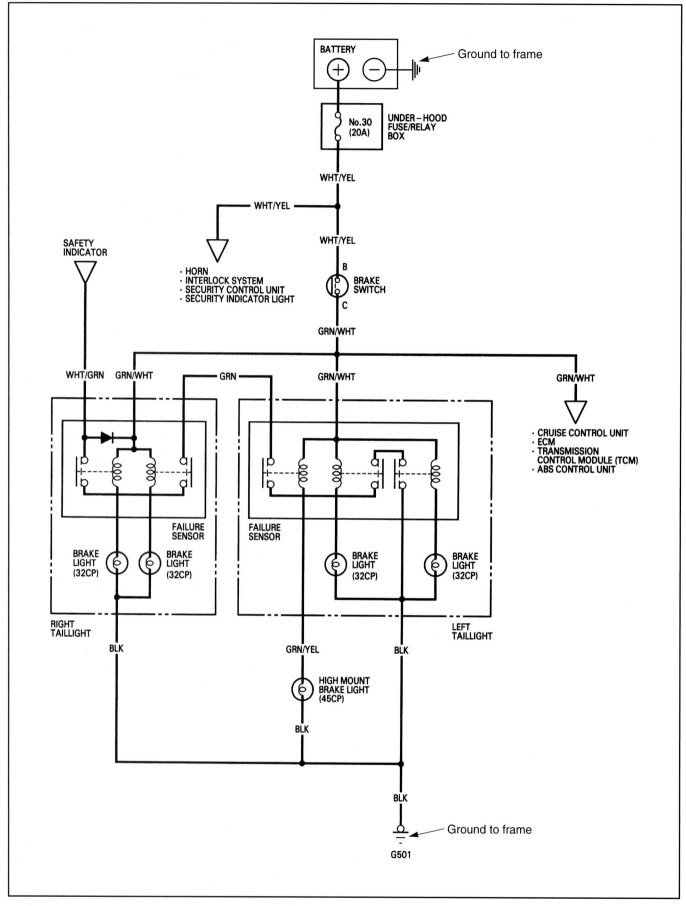

Figure 20-4. *A typical automotive brake light circuit wiring schematic. The brake switch is off; no electrical current is traveling to the brake lights. (Honda)*

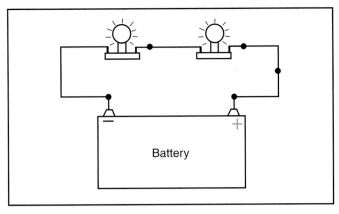

Figure 20-5. *A series circuit. If an ammeter was inserted in the circuit, all the readings would be the same.*

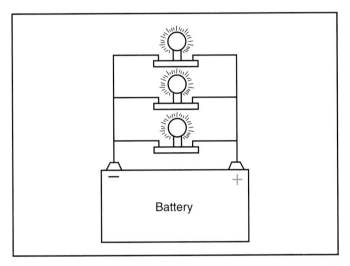

Figure 20-6. *A parallel circuit. Current flow in amperes would depend upon the resistance unit at that point in the circuit.*

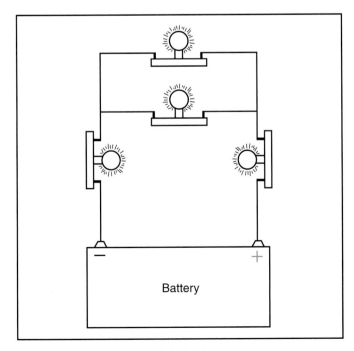

Figure 20-7. *A series-parallel circuit.*

some parts of the circuit, while in other parts, the flow is split between two electrical circuits in parallel. An example of the series-parallel circuit is the brake light switch and brake lights. All brake light current passes through the fuse and switch, but only part of the current passes through each bulb.

Circuit Defects

There are three types of wiring defects: shorts, grounds, and opens. A ***short circuit*** is caused when the wire insulation fails or is removed, and the wires contact each other, the frame, body, or another grounded part of the vehicle, **Figure 20-8.**

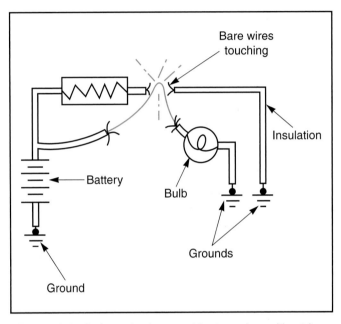

Figure 20-8. *A short circuit caused by two wires without insulation, coming into contact with each other. (General Motors)*

A ***grounded circuit*** is like the short circuit, but the current flows directly into a ground that is not part of the original circuit. This can be caused by a wire with damaged insulation touching the frame or body. A grounded circuit may also be caused by oil, dirt, or moisture around connections and/or terminals, which provide a good path to ground. A grounded circuit is shown in **Figure 20-9.** If a fuse or circuit breaker, does not protect the circuit, a short could lead to burned wiring, damaged electrical components, or a fire.

An ***open circuit*** is a circuit that is not complete. Current cannot flow in an open circuit, **Figure 20-10.** Common causes of open circuits are loose or corroded connections, disconnected wires, and defects in electrical components such as switches, bulbs, and fuses. A related problem is a high resistance electrical connection, usually caused by corrosion or overheating. The high resistance may cause the circuit to stop operating, or possibly catch fire at the connection.

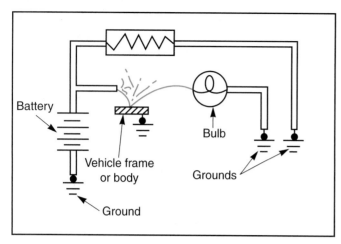

Figure 20-9. *A grounded circuit. The wire insulation has failed, allowing the bare wire to come into contact with the vehicle body or frame. (General Motors)*

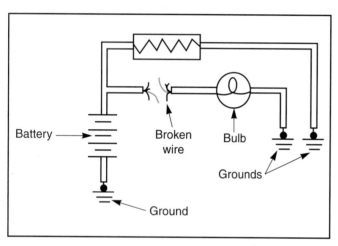

Figure 20-10. *An open circuit happens whenever there is a break in the circuit. The break may be corrosion at a connector, a broken wire, or a wire that burned in two from too much electrical current. Switches and circuit breakers that fail can also create an open circuit. (General Motors)*

Vehicle Wiring and Components

The modern automotive electrical system is a complex arrangement of wiring and electrical components. The electrical system must produce electricity and deliver it to the proper places; protect circuits from damage; reduce or increase voltage; change electricity into light, motion, or heat; and use electricity to control the movement of liquids and gases. The construction, operation, and function of electronic devices to accomplish this will be discussed later in this chapter.

Wiring

Most automotive wiring is made of copper, aluminum, or aluminum coated with copper. Wires are plastic coated and installed into wrapped assemblies called **harnesses.** For easy circuit tracing, automotive wiring is **color coded,** **Figure 20-11.** This is done by giving the insulation of every vehicle wire a specific color. Modern vehicles have many wires, which has made it necessary to increase the number of available colors by adding a stripe or stripes of contrasting colors to the original insulator color.

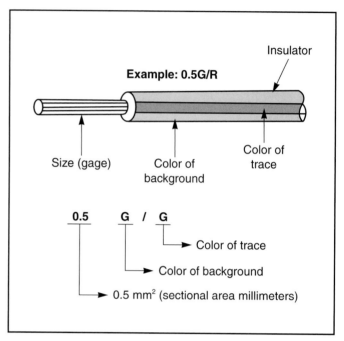

Figure 20-11. *An example of a color-coded piece of wire. Colors vary between vehicle manufacturers. (Chevrolet)*

On most vehicles, there is no return wiring from various electrical units back to the battery. The vehicle frame or body forms the return, or **ground wire.** On all modern vehicles, the negative terminal is the ground terminal. On most vehicles, the battery negative cable is attached to the engine block. The body and chassis may ground directly to the battery through a smaller cable attached to the negative post, or to the engine block by one or more straps.

Wire Size

All wires, no matter how well they conduct current, have some resistance, and lose some electrical power as heat. It is important the wire in a circuit be large enough to carry the rated amperage without overheating. At the same time, the wire should not be unnecessarily large or long as to increase weight, bulk, and cost. **Wire gage** is the rating system for wire diameter. The larger the gage number, the smaller the wire. Wire gage is measured in AWG (American Wire Gage) or metric sizes. Typical wire gages are shown in **Figure 20-12.** The wire gage shows the highest amperage that can be carried without overheating. To further reduce resistance losses, wires carrying high amperages are designed to be as short as possible.

Metric Wire Sizes (mm²)	AWG Sizes
.22	24
.35	22
.5	20
.8	18
1.0	16
2.0	14
3.0	12
5.0	10
8.0	8
13.0	6
19.0	4
32.0	2

Light wire → Heavy wire

Figure 20-12. *A typical AWG (American Wire Gage) chart. (Cadillac)*

Shielded Wiring

Many wires on modern vehicles carry extremely low voltages. Low voltages are used as signals between various electronic components. Low voltage signal wires are extremely vulnerable to magnetic fields generated by current flow in nearby wires and devices. Heavy current flow in a wire next to a sensor wire can cause inaccurate readings.

To reduce the chance of stray magnetic fields affecting low voltage wires, all chassis wiring is carefully routed to prevent crossed wiring. Manufacturers also use **shielded wiring** in circuits that are particularly sensitive to magnetic fields, such as anti-lock brake systems.

Most shielded wires are made by surrounding the signal carrying wire with Mylar tape. **Mylar tape** is a tough, flexible plastic tape that has an outer coating of aluminum. A grounding wire, known as a **drain wire,** is wrapped around the Mylar tape, then the entire wire is wrapped with the same type of plastic insulation used on non-shielded wires, **Figure 20-13.** The aluminum coating on the Mylar tape absorbs any stray magnetic fields. Any voltage produced by the field is discharged through the drain wire to chassis ground.

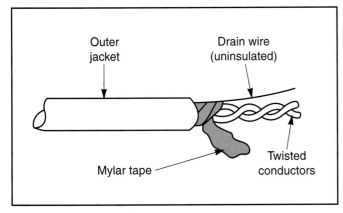

Figure 20-13. *Cross-section of shielded wire as used with most anti-lock brake systems. (General Motors)*

If one heavy current wire has the potential to affect several sensor wires, then the high current wire is shielded. The shield could be Mylar tape, fine wire braided around the high voltage wire, or a metal tube that surrounds the wire for all or part of its length. The shield is grounded to the vehicle chassis. Magnetic fields are absorbed by the shielding, and do not leave the immediate vicinity of the current carrying wire.

Plug-In Connectors

The majority of connectors used in modern vehicles are plug-in types. A **plug-in connector** has male and female ends which are plugged into each other. Connectors having more than one wire are called multiple connectors. Many service manuals refer to a specific connector by the number of wires it contains, such as 12 wire connectors, 23 wire connectors, and so on. Modern plug-in connectors cannot be assembled incorrectly, due to the shape of the connector, or by the use of special aligning lugs and slots on each side of the connector.

Modern plug-in connectors are often thoroughly sealed to keep out moisture and corrosion, **Figure 20-14.** Since many vehicle electronic components operate on very low voltages, a small increase in resistance due to moisture and corrosion can affect circuit operation.

A few connectors may be screw terminals, or bolts which pass through a terminal eye ring to form a connector, **Figure 20-15.** These connectors are usually used to connect a ground strap to the vehicle body, or on circuits with very high current loads, such as the starter and charging system.

Schematics

A wiring diagram or **schematic,** is a drawing showing electrical units and the wires connecting them. Schematics also show wire colors and terminal types. Schematics allow technicians to trace out defective wiring and components in the electrical system. Many vehicle manufacturers break down the overall vehicle wiring into separate circuit diagrams, as shown in **Figure 20-16.**

Schematics use symbols to represent electrical devices. There is some variation in the use of these symbols. Some schematics have a combination of company-specific and standardized symbols. **Figure 20-17** illustrates some symbols widely used in automotive electrical diagrams.

Circuit Protection Devices

To protect vehicle circuits from damage due to excessive current flows, **circuit protection devices** are used. These include fuses, fusible links, and circuit breakers. Short circuits or defective components that draw too much current can cause excessive current flows. All electrical circuits except the starter and alternator output will have a circuit protection device.

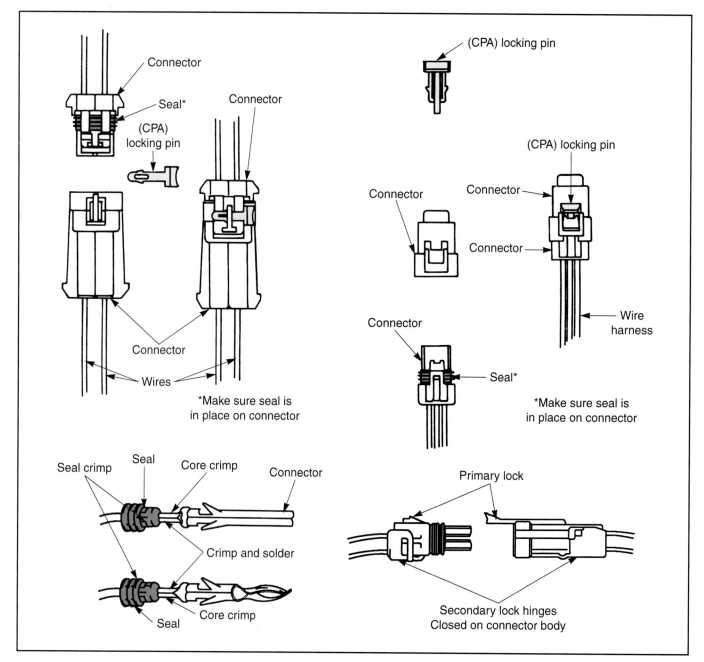

Figure 20-14. *An assortment of plug-in connectors which use a seal to keep out moisture. These are used with anti-lock brakes. Note the CPA (connector position assurance) locking pin. This pin prevents the connectors from coming apart. Trying to separate this type of connector will lead to wire and/or connector damage if the pin is not removed first. (General Motors)*

Figure 20-15. *An engine compartment wiring harness illustrating eye or "ring" wiring terminals. (Jack Klasey)*

Fuses

Fuses are made of a soft metal that melts when excess current flows through them, before the current can damage other components or circuit wiring. The majority of fuses are installed in a *fuse block,* which is usually located under the dashboard or in the glove compartment. A melted or blown fuse must be replaced. Most fuses are color-coded to indicate their current rating, **Figure 20-18.**

Fusible Links

A *fusible link* is a length of wire made of soft metal. It operates in the same manner as a fuse, melting when exposed to excess current. They are rated in amperes and

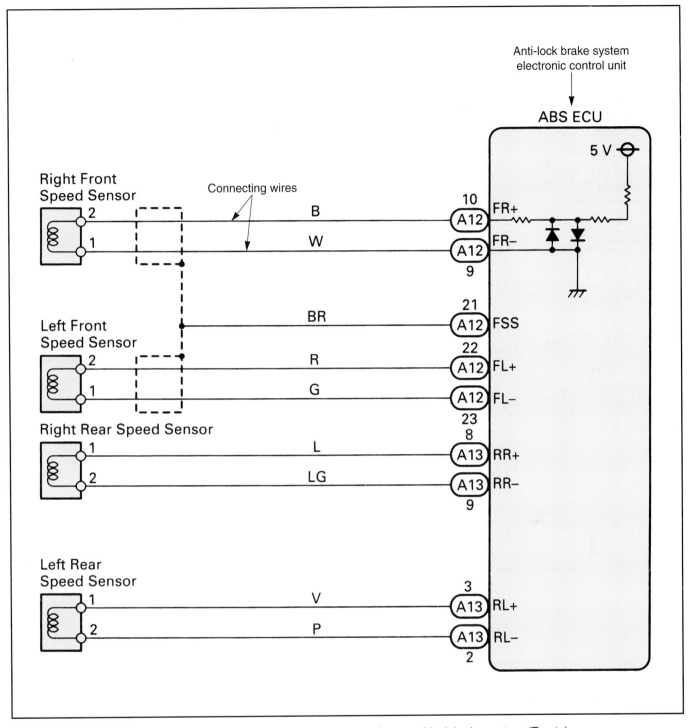

Figure 20-16. *Wiring harness schematic for the wheel speed sensors in an anti-lock brake system. (Toyota)*

may be color-coded. Fusible links are usually installed in the main wiring leading from the battery or starter solenoid to the main electrical circuits. A circuit may be protected by both a fuse in the fuse box and a fusible link ahead of the box, **Figure 20-19.** In newer vehicles, fusible links have been replaced by high amperage Pacific or Maxifuses.

Circuit Breakers

Circuit breakers consist of a contact point set attached to a bimetallic strip. The bimetallic strip will heat and bend as current flows. When it becomes too hot, it bends enough to open the point set, which opens the circuit. When the strip cools off, it straightens out and allows the points to close. The advantage of the circuit breaker is that it can reset itself. A circuit breaker is shown in **Figure 20-20.**

A second type of circuit breaker is solid state and is called a positive temperature coefficient (PTC) unit. This circuit breaker builds (increases) its resistance to current flow as it heats. When the resistance reaches a predetermined level, the PTC unit opens. This breaker will reset in 1-2 seconds, but not until voltage is removed from the circuit.

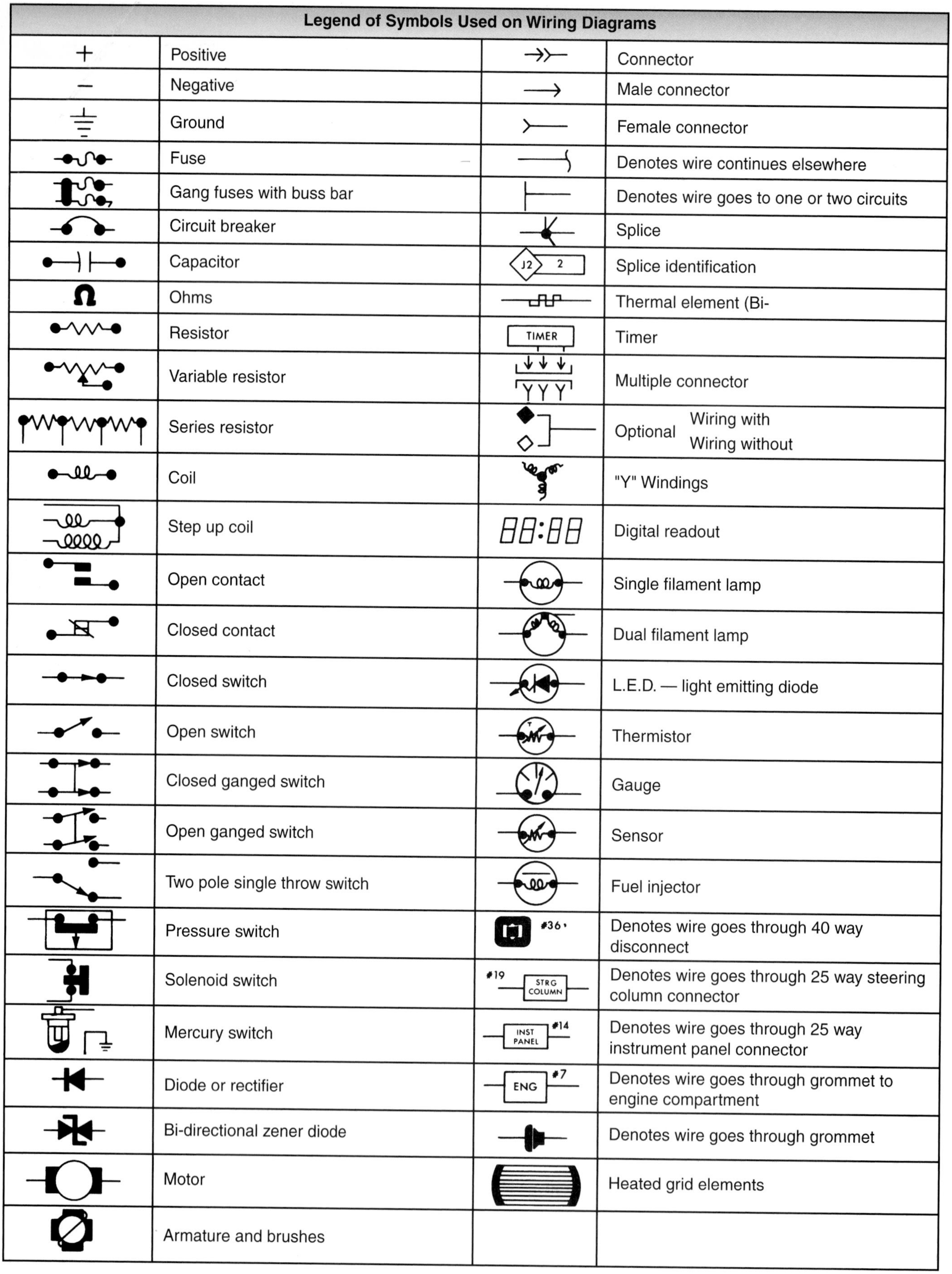

	Legend of Symbols Used on Wiring Diagrams			
+	Positive	⟫—	Connector	
−	Negative	⟶	Male connector	
⏚	Ground	⟩—	Female connector	
Fuse symbol	Fuse	⟶⌐	Denotes wire continues elsewhere	
Gang fuses symbol	Gang fuses with buss bar	⊢	Denotes wire goes to one or two circuits	
Circuit breaker symbol	Circuit breaker	Splice symbol	Splice	
Capacitor symbol	Capacitor	J2 2	Splice identification	
Ω	Ohms	Thermal element symbol	Thermal element (Bi-	
Resistor symbol	Resistor	TIMER	Timer	
Variable resistor symbol	Variable resistor	↓↓↓	Multiple connector	
Series resistor symbol	Series resistor	◆ ◇	Optional	Wiring with / Wiring without
Coil symbol	Coil	"Y" symbol	"Y" Windings	
Step up coil symbol	Step up coil	88:88	Digital readout	
Open contact symbol	Open contact	Single filament lamp symbol	Single filament lamp	
Closed contact symbol	Closed contact	Dual filament lamp symbol	Dual filament lamp	
Closed switch symbol	Closed switch	L.E.D. symbol	L.E.D. — light emitting diode	
Open switch symbol	Open switch	Thermistor symbol	Thermistor	
Closed ganged switch symbol	Closed ganged switch	Gauge symbol	Gauge	
Open ganged switch symbol	Open ganged switch	Sensor symbol	Sensor	
Two pole single throw switch symbol	Two pole single throw switch	Fuel injector symbol	Fuel injector	
Pressure switch symbol	Pressure switch	#36	Denotes wire goes through 40 way disconnect	
Solenoid switch symbol	Solenoid switch	#19 STRG COLUMN	Denotes wire goes through 25 way steering column connector	
Mercury switch symbol	Mercury switch	INST PANEL #14	Denotes wire goes through 25 way instrument panel connector	
Diode symbol	Diode or rectifier	ENG #7	Denotes wire goes through grommet to engine compartment	
Bi-directional zener diode symbol	Bi-directional zener diode	Grommet symbol	Denotes wire goes through grommet	
Motor symbol	Motor	Heated grid symbol	Heated grid elements	
Armature symbol	Armature and brushes			

Figure 20-17. *An assortment of symbols which are used to represent electrical devices within one manufacturer's service manual. (Chrysler)*

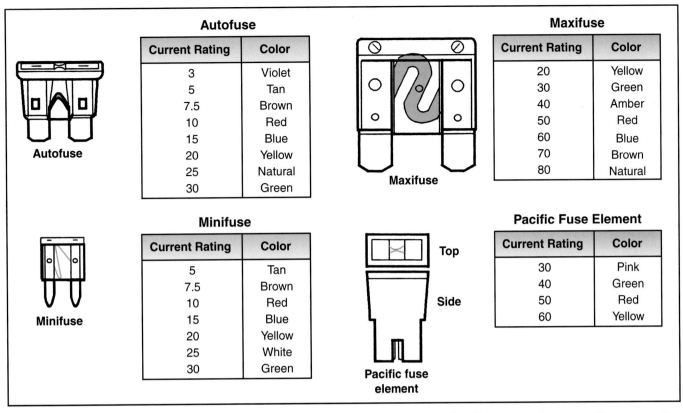

Figure 20-18. *An assortment of common fuses, their shapes and sizes. Note that each individual type has a specific current rating in amperes and is color-coded. The autofuse is the most common fuse in use. The Maxifuse is used to replace the fusible link and is designed to protect cables that normally go between the battery and fuse block. It will protect the circuit from a direct short, or a resistive short. The Maxifuse has a higher amperage rating than the autofuse. The minifuse is a smaller version of the autofuse, but provides the same level of circuit protection as the autofuse. The Pacific fuse element is designed to be a replacement for the fusible link wire. They protect the wiring from a direct short. This style makes them easier to inspect and service than the standard wire type fusible link. (Pontiac)*

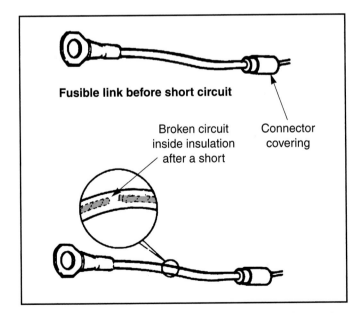

Figure 20-19. *Fusible links are a one time protection device. The fusible link is about four wire gages (sizes) smaller than the wire it is to protect. The thick insulation makes the fusible link look larger than it really is. The links are generally marked on the insulation with their wire gage size. When replacement is needed, always use a new link with the same wire size as the old defective one. (General Motors)*

Switches

To control the flow of electricity through a circuit, some sort of switch is used. Some of these switches have simple on-off positions, while others have several positions to place the circuit in varying operating modes. The vehicle driver manually operates many switches. Other switches are operated as a byproduct of other driver actions, such as the brake and backup light switches.

Brake Light Electrical Diagnosis

In most cases, brake system electrical diagnosis consists of checking for malfunctioning lights. Two types of lights are associated with the brake system: lights on the rear of the vehicle that illuminate when the brake pedal is pressed, and dashboard lights that warn the driver when a problem exists. This section deals with the lights at the rear of the vehicle.

Note: For information on diagnosing and repairing electrical devices associated with anti-lock brake and traction control systems, see Chapters 21 and 22.

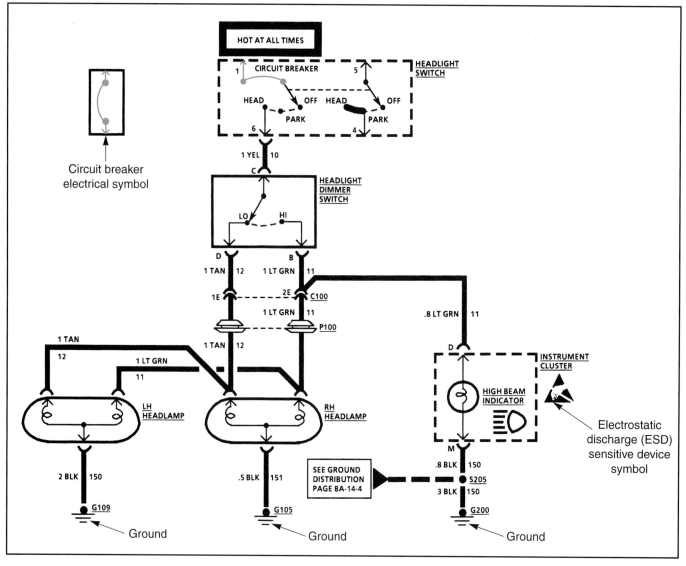

Figure 20-20. *A typical headlight electrical circuit schematic illustrating a circuit breaker. (Chevrolet)*

The wire in the bulb, called a **filament,** is made of a type of resistance wire that glows as electrons pass through. The other wiring is made of metals that allow the electrons to pass with very little resistance. Before proceeding to check a light problem, be sure to remember what is necessary for a bulb or any electrical device to operate properly:

❑ There must be a source of electricity (voltage).
❑ The electrical device must be in operating condition.
❑ The electrical device must be properly grounded.

Diagnosing Brake Light Problems

Rear brake light problems are rarely difficult to find and repair. Most rear brake light problems fall into four categories:

❑ Some brake lights do not light when the brake pedal is depressed.
❑ No brake lights come on when the brake pedal is depressed.
❑ The brake lights are on at all times.
❑ The front parking lights flash when the brakes are applied.

Some Brake Lights Do Not Light

When some brake lights do not light, the most common cause is a burned out bulb. Before proceeding, make sure the light is supposed to come on when the pedal is pressed. Many rear lights are single filament types that are used as taillights only.

To check for a burned out bulb, remove the bulb and inspect it. Most burned out bulbs will be obviously bad. Also, a defective filament will vibrate when the bulb is lightly shaken. Double-filament bulbs used in older vehicles may have a burned out brake filament while the taillight filament is still good. If either filament is broken, replace the bulb. If the bulb appears to be good, check for power at the center terminal using a non-powered test light, **Figure 20-21.** If power is reaching the bulb, and it still does not light, check the ground. Also check for grease or rust in the bulb socket. On some older vehicles, the turn signal switch may cause the lights to be off on one side of the vehicle.

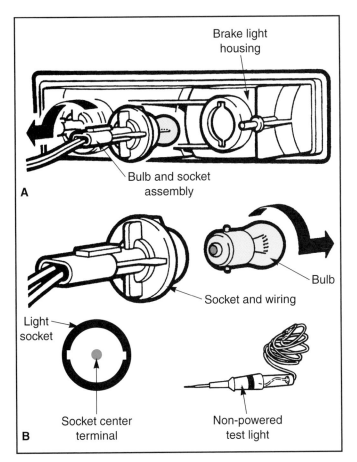

Figure 20-21. *A—A brake light housing along with bulb and socket assembly. B—Use a non-powered test light to check for voltage. Touch the test light point to the center wire terminal of the socket. Be sure you have a good ground for the test light. (General Motors and Land Rover)*

All Brake Lights Are Off

When no brake lights come on when the pedal is pressed, a blown fuse or a defective brake light switch is the usual cause. Check the fuse first. If other electrical devices protected by the brake light fuse are also inoperative, you can conclude the fuse is probably blown before looking at it. If the fuse is blown, you must determine what caused the electrical overload. Do not assume the overload is caused by the brake light circuit; check all the circuits protected by the blown fuse. Also, do not simply replace the fuse and return the vehicle to the owner.

If the fuse is ok, check the brake switch, **Figure 20-22.** The switch may be defective, or it may simply be misadjusted. You can use an ohmmeter or a powered test light to check switch operation. The switch should show infinite resistance when the pedal is in the unapplied position, and low resistance when the brake pedal is lightly pressed. If the switch does not give these readings, it should be adjusted or replaced as necessary.

If the switch checks out good, look for disconnected, broken, or shorted wiring. In many instances, the wiring connector has been knocked off the brake switch itself. However, all modern vehicles have several connectors, and you may need to make a search of the entire wiring harness.

Brake Lights On at All Times

If the brake lights are on at all times, the most likely cause is a stuck brake light switch. Check the switch by unplugging it. If the lights go off, you have located the problem. The switch may be misadjusted or may have failed internally. If the light remains on when the brake switch is unplugged, check for a grounded wire.

Front Parking Lights Flash when Brakes Are Applied

If the front parking lights come on when the brake is applied, electricity is being fed to the parking lights by one of three paths:

❏ A broken filament in a dual filament bulb is touching the other filament.
❏ There is a short between a brake light wire and a parking light wire.
❏ There is a defect in the turn signal switch.

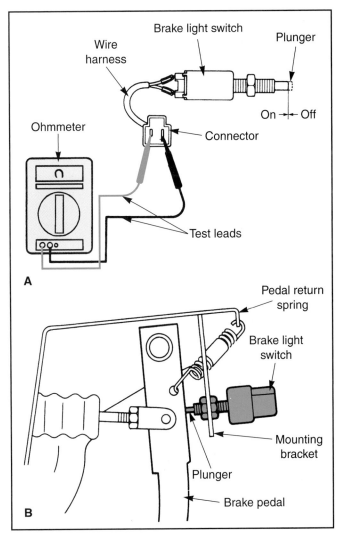

Figure 20-22. *A—Testing a brake light switch for proper operation with a ohmmeter. This switch was removed from its mounting bracket for easier testing. B—Brake switch and housing are placed back in the mounting bracket. The wire harness (not shown) will be connected to the switch next. (General Motors)*

The most common cause of this problem is filament contact inside a dual filament bulb. Shorts between wires or a defective turn signal switch are rare. Begin checking by observing the brake lightbulbs as an assistant presses on the brake pedal. Any bulb that is weak or out completely is probably the source of the problem. Remove and check the bulb. If the bulbs are ok, check for a wire short, and finally, unplug the turn signal harness and press on the brake pedal. If the brake lights do not come on, brake current is being routed through the turn signal harness and the turn signal switch is defective.

Diagnosing Warning Light Problems

The typical red Brake warning light tells the driver the parking brake is applied, the pressure differential switch is out of position, or that the pad wear indicator (when used) is grounded. There are two kinds of warning light problems:

❑ The warning light does not light at any time.
❑ The warning light is on at all times.

Diagnosis and service of these types of warning light problems is discussed in the following paragraphs. The ABS and TCS lights are computer-controlled and diagnosis of these lights is located in Chapter 22.

Warning Light Does Not Light

If the red Brake warning light does not come on when the parking brake is applied, check the fuse. If the fuse is good, check the parking brake electrical switch. A typical switch is shown in **Figure 20-23**. The switch may be out of adjustment, may be defective, or the electrical connector may be disconnected. Adjust or repair as necessary. The switch can be checked with a non-powered test light.

If the fuse and switch are ok, check the warning light bulb. To remove the bulb, it may be necessary to remove the instrument cluster, air conditioner ducts, or partially disassemble the dashboard, **Figure 20-24**. Consult the proper service manual for these procedures. The bulb may be burned out, or may have fallen out of its socket. In some cases, someone has removed the bulb because it was on at all times. Install a new bulb and find out why it is on at all times.

Warning Light Is On at All Times

If the warning light is on at all times, first determine which vehicle systems operate the light. On most vehicles, the parking brake and pressure differential switches can turn on the red Brake warning light. On a few vehicles, the disc brake pad wear sensors also operate this light. Begin diagnosis by unplugging the pressure differential switch, **Figure 20-25**. If the light goes out, service the pressure differential switch or the brake hydraulic system as needed.

If the light stays on, reattach the pressure differential switch and unplug the parking brake switch. If the light goes out, adjust or replace the parking brake switch as necessary. If the light does not go off, check the pad wear indicators, if used. In most cases, the pad wear indicators are grounded because the pads are worn out. Service the brakes as outlined in Chapter 13. If the light remains on when all switches have been disconnected, look for a wire that has grounded against the vehicle body.

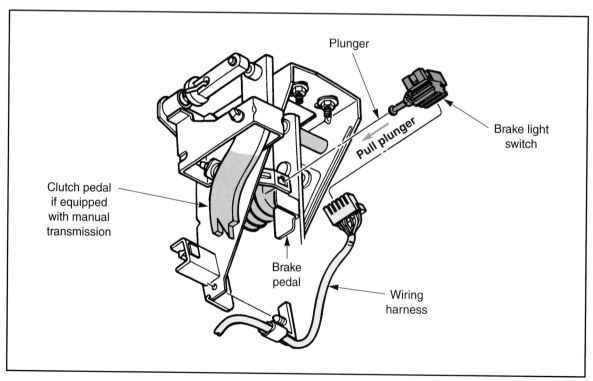

Figure 20-23. *One particular brake light switch and wire harness. Switch styles and locations will vary from one vehicle to another. (Chrysler)*

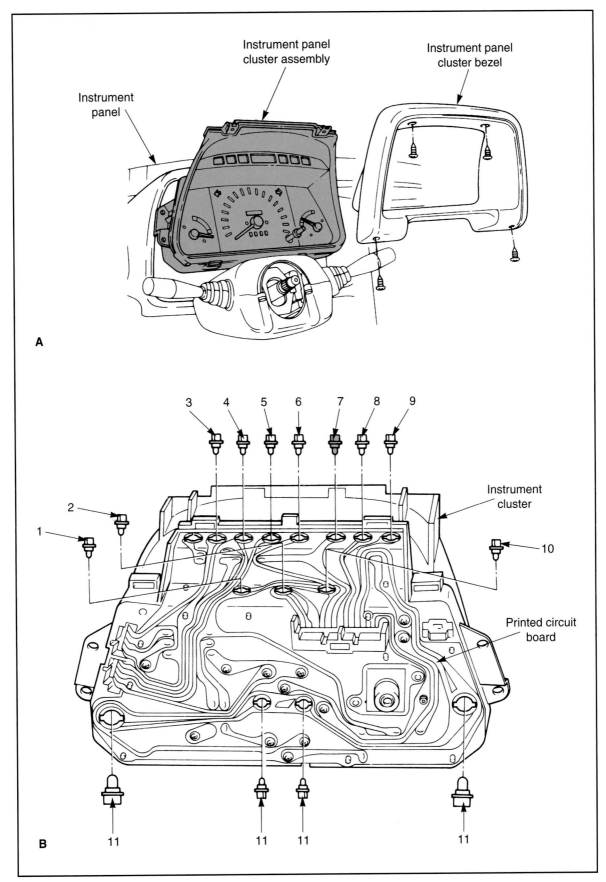

Figure 20-24. *A—An instrument panel disassembly on one car. B—This shows the backside of the instrument cluster assembly and the various lightbulb locations. 1—Right-hand turn indicator. 2—High-beam indicator. 3—Seat belt indicator. 4—Charge indicator. 5—Daytime running lights indicator. 6—Four-wheel drive indicator. 7—Brake indicator. 8—Oil pressure indicator. 9—Malfunction indicator light (MIL). 10—Left-hand turn indicator. 11—Instrument cluster illumination lights. (General Motors)*

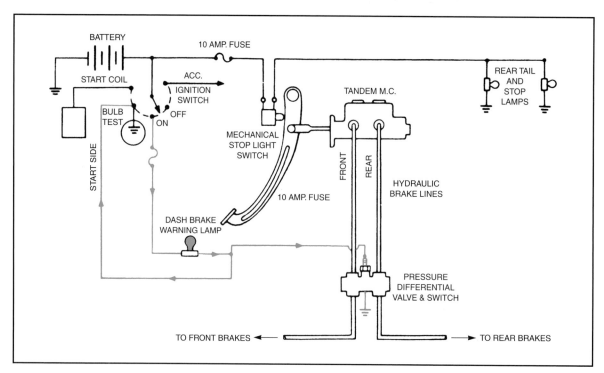

Figure 20-25. *A wiring schematic for a base brake system incorporating a pressure differential valve and warning light. (FMC)*

Electromagnetism

There is a well-defined relationship between electricity and magnetism. When a material is *magnetic,* the electrical charges of its electrons are aligned to create a force that extends outward from the material. A magnetic material attracts metals which contain iron. Some materials are naturally magnetic, while others can be magnetized by electricity. Magnetic fields have definite North and South poles. This property of magnetic fields is called *polarity.* Like poles repel each other and unlike poles attract. The relationship between electricity and magnetism is called **electromagnetism.** Electromagnetism is used to operate many electrical devices on the vehicle.

When current flows in a wire, the electrons start moving in the same direction. This alignment of electrons creates a magnetic field around the wire as long as current is flowing. When the wire is wound into a coil, the magnetic fields of each wire loop combine to create a very strong magnetic field. The combination of the coil winding and an iron core is called an **electromagnet,** **Figure 20-26.** The iron core helps to increase field strength, and may be movable. The magnetic field created may be used to move linkages or open electrical contacts. Electromagnets are the basic component of solenoids, relays, and starters.

When a wire moves through a magnetic field, or a magnetic field moves through a wire, the effect of the field on the electrons in the wire causes them to begin moving. This causes current to flow in the wire. This process is called **induction,** and it is how current is produced in wheel speed sensors.

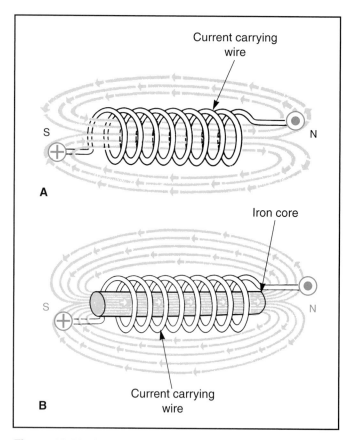

Figure 20-26. *An electromagnet. A—The conductor in several loops multiplies the magnetic field. B—Placing an iron core within the windings increases field strength and will form a stronger electromagnet. The strength of the magnetic field at the N and S poles is increased greatly by adding the iron core. The basic reason for this increase is air itself is a very bad conductor of magnetism while iron is a very good conductor. (Deere & Co.)*

Solenoids and Relays

Some electrical components are electromagnetic control devices. Electricity creates a magnetic field which causes movement of a metal part. A *relay* uses a magnetic field that closes one or more sets of electrical contacts, causing electrical flow in a circuit. See **Figure 20-27.** This is useful when switching high current devices, such as fan motors, without excessive lengths of heavy wire. In a *solenoid,* the magnetic field performs a mechanical task, such as opening or closing anti-lock brake actuator valves, **Figure 20-28.**

Motors

To turn electricity into rotation, *motors* are needed. The most commonly used motors consist of a central *armature* made of many loops of wire. Surrounding the armature are *field coils*. The field coils produce a magnetic field. Current flows through the armature windings. By controlling the direction of current flow through the windings, the armature can be made to turn. Current direction is usually controlled by the use of a commutator and brushes. Refer to **Figure 20-29.**

Figure 20-27. *Relays are usually located in junction or fuse blocks located in the engine compartment. (Jack Klasey)*

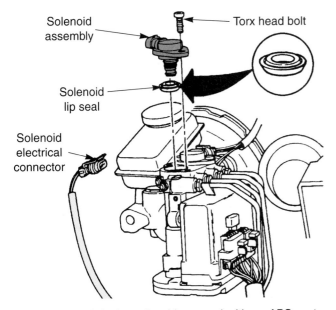

Figure 20-28. *A brake solenoid as used with an ABS system. Most ABS systems have 3-4 solenoids. (General Motors)*

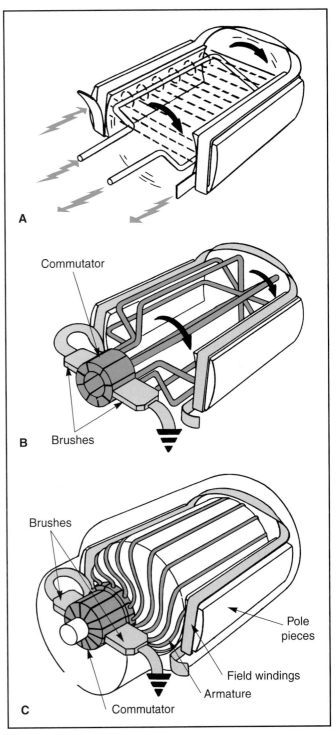

Figure 20-29. *Changing electricity into rotation. If you put a loop of wire in the magnetic field between pole pieces, and send electrical current through the loop, the result will be a simple armature. A—The magnetic field that is around the loop and the field located between the pole pieces will repel one another, causing the loop to turn. B—This shows an actual (normal) armature with many more loops. If you attach individual metal segments to the ends on each loop, a simple contact surface (commutator) is formed. As electrical current is sent to the commutator by sliding contacts (brushes), the repelling action causes a continuous rotation. Starter motors use this principle to generate mechanical energy. C—Most starter motors have brushes, armature, field windings, pole pieces, and a mechanical drive mechanism. (Deere & Co.)*

Electric motors are found throughout the vehicle, including the starter motor, heater blower motor, windshield wiper motor, power window, seat, and antenna motors, and small motors which provide hydraulic pressure for the ABS system.

Brake Motors

Some anti-lock brake systems require precise hydraulic control. To do this, the motors must stop immediately when electrical power is removed. To stop motor rotation as soon as power is removed, **brake motors** are used. There are two types of brake motors in use.

One type is used on a few systems. This type has an electromagnetic brake assembly, **Figure 20-30A.** Ground for the brake assembly is provided through the ABS controller. When the motor is turning, the clutch is energized and retracts the spring. When the motor stops, the clutch is de-energized, and the spring expands, stopping the motor immediately.

The other type consists of an expansion spring that connects the motor shafts and the piston drive gear, **Figure 20-30B.** In its normal, non-moving position, the spring expands against the sides of its housing, locking it in place. The spring is connected to the motor shafts so that any movement causes it to be contracted. This contraction frees the spring from the sides of the housing, allowing it and the shafts to turn. As soon as the motor stops, the spring expands and locks itself to the housing, stopping the shaft.

Static Electricity

While most electronic components used in vehicles are designed to withstand heat and some physical shock and abuse, other components, especially computer circuits, are sensitive to damage by **static electricity.** You can become charged with static electricity simply by sliding across a seat, which is a common occurrence when servicing most vehicles.

If you become charged with static electricity and touch a computer or other static sensitive component, the charge will arc to ground through the component, destroying it in the process. Fortunately, static electricity can be discharged simply by grounding yourself on the vehicle chassis or any metal part connected to the vehicle chassis before handling any static sensitive component. Most of these components and schematics that contain static sensitive components are marked, **Figure 20-31.**

Radio Frequency and Electromagnetic Interference

With the ever-increasing number of wires and circuits in the modern vehicle, the chances for problems increase. A problem that has existed since the introduction of electronic ignition systems and has gone unnoticed for the most part is starting to occur more frequently in newer

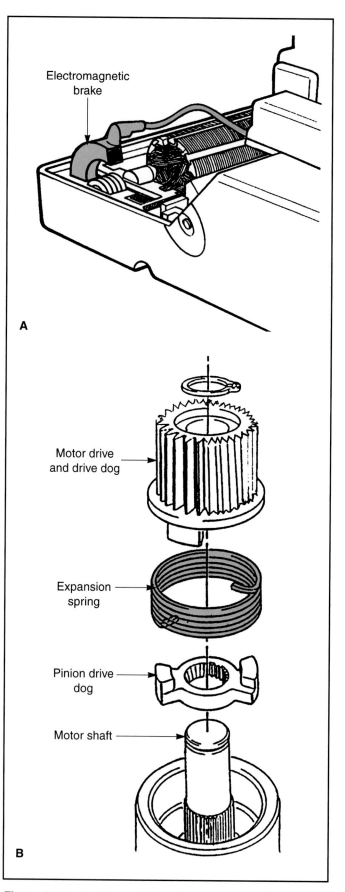

Figure 20-30. A—Some ABS systems use specialized motors with braking features. This motor has an electromagnetic brake. B—This motor uses a spring-loaded brake feature. (General Motors)

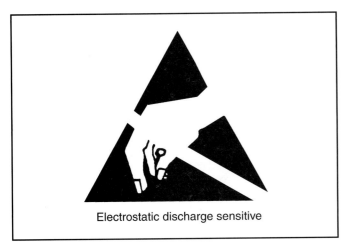

Electrostatic discharge sensitive

Figure 20-31. *Symbol used to indicate circuits or components that are electrostatic sensitive. (General Motors)*

vehicles. This problem is electrical malfunctions caused by *radio frequency interference (RFI)* and *electromagnetic interference (EMI)*.

RFI and EMI interference is usually caused by voltage spikes and stray magnetic fields created by defects in such circuits as the ignition and charging systems. Devices such as police and CB radios and other electronic accessories can also generate RFI and EMI during normal operation. In the past, this was limited to background noise that was sometimes heard through the radio speakers. In the modern vehicle, RFI and EMI can create false signals, resulting in sensor misreadings, unusual output device operation, and inaccurate diagnostic codes.

Vehicle Computers

The modern automotive computer is a collection of electronic circuits that can operate many vehicle systems. All modern cars and trucks have at least one computer, and most have several. The following information applies to all vehicle computers, but will be most useful when working on ABS or TCS computers. All vehicle computers have some common operating features.

Semiconductor Materials

All modern automotive computers and other electronic devices depend on the use of materials known as *semiconductors.* Semiconductors are made of silicon or germanium with a small amount of impurities to cause them to become either conductors or insulators, depending on how voltage is directed into them. Semiconductors change at the atomic level, depending on how their electrons are affected. Common semiconductors are diodes and transistors. These devices are extremely small and can be combined into compact computers, performing the same number of calculations as a room full of older electronic devices.

Diodes

Diodes are semiconductors that allow current to flow in only one direction. Diodes are one-way check valves for electricity, becoming a conductor or an insulator, depending on which way the current tries to flow, **Figure 20-32.** Diodes are a familiar part of alternators, where the alternating current must be *rectified,* that is changed, to direct current to charge the battery. For this reason, diodes are sometimes called rectifiers. Diodes are used to prevent voltage surges, either by being installed inside the computer or installed in the wiring harness to some high current devices. Diodes are used in the wiring harness of ABS systems to prevent reverse current surges.

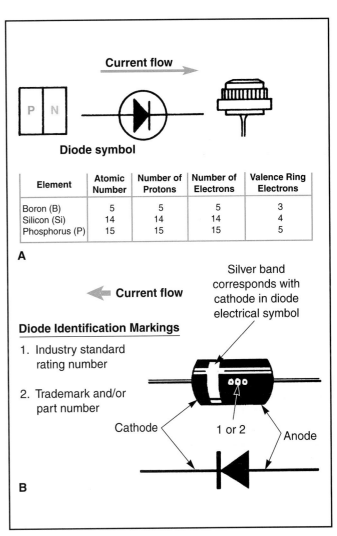

Figure 20-32. *A—A diode is an electrical device which will allow electrical current to pass through itself in only one direction. A diode is formed when two semiconductor materials are joined, one of N-type material, and the other P-type material. The N and P materials will attract one another, but are stabilized by the positive and negative ions on each side, An ion is an atom which has a shortage or excess of electrons. B—Another style of diode and construction. Diodes are used to help protect certain electrical components from the surges or spikes (higher voltage than the part or circuit was designed to handle), and to also isolate circuits. (Deere & Co. and General Motors)*

Computer Circuits

Many diodes and transistors, as well as chip capacitors, resistors, and other parts are combined onto a large complex electronic circuit by etching the circuitry on small pieces of semiconductor material. A circuit made in this manner is called an *integrated circuit,* or *IC.* The IC also contains resistors, capacitors, and other electronic devices. Modern ICs, sometimes called chips or microprocessors, can control vehicle functions formerly done by mechanical devices, such as controlling engine timing. Microprocessors have made some systems, such as anti-lock brakes, not only possible but also available in even low-priced vehicles.

Control Loops

The microprocessors that make up a modern computer can perform control operations based on their ability to process input messages and issue output commands. This ability allows them to operate as part of a control loop. A *control loop* can be thought of as an endless circle of causes and effects, which are used to operate part of the vehicle, such as the ABS system, **Figure 20-33.**

When the control loop is operating, the *input sensors* furnish information to the computer, which makes output decisions and sends commands to the *output devices.* The operation of the output devices affects the operation of the vehicle system, causing changes in the readings furnished by the input sensors. There is a continuous loop of information, from the input sensors, to the computer, to the output devices, to the vehicle system, and back to the input sensors. Some computers operate many control loops at the same time.

Main Computer Sections

All automotive computers contain two main sections: the central processing unit, or CPU, and the memory. The CPU and memory are built into the control module, and the entire module must be replaced if either is defective.

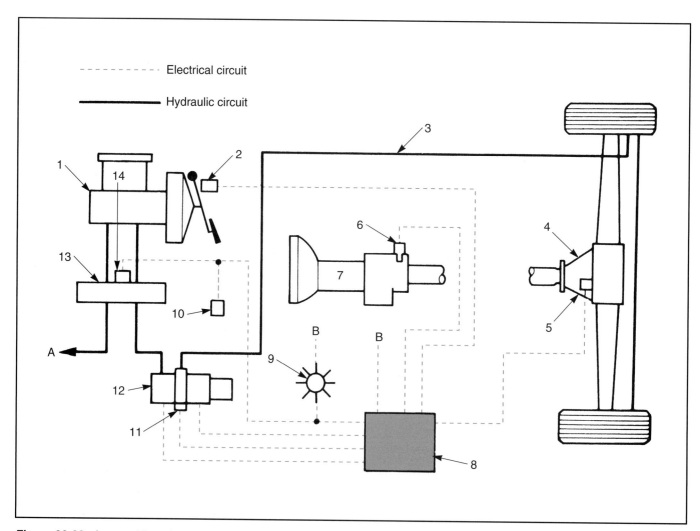

Figure 20-33. *A control loop for one particular rear wheel anti-lock brake system (RWAL). Note that all the various parts that send an electrical signal to the electronic brake control module (EBCM). A—To front brakes. B—To 12-volt ignition. 1—Master cylinder. 2—Stoplight switch. 3—Rear brake circuit. 4—Rear differential. 5—Rear wheel speed sensor. 6—Four-wheel drive switch. 7—Transmission. 8—Electronic brake control module (EBCM). 9—Brake warning light. 10—Parking brake switch. 11—Reset switch. 12—Pressure limiting valve. 13—Proportioning/differential valve. 14—Differential switch. (General Motors)*

Central Processing Unit

The **central processing unit,** or **CPU** is the section of the computer that receives the sensor information, compares this input with the information stored in memory, performs calculations, and makes output decisions. The CPU may contain several microprocessors, as well as other electronic parts.

Memory

The two basic types of computer **memory** used in ABS and TCS control modules are **read only memory (ROM)** and **random access memory (RAM).** The ROM and RAM contain microprocessors and other solid state parts. The read only memory (ROM) contains permanent programs that tell the computer what to do under various operating conditions. The ROM also contains the operating instructions for the module and the system. Information is programmed in the ROM at the time of manufacture, and cannot be modified in the field. The ROM is permanent, or nonvolatile memory. This means the information stored in ROM will remain if the battery is disconnected or the system fuse is removed.

The random access memory (RAM) is a temporary storage place for data from the input sensors. As information is received from the sensors, it is temporarily stored in RAM, overwriting any old information. The computer constantly receives new signals from the sensors as the ABS or TCS system operates. Any trouble codes generated by a system defect is stored in RAM. The RAM is temporary, or volatile memory. If the battery cable or system fuse is removed, all data stored in RAM will be erased, including any trouble codes.

 Note: The memory in some newer ABS/TCS control modules cannot be cleared by removing battery power.

Computer Diagnostic Outputs

In addition to operating the output devices, the computer provides a diagnostic output to the technician. This consists of a dashboard mounted light to indicate that a problem is present, and a data link connector for retrieving diagnostic information. When a system defect occurs, the control module saves information about the defect in memory. This information takes the form of a **trouble code.** On vehicles up to 1995, trouble codes were 2- or 3-digit numbers corresponding to a specific defect. In vehicles equipped with OBD II, the trouble code is in alpha-numeric format, consisting of one letter and four numbers, **Figure 20-34.** To determine whether any trouble codes are present, the technician must use a **scan tool** to cause the module to enter the **self-diagnosis** mode and retrieve trouble codes.

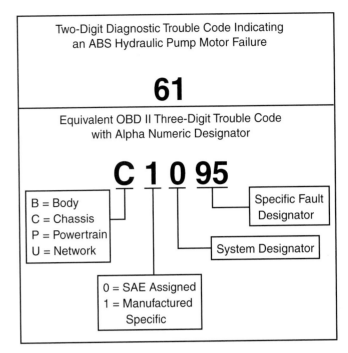

Figure 20-34. *Trouble code format used in loder vehicles had 2-3 numbers. Vehicles built in 1996 and after use the OBD II system, which contains an alpha numeric designator, along with a three digit code number indicating the system and the specific problem.*

Summary

Every atom has a positively charged center. Negatively charged electrons revolve around the center. Electricity is the movement of electrons from atom to atom. Materials whose atoms easily give up or receive electrons are conductors. Materials whose atoms resist giving up or accepting atoms are insulators. The path through which the electrons move is called a circuit. Electrons will not flow in a circuit unless there are more electrons in one place than another and if there is a path between the two places. The three basic electrical properties are amperage, voltage, and resistance. Unknown electrical properties can be calculated using Ohm's law.

The two types of current flow are alternating and direct. Automotive electrical systems are always direct current systems. Alternating current is used in some sensors. The three types of automotive circuits are series, parallel, and series-parallel. Magnetism can be used to make electricity and electricity can be used to create magnetism. This relationship between electricity and magnetism is called electromagnetism.

Automotive wiring can be copper, aluminum, or copper coated aluminum, plastic coated, color coded, and installed into harnesses. Wire gage is the rating system for wire diameter. Most electrical connectors are plug-in types. A connector with more than one wire is called a multiple connector. A schematic, or wiring diagram, is a drawing of electrical units and connecting wires.

Schematics allow the technician to trace out defective components in the wiring system.

Circuit protection devices include fuses, fusible links, and circuit breakers. Switches control the flow of electricity through a circuit. Relays and solenoids are electromagnetic control devices. Relays control electrical flow, while solenoids cause physical movement. Motors turn electricity into rotation. Resistors are used to reduce current flow through circuits.

All modern automotive computers depend on the use of materials known as semiconductors. Semiconductors can be conductors or insulators, depending on how voltage is directed into them. Common semiconductors are diodes and transistors. Diodes allow current to flow in one direction only. Transistors are used as switches or amplifiers. Transistors carry large amounts of current, but are operated by low currents. Transistors can perform switching operations with no moving parts. Transistors, diodes, capacitors, and resistors are etched onto a small piece of semiconductor material to form an integrated circuit, or IC. ICs are also called chips or microprocessors.

Computers are collections of microprocessors and other components. They form control loops, using the inputs and outputs of the system and internal controls. The computer processes inputs from sensors and issues commands to output devices. Computer internal controls are the central processing unit, or CPU, and the memory circuits. The two types of memory are read only memory or ROM, which consists of permanent settings, and random access memory, or RAM, which changes as the vehicle operates. Computers, in addition to controlling the operation of a system, can produce diagnostic codes for aid in troubleshooting problems.

Review Questions—Chapter 20

Please do not write in this text. Write your answers on a separate sheet of paper.

1. All modern vehicles use a _____ volt electrical system.
2. Everything is made of _____.
3. Match the material with its electrical function.
   _____ Copper   (A) conductor
   _____ Plastic   (B) insulator
   _____ Aluminum
   _____ Glass
4. Electrical pressure is called _____.
5. How many terminals do a vehicle battery and alternator always have?
6. Failed insulation is a common cause of _____ circuits.
7. An electromagnet is a combination of a coil _____ and a metal _____.

8. Electromagnets are the basic components of three electrical devices used in brake systems. Name them.
9. Electrical connectors with more than one wire are called _____ connectors.
10. One type of circuit protection device resets itself. This is called a _____.
11. How do resistors affect the current flowing in a circuit?
12. The brake technician will usually be called on to perform basic electrical diagnosis when one or more brake _____ fail.
13. What three things are necessary before an electrical device will operate properly?
14. The two main sections of an automotive computer are the _____ and _____.
15. How does the automotive computer warn the driver of a problem?

ASE Certification-Type Questions

1. Technician A says that glass easily gives up electrons. Technician B says that aluminum and copper are good conductors. Who is right?
   (A) A only.
   (B) B only.
   (C) Both A & B.
   (D) Neither A nor B.
2. If there is an excess of electrons in one part of a complete circuit and a lack of electrons in another part, what will happen?
   (A) Electrons will flow through the circuit.
   (B) Atoms will flow through the circuit.
   (C) Protons will flow through the circuit.
   (D) Neutrons will flow through the circuit.
3. Amperage is the number of _____ flowing past a point in a circuit.
   (A) volts
   (B) protons
   (C) electrons
   (D) atoms
4. If voltage and resistance are known, you can use Ohm's law to find _____.
   (A) horsepower
   (B) heat
   (C) wattage
   (D) amperage

5. In a _____ circuit, only some of the current flows through parts of the circuit.

(A) series

(B) parallel

(C) short

(D) open

6. Wire gage is the measurement of wire _____.

(A) length

(B) material

(C) insulation

(D) diameter

7. Motors turn electricity into _____.

(A) magnetism

(B) rotation

(C) heat

(D) light

8. Which of the following is *not* a circuit protection device?

(A) Fuses.

(B) Circuit breakers.

(C) Resistors.

(D) Fusible links.

9. Technician A says that diodes are used to reduce voltage surges. Technician B says that diodes are used to reduce current flow. Who is right?

(A) A only.

(B) B only.

(C) Both A & B.

(D) Neither A nor B.

10. The left side brake lights on a vehicle do not work. Technician A says that the problem could be a defective brake switch. Technician B says that the problem could be a bad ground on the left side. Who is right?

(A) A only.

(B) B only.

(C) Both A & B.

(D) Neither A nor B.

11. None of the brake lights on a late-model vehicle operate when the brake pedal is pressed. Which of the following is the *least* likely cause?

(A) Defective brake switch.

(B) Blown brake light fuse.

(C) Disconnected brake switch connector.

(D) All bulbs burned out.

12. The front parking lights flash when the brake pedal is pressed. All of the following could cause this problem, EXCEPT:

(A) brake pedal switch.

(B) turn signal switch.

(C) shorted wiring.

(D) broken bulb filament.

13. A brake warning light does not come on when the parking brake is depressed. Technician A says that the problem could be a blown fuse. Technician B says that the problem could be a burned out bulb. Who is right?

(A) A only.

(B) B only.

(C) Both A & B.

(D) Neither A nor B.

14. Which of the following defects will *not* cause the dashboard warning light to come on?

(A) Defective parking brake switch.

(B) Defective brake light switch.

(C) Loss of hydraulic pressure on one side of the hydraulic system.

(D) Grounded pad wear sensor.

15. The computer diagnostic output system consists of all of the following, EXCEPT:

(A) dashboard warning light.

(B) scan tool.

(C) data link connector.

(D) internal computer circuits.

A compact anti-lock brake hydraulic actuator. The small size of this unit allows it to be mounted almost anyplace on the vehicle. (Continental Teves)

Chapter 21

Anti-Lock Brake and Traction Control System Components and Operation

After studying this chapter, you will be able to:

- ❑ Explain why anti-lock brake systems were developed.
- ❑ Explain anti-lock brake system operation.
- ❑ Identify and explain the purpose of anti-lock brake system components.
- ❑ Explain traction control system operation.
- ❑ Identify and explain the purpose of traction control system components.
- ❑ Identify common components of anti-lock brake systems and traction control systems.
- ❑ Identify major manufacturers of anti-lock brake systems and traction control systems.

Important Terms

Tire slip

Anti-lock brake system (ABS)

Base brakes

Traction control systems (TCS)

Pedal feedback

Riding the brakes

Two-wheel systems

Four-wheel systems

Integral hydraulic systems

Non-integral hydraulic systems

Channel

Input sensors

Wheel speed sensors

Tone wheel

G-force sensor

Lateral acceleration sensor

Brake light switch

Pedal travel switch

Control module

Serial data

Universal asynchronous receive and transmit (UART)

Class 2 serial communications

Hydraulic actuator

Hydraulic modulator

Lamp driver module

Society of Automotive Engineers (SAE)

Engine torque management

This chapter discusses the design and operation of anti-lock brake and traction control systems. These two systems are similar since they use many of the same input sensors, electronic control modules, and hydraulic actuators. All anti-lock brake and traction control systems work with the conventional brake hydraulic and friction elements. The anti-lock brake system is activated to reduce slipping during braking. The traction control system works to reduce slipping during acceleration.

 Note: In this and Chapter 22, the abbreviations ABS and TCS will be used frequently. ABS stands for anti-lock brake system. TCS stands for traction control system. They will sometimes be grouped together as ABS/TCS, which is an abbreviation for anti-lock brake and/or traction control system.

Tire Slip

When confronted with an emergency, most drivers respond by pushing on the brake pedal as hard as possible. This is a natural reaction, but in many situations, it is the wrong thing to do. Heavy brake pedal pressure may cause one or more wheels to lock up, or stop turning. A locked-up wheel is said to be **skidding**, or sliding along the pavement instead of rolling over it. A skidding wheel does not contribute to slowing or steering the vehicle.

At one time or another, every driver has also experienced **wheel spin** when accelerating on a slippery surface. Not only is this annoying, the resulting loss of control can cause an accident. Tire skidding and spinning are referred to by the technical term **tire slip**. Tire slip is measured between 0-100% and referred to in terms of negative and positive slip.

At 0% negative slip, the tire is turning and at 100% negative slip, the wheel is locked and skidding. At 100% negative slip, the brakes stop the tire, but not the vehicle. Maximum braking force occurs between 10-20% slip, **Figure 21-1**. Therefore, some tire slip is needed for maximum braking. Conversely, a tire with 0% positive slip is gripping the road and a tire with 100% positive slip is spinning over the road surface with no traction. In this condition, less tire slip is desirable.

Purpose of Anti-Lock Brake Systems (ABS)

When the road is dry, wheel lockup causes tremendous friction between the tire tread and the pavement. This produces excessive heat which can melt the tire's rubber. The tire then skids along on the layer of melted rubber. The skid marks seen on almost every road is actually melted rubber from panic stops. When the pavement is wet or icy,

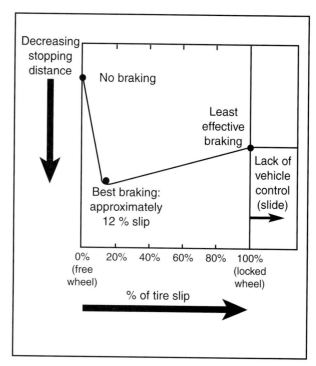

Figure 21-1. *Too much negative tire slip will result in skidding. The most effective braking occurs at approximately 12% slip. (General Motors)*

hard braking causes skidding on the moisture between the tires and pavement. The moisture acts as a lubricant. A skid on a wet or icy road can cause a driver to lose vehicle control completely, possibly resulting in an accident.

In the past, drivers avoided skids in wet or icy conditions by **pumping** (alternately applying and releasing) the brake pedal. Pumping the pedal applies and releases the hydraulic pressure in the system, allowing the friction members to apply and release, reducing the chance of wheel lockup. However, drivers often forget to pump the brakes or fail to release the brake pedal once a skid begins, especially during a panic stop.

It is impossible to manually pump the brakes quickly enough to prevent some skidding. The purpose of the **anti-lock brake system (ABS)** is to perform the pumping action for the driver. Anti-lock brake systems respond quickly to wheel lockup and pump the brakes at a much faster rate, up to 15 times per second or more. Since the anti-lock system only operates during very hard stops, normal braking is unaffected.

 Note: Some manufacturers originally called their ABS systems anti-skid systems.

In addition to reducing the chance of skidding and improving vehicle control during hard braking, the anti-lock brake system reduces tire wear and flat spotting caused by panic stops. Whenever a tire is not turning on a moving vehicle, friction will melt a flat spot in the bottom of the tire.

Even when skidding on wet roads, some friction remains to grind a flat spot into the tire. If the tire can continue to rotate as the vehicle is stopped, tire wear will be even.

Most anti-lock brake systems used on automobiles and light trucks are similar to the one shown in **Figure 21-2**. This ABS system controls hydraulic pressure to all four wheels. ABS systems used on some older light trucks and small cars control only the rear wheels, **Figure 21-3**.

The conventional brake parts you studied earlier are called the **base brakes** or *foundation brakes*. The friction and hydraulic components that make up the base brakes are identical in design and function to those used on vehicles without ABS.

Purpose of Traction Control Systems (TCS)

To reduce wheel spin on acceleration, manufacturers have used various means of traction control. In the past, the most common means of traction control was the

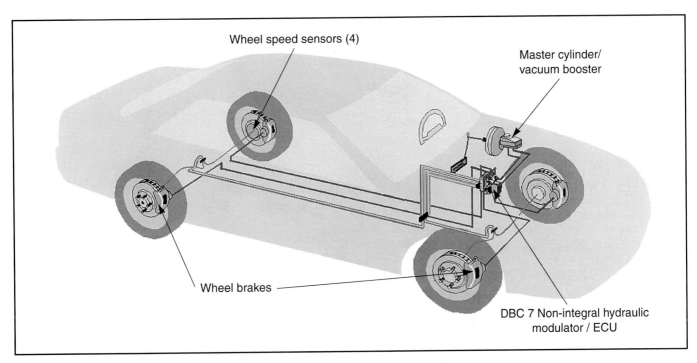

Figure 21-2. *A four-wheel anti-lock brake system. Study carefully. (General Motors)*

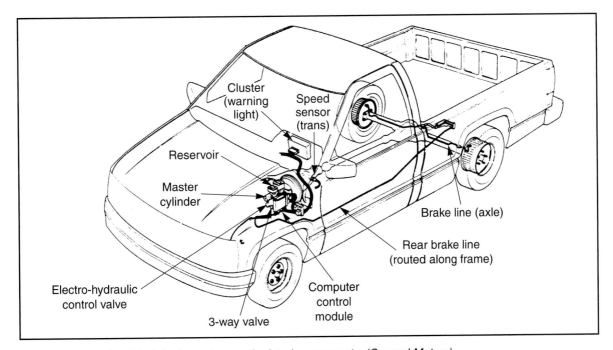

Figure 21-3. *A rear-wheel anti-lock brake system and related components. (General Motors)*

limited-slip differential. However, limited-slip differentials cannot precisely control power transfer, and can only react to wheel spin rather than anticipating it. In addition, limited-slip differentials are ineffective when both drive wheels are on a slippery surface.

To assist vehicles in achieving better acceleration control, more and more vehicles are equipped with electronic *traction control systems (TCS).* These systems react quickly to reduce engine torque and operate the drive wheel brakes before wheel spin becomes excessive. Some traction control systems are designed to operate during hard cornering to reduce body lean and sideslip. **Figure 21-4** shows the layout of a typical traction control system.

 Note: Some manufacturers refer to their TCS systems by the name *traction assist.*

If the traction control system determines that all drive wheels are spinning excessively, it can reduce engine torque to prevent further spinning. If reducing engine torque fails to eliminate excessive wheel spin, the traction control system begins to apply the brakes. On a two-wheel drive vehicle, the traction control system applies the brakes on one or both wheels. If the system detects a drive wheel spinning at a faster rate than the others, it will apply the appropriate

amount of braking force to slow the wheel to match the speed of the other wheel. On a full-time four-wheel drive system, it can apply any combination of the four brakes.

When the excessive wheel spin is eliminated, normal engine operation is resumed. The traction control system is usually designed to suspend operation above a certain vehicle speed or if the system has been used excessively.

Some vehicles use a traction control system without ABS, or a traction control system separate from the ABS system. On most newer vehicles, however, the ABS and TCS systems are paired together, sometimes in the same housing, and operated by a single electronic control module. Systems with one electronic module may have a single hydraulic actuator, or separate actuators for braking and traction control functions.

Many traction control systems have an on-off switch, for situations where some slippage may be desirable, for example rocking the vehicle out of a deep snowbank or mudhole.

Proper Use of ABS and TCS Systems

Reports have indicated that many drivers do not know how to get the maximum benefit from their anti-lock brake and traction control systems. Many drivers were taught to

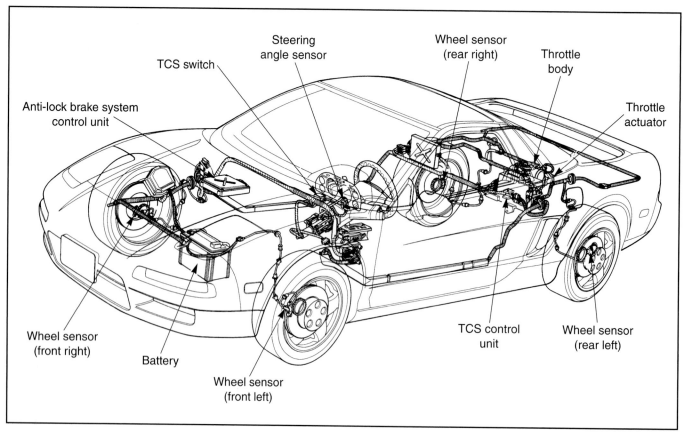

Figure 21-4. *The overall layout of one particular combination anti-lock brake and traction control system (TCS). This system will automatically become "ready" each time the engine is started. If not desired, the traction control system may be turned off by the driver. (Honda)*

pump the brakes during emergency braking situations. However, pumping the brakes turns the ABS system on and off, keeping it from working properly. If the brakes are pumped in situations when the ABS would be in operation, the overall braking effect is actually worse than it would be on a non-ABS vehicle. In an emergency calling for hard braking, always keep steady pressure on the brake pedal until the emergency is over.

A common belief of TCS systems is they will allow you to drive a vehicle out of any reduced traction situation. The truth is, TCS systems only provide additional acceleration control in situations where traction is limited. TCS provides additional traction on icy, wet, or sandy road surfaces. However, TCS will have minimal or no effect in deep snow or mud.

Pedal Feedback

One reason that drivers give for lifting their foot off the brake pedal during an ABS stop is pedal feedback. *Pedal feedback* is an upward pressure or pulsation of the brake pedal against the driver's foot that occurs during ABS braking. This feedback should not be confused with the pulsation that occurs due to an out-of-round rotor or drum. An out-of-round rotor or drum will cause pulsation that can be felt in the steering wheel or the seat of the pants during almost any level of braking. Pedal feedback from the ABS system should only occur during an ABS-assisted stop. The pedal feedback is usually accompanied by an audible clicking, humming, or buzzing, which is the operation of the ABS pump and/or valves.

Pedal feedback varies between ABS systems, from very firm to very light. Newer systems have much less pedal feedback than older systems. To a driver not familiar with ABS, the pulsation and noise can raise fear that there is something wrong with the brake system. Education in the proper use of ABS has reduced the number of complaints regarding pedal feedback.

Riding the Brakes

Another common problem is drivers resting their foot on the brake pedal. This is commonly referred to as *riding the brakes.* Riding the brake pedal can cause several conditions to occur:

❏ Can cause the base brakes to overheat and wear prematurely.
❏ Sets a diagnostic code and disables the ABS system.
❏ The ABS system will not self-test after start-up until the brake pedal is released.
❏ Causes the control module to disable the TCS system on some vehicles.

Drivers who operate the accelerator and brake using one foot on each pedal usually are guilty of riding the brake pedal. These drivers should be instructed that riding the brake pedal will reduce or eliminate any benefits of the ABS or TCS system.

ABS Malfunction Effect on Base Brakes

One belief of some drivers is that any ABS malfunction will affect base brake operation. The chance of this occurring was much greater in older ABS systems. In some cases, a control module or hydraulic pump malfunction could affect base braking. The chance of an ABS system malfunction in newer systems affecting the base brakes has been greatly reduced.

In most cases, an ABS malfunction will have no effect on the base brake system. Usually, if a malfunction in the ABS system occurs, base braking will still be available. However, a base brake problem will affect the ABS system. For example, a leaking brake line will affect both base and ABS brake systems while a defective ABS wheel speed sensor will set a trouble code, but will not affect base brake operation.

Types of ABS/TCS Systems

There are a variety of anti-lock brake and traction control systems in use by vehicle manufacturers. They can vary in operation by model and sometimes by model year, so it is important that you use the most up-to-date service information. This section gives the various general classifications of ABS/TCS systems. Information about specific systems is given later in this chapter.

Two- and Four-Wheel Systems

There are two general types of systems used, depending on the number of wheels controlled. They are two-wheel and four-wheel. *Two-wheel systems* or *rear-wheel anti-lock (RWAL)*, provides stopping assistance for the rear wheels only, and is primarily used on trucks. *Four-wheel systems* or *four-wheel anti-lock (4WAL)*, controls brake actuation on all four wheels. ABS and TCS systems can be further broken down into one of two general classifications of hydraulic modulator—integral and non-integral.

Integral Hydraulic Systems

Integral hydraulic systems were the first ABS systems installed on vehicles. They were large, complex units that operated at pressures over 2000 PSI (13 790 kPa). Integral systems took the place of the conventional master cylinder and power booster, **Figure 21-5**. The master cylinder, accumulator, hydraulic valves, pressure pump, and other components were all part of the assembly. One noticeable drawback to integral systems was they generated high brake pedal feedback during an ABS stop. These systems had other disadvantages:

❏ Expensive to build and service.
❏ Specialized tools often required to perform even the simplest repairs.
❏ Some integral systems had very limited or no self-diagnostics.

❑ High pressures made system work potentially dangerous for technicians.

❑ Likelihood of an ABS malfunction affecting base braking greater.

❑ Most systems did not have traction control capability.

However, the high operating pressures of integral systems often made bleeding an easy or in some cases, an unnecessary task.

Figure 21-5. *An integral anti-lock brake system as installed in the vehicle. (General Motors)*

Non-Integral Hydraulic Systems

In the late 1980s, manufacturers began to install **non-integral hydraulic systems** rather than integral systems. These systems are used with a conventional brake master cylinder and power booster, **Figure 21-6.** Often, they take the place of the proportioning/metering valve. Some non-integral systems are connected directly to the master cylinder. These systems have many advantages over integral systems, including:

❑ Lower cost when compared to integral systems.

❑ Modern bi-directional self-diagnostics similar to powertrain diagnostics.

❑ Fewer specialized tools required for service.

❑ Lower operating pressures minimize brake pedal feedback, noise, and hazard to technicians.

❑ Less likelihood of an ABS malfunction affecting base braking.

❑ Provisions for traction control.

Bleeding most non-integral systems is similar to bleeding a base brake hydraulic part, such as a proportioning valve. However, some non-integral systems require a special bleed procedure.

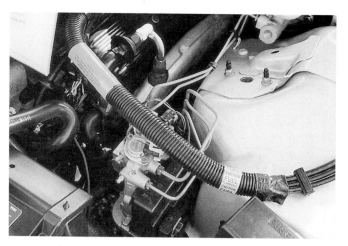

Figure 21-6. *A non-integral ABS system. Non-integral systems can be located almost anywhere in the engine compartment or on the vehicle.*

ABS/TCS System Configurations

The number of wheel speed sensors and hydraulic controls an ABS system uses is referred to as channels. A **channel** is a single system of ABS sensor input and hydraulic output. There are a maximum of four channels, one for each wheel. One-, two-, and three-channel systems are also used.

One- and Two-Channel Systems

The one-channel system was used on pickup trucks equipped with RWAL, **Figure 21-7.** One-channel ABS consists of a single sensor mounted in the transmission or differential. Two channels were used with some of the first anti-lock brake systems. They consisted of two sensors, one at each rear wheel. The input to the control module allowed individual control of each wheel.

Three- and Four-Channel Systems

Many cars and trucks made in the last 10 years are equipped with three-channel systems, **Figure 21-8.** Three-channel systems control each of the front wheels individually, while using one circuit to control both rear wheels. In effect, it works like a combination one- and two-channel system, with the two-channel system used on the front wheels. Most modern vehicles use four-channel systems, that allow for individual wheel control, **Figure 21-9.** Any rear-wheel drive vehicles with traction control and many front-wheel drive vehicles are equipped with four-channel systems.

ABS and TCS System Components

ABS and TCS systems use an electronic control system to monitor and modify the operation of the base brake hydraulic system as needed. The electronic and hydraulic components work together to prevent wheel lockup during

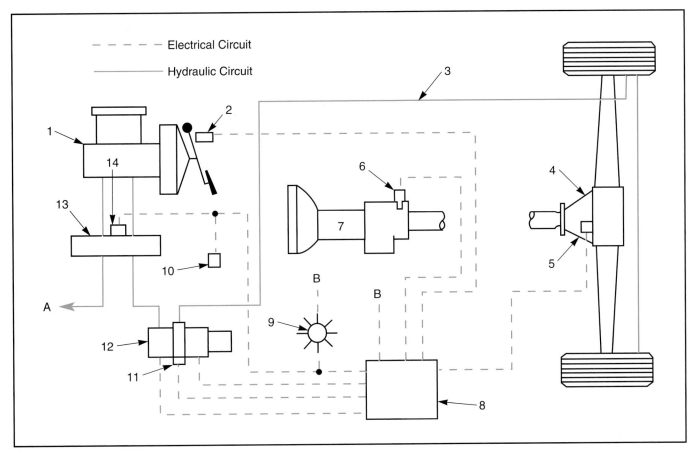

Figure 21-7. *A rear-wheel anti-lock brake system and related components. This system has one channel. A—To front brakes. B—To 12 volt ignition. 1—Master cylinder. 2—Stoplamp switch. 3—Rear brake circuit. 4—Rear differential. 5—Rear wheel speed sensor. 6—Four-wheel drive switch. 7—Transmission. 8—Electronic brake control module (EBCM). 9—Brake warning lamp. 10—Parking brake switch. 11—Reset switch. 12—Pressure limiting valve. 13—Proportioning/differential valve. 14—Differential switch. (General Motors)*

periods of hard braking or wheel spin during acceleration. This process uses control loops as was discussed in Chapter 20.

Anti-lock brake and traction control systems function by controlling the operation of the conventional brake system hydraulic and friction parts. Regardless of manufacturer, all ABS and TCS systems contain three common classes of components. These components include input sensors, the control module, and output devices.

Input Sensors

Input sensors provide information to the control module to allow it to make decisions and issue output commands. ABS and TCS systems both use wheel speed sensors as inputs to the control module. In addition, some ABS units use G-force sensors, brake pedal switches, or pressure switches as inputs.

Wheel Speed Sensors

To determine wheel rotation speed, **wheel speed sensors** are used. Most wheel speed sensor units consist of a toothed rotor or **tone wheel,** and a sensing unit, such as the one shown in **Figure 21-10.** The tone wheel can be

installed with the sensor inside a sealed wheel bearing, on the wheel hub, transfer case, CV joint housing, brake rotor, or inside the rear axle or transmission extension housing. Typical locations of wheel speed sensors are shown in **Figure 21-11.**

The wheel speed sensor, contains a small permanent magnet, which in operation, acts like a small electrical generator. As the tone wheel spins, it cuts through the sensor's magnetic field. The sensor converts the magnetic field into an ac (alternating current) voltage signal. **Figure 21-12** shows how the field is created and how the variations in field strength are converted into a sine wave ac signal. The frequency (rate of occurrence) of this signal varies in relation to the speed of the wheel. The control module reads this signal and converts it to a wheel rotation rate.

G-Force Sensor/Lateral Acceleration Sensor

Some ABS and traction control systems contain a **G-force sensor** which measures the rate of deceleration by comparing vehicle tilt during braking to the normal ride position. Most G-force sensors are mercury switches, such as the one shown in **Figure 21-13.** Acceleration and deceleration cause the mercury in the switch to

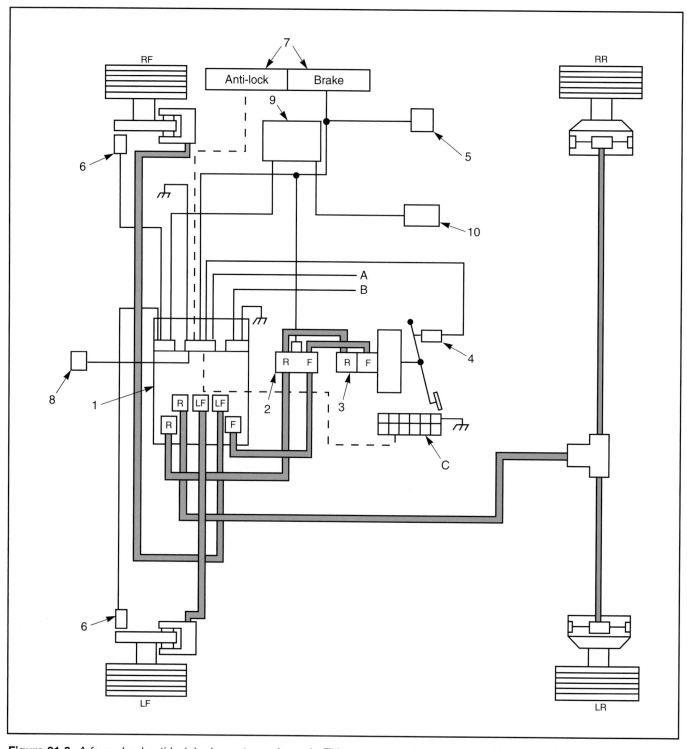

Figure 21-8. *A four-wheel anti-lock brake system schematic. This system has three channels for brake fluid to flow. A—To ignition switch (B+). B—To battery (B+). C—Data link connector (DLC). 1—Brake pressure modulator valve. 2—Combination valve. 3—Master cylinder. 4—Brake pedal switch. 5—Parking brake switch. 6—Wheel speed sensor. 7—Warning lamps. 8—Four-wheel drive switch. 9—Vehicle speed sensor (VSS) buffer. 10—Vehicle speed sensor (in transmission). (General Motors)*

flow between two electrical contacts. This completes the electrical circuit and signals the control module that a certain rate of acceleration or deceleration has been exceeded.

Some G-force sensors consist of a weight and a piezoelectric crystal. The piezoelectric crystal is a semiconductor device which produces an electrical signal when pressure is applied. This sensor is installed so the weight pushes on the piezoelectric crystal when the vehicle is braking. The resulting pressure produces an electrical signal which is sent to the control module. The control module compares the deceleration rate with the wheel rotation rates to effectively control brake application.

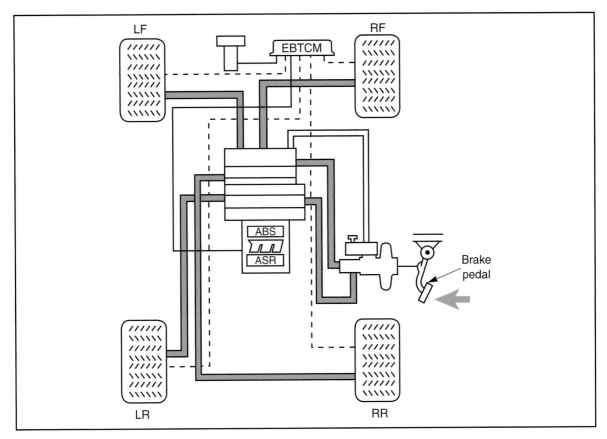

Figure 21-9. *Four-wheel anti-lock brake system with four channels. (General Motors)*

Figure 21-10. *A front wheel speed sensor assembly. The toothed rotor is an integral part of the CV joint.*

G-force sensors are usually installed in the passenger compartment to reduce the effect of cornering forces and body dive, **Figure 21-14.** Some ABS designs use a variation of the G-force sensor called a ***lateral acceleration sensor.*** The lateral acceleration sensor is used to sense cornering speed. These units contain two mercury switches, one for each side direction. G-force sensors are sealed units, and must be replaced if defective.

Some G-force and lateral acceleration sensors are also part of a ***handling control system.*** The handling control system uses the G-force and lateral acceleration sensors to ensure the vehicle does not spin out of control during hard braking or acceleration. The handling control system is usually integrated into the ABS/TCS system.

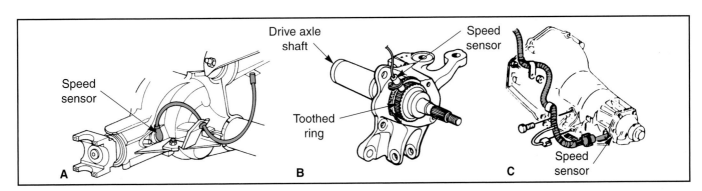

Figure 21-11. *Several common wheel speed sensor locations. A—Inside the rear axle housing. B—At the wheel. C—Inside a transmission extension housing. (Wagner)*

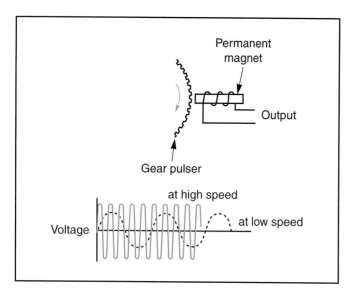

Figure 21-12. *Wheel speed sensor assembly produces an alternating current (voltage) signal. (Honda)*

Brake Light Switch

ABS and TCS control modules monitor the **brake light switch** to detect when the driver depresses the brake pedal. The signal from the switch will reset the ABS control system if the driver pumps the brakes during an ABS stop. The TCS system on many vehicles will be disabled if the brake pedal is depressed. In many cases, the brake light switch resembles the button switches mounted on non-ABS equipped vehicles. These switches may have an additional terminal for the ABS plug-in connector. Switch adjustment is crucial and is one of the most common causes of ABS and TCS problems.

Pedal Travel Switch

On some vehicles with ABS systems, a **pedal travel switch** is installed on the brake pedal assembly. **Figure 21-15** shows a pedal travel switch mounted on the brake pedal assembly frame. The switch is able to measure pedal travel and determine how much pedal pulsation (up and down movement) is occurring per second. This information is sent to the control module. The switch has no function when the ABS is not in operation, even when the brake pedal is depressed.

When the ABS system is operating, the pedal travel switch signal alerts the control module when pedal pulsation becomes excessive. The control module modifies fluid flow through the hydraulic actuator to reduce pulsation. Most pedal travel switches have a provision for

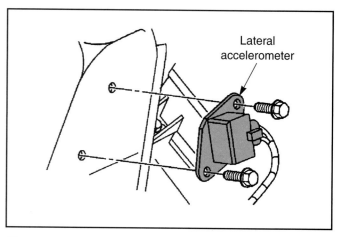

Figure 21-14. *A piezoelectric G-force type sensor. Handle with care. (Cadillac)*

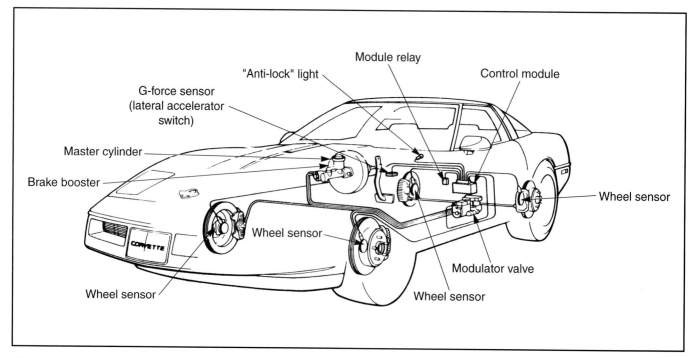

Figure 21-13. *A vehicle which uses anti-lock brakes and traction control. Two G-force sensors (mercury switches) are used to help maintain vehicle control. These units are not serviceable. Use care when handling, as they are easily damaged. (General Motors)*

adjustment. They may require adjustment when the brake pedal height is changed by the installation of a new master cylinder or other brake pedal component. Manufacturers' procedures must be followed when adjusting this switch.

Note: This switch is part of the brake light switch assembly on many vehicles. Many newer ABS systems do not use a travel switch.

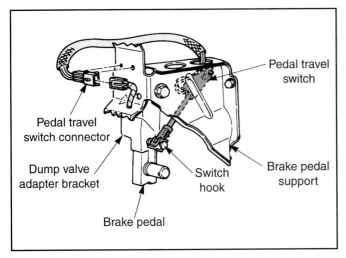

Figure 21-15. *A brake pedal travel switch assembly. (Ford)*

Fluid Level Sensor

The control module also monitors the fluid level in the hydraulic system. It often uses the same type of sensor usually mounted in the proportioning valve of a non-ABS vehicle. If the control module detects a low fluid condition, it disables the ABS system and illuminates the amber ABS dashlight, along with the red Brake dash light.

Pressure Switches

Most ABS and TCS systems have one or more **pressure switches.** A typical switch location is shown in **Figure 21-16.** The function of these switches varies by system and manufacturer. Some switches prevent overpressurizing the system, while others send a message to the control module that system pressures are excessively low. Other switches directly energize the hydraulic pump motor when system pressure drops too low. Some switches are designed to illuminate the warning light when a low pressure condition is detected. On some systems, the pressure switch may perform several of these functions.

Control Module

The **control module** is a computer that controls the operation of the ABS and/or TCS system, based on sensor inputs. All ABS and TCS control modules are computers,

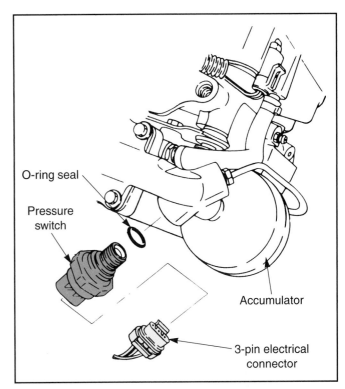

Figure 21-16. *One particular electrically operated pressure switch used on an integral ABS assembly. (Pontiac)*

similar to those that control the engine and other vehicle systems. Names for the ABS or TCS control module include:

❑ *Electronic Brake Control Module (EBCM).*
❑ *Electronic Brake and Traction Control Module (EBTCM).*
❑ *ABS/TCS controller.*
❑ *Body Control Module (BCM).*

Most modules are usually installed in the vehicle passenger compartment, trunk, or other location away from engine heat and vibration. **Figure 21-17** shows a system with the module installed in the trunk. Some modules are part of the hydraulic actuator and all components are installed under the hood at or near the master cylinder, as shown in **Figure 21-18.** On some newer vehicles, the ABS/TCS control functions are built into the powertrain control computer, which is referred to as a **vehicle control module (VCM), Figure 21-19.**

Serial Data

On newer vehicles, the control module receives and transmits **serial data** to other on-board computers. Serial data is simply sensor and actuator information that is shared by the various vehicle on-board computers. By sharing data, it eliminates the need for duplicate sensors for each system. This information is transmitted over wiring referred to as a *data bus.* In the near future, vehicles may use fiber optics as the data bus.

ECM External Communications

There are two types of external serial communication used in automotive computer systems. Most modern vehicles

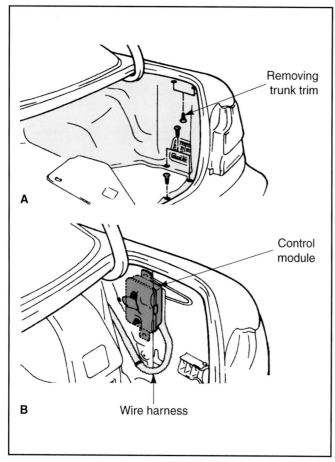

Figure 21-17. *A control module mounted in the car trunk. A—Access to the module requires removing some of the trunk trim. B—Trim has been removed, exposing the control module and its wiring harness. (Hyundai)*

use a system called ***universal asynchronous receive and transmit (UART).*** This signal is used for communication between the ECM, off-board diagnostic equipment, and other control modules. UART is a data line that varies voltage between 0-5 volts at a fixed pulse width rate. Some of the newest vehicles use UART, but also depend on the use of ***Class 2 serial communications.*** Class 2 data is transferred by toggling the line voltage from 0-7 volts, with 0 being the rest voltage, and by varying the pulse width. The variable pulse width and higher voltage allows Class 2 data communications to better utilize the data bus, **Figure 21-20.**

Control Loop

The ABS control module is part of a ***control loop,*** explained in Chapter 20. In a functioning control loop, the input sensors furnish information to the control module, which makes decisions and sends commands to the output devices. The operation of the output devices affects brake operation, causing changes in the readings furnished by the input sensors. A continuous loop of information, from the input sensors, to the control module, to the output devices, to the brake system, and back to the input sensors is constantly in operation. Depending on the system design, the control module may have as many as four different control loops operating at each time, depending on the number of wheel speed sensors and the design of the hydraulic actuator.

The control module also monitors wheel speed, and will not operate the ABS system under a certain minimum speed. On most systems, this minimum speed is between 6-10 mph (8-14 kph). This prevents ABS operation at speeds where it is not really needed.

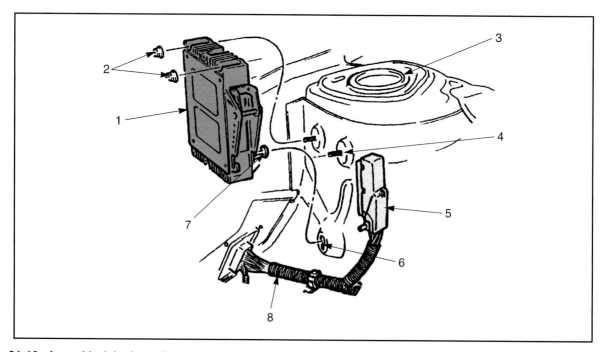

Figure 21-18. *An anti-lock brake and traction control module which is installed in the engine compartment and is bolted to the left-hand suspension strut tower. 1—Module. 2—Nuts. 3—Left-hand strut tower. 4—Studs. 5—Module connector. 6—Lating slot. 7—Locating tab. 8—Wire harness. (Pontiac)*

ABS Self-Test

Most ABS systems will perform a short **self-test** as the vehicle begins to move. This tests normally consists of cycling the valves in the hydraulic actuator for a period of 1-2 seconds. This self-test occurs when a vehicle equipped with ABS reaches approximately 3-9 mph (4-12 kph) after start-up. The purpose of this test is to ensure the hydraulic actuator, as well as its pump and relays are functioning properly. The driver may hear a slightly audible buzz or clicking; this noise is normal and is similar to the noise the actuator will make during an ABS stop. Not all ABS systems perform this self-test.

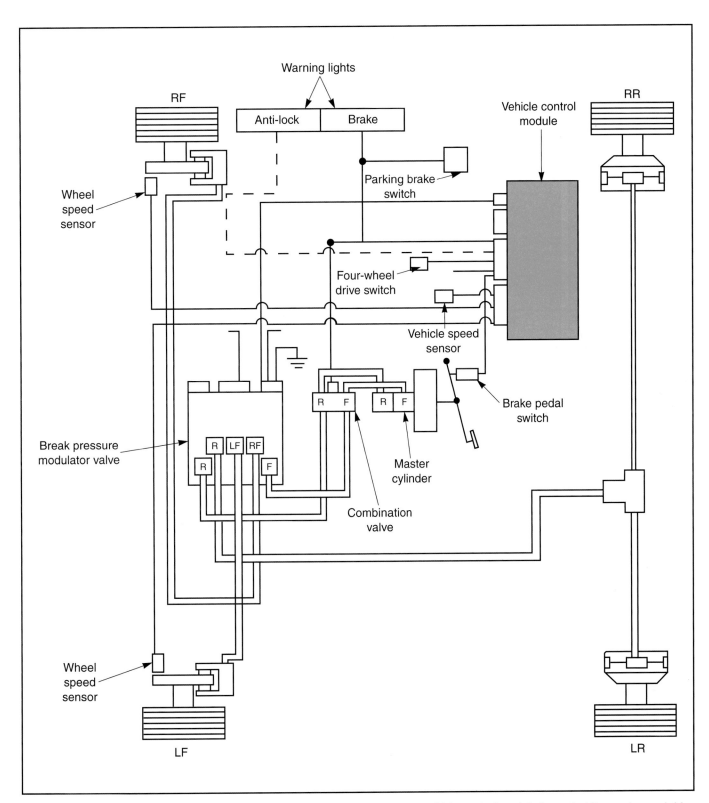

Figure 21-19. *A schematic of a four-wheel anti-lock brake system that uses a vehicle control module to control the engine and drive train (brakes). This is referred to as the vehicle control module/four-wheel anti-lock brake system (VCM4WAL). (Chevrolet)*

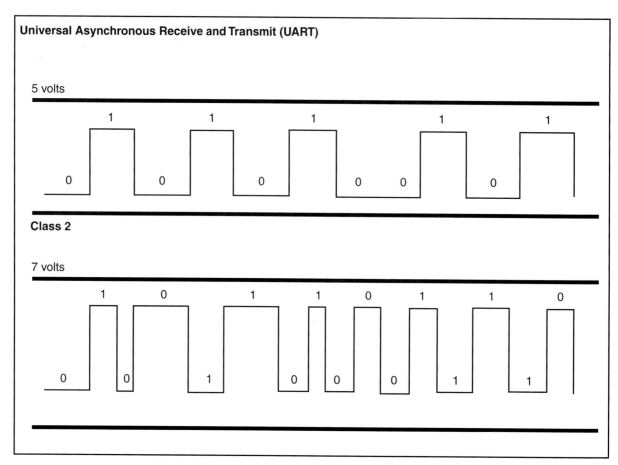

Figure 21-20. *Computers communicate using one of two forms of data transmission. Older systems use UART, while newer vehicles use Class 2.*

Electrical Inputs to the Control Module

The control module receives electric power through more than one circuit. Typically, there are two separate circuits. One circuit is powered at all times. This circuit allows the control module to maintain any information in RAM, such as operating data and trouble codes. The other circuit is energized through the ignition switch. When this circuit is energized, the control module can operate the hydraulic actuator and illuminate dashboard lights. Both circuits are wired through fuses or fusible links. Some ABS/TCS systems have as many as four separate electrical inputs to the control module.

ABS/TCS control modules monitor battery voltage to ensure proper operation and to prevent system damage. Most ABS/TCS systems will be disabled automatically and set a diagnostic code if system voltage drops below 9.5 volts. It will also disable the ABS system to prevent damage and set a code if battery voltage rises above 14.5-15 volts.

ABS Diode

Some systems contain **diodes** in the wiring harness to reduce back electrical charges from the red Brake warning light when it is illuminated by application of the parking brake. This prevents Brake warning light voltage from causing the control module to falsely illuminate the ABS warning light and disabling the anti-lock brake system.

Output Devices

Output devices consist of the hydraulic actuators and dashboard mounted information and warning lights. On traction control systems, some outputs to the electronic control module reduce engine power when necessary. Hydraulic actuators regulate the pressure delivered to the brakes, based on commands from the control module. Dashboard lights indicate system operating modes and/or system problems. There are two major types of hydraulic actuators and several kinds of dashboard lights and engine torque controls.

Electrical Relays

Hydraulic actuators are often controlled through electrical **relays.** This is done so the total amount of current needed to operate the hydraulic module devices does not flow through the control module. Instead, the control module uses a small current to energize the relay, which then directs heavier current to the hydraulic actuator electrical components.

Many relays are electromechanical devices in which the small current from the control module creates a magnetic field that causes a set of contact points to close. Some relays are electronic, using transistors in place of mechanical points. These relays may be called *system modules* or *drivers*.

Hydraulic Actuators

As you have learned in earlier chapters, pushing the brake pedal creates pressure in the master cylinder. This pressure is transmitted through lines and hoses to the brake caliper pistons and/or wheel cylinders to apply the brake friction members. On all ABS designs, the same basic hydraulic system is used. However, the ABS adds additional hydraulic components. These components do two things:

❏ They maintain, increase, and reduce brake pressure to the wheel hydraulic units.

❏ They develop additional pressure in the system when needed.

The *hydraulic actuator* reduces pressure to individual wheels that are beginning to lock up. This is done by bleeding fluid, and therefore pressure, from the affected brake line(s). The hydraulic actuator can increase and also create additional pressure when the original pressure drops too much. Therefore, the purpose of the hydraulic actuator can be described as that of controlling and creating hydraulic pressure in the brake system. The two main types of hydraulic actuators are discussed later.

On most vehicles, power brake boosters are separate units operated by engine vacuum or power steering pressure. Therefore, the additional braking power supplied by these units is independent of the anti-lock brake system. Most ABS systems use an electric pump which provides hydraulic pressure. A schematic of this system is shown in **Figure 21-21.**

Pump Operated Hydraulic Actuator

The *pump operated hydraulic actuator* consists of the solenoid valves, a hydraulic pump, an accumulator, and various tubing connections and electrical connectors. Most modern hydraulic actuators have all components contained in a single unit, as shown in **Figure 21-22.** On integral systems, they are combined with the master cylinder. Some older systems have separate actuator components connected by high pressure hoses. The hydraulic actuator on a two-wheel system has two solenoid valves. Most four-wheel ABS hydraulic actuators have four solenoid valves. Some four-wheel ABS actuators have three solenoid valves. There is one solenoid for each front wheel, and a single solenoid for the rear wheels.

Hydraulic Pump, Pressure Switch, and Relay

The electrically operated **hydraulic pump** is usually a one-piece assembly. It is controlled through a relay operated by a pressure switch. Hydraulic pumps can produce thousands of pounds or kilopascals of pressure, and may draw 20 amps or more of power. As an additional safety factor, the pump motor is usually operated by a pressure switch and relay that are separate from the ABS controller. This design allows the pump to provide power assist, even when another component of the ABS system has a defect, **Figure 21-23.**

The **pressure switch** closes when the accumulator pressure falls below a set value. When the pressure switch closes, current flows to the pump relay, energizing internal

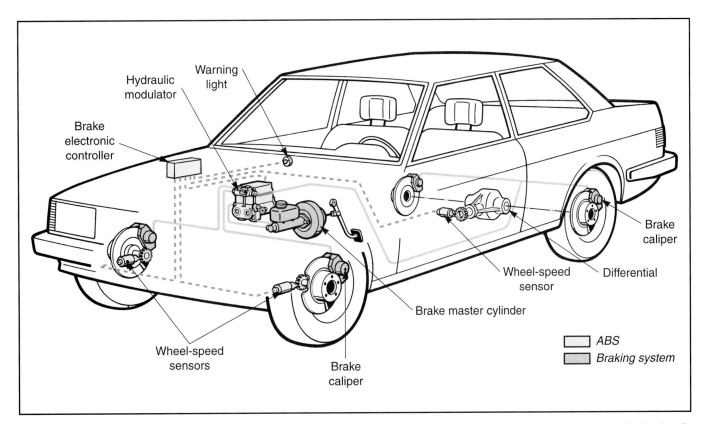

Figure 21-21. *A schematic of a four-wheel anti-lock brake system which uses an electric pump to supply pressure to the hydraulic actuator. Note that the rear wheel speed sensor is located on the rear differential assembly. (Bosch)*

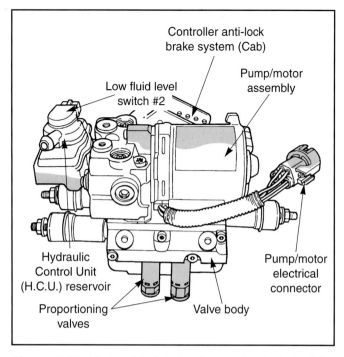

Figure 21-22. *Most anti-lock hydraulic actuators include all the parts. Make sure you inspect these whenever the actuator is replaced. (Dodge)*

the relay coil. When the coil is energized, the relay closes a set of contact points, which sends battery current into the pump motor. The pump motor runs until pressure reaches preset values, and the pressure switch de-energizes the pump relay.

Accumulators

Most hydraulic actuators also contain an **accumulator, Figure 21-24.** The accumulator absorbs extra fluid when the actuator valves are bleeding off pressure. The accumulator also holds fluid pressure in reserve to allow for brake operation if the foot pressure and/or hydraulic pump is unable to keep up with system pressure needs.

Piston Operated Hydraulic Modulator

The **piston operated hydraulic modulator** does not have a motor driven hydraulic pump. Movable pistons vary the pressure. The pistons are operated by small electric motors through reduction gears. Refer to the cutaway view of the actuator in **Figure 21-25.** Unlike the pump operated actuator, this system is referred to as a **hydraulic modulator.** The hydraulic modulator receives electrical signals from the control module. Although the modulator assembly regulates the pressure delivered to the brakes,

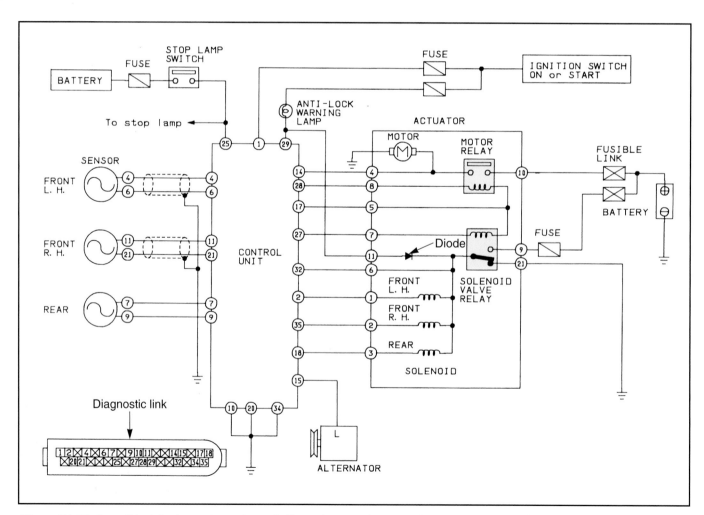

Figure 21-23. *An anti-lock brake system schematic illustrating two electromechanical relays. (Nissan)*

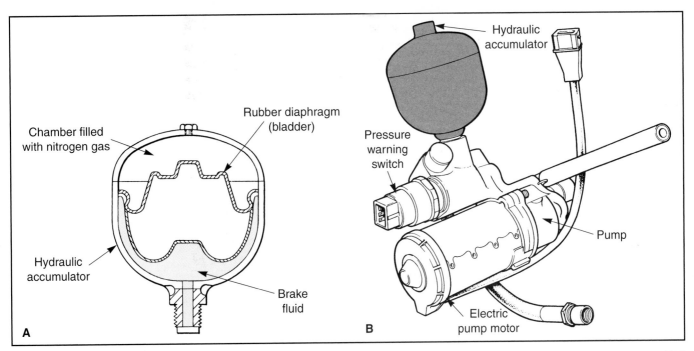

Figure 21-24. *A—A cross-sectional view of an accumulator assembly. B—Accumulator mounting position on one particular ABS motor and pump assembly. Accumulator styles will vary, but they perform the same basic functions. (Jaguar)*

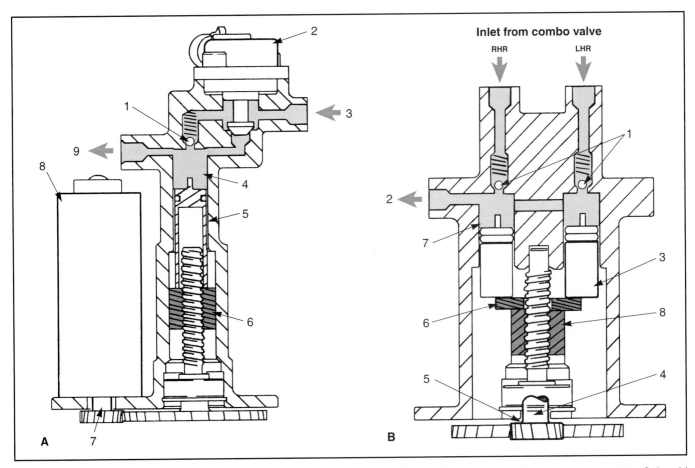

Figure 21-25. *Piston operated, motor driven hydraulic actuator. A—Front ABS braking mode. 1—Check valve closed. 2—Solenoid valve closed 3—Inlet from combo valve. 4—Modulation chamber. 5—Piston in "modulation" position. 6—Ball screw in "modulation" position. 7—Expansion spring brake (ESB). 8—Front modulator motor. 9—Modulated ABS pressure to front brake. B—Rear ABS braking mode. 1—Check valves closed. 2—Modulated pressure to rear brakes. 3—Piston lowered to reduce pressure. 4—Expansion spring brake (ESB) location. 5—Motor pinion. 6—Yoke on ball screw drives both rear circuit pistons. 7—Modulation chamber. 8—Ball screw. (General Motors)*

each solenoid has only one function: to isolate, or seal the passage between the master cylinder and the affected wheel cylinder or caliper. Each front wheel piston has a separate ball screw, while the rear pistons are both operated by a single ball screw.

If additional pressure is needed during ABS operation, the pistons are moved up or down in their bores to increase or reduce the hydraulic pressure to a particular brake line. Motor-driven screw assemblies called **ball screws** operate the pistons. As the motor turns, the ball screw is moved up or down on the threaded rods. This operation is similar to a bench vise, where the movable jaw travels in or out by turning a threaded rod. The motor pack in some systems uses an electromagnetic brake, **Figure 21-26A,** while others have an expansion spring brake to control motor action, **Figure 21-26B.**

The motor and ball screw assembly can drive the pistons forward to increase pressure, backward to reduce pressure, or hold the pistons in position to maintain pressure. The pistons also serve as accumulators for extra pressure. Piston position is precisely controlled by extremely close tolerances in the ball screws and reduction gears. Brakes on the drive motors cause them to instantly stop turning when current is turned off. The motor brakes can be electromagnetic or spring loaded types. A typical drive motor pack and hydraulic modulator assembly is shown in **Figure 21-27.**

Warning Lights

In addition to operating the hydraulic actuator, the control module operates various dashboard lights. The purpose of these lights is to indicate when the system is in operation, in self-diagnostics, or to warn the driver when the system develops a defect.

The control module will illuminate an amber warning light on the vehicle dashboard if there is a malfunction in the ABS system. The amber light also illuminates during and shortly after the engine starts. This is done to check the ABS electronic and hydraulic components, bulb, and wiring. On some systems, the light is used to retrieve trouble codes from the control module. Some systems use a **lamp driver module** that is separate from the control module.

On some traction control systems, a green indicator lamp alerts the driver when the traction control system is operating. Some traction control systems also have a separate warning light to warn of a traction control system malfunction.

The ABS or traction control lights should not be confused with the brake system warning light. ABS lights are amber and traction control lights are blue or green, and indicate a problem with these systems and not with the base brakes. On some vehicles, the light is designed to spell out Brake, Traction, or ABS, or show a symbol indicating which system it monitors.

The Brake warning light is always red, and warns of a pressure loss in one side of the dual brake system. Some red Brake warning lights are also connected to a fluid level switch

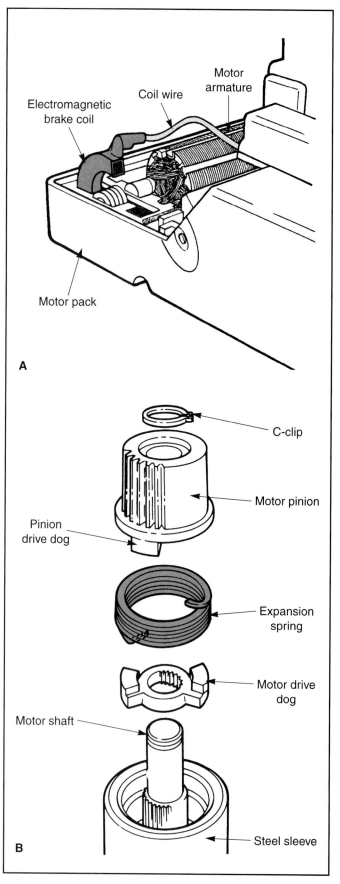

Figure 21-26. A— A cutaway view of a front motor pack showing the electromagnetic brake (EMB). These brakes are used to hold the front piston in the resting or home position. B—An exploded view of a expansion spring brake (ESB). (General Motors)

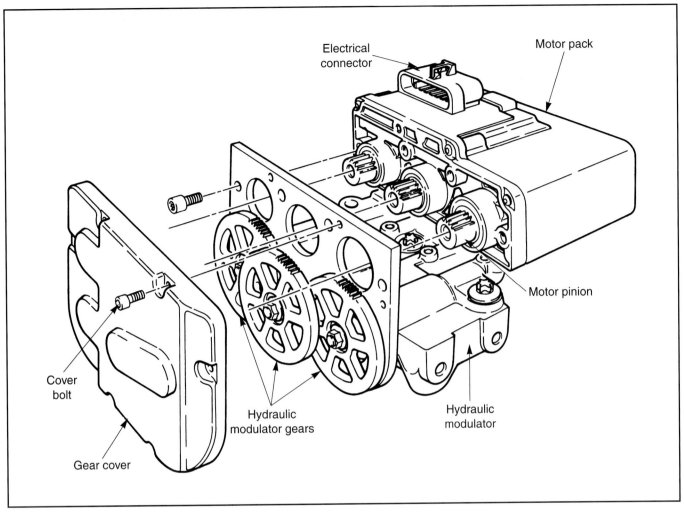

Figure 21-27. *A partially exploded view of a electric motor pack and the hydraulic modulator assembly. (General Motors)*

in the master cylinder reservoir, a brake pad wear switch, or to a switch installed on the parking brake. Typical system lights are shown in **Figure 21-28.** On many later vehicles, the control module may operate all brake related warning lights

and illuminate the red Brake light when an ABS malfunction occurs. Some early rear-wheel ABS systems use the red Brake light to indicate the presence of a problem.

Data Link Connector

The use of on-board diagnostics simplifies the ABS troubleshooting process. When a system defect occurs, the control module saves information about the defect in memory. This information takes the form of a *trouble code*. Trouble codes can be accessed using the same methods used to retrieve powertrain trouble codes. In most cases, the same data link connector is used to retrieve ABS information. Trouble codes on older vehicles were designated by each vehicle manufacturer, and in most cases, were not the same between different manufacturers.

To determine whether any trouble codes are present, the technician must cause the module to enter the *self-diagnostic* mode. In the self-diagnostic mode, trouble codes can be retrieved by one of two methods: observing the sequence of dashboard light flashes; or attaching a scan tool to the system's data link connector and reading the codes on the tool screen. On vehicles built before 1996, trouble codes are 2- or 3-digit numbers which correspond to a specific defect.

Figure 21-28. *Typical symbols and words used on an instrument cluster for the anti-lock brake system (ABS), and traction control lights (TRAC). (Lexus)*

If the vehicle was made after 1995, the vehicle will be equipped with the OBD II *(on-board diagnostics generation two)* diagnostic system. In the OBD II diagnostic system, the trouble code format has been standardized by the **Society of Automotive Engineers (SAE),** and is in alphanumeric form consisting of one letter and four numbers. All OBD II systems use the same 16-pin data link connector, **Figure 21-29.**

A universal scan tool must be used to retrieve ABS trouble codes from any vehicle using the OBD II system. Many tool companies make scan tools capable of retrieving trouble codes from more than one make of vehicle. Universal scan tools are available from various tool suppliers. Code retrieval and other diagnosis procedures will be discussed in more detail in Chapter 22.

System Operation

The following is a brief review of overall system operation during ABS and TCS modes. Note that during normal operation, the wheel speed sensors are generating a signal. However, the control module does not respond unless a severe braking or traction loss situation occurs.

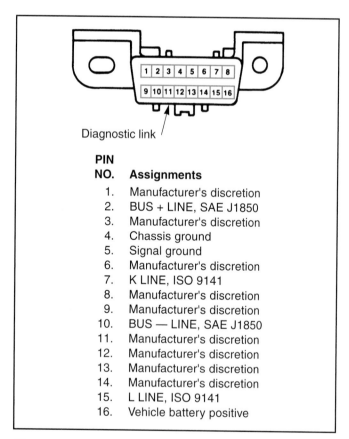

Diagnostic link

PIN
NO.   Assignments
1.   Manufacturer's discretion
2.   BUS + LINE, SAE J1850
3.   Manufacturer's discretion
4.   Chassis ground
5.   Signal ground
6.   Manufacturer's discretion
7.   K LINE, ISO 9141
8.   Manufacturer's discretion
9.   Manufacturer's discretion
10. BUS — LINE, SAE J1850
11. Manufacturer's discretion
12. Manufacturer's discretion
13. Manufacturer's discretion
14. Manufacturer's discretion
15. L LINE, ISO 9141
16. Vehicle battery positive

Figure 21-29. *A frontal view of one particular 16-pin data link connector (DLC), as used with the OBD II diagnostic system. These test links are generally located in the passenger compartment under the instrument panel on the driver's side. The connector is somewhat "D-shaped" and keyed so the data link and corresponding scan tool adapter can only mate in one direction. Note the pin numbers and their assignments. (Snap-On)*

Hydraulic Control during ABS Braking

Any time the brake pedal is applied, the control module interprets signals from the wheel speed sensors. The actuator controls the hydraulic system by positioning solenoids in three modes, according to the presence and severity of a lockup condition.

Pressure Maintain

When the wheel speed sensors indicate that one or more wheels are nearing lockup condition, the control module signals the appropriate solenoid valve(s) in the hydraulic actuator. The solenoids are first positioned to seal the passage between the master cylinder and brake lines to prevent additional fluid pressure from reaching the affected wheel cylinder or caliper, **Figure 21-30.** Pressure does not increase, but remains at a constant level.

Pressure Decrease

If the control module receives input from the wheel speed sensor indicating severe or complete wheel lockup, it commands the hydraulic actuator to position the appropriate solenoid valve(s) to bleed off pressure to the affected caliper(s) and/or wheel cylinder(s). Brake fluid from the affected wheel(s) flows into the accumulator, reservoir, or pump intake, depending on the design of the system. This is shown in **Figure 21-31.** This fluid removal reduces, or bleeds off, hydraulic pressure going to the individual wheel cylinder or caliper piston. If the hydraulic modulator is a piston type, the solenoid valves isolate the circuit from the master cylinder, and the internal pistons are moved to reduce pressure.

Each solenoid operates independently, and hydraulic pressure to one wheel may be bled off to a low value while another is receiving full pressure. The operation of each solenoid is based on wheel sensor inputs to the control module. The control module uses these inputs to decide how much pressure should be applied to each wheel.

Pressure Increase

When the wheel speeds up, pressure is increased, either by supplying pressure from the accumulator or moving the pistons to increase pressure on the piston. This allows the brake system to continue the process of slowing the vehicle. This is shown in **Figure 21-32.** If the control module senses excessive wheel lockup after pressure increase has begun, it will move the appropriate solenoids to first maintain pressure and then, if needed, reduce pressure.

During a normal ABS-assisted stop, hydraulic system pressure is constantly maintained, reduced, and increased to provide maximum braking without wheel lock. If the ABS is engaged for long periods, the original pedal pressure may not be sufficient to apply the brakes. The pump built into many hydraulic actuators provides makeup pressure to the hydraulic system.

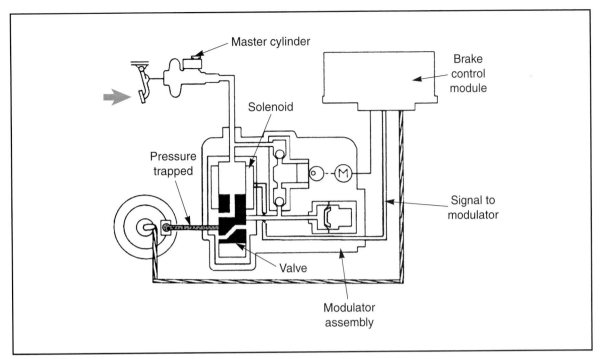

Figure 21-30. *A schematic of a four-wheel, anti-lock brake system during operation which shows system operation when wheel lockup is about to occur. The solenoid valve has been positioned to trap hydraulic pressure traveling to the affected wheel. This will prevent a pressure increase, which would cause the wheel to lock and skid. (General Motors)*

Systems with a hydraulic pump start to increase pressure whenever the pressure switch indicates that available pressure is dropping excessively. The intake on some hydraulic pumps is piped through the solenoids and is used to quickly draw off pressure when a particular wheel must be depressurized quickly. This process is controlled by the solenoids. This pressure is directed either to the hydraulic actuator circuits or the accumulator.

When all the vehicle's wheels are again rotating normally, the speed sensor input indicates to the control module that ABS operation is no longer needed. The control module then de-energizes all solenoid valves in the hydraulic actuator. When the solenoid valves return to their original position, the base brake system takes over. After the ABS is disengaged, the hydraulic actuator pump used on some systems reduces system pressure by diverting any excess fluid from the accumulator into the master cylinder.

This cycle is repeated over and over as long as the possibility of wheel slippage exists. The cycle can occur 15 times or more per second. The slight pulsation felt in the brake pedal during ABS operation is caused by the system raising and lowering hydraulic pressure at an extremely fast rate.

Hydraulic Control during TCS Assisted Acceleration

During acceleration, the traction control system monitors wheel speed. When vehicle acceleration does not result in loss of traction, the traction control system is inactive. If the control module detects tire slip on a drive wheel, it can reduce wheel slip by either reducing engine torque or directing the hydraulic actuator to apply the brake on the spinning wheel(s). Since the driver will not

Figure 21-31. *A schematic illustrating how the modulator assembly will help to prevent severe wheel lockup. The solenoid valve has been positioned to allow excessive fluid pressure to travel to the accumulator, where it will be stored briefly. (General Motors)*

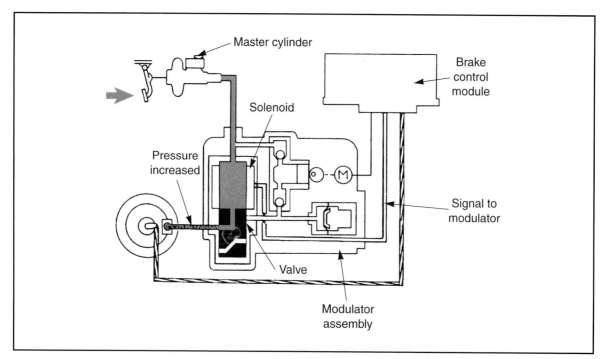

Figure 21-32. *Modulator position when pressure to the brake assembly is increased. (General Motors)*

normally be pressing on the brake pedal during accelera-tion, the pump or accumulator supplies hydraulic pressure to the spinning wheel(s). Traction control is usually active at vehicle speeds below 30 mph (48 kph).

Engine Torque Management

In some newer TCS-equipped vehicles, *engine torque management* is used first to reduce wheel spin before the TCS control module energizes the hydraulic actuator. On these traction control systems, the TCS control module works and in some cases, is part of the engine control com-puter. Several means are used to reduce engine torque. Some vehicles use more than one method. Not all vehicles use the same method:

❑ Enleaned air-fuel mixture.
❑ Retarding ignition timing.
❑ EGR valve open 100%.
❑ Disabling one or more cylinders.
❑ Reducing throttle angle.
❑ Upshifting transmission.
❑ Opens turbocharger wastegate or disables super-charger (if equipped).

The first step taken by the ECM in torque management is to enlean the air-fuel ratio. This is done by reducing the amount of fuel injected into the engine. In slight tire slip situations, this is often sufficient torque management to maintain traction. If the tires continue to slip, the ECM will retard engine timing next.

The ECM in most TCS-equipped vehicles retards igni-tion timing to reduce power. This is done by the computer through the ignition module in the same manner as during normal engine operation. Retarding the ignition timing reduces engine torque by reducing combustion efficiency. However, spark retard, cannot be decreased below top

dead center, and is limited to a few seconds per traction control episode.

Some ECMs will order the exhaust gas recirculation (EGR) valve 100% open to route excess exhaust gas back into the engine. A fully open EGR valve dilutes the engine air-fuel mixture with recirculated exhaust gas, thereby reducing engine torque. Like spark retard, this can only be done for a few seconds per traction control episode.

On some modern systems, the control computer causes the automatic transmission to upshift at lower than normal speeds. This reduces transmission torque multipli-cation, lowering power output to the drive wheels. ECMs on vehicles equipped with turbochargers may open the wastegate to reduce power. If the vehicle has a super-charger, the ECM may disable supercharger operation dur-ing periods of excessive wheel slip.

The last step in engine torque management on some engines is to selectively disable one or more cylinders. The ECM does not open the fuel injector and/or commands the ignition system not to fire on those cylinders. This system can disable up to half the engine cylinders at once.

On some engines, the ECM uses a throttle control sys-tem that has a stepper motor to close the throttle plate slightly, reducing airflow into the engine. This reduces engine torque output. This motor overrides the driver foot pressure when necessary, even when the driver is attempt-ing to open the throttle further. Systems which reduce throttle opening may also cause a reverse push on the accelerator pedal to inform the driver the traction control system is operating. Some motors use this stepper motor to move the throttle plate during normal and cruise control driving. Most vehicles now use the throttle control system, since retarding the ignition timing increases exhaust emis-sions and reduces fuel mileage.

Hydraulic Actuator Modes In Traction Control

The hydraulic actuator develops pressure using the hydraulic pump or pistons, and uses the solenoids in the hydraulic actuator to apply this pressure to the caliper or wheel cylinder at the wheel(s) that are slipping. As with the anti-lock brake system, the traction control can vary this pressure as much as 15 times per second.

When wheel speed sensor input indicates that slippage has been reduced or eliminated, the traction control system becomes inactive, and the control module stops activating the hydraulic actuator and pump. Hydraulic pressure on the affected wheel drops to zero and, if necessary, the control module signals the ECM to restore normal engine torque output.

Manufacturers of Anti-Lock Brake and Traction Control Systems

The various types of ABS and TCS systems are similar in design and operation, but a few differences can be found between manufacturers. The major differences involve:

❑ Location of the control module and hydraulic actuator in the vehicle.
❑ Interconnections with other vehicle computers.
❑ Use of peripheral devices such as brake pedal switches and G-force sensors.

These differences are important when diagnosing and servicing ABS systems, and will be covered in Chapter 22. In addition, you should *always* obtain and use the proper service manual for the system being serviced.

Common ABS and TCS Systems

There are a relatively small number of ABS/TCS system manufacturers. However, there are many different systems and vehicles on which ABS and TCS systems are used. Both vehicle manufacturers and parts suppliers make ABS and TCS systems. Many systems, such as those made by Bosch, Kelsey-Hayes, and Teves, are used on vehicles made by different manufacturers. This can cause confusion about what type of ABS and TCS system is used on a particular car or truck. The following is a brief description of the more common ABS and TCS systems, and where they are used. The best source of information is a manual and other service literature.

 Note: ABS manufacturers modify some actuators slightly to fit each vehicle manufacturer's application. The actuators shown here may appear slightly different from those you actually encounter in the vehicle.

Bendix 3

Type of system: Non-integral four-wheel. Also known as the Bendix LC4 (LC for "Low Cost").

This is a compact version of the Bendix 6. It is used on newer Chrysler cars and minivans. It can be identified by the presence of two accumulators and six bleed screws on the modulator. This system uses four channels to control each brake separately.

Bendix ABX-4

Type of system: Non-integral four-wheel.

Used on Chrysler's Neon, Cirrus, and Stratus models. Compact, four-channel system. The modulator is located below the master cylinder and can be identified by its lack of an accumulator, **Figure 21-33.** The master cylinder on cars equipped with this system have only two lines, while non-ABS models will have four lines.

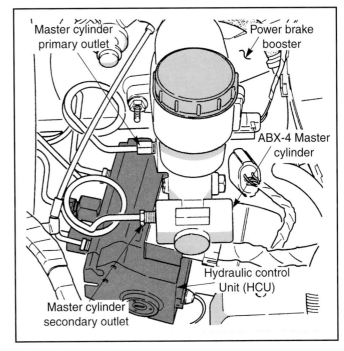

Figure 21-33. *Location of the Bendix ABX-4 hydraulic actuator in one vehicle. (Chrysler)*

Bendix 6

Type of system: Non-integral four-wheel.

Used on many small and midsize Chrysler automobiles and some Jeeps. Three-channel system that uses six ABS solenoids. It has dual accumulators like other non-integral Bendix systems. The hydraulic modulator is usually located on the left frame rail.

Bendix 9

Type of system: Integral four-wheel.

Used on the Jeep Cherokee and Wagoneer from the late 1980s to early 1990s. It can be identified by its separate pump and modulator assembly. The Bendix 9 is a three-channel system. Located under the back seat attached to the floor pan. This system uses two accumulators. Does not become functional until the vehicle reaches 12-15 mph (7.5-25 kph).

Bendix 10

Type of system: Integral four-wheel.

This system is used on Chrysler's larger cars and minivans. It is similar to the Bendix 9, with the only difference being the use of 10 ABS solenoid valves. This system has less pedal feedback than the Bendix 9.

Bendix Mecatronic II

Type of system: Non-integral four-wheel.

Used on some small Ford and Mercury cars. This system is located on the frame, below the master cylinder. The hydraulic modulator holds the pump, relays, and control module. This system uses two microprocessors to separately control the modulator and monitor vehicle speed and overall ABS operation. This system also has a traction control option.

Bosch 2, 2E, 2S, 2S Micro, and 2U

Type of system: Non-integral four-wheel.

Widely used on General Motors, Ford, Chrysler, European, and Asian vehicles. It has a one piece relay, motor, and hydraulic unit, usually installed under the master cylinder or on the frame. A three-channel system, it can be identified by four to five lines that exit the hydraulic unit at the side, **Figure 21-34.** System also has traction control option.

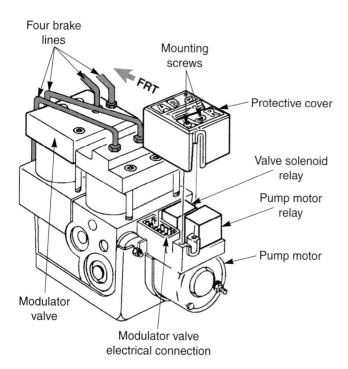

Four brake lines

Mounting screws

FRT

Protective cover

Valve solenoid relay

Pump motor relay

Pump motor

Modulator valve

Modulator valve electrical connection

Figure 21-34. *Overall view of a Bosch 2 ABS unit. Note the one piece relay, motor and hydraulic unit. Four hydraulic brake lines exit the top of the modulator valve assembly. (General Motors)*

Bosch 3

Type of system: Integral four-wheel.

Four-channel system used on some General Motors and Chrysler vehicles. The hydraulic modulator is built into the master cylinder and is connected to the pump by high pressure hoses. A special tester is required to diagnose this system. A variation of this system equipped with a separate traction control unit was used on the Cadillac Allanté in the early 1990s.

Bosch RWAL

Type of system: Non-integral rear-wheel.

Used on Chrysler imports and some Mitsubishi vehicles. It can be identified by the presence of a large separately mounted accumulator referred to as a modulator.

Bosch 5 ABS/ASR

Type of system: Non-integral four-wheel.

Used on Porsche, Chevrolet Corvette and General Motors luxury vehicles. This system has integral traction control. One of the first systems to use both engine torque management and braking to limit tire slip. Three-channel system.

Bosch 5.3 ABS

Type of system: Non-integral four-wheel.

Used in late-model General Motors, Toyota, and Subaru vehicles. Four-channel system with built-in traction control.

Bosch VDC

Type of system: Non-integral four-wheel.

Used on late-model Mercedes Benz cars. Four-channel system. Has traction control option.

Delco Moraine III

Type of system: Integral four-wheel. Also known as the Powermaster III.

Used exclusively on General Motors' W-body vehicles from the late 1980s to the early 1990s. This unit can be identified by its unusually large reservoir top, **Figure 21-35.** One of the first ABS systems produced by a vehicle manufacturer, it has no traction control function. This system was discontinued in favor of the Delphi Chassis VI.

Delphi Chassis VI

Type of system: Non-integral four-wheel. Also known as the Delco Moraine ABS-VI system.

Used on small to mid-size General Motors cars and minivans and can be identified by the solenoid electrical connectors on top of the control unit. Shown in **Figure 21-36**, this system has no hydraulic motor or accumulator. Operates at low pressure, resulting in almost no pedal feedback during ABS-assisted stops. Some systems built after 1994 have a separate traction control modulator.

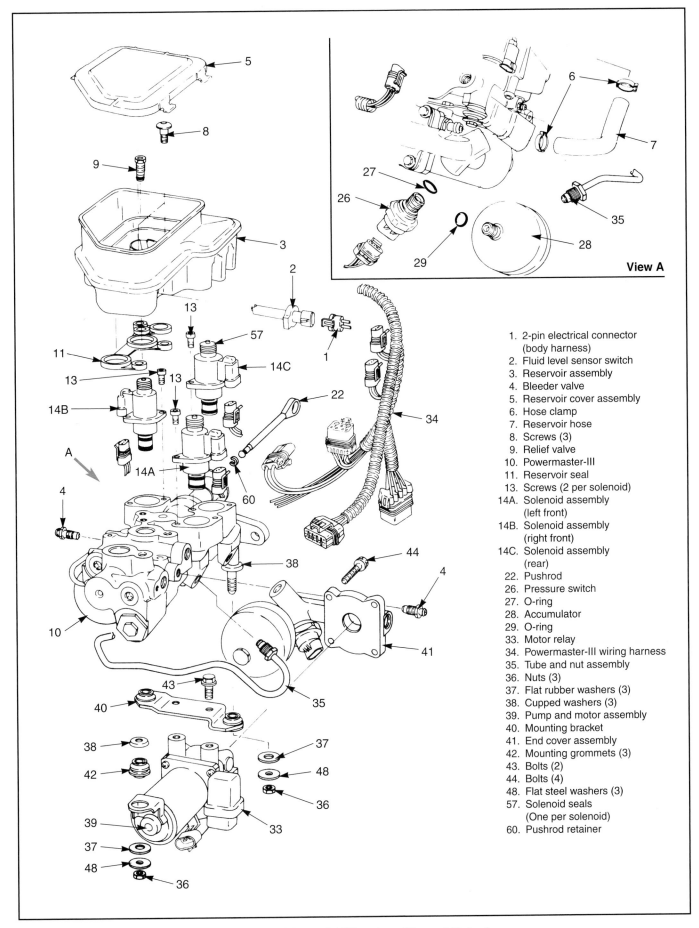

1. 2-pin electrical connector
   (body harness)
2. Fluid level sensor switch
3. Reservoir assembly
4. Bleeder valve
5. Reservoir cover assembly
6. Hose clamp
7. Reservoir hose
8. Screws (3)
9. Relief valve
10. Powermaster-III
11. Reservoir seal
13. Screws (2 per solenoid)
14A. Solenoid assembly
    (left front)
14B. Solenoid assembly
    (right front)
14C. Solenoid assembly
    (rear)
22. Pushrod
26. Pressure switch
27. O-ring
28. Accumulator
29. O-ring
33. Motor relay
34. Powermaster-III wiring harness
35. Tube and nut assembly
36. Nuts (3)
37. Flat rubber washers (3)
38. Cupped washers (3)
39. Pump and motor assembly
40. Mounting bracket
41. End cover assembly
42. Mounting grommets (3)
43. Bolts (2)
44. Bolts (4)
48. Flat steel washers (3)
57. Solenoid seals
    (One per solenoid)
60. Pushrod retainer

Figure 21-35. *Exploded view of a Delco Moraine III integral ABS system. (General Motors)*

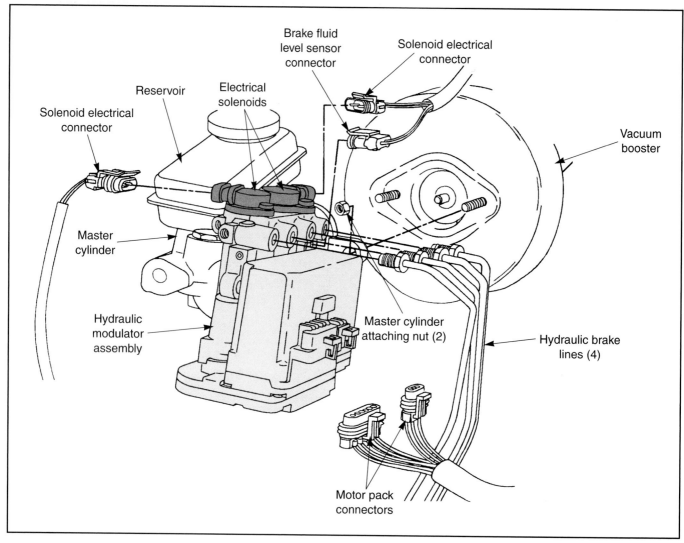

Figure 21-36. *Overall view of a Delphi Chassis VI ABS assembly. Note that it can be identified by the electric solenoids mounted on top of the hydraulic modulator unit. This system is also referred to as a Delco Moraine VI system. (Pontiac)*

DBC 7

Type of system: Non-integral four-wheel.

Built by Delphi and used on General Motors vehicles. Can be identified by its integral hydraulic actuator/EBCM assembly, **Figure 21-37.** Has internal accumulator, relays, and motor; very compact. Located on the lower frame rail, near the front of the engine compartment. Unit has optional traction control capability.

Honda ABS

Type of system: Non-integral four-wheel.

Used on Honda and Acura cars. Uses a conventional master cylinder, but has a separate reservoir tank on the hydraulic modulator, **Figure 21-38.** Also has a high pressure accumulator and hydraulic motor. It is a three-channel system.

Kelsey-Hayes RWAL

Type of system: Non-integral rear-wheel. Also known as EBC2, RWAL II, ZPRWAL, RABS, RABS II, and ZPRABS.

Figure 21-37. *Exploded view of a DBC 7 anti-lock brake system. (General Motors)*

Used on light trucks. All North American manufacturers and many Asian manufacturers used RWAL. It can be identified by the presence of a relatively small

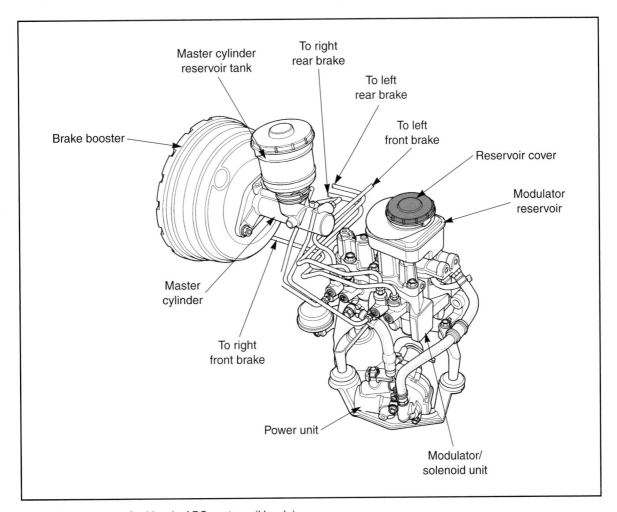

Figure 21-38. *Components of a Honda ABS system. (Honda)*

hydraulic unit with only two hydraulic lines, **Figure 21-39.** The hydraulic unit is mounted below the master cylinder or on the vehicle frame. Many manufacturers have discontinued using RWAL in favor of four-wheel systems. This system has no traction control ability or hydraulic pump and very limited self-diagnostics.

Kelsey-Hayes 4WAL

Type of system: Non-integral four-wheel. Also known as the EBC4.

Four-wheel ABS system used on some General Motors, Isuzu, and Kia trucks, vans, and sport utility vehicles. It is a large hydraulic unit with five hydraulic lines and two accumulators. The unit is mounted either on the bulkhead below the master cylinder or on the left side of the engine compartment, **Figure 21-40.** On most vehicles, the control module is built into the hydraulic unit. Some newer vehicles with the OBD II diagnostic system use a VCM to control the modulator. It has bidirectional self-diagnostics, however, it requires specialized tools and procedures for proper service. Does not have traction control features.

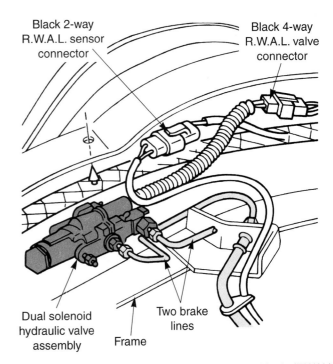

Figure 21-39. *A Kelsey-Hayes rear wheel anti-lock (RWAL) hydraulic unit, wiring harness, and the two brake lines. These units can also be mounted on a bracket located next to the master cylinder. (Dodge)*

Figure 21-40. *Kelsey-Hayes 4WAL anti-lock brake system. This system houses the actuator and control module as one unit.*

Kelsey-Hayes EBC5H and EBC5U

Type of system: Non-integral four-wheel.

Used on Ford and Chrysler trucks. The 5H system is installed on Chrysler trucks and is a standard Kelsey-Hayes RWAL system with a separate modulator for the front wheels. The 5U system is unitized and is installed on Fords truck. Both systems are three-channel and are mounted on the left side of their respective vehicles, at or near the frame rail.

Kelsey-Hayes EBC310

Type of system: Non-integral four-wheel.

Used on large General Motors trucks. It is a three-channel system that uses six ABS valves. This system is similar to the EBC5U system, however, the parts are not interchangeable. This system may have the combination valve bolted to the actuator.

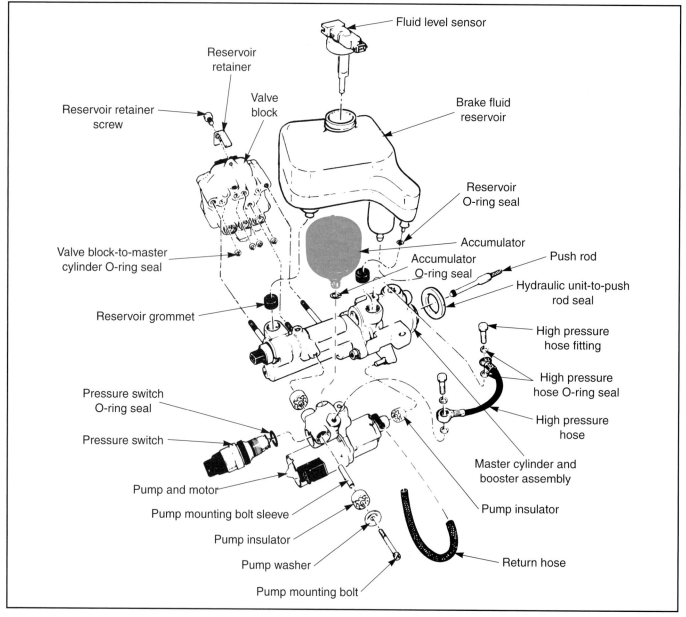

Figure 21-41. *A Teves II anti-lock brake (ABS) system. This exploded view shows the accumulator and other related components. Study carefully. (Wagner)*

Kelsey-Hayes EBC410

Type of system: Non-integral four-wheel.

Used on Ford minivans. It is similar to the EBC310 except that it uses eight ABS solenoids and has four-channels. This system is unique in the K-H line in that it has a separate control module. Has full bidirectional self-diagnostics.

Nippondenso ABS

Type of system: Non-integral four-wheel.

Used on some Infiniti, Toyota, Lexus, and Subaru models. Has self-diagnostics. Located on right front of the engine compartment on most models. On Lexus vehicles, this system is coupled with a separate Vehicle Skid Control (VSC) system.

Sumitomo ABS

Type of system: Non-integral four-wheel.

Used on some Mazda, Honda, and Ford vehicles. This unit can be identified by its separate fluid reservoir mounted on top of the hydraulic control unit. The actuator unit is located on the right front of most vehicles equipped with this system. Can be three- or four-channel system depending on the presence of traction control and the application.

Teves Mark II

Type of system: Integral four-wheel.

Used in Ford, General Motors, and European vehicles. One of the earliest integral systems installed on production vehicles. This system can be easily identified by the presence of a large accumulator at the top of the hydraulic unit, **Figure 21-41**. It has no TCS version and, depending on the year of production, very limited or no self-diagnostics. Early models require the use of a breakout box for in-depth diagnostics.

Teves Mark IV

Type of system: Non-integral four-wheel.

Used on many larger Ford, General Motors, Chrysler, and Jeep vehicles. The Mark IV can be identified by four hydraulic lines attached to the hydraulic control unit. The hydraulic unit is usually on the left side of the vehicle, mounted on the frame rail, **Figure 21-42**. This system has modern self-diagnostics, which can be accessed with a scan tool. Newer systems can be equipped with traction control.

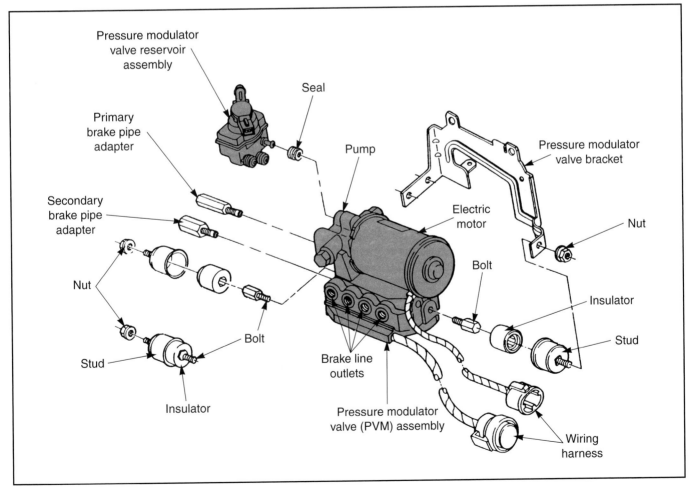

Figure 21-42. *A Teves Mark IV ABS pump and pressure modulator valve (PMV) assembly. Each individual unit can vary somewhat from year-to year, as improvements continue. (General Motors)*

Teves Mark 20

Type of system: Non-integral four-wheel.

Used on late-model BMW, Chrysler, Ford, and Honda vehicles. Very compact unit that has the pump, electric motor, valve body, and electronic controls all in one assembly. Has traction control option. Can be three- or four-channel, depending on whether traction control is installed.

Toyota ABS

Type of system: Non-integral rear-wheel.

Used on Toyota trucks and sport utility vehicles. This system is unique as it uses the power steering pump to generate additional brake hydraulic pressure to the rear wheels. The hydraulic actuator is located on the right side of the vehicle, near the power steering pump.

Summary

The purpose of the ABS system is to alternately apply and release, or pump, the brakes. It can do this at a much faster rate than the vehicle driver is able to do. Traction controls use the brake system and engine controls to reduce excessive wheel spin during acceleration. To do this, anti-lock brake and traction control systems use electronic and hydraulic components. The purpose of these devices is to increase vehicle control during braking and acceleration. Modern anti-lock brake systems used on most automobiles control all four wheels. Older systems, especially those used on small cars and light trucks, control the rear wheels only.

All ABS and traction control systems operate by controlling the brake hydraulic system. The basic components and operating principles of different brands of anti-lock brake and traction control systems are similar. These components include input devices, the control module, and output devices. On new vehicles, the anti-lock brake and traction control systems are combined, and are operated by a single electronic control module.

Common ABS and traction control inputs include wheel speed sensors, G-force sensors, and brake travel switches. The control module processes these inputs, comparing them to the information stored in its memory. The control module then issues commands to the output devices. The continuous interaction of the input sensors, control module, and output devices is called a control loop.

The most prominent output device is the hydraulic actuator. The actuator consists of electrically operated solenoids which control hydraulic fluid pressures when the ABS or traction control is operating. In addition, most hydraulic actuators have a hydraulic pump and accumulator. All anti-lock brake and traction control systems have amber or green dashboard lights, in addition to the red Brake warning light used on all vehicles. The control

module may operate all of these lights. The traction control is interconnected to the engine control computer. When necessary, the engine control computer reduces engine torque to eliminate traction loss.

When the ABS is operating, the wheel speed sensors tell the control module that a wheel is skidding. The control module responds to this by preventing excess pressure from reaching the affected wheel. In cases of severe lockup, fluid is bled from the hydraulic line. The accumulator and pump increase available pressure when necessary.

When the traction control is operating, the wheel speed sensors tell the control module that traction has been lost. The module then directs the hydraulic actuator to apply the brakes on the spinning wheel, and will reduce engine torque if slippage is occurring on all drive wheels.

Although all traction controls operate in a similar manner, there are some differences between individual manufacturers. Most ABS and traction controls installed on modern vehicles are made by a few manufacturers. The modern ABS control module interfaces with the drivetrain computer. There may not be a separate ABS computer.

Review Questions—Chapter 21

Please do not write in this text. Write your answers on a separate sheet of paper.

1. When a tire skids on a wet surface, the water is acting as a _____.

2. Wheel spin when the vehicle is being accelerated can only occur on a _____ wheel.

3. A traction control system may reduce _____ of all drive wheels that are spinning.

4. On older vehicles, the anti-lock and traction control systems may be _____.

5. A pedal travel switch is sometimes installed to sense excessive pedal _____.

6. Problems in the ABS or traction control system are saved by the computer memory in the form of a _____.

7. On most ABS systems, extra pressure is developed by an electric _____.

8. The hydraulic actuator contains valves which control pressure to the brake units at each _____.

9. What is the purpose of the accumulator?

10. To reduce engine output when the drive wheels are slipping, the control module may close the _____ or retard the _____.

ASE Certification-Type Questions

1. Technician A says that an ABS system can pump the brakes at a much faster rate than the driver can. Technician B says that there is some confusion about the proper use of ABS. Who is right?
   (A) A only.
   (B) B only.
   (C) Both A & B.
   (D) Neither A nor B.

2. Use of ABS reduces wear of all of the following, EXCEPT:
   (A) brake pads.
   (B) brake shoes.
   (C) master cylinder seals.
   (D) tires.

3. Technician A says that a skidding wheel can slide on ice or water. Technician B says that a skidding wheel can slide on melted rubber. Who is right?
   (A) A only.
   (B) B only.
   (C) Both A & B.
   (D) Neither A nor B.

4. ABS wheel speed sensors tell the control module which of the following?
   (A) How fast the wheel is turning.
   (B) The wrong size wheel is being used.
   (C) A tire is underinflated.
   (D) A tire is overinflated.

5. All of the following statements about ABS pressure switches are true, EXCEPT:
   (A) the function of these switches varies with system manufacturers.
   (B) some switches prevent over or under pressure in the system.
   (C) some pressure switches energize the control valves directly.
   (D) some switches illuminate a warning light.

6. All of the following are components used by both the anti-lock braking and traction control systems, EXCEPT:
   (A) speed sensor.
   (B) control module.
   (C) throttle control.
   (D) hydraulic actuator.

7. Technician A says that the G-force sensor tells the control module how fast the vehicle is accelerating. Technician B says that the brake pedal travel switch tells the control module when the driver releases the brake. Who is right?
   (A) A only.
   (B) B only.
   (C) Both A & B.
   (D) Neither A nor B.

8. The control module compares wheel sensor input to information stored in ABS system _____.
   (A) memory
   (B) output devices
   (C) service literature
   (D) scan tools

9. Which of the following is *not* a control module output?
   (A) Hydraulic actuator.
   (B) G-force sensor.
   (C) Warning light.
   (D) Pressure control solenoid.

10. The accumulator stores pressurized _____.
    (A) air
    (B) brake fluid
    (C) power steering fluid
    (D) engine oil

11. All of the following lights indicate a problem in the ABS or traction control system, EXCEPT:
    (A) amber Brake light.
    (B) green Traction light.
    (C) amber Check Engine light.
    (D) red Brake light.

12. On a piston operated hydraulic modulator, the pistons are operated by reversible _____.
    (A) motors
    (B) accumulators
    (C) solenoids
    (D) Both A & B.

13. The ABS system receives and monitors inputs from the wheel speed sensors when the _____.
    (A) brake pedal is depressed
    (B) vehicle starts to skid
    (C) vehicle is decelerating in drive
    (D) dashboard warning light is on

14. Technician A says that a common ABS system may be used on several makes of cars and trucks. Technician B says that the vehicle trouble codes are the same for all manufacturers when the same ABS system is used. Who is right?
    (A) A only.
    (B) B only.
    (C) Both A & B.
    (D) Neither A nor B.

15. To check ABS operation, the scan tool must be able to connect with the _____ computer.
    (A) vehicle
    (B) ABS system
    (C) shop PC
    (D) manufacturing plant

Most ABS systems are easy to locate and service. Newer vehicles have their ABS systems located away from the master cylinder, somtimes on the vehicle frame, such as this unit.

Chapter 22

Anti-Lock Brake and Traction Control System Service

After studying this chapter, you will be able to:

❑ Perform anti-lock brake system maintenance.
❑ Use scan tools to retrieve trouble codes from an anti-lock brake or traction control system.
❑ Use scan tools to diagnose problems in an anti-lock brake or traction control system.
❑ Check electrical and electronic components of ABS/TCS systems.
❑ Make pressure checks of ABS/TCS hydraulic components.
❑ Diagnose anti-lock brake system problems.
❑ Diagnose traction control system problems.
❑ Isolate defective anti-lock brake system components.
❑ Isolate defective traction control system components.
❑ Adjust ABS/TCS wheel sensors and brake travel switches.
❑ Replace defective anti-lock brake system components.
❑ Replace defective traction control system components.

Important Terms

Self test

Tire chirp

Electromagnetic interference (EMI)

Radio frequency interference (RFI)

Trouble code

Scan tool

Snapshop

Freeze frame

Hard codes

Soft codes

This chapter covers the diagnosis and repair of anti-lock brake and traction control systems. As you learned in Chapter 21, ABS and TCS systems work with the conventional brake hydraulic and friction elements, usually called the foundation or base brakes. You also learned there are two different designs of ABS and TCS systems, integral and non-integral. Design and operation of the two systems are similar. After completing this chapter, you will understand the concepts of ABS/TCS diagnosis and repair.

Caution: When working on ABS and TCS systems, make sure you do not connect test equipment to the air bag circuits. The air bag circuit wiring is generally yellow in color. Accidental deployment of the air bags and/or circuit damage could result.

Working on ABS and TCS Systems

There are several rules you need to keep in mind while working on ABS/TCS systems. Some of these were discussed in Chapter 21. These rules and others will be stated again later in this chapter as needed.

❏ If the red Brake light is illuminated, you must check the base brake system first, even if the ABS or TCS light is on. There are no exceptions to this rule.

❏ Do not assume that an ABS/TCS problem is in the ABS

or TCS system without thoroughly checking all base brake components. You should proceed to test the ABS or TCS system components only after the base brakes have been checked.

❏ Most base brake hydraulic components and some friction components (drums, rotors, pads, shoes, etc.) used on ABS and TCS equipped vehicles are different from those in non-ABS/TCS vehicles.

❏ Most integral and a few non-integral ABS systems operate at pressures over 2000 psi (13 790 kPa). Be sure to depressurize the hydraulic system before performing any work, even when checking the fluid level.

❏ Always remember to check the simple things first, such as fluid level and fuses. Many technicians have wasted time and energy trying to diagnose ABS/TCS problems caused by a missing or blown fuse.

❏ Many ABS and TCS system parts are very expensive. Do not simply throw parts on a vehicle in an attempt to fix the problem. *You must diagnose the system.*

❏ Work using a logical, step-by-step process. A good process to follow is discussed in Chapter 23.

ABS and TCS System Maintenance

ABS and TCS systems require very little periodic maintenance. Whenever the vehicle is raised on a lift for service, the wheel sensors and tone wheels should be checked for damage or debris. Any debris, such as chassis

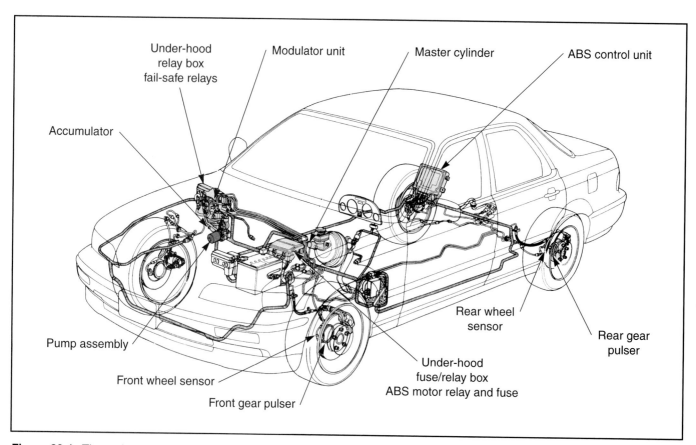

Figure 22-1. *The various parts used on one vehicle equipped with anti-lock brakes. (Honda)*

grease or road tar, should be removed from the sensor and trigger wheel using brake cleaner or other non-damaging solvent. Check the wiring for damage and proper routing. A quick check of all system components should be done at each service, **Figure 22-1.**

The level in the ABS brake master cylinder, **Figure 22-2,** should be checked periodically, just as on a conventional brake system. Check the manufacturer's fluid requirements before adding brake fluid. Components and seals in most

anti-lock brake and traction control systems will be damaged if DOT 5 silicone brake fluid is used. You may need to depressurize some integral ABS systems before checking the fluid level.

 Caution: When adding brake fluid to a vehicle with an ABS and/or TCS system, always use fluid from a newly or recently opened (less than 48 hours) container. This will minimize the chance of system damage from contaminated brake fluid.

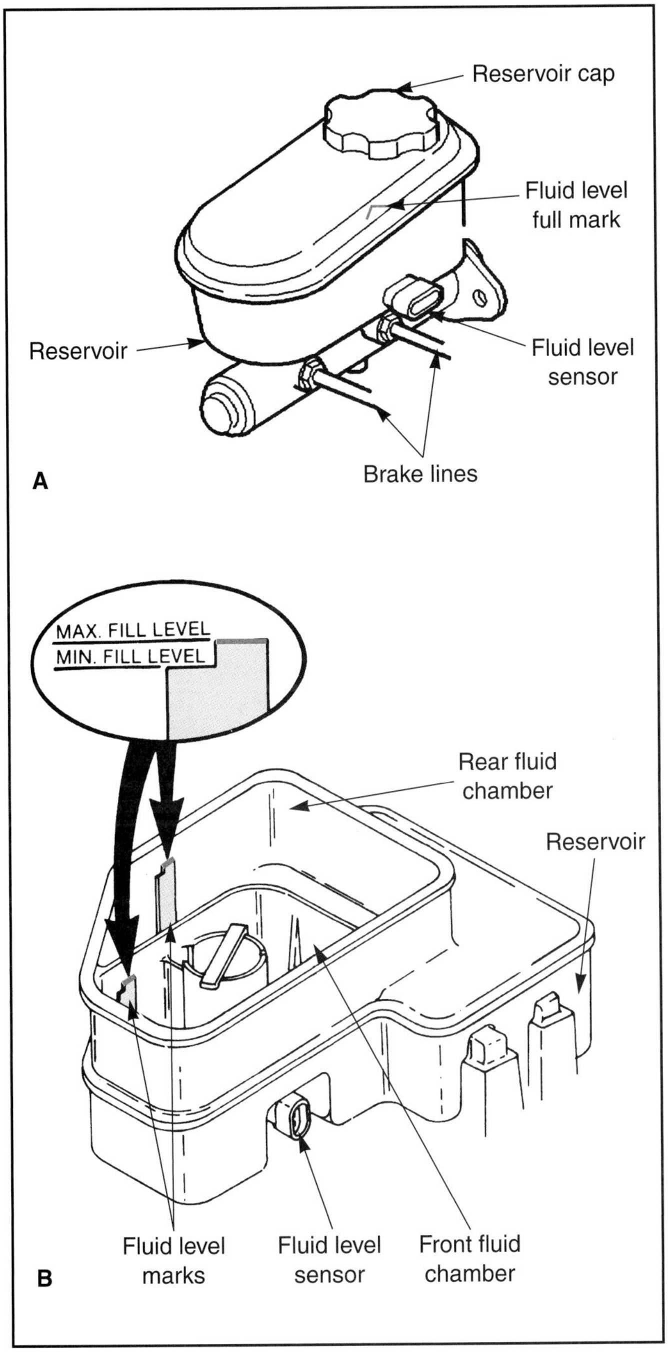

Manufacturers, in increasing numbers, are recommending periodic flushing of the ABS system. This may be performed as a maintenance service, or as a way to cure problems caused by a buildup of dirt and particles in the system. Always consult the service manual for exact flushing procedures.

On vehicles with a pedal travel switch, the switch may require adjustment to compensate for pedal wear, or after a brake pedal component is replaced. Under normal circumstances, however, the switch should retain its adjustment.

Performing Brake Service on ABS/TCS Equipped Vehicles

Many of the most common brake service procedures, such as pad and shoe replacement, rotor and drum service, and wheel bearing replacement, are not affected by the presence of anti-lock brake or traction control systems. If the lining or friction service replacement procedures involve the wheel speed sensors, treat them gently, and recheck the gap where applicable. Remember not to drop or hammer on the sensor rings, or use them to pry on other components.

Some manufacturers recommend opening the bleeder screw to allow fluid to escape, rather than pushing it back into the hydraulic actuator and master cylinder. This minimizes the chance of contamination. Do not replace any system hoses with standard (non-ABS) hoses. The higher pressures in these systems can rupture a standard brake hose.

Preparing to Diagnose ABS and TCS Problems

Due to the relative newness of ABS and TCS systems and their interaction with the base brakes, diagnosis may be complicated by the driver's unfamiliarity with the system. Therefore, you must separate the driver's reaction to a normal system condition from real problems. A properly operating ABS or TCS system will have no effect on normal vehicle performance. Actual anti-lock brake and

Figure 22-2. *Checking the brake fluid level in an ABS master cylinder reservoir. A—Reservoir for a non-integral system with fill indicator mark on the outside. B—Integral ABS reservoir with fluid level marks molded on the inside. These marks are exposed for easier checking when the reservoir cover is removed. (Pontiac)*

traction control system problems fall into two general categories:

- ❏ Failure of the system to operate (wheel lockup on braking and/or lack of traction assist during acceleration).
- ❏ Warning light illumination, either steady or intermittent.

Sometimes, what appears to be a problem will actually be normal operation. For instance, brake pedal pulsation during hard braking is a normal condition, which may appear to the driver to be a defect. Also, the TCS light on some systems will illuminate when the traction control system is operating, but the driver may interpret this light illumination as a warning of possible trouble.

Performing a Road Test

If possible, begin ABS system diagnosis by interviewing the vehicle driver. It may be helpful to fill out a check sheet while talking to the driver, **Figure 22-3**. Once the actual driver complaint has been determined, proceed to check the indicated areas. The first step is to perform a road test. In some cases, what the driver perceives as a problem may be normal ABS or TCS operation. For example, if the driver says the brakes lock up at low speeds, (roughly 6-10 mph or 9.6-16 kph) the ABS system may be operating normally. Before deciding the ABS system is defective, consult the service manual to determine the exact speed at which the system is deactivated. Lockup below the deactivation speed is not a defect.

The first check should be done as soon as the engine is started. If the red Brake light is illuminated, *do not test drive the vehicle.* Carefully move the vehicle into the shop and inspect the base brake system. Under normal operating conditions, the ABS and TCS lights should remain illuminated for 2-5 seconds after engine startup. If either light stays on for an extended period before going out, the hydraulic system, pump, or accumulator may be defective or leaking. If either light stays on constantly, there is a problem in the system.

If the red Brake, amber ABS, or TCS lights go out after start-up, check the brake pedal for firmness. If the pedal is low, or there appears to be little or no brake action, take the vehicle in the shop and inspect the base brake system. If the pedal feels ok, at this point you can begin the road test.

ABS System Feedback

On most systems, once the vehicle reaches between 3-9 mph (4.8-14.4 kph), the ABS or TCS system will perform a **self test** by cycling the hydraulic motor(s) and relays. When this test is performed, you may hear an audible buzz or clicking; this sound is normal. The purpose of this self test is two-fold: it allows the ABS control module to check the actuator and its relays for faults, and it prepares the system for operation. Observe all traffic rules when testing the antilock brake and traction control system. Perform all road testing in a safe area, away from other vehicles.

As the first check of anti-lock brake operation, attempt an ABS-assisted stop in a safe area, away from traffic and parked vehicles. The vehicle should stop straight, with no pulling or excess wheel lock-up. Remember that during hard vehicle braking, the brake pedal will be felt to pulsate slightly. This is a normal result of hydraulic pressure variations as the ABS system operates. If you cannot detect any pulsation during hard braking, the ABS may not be operating. Pedal pulsation during a hard stop means the ABS system is doing its job.

 Note: Pedal pulsation in some vehicles is very difficult to detect. You may need to listen for system operation.

Another indicator of system operation during an ABS stop is the tires will make noise in the form of a "chirp." Some **tire chirp** during an ABS-assisted stop is considered normal as it is the noise a tire makes just before wheel lockup, which in a non-ABS vehicle, would be lost in the sound of skidding tires.

TCS System Feedback

The TCS light on some systems may illuminate when the traction control is operating. This is considered part of normal system operation. The most obvious sign of traction control failure is wheel spin during acceleration. If the traction control system seems to be completely inoperative, first check to see if the control switch is in the *on* position. This is a cause of some TCS inoperative complaints.

To test the traction control system, find a paved location at your shop away from traffic and parked vehicles. Spray some water on the ground around the rear wheels; also soak the rear tires with water. Accelerate the vehicle, taking note if there is excessive wheel spin.

In some cases, the TCS system's normal operation may be misinterpreted as an engine performance problem. Some drivers may interpret the lack of wheel spin as low engine power. In some cases, drivers may also complain of feedback through the accelerator pedal (similar to brake pedal feedback during ABS operation). Remember, the first step most TCS systems take to reduce wheel spin is to reduce engine power. The TCS system does this in some cases by closing the throttle plate slightly or by disabling one or more engine cylinders. This temporary reduction in engine power and accelerator pedal feedback are both considered part of normal TCS system operation.

If the TCS light stays on constantly, or the TCS goes out, but excessive tire slip still occurs, the system is malfunctioning. If a light does not come on at any time, the problem may be as simple as a blown fuse or burned out bulb, or may indicate a control module problem.

Other ABS/TCS Problems

Very rarely does an ABS or TCS problem affect normal brake system operation. On some occasions, a valve body solenoid will cause pulling during normal braking.

Customer Problem Analysis

ABS Check Sheet	Inspector's Name

		Registration No.	
Customer's Name		**Registration Year**	/ /
		Frame No.	
Date vehicle brought in	/ /	**Odometer Reading**	km Mile

Condition of problem occurrence	Date of problem occurrence	/ /
	How often does problem occur?	☐ Continuous ☐ Intermittent (times a day)

Symptoms	☐ ABS does not operate.
	☐ ABS does not operate efficiently.
	ANTI-LOCK warning light abnormal ☐ Remain ON ☐ Does not light up

Check item	TRAC indicator light	☐ Normal	☐ Does not light up

Diagnostic trouble code check	1st time	☐ Normal code	☐ Malfunction Code (Code)
	2nd time	☐ Normal code	☐ Malfunction Code (Code)

Figure 22-3. *Check sheet used in diagnosing ABS problems. (Toyota)*

Some RWAL and non-integral four-wheel ABS systems can develop a soft or falling pedal due to a defective accumulator. Many of these problems can be diagnosed as part of the general system inspection. Some of these problems may be addressed in technical service bulletins, issued by the vehicle's manufacturer.

 Note: RWAL brake systems in some part-time four-wheel drive light trucks will be deactivated during four-wheel drive operation. This is a normal condition and is not caused by a system defect. When testing an RWAL system in part-time four-wheel drive vehicles, make sure the transfer case is in two-wheel high.

Using Service Information

Many vehicle and parts manufacturers make anti-lock brake systems. Major producers, as discussed in Chapter 21, include Bendix, Bosch, Delphi, Kelsey-Hayes, and Teves. These systems are similar in basic operation. However, they vary greatly in the design of their hydraulic and electronic components, and require different troubleshooting and service procedures.

In addition, one type of system may be installed on several makes of vehicle, or the same manufacturer's line of vehicles may have different systems on each model. Therefore, it is important to obtain and use the correct, and most recent service manual before beginning any repairs on ABS or TCS systems.

A factory service manual is the best choice for a particular vehicle. A specialized manual covering all makes of ABS and TCS systems allows you to access service information for many vehicles. You should also have access to any technical service bulletins (TSBs) which detail problems or update service procedures for the particular system.

Visual Inspection

Many preliminary checks can and must be made visually. During the road test, you should have checked to make sure the Brake, ABS, and TCS lights are working. On many systems, trouble codes cannot be retrieved unless the ABS or TCS lights are operating. If any of the trouble lights do not work, check for a burned out bulb and related fuses and wiring. Once you are certain the dash lights are in working order, you can begin to visually inspect the vehicle and ABS/TCS system for problems.

Mismatched Tires

If the problem appears to be related to vehicle speed, check for mismatched tires first. Incorrect tire sizes or mismatched tires will affect the wheel rotation rate as read by the wheel sensors. This will confuse the ABS controller, leading to erratic system operation. This should be checked before the test drive. Usually, mismatched tires will have an obvious effect on the vehicle's ride height.

ABS problems often occur after the owner changes to aftermarket rims and/or low profile or extended scrub radius tires. For best system operation, tires should be roughly the same height as original equipment. Check that all tires on the vehicle are the same size and type. Tread width can be greater, as long as the height is not changed. However, mismatched tires (for instance, one P205/70 and three P195/75s) can cause problems.

 Note: It is possible to change the tire size parameters in the ABS control modules on some newer vehicles.

Brake Fluid Level and Base Brake Hydraulic Problems

Low brake fluid level can cause the red Brake warning light to come on, and may also illuminate the ABS light. It is normal for the fluid level to gradually lower as the brake pads wear, causing the caliper pistons to extend further. More fluid is retained inside the caliper, instead of returning to the master cylinder. However, a very low fluid level is usually caused by severely worn pads, rotors, or by external fluid leaks. Leaks can occur at the ABS/TCS hydraulic components or at any point in the base brakes. A severe leak will quickly empty the reservoir, and may cause erratic pressure application. General brake hydraulic system service was covered in Chapters 4-6.

If suspected, check for hoses that may be swollen shut, kinked or crushed steel lines, and stuck calipers or wheel cylinders. Any one of these can cause poor brake application or pulling. A pressure differential valve that sticks in any spot but the center position will illuminate the red Brake warning light. Remember, if the red Brake light is on, you must check the base brake system first before the ABS/TCS system.

Power Brake Booster

If the vehicle is equipped with an integral ABS system, skip this step. A defect in the power booster of vehicles equipped with non-integral ABS can cause a hard pedal or slow brake application. Check for a collapsed or leaking hose to the vacuum unit and low oil level on a hydraulic booster. For other diagnosis and service procedures to power brake units, see Chapters 7 and 8.

Worn or Misadjusted Friction Members

Worn disc brake pads, rotors, drums, or misadjusted shoes will force the hydraulic system to provide more fluid than normal to the caliper pistons or wheel cylinders. This can upset pressure development in the base brake hydraulic system that will affect ABS operation. Out-of-round rotors or drums that set up fluctuations in the hydraulic system may also affect pressure development. A stuck or misadjusted parking brake cable will cause partial application of the parking brakes, resulting in hydraulic system or sensor malfunctions.

Some brake pads have a wear sensor that will turn on the red Brake light when the pads wear past a certain point. No matter how well the ABS system works, worn, out-of-round, or misadjusted friction members cannot do a good job of stopping the vehicle. If necessary, review the base brake service information in earlier chapters.

Vehicle Charging System Problems

High or low charging system voltage will cause the control module to operate incorrectly. A quick check of charging voltage can be made with the engine idling in neutral. Make sure that any electrical equipment which draws heavy current (such as the headlights, blower motor, and rear window defroster) are *off*. Minimum charging system voltage should be about 13 volts at idle and not go over roughly 14.5 volts when the throttle is wide open. Most ABS or TCS systems will not operate and in many cases, will set a trouble code if system voltage drops below 9.5-10 volts or goes above 14.5-15 volts.

If voltage is low, check for a slipping alternator drive belt and a defective voltage regulator. If voltage is too high, check the voltage regulator for proper grounding and internal defects. When necessary, make further checks using a starting and charging system tester.

Wiring Problems and EMI/RFI Interference

After the road test, check all vehicle wiring, especially the ABS and TCS system wiring, for proper connections and routing. This includes the wiring from the fuse block to the relays, hydraulic actuator, control module, wheel speed sensors, etc. One poor connection can cause a multitude of problems.

It is very important to ensure that all ABS and TCS wiring is properly routed and/or shielded. ABS and TCS systems are *extremely sensitive* to **electromagnetic interference (EMI)** and **radio frequency interference (RFI).** If an ABS or TCS problem occurs intermittently, especially when a particular vehicle or aftermarket system is operating, EMI or RFI interference may be the cause. Sources of EMI and RFI interference include:

❑ Misrouted vehicle system or ABS/TCS wiring.
❑ Defective (leaking or grounded) ignition secondary wiring.
❑ Defective alternator diodes.
❑ Police and CB radios (especially units that use signal amplifying circuitry).
❑ Improperly installed aftermarket components.
❑ Aftermarket equipment that generates an electromagnetic field or radio frequency signal.

A misrouted wheel speed sensor harness that crosses another wire, even one in a low power circuit, can cause problems in some ABS/TCS systems. Shielding is used on some ABS and TCS wiring to reduce the chance of problems caused by EMI and RFI interference. In some cases, aftermarket component relocation or additional wire shielding may be needed to correct the problem.

 Warning: It is extremely important that all wiring be properly routed and/or shielded on vehicles equipped with ABS systems. EMI/RFI interference can cause a condition in some systems that may result in the complete loss of both anti-lock and base braking capability.

Note that on some vehicles, seemingly unrelated areas, such as parking brake or brake light switch adjustment, brake pedal height, loose or worn front suspension parts, or defects in drive axles, can also affect system operation.

ABS and TCS System Diagnosis

Once all preliminary checks have been made, the system itself can be checked. Most anti-lock brake and traction control systems require special testers to pinpoint internal system troubles. On late-model vehicles, a scan tool, **Figure 22-4,** must be used to check the system for trouble codes and to run other diagnostic procedures. Some of the first vehicles with anti-lock brakes require the use of a breakout box or a system specific tester, such as the one shown in **Figure 22-5,** to check ABS components.

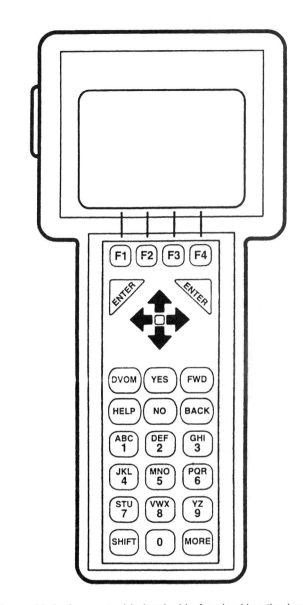

Figure 22-4. *A scan tool is invaluable for checking the brake system for trouble codes as well as diagnostics and bleeding.*

Trouble Code Retrieval

When an ABS or TCS system problem occurs on most vehicles, the computer stores information about the problem as a **trouble code.** Trouble codes identify the general area of a problem, or may identify the defective component. On older systems, the problem identified by the trouble code varied with the system and vehicle manufacturer. Newer systems use a standardized trouble code system. A few early integral ABS systems will produce no trouble codes, and must be diagnosed using a breakout box.

Methods of retrieving trouble codes are similar to those used to retrieve trouble codes from the powertrain computer. In many cases, the same data link connector is used. The most common way to retrieve trouble codes from modern vehicles is by using a scan tool. The retrieved trouble codes should then be compared to the appropriate trouble code chart to determine potential problems. Always consult the vehicle service manual for the correct code retrieval method.

Note: On some vehicles, the control module can only store a limited number of trouble codes. After servicing the system, drive the vehicle and observe the trouble light to ensure that no additional trouble codes occur.

Reading Codes Using Dashboard Lights

The most common method of retrieving trouble codes is to ground a terminal at the data link connector. Once the terminal is grounded, the technician can observe a series of flashes from the dashboard indicator light. There are two variations of this method: observing the flashes of an LED (light emitting diode) installed on the control module after entering the diagnostic mode, **Figure 22-6;** and accessing trouble codes through the digital air conditioner control panel. **Figure 22-7** shows some methods of using the dashboard lights to obtain trouble codes.

Another method of placing the control module in the self-diagnosis mode is by turning the ignition switch to the *on* position without starting the engine, or by depressing the brake pedal for a certain number of seconds with the ignition on. On these systems, the codes are read from the dashboard trouble light.

Reading Codes Using a Voltmeter

Another method of accessing trouble codes is to ground the proper terminal and connect a voltmeter across two other terminals. Then the voltage pulses created by the control module are read to determine the codes. This method is used on only a few vehicles.

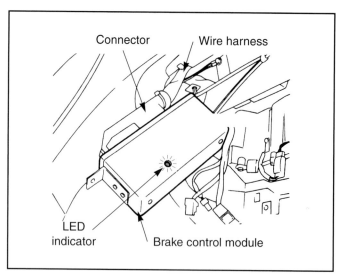

Figure 22-6. *A light emitting diode (LED) placed on the brake control module cover.*

Using Scan Tools

The best way to access the self-diagnostic capabilities of an anti-lock brake or traction control system is to use a *scan tool,* **Figure 22-8.** Scan tools are commonly used to diagnose problems in the engine and drivetrain. Less well known is their ability to assist in troubleshooting and servicing the brake system.

The following sections explain how to use a scan tool to retrieve trouble codes and perform other ABS/TCS system diagnosis. Always refer to the scan tool manufacturer's instructions for exact procedures.

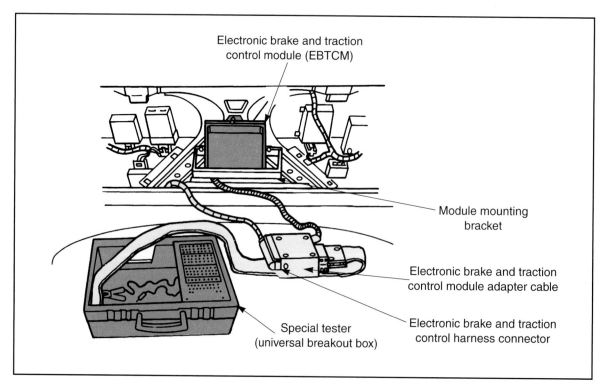

Figure 22-5. *Some ABS systems require the use of a breakout box and other tools for proper diagnostics. This is primarily on older ABS equipped vehicles. (Cadillac)*

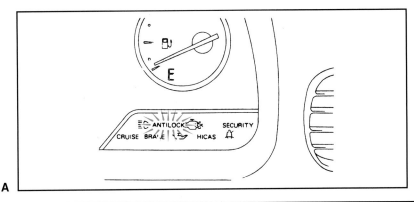

Code No.	Flashing Pattern	Malfunction
11	ON ⊓_⊓_⊓ OFF	Solenoid relay circuit wire disconnected.
12	⊓_⊓⊓	Solenoid relay circuit wire shorted.
13	⊓_⊓⊓⊓	Pump motor relay circuit wire disconnected.
14	⊓_⊓⊓⊓⊓	Pump motor relay circuit wire shorted.
21	⊓⊓_⊓⊓	Right front wheel 3-position solenoid circuit wire disconnected or shorted.
22	⊓⊓_⊓⊓	Left front wheel 3-position solenoid circuit wire disconnected or shorted.
23	⊓⊓_⊓⊓⊓	Right rear wheel 3-position solenoid circuit wire disconnected or shorted.
24	⊓⊓_⊓⊓⊓⊓	Left rear wheel 3-position solenoid circuit wire disconnected or shorted.
31	⊓⊓⊓_⊓	Right front wheel speed sensor circuit signal malfunction.
32	⊓⊓⊓_⊓⊓	Left front wheel speed sensor circuit signal malfunction.
33	⊓⊓⊓_⊓⊓⊓	Rear wheel speed sensor circuit wire malfunction.
34	⊓⊓⊓_⊓⊓⊓⊓	Front wheel speed sensor circuit wire severed.
35	⊓⊓⊓_⊓⊓⊓⊓⊓	Left front or right rear wheel speed sensor circuit wire severed.
36	⊓⊓⊓_⊓⊓⊓⊓⊓⊓	Right front or left rear wheel speed sensor circuit wire severed.
37	⊓⊓⊓_⊓⊓⊓⊓⊓⊓⊓	Wrong left and right rear axle hubs
41	⊓⊓⊓⊓⊓_⊓	Battery voltage low (9.5V or lower)
42	⊓⊓⊓⊓⊓_⊓⊓	Abnormally high battery voltage (16.2V or higher).
51	⊓⊓⊓⊓⊓⊓_⊓	Pump motor locked.
Always On	⌐‾‾‾	Computer malfunction.

Figure 22-7. A—Using the dashboard ABS light to retrieve trouble codes. B—This shows the code number, light flashing pattern, and the malfunction. If there is more than one malfunction code, the code with the lowest number will flash first.

⚠ **Warning: Some ABS systems are disabled anytime a scan tool is connected to the data link connector. Do not perform any actuator or scan tool intrusive tests while driving the vehicle. Doing so may disable the brake system.**

Reading Trouble Codes
To retrieve trouble codes using a scan tool, first ensure the correct software cartridge is installed in the tool, if applicable. Next, locate the data link connector. Most modern vehicles use the same data link connector to access both engine and ABS/TCS codes. **Figure 22-9** shows some typical data link connector locations.

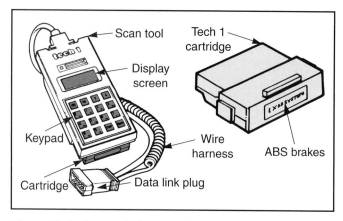

Figure 22-8. *One particular type of scan tool along with an anti-lock brake system (ABS) cartridge. Handle these tools with care. (Pontiac)*

Make sure the ignition is in the off position and connect the tool to the data link connector using the proper adapter. A typical connection is shown in **Figure 22-10.** If necessary, connect the scan tool power adapter to the cigarette lighter or battery terminals as necessary. Some scan tools are powered through the data link connector.

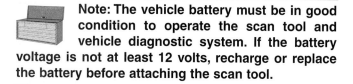 **Note: The vehicle battery must be in good condition to operate the scan tool and vehicle diagnostic system. If the battery voltage is not at least 12 volts, recharge or replace the battery before attaching the scan tool.**

After it is connected to a power source, the scan tool will perform a self test. Most scan tools will then prompt you to enter the vehicle identification number (VIN) or otherwise identify the type of vehicle or system being worked on. Once the date has been entered, the display screen will prompt you to select the system to be diagnosed, in this case, the ABS or TCS system. The ignition key may have to be turned to the *on* position to retrieve codes. Check the vehicle manufacturer's service instructions.

Next the display screen will prompt you to retrieve trouble codes. Once you select this option, the scan tool will retrieve and display the trouble codes. Older ABS systems may have one of several types of trouble code formats. Later vehicles with the OBD II system will use a standardized code. Many of the latest scan tools have a **snapshot** or **freeze frame** mode, which will enable the control module to display the system readings present at the time the malfunction occurred.

In some cases, the scan tool will read all vehicle codes, and it will be necessary to separate the ABS or TCS codes from the engine codes. In many cases, information on engine and drivetrain problems can be helpful in diagnosing brake problems.

 Note: Always retrieve trouble codes before performing any other diagnostic routines.

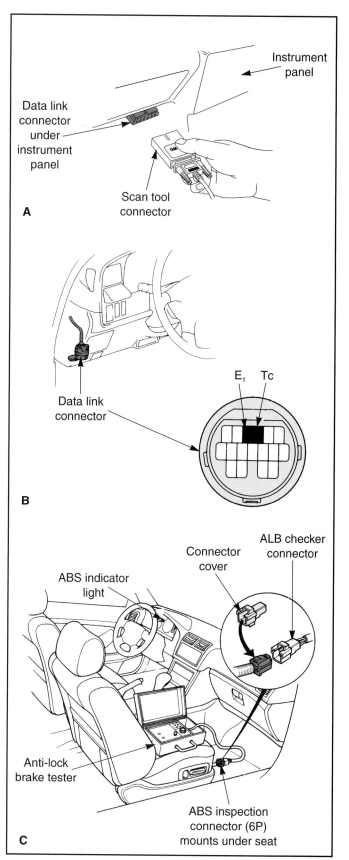

Figure 22-9. *A—Underdash location for the data link connector (DLC). This is the DLC location on OBD II vehicles. B—Some older vehicles have separate DLC connectors for the ABS system. C—Some connectors are located under the seats, carpet, in the engine compartment, or in the trunk. On vehicles built before 1996, connector shapes will vary between car manufacturers.*

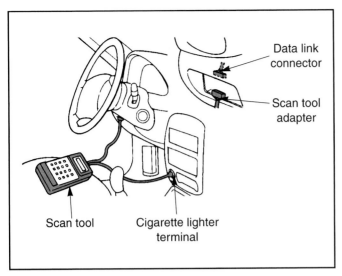

Figure 22-10. *A typical scan tool connection. Be sure the connectors are properly engaged.*

Hard and Soft Codes

Two types of codes are stored in the ECM's volatile memory. They are called hard codes and soft codes. **Hard codes** (permanent) indicate an on-going problem that still exists in the vehicle. They are often the easiest to track down. Soft codes or *intermittent codes* (temporary), have been set by problems that occur occasionally, or only once, and are harder to isolate.

To properly diagnose an ABS/TCS system with multiple stored trouble codes, you must separate hard codes from soft codes. First, record all codes, then use the scan tool or remove electrical power from the control module to erase the stored codes. Restart the engine and, if needed, drive the vehicle for a period of time as specified by the manufacturer. After the vehicle has been driven long enough, stop it and re-enter the diagnostic mode.

Hard Codes

Hard codes will usually reset almost immediately after the key is turned to run, after the engine is started, or when the control module requests information from a sensor or tries to activate an output device. Hard codes are usually caused by a defective component, open or shorted circuit, or connector. Hard codes make it relatively easy to isolate the defect to a specific area, since the problem is always there.

Soft Codes

Soft codes will only turn the light on while the problem is occurring, after which the light will go out. In many cases, a soft code may not mean the sensor, actuator, or other device is defective, only that it is responding to another problem. Problems that set soft codes usually take longer to diagnose.

Keep in mind that sensors and switches can go slightly out-of-range without setting a trouble code. Lack of a trouble code in any computer control system is not absolute proof that a sensor is good.

Checking for Vehicle Electrical Problems

Once any trouble codes have been retrieved and verified, the system circuits and components can be checked for defects. A scan tool can be used to accomplish some component testing.

Sometimes when the trouble code(s) indicate an electrical problem, checking for obvious electrical defects can save time. The most common electrical system problems are low charging system voltage, blown fuses or fusible links, corroded or overheated connections, defective power relays, and disconnected or high resistance wiring. Charging system voltage checking was discussed earlier in this chapter.

Remember that various circuits in the control module may be powered through more than one fuse, fusible link, or relay. Notice the dual feeds in the schematic in **Figure 22-11**. For this reason, it is vital to obtain the correct system schematics and determine what circuits and devices directly affect the control module.

 Note: Do not remove any fuses before attempting to retrieve trouble codes. Removing fuses may erase stored trouble codes from the control module memory.

After checking the control module fuses, proceed to check all other vehicle fuses. A blown fuse in an unrelated circuit may be caused by current attempting to ground through the dashboard wiring. This could affect control module operation or illuminate the ABS or TCS light.

Electrical relays that are operated by or deliver power to the system may be located in a panel with other relays, or separately mounted almost anywhere on the vehicle. **Figure 22-12** shows a typical power distribution center containing relays.

After locating the appropriate relay(s) using the correct schematic, check that the relay electrical sockets and connectors are not overheated, corroded, or otherwise damaged. Some relays can be checked by operating them with jumper wires from the vehicle battery, sometimes called bypassing. Before bypassing any relay, disconnect it from its electrical connector. Do not try to bypass any electronic relay or control module. Carefully check all electrical connections, especially those under the vehicle. Also check for disconnected or corroded ground wire connections, or other evidence of damage.

Finally, do not forget to check the brake lights. If the brake light switch is defective, this can set a trouble code. On a few vehicles, a defect in the turn signal switch or flasher relay can cause the brake lights to become inoperable and set a code.

Checking Wheel Speed Sensors

If the trouble codes or scan tool readings indicate a wheel speed sensor problem, the affected sensors should be visually inspected before proceeding. Check for buildup of dirt or grease between the tone wheel teeth or on the sensor.

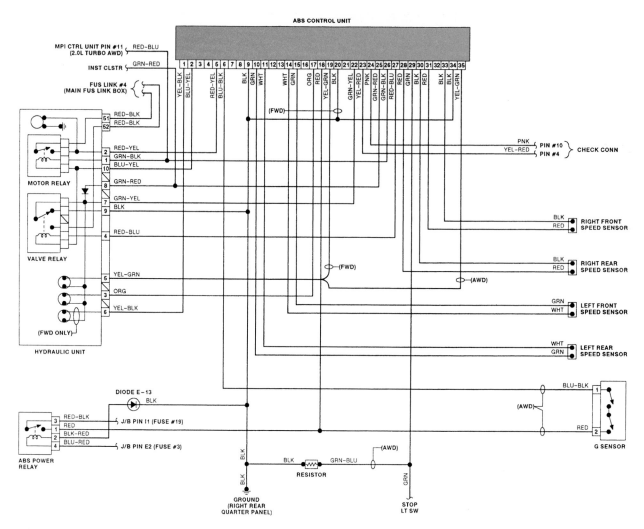

Figure 22-11. *An anti-lock brake system (ABS) control unit and its various electrical feeds shown in a schematic form.*

Check the tone wheel for missing teeth or other obvious damage. Check the sensor mounting hardware for looseness or damage. Always consult the proper service manual for exact sensor checking procedures and specifications.

Check the tone wheel-to-sensor air gap using a brass feeler gauge, as shown in **Figure 22-13.** Remember to rotate the wheel and check the gap at several spots. If readings vary widely as the wheel is rotated, check for overtorqued lug nuts or rotor damage. A warped rotor can cause an erratic signal to be produced by a good sensor. If gap specifications are not available, check the gaps on all wheel sensors, and compare the readings. If one gap is much larger or smaller than the others, check for bent or loose parts. Note that incorrect gap is seldom a problem with sensors mounted on the transmission or differential carrier.

If the gap is adjustable, follow the manufacturer's procedures for adjustment. When adjusting the gap, be sure to use a brass feeler gauge. A steel gauge will drag on the sensor magnet, making accurate adjustment difficult.

Also check for careless tone wheel handling or installation by previous service technicians. The sensor tone wheels may be damaged if they were heated, dropped, or hammered. Finally, make sure the tone wheel has the correct number of teeth.

Figure 22-12. *A power center which houses the relays, and fuses for one particular vehicle.*

Making Sensor Electrical Checks

Various electrical checks can be made to the sensor before deciding that it is defective. On some systems, an ohmmeter can be used to determine sensor winding resistance, **Figure 22-14.** Remove the sensor harness connector

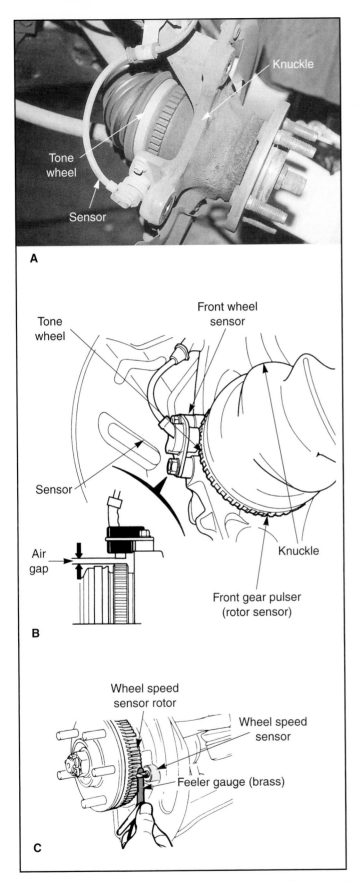

Figure 22-13. *A—Wheel speed sensor location on a front-wheel drive car. B—After installing a new speed sensor, bring the air gap back to proper operating specifications. C—Checking the tone wheel-to-sensor air gap with a brass or other nonmagnetic feeler gauge. (Chrysler, Honda)*

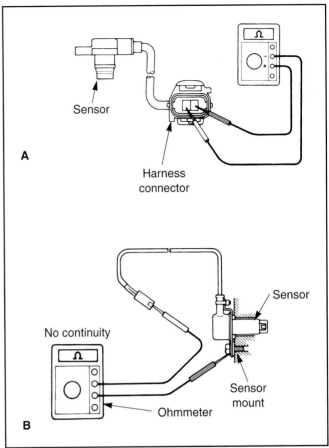

Figure 22-14. *A—Using an ohmmeter to check the wiring harness connector for sensor winding resistance. B—Sensor showing no continuity when tested with the ohmmeter. No continuity indicates that the sensor and wiring are fine.*

and connect an ohmmeter across the winding at the input leads. If the resistance reading does not fall within specifications, the sensor should be replaced. General sensor resistance should be around 1000 ohms. A sensor winding which indicates zero or infinite resistance can be assumed to be defective.

If a scopemeter is available, the wheel sensors can be checked to determine whether they are creating the proper waveform output. Begin by connecting the scope or meter to the wheel sensor. If the scan tool is capable of displaying this type of reading, use its menu provision to access wheel sensor output. Then raise the wheel from the ground and spin it. An ac sine wave should be seen as the wheel rotates. Typical sine wave patterns are shown in **Figure 22-15.** If an ac sine wave cannot be obtained, replace the sensor.

Before deciding a sensor has no electrical problems, check the sensor's wiring. If the sensor wire passes too closely to electrical devices that produce strong magnetic fields, a false signal may be generated. Keep sensor wires away from the exhaust system, ignition wires, starter cables, and fuel pump wires. Also check the wiring where it passes through clips and body openings to ensure that it has not been cut or chafed by vehicle movement. Repair any damaged wires before proceeding with diagnosis.

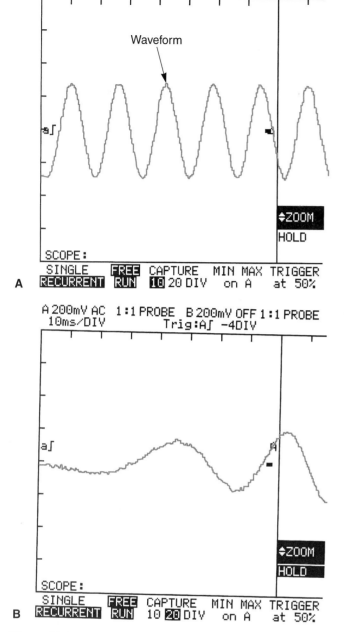

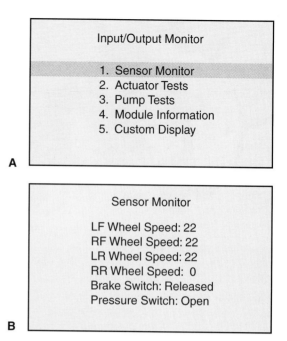

Figure 22-16. A—Scan tool menu for testing ABS/TCS systems. B—Scan tool screen showing sensor readings from the ABS system.

Figure 22-15. Using a scopemeter to check wheel speed sensor waveforms. A—Ac voltage is produced at the wheel speed sensor, while the wheel is being rotated at a constant speed. B—All wheel speed sensors are generally the same—varying voltage and frequency will depend on road speed. Rapid wheel acceleration is shown.

Checking Wheel Speed and Other Sensors Using a Scan Tool

To check the condition of sensors, use the scan tool menu to access the proper diagnostic routine. On most menus, this routine is called the test function or a similar name, **Figure 22-16.** Follow menu directions to check the voltage and/or waveforms produced by the input sensors. Compare the readings with manufacturer's specifications. The scan tool may have these specifications in its memory.

Checking and Adjusting the Brake Light and Pedal Travel Switches

Before checking the brake light or pedal travel switches, check the service manual to determine whether there is a specific procedure that should be used. If necessary, check and adjust the pedal height. The adjustment on most brake light switches can be checked simply by pulling up on the brake pedal. There should be no clicking heard. If two or more clicks are heard when the pedal is pulled up, the brake light switch was probably misadjusted.

 Note: A misadjusted brake light switch is a common cause of ABS problems in new vehicles delivered from the factory.

Once the brake light switch is adjusted, the pedal travel switch can be checked, if the vehicle has one. Keep in mind not all vehicles have a pedal travel switch. Most travel switches have a 2-pin connector. When troubleshooting indicates improper pedal travel, remove the 2-wire connector and check the switch for continuity with a powered test light or ohmmeter. See **Figure 22-17.** Both pins should have continuity to ground with the pedal in the released position, and no continuity with ground with the pedal firmly applied. If these results are not obtained, adjust the switch. If adjustment does not result in proper continuity readings, replace the switch.

On most vehicles, the pedal travel switch is self-adjusting. To ensure that adjustment is correct, remove the switch plunger from its mounting clip, **Figure 22-18.** Then gently pull the plunger to the end of its outward travel. Then reinstall the switch and press the brake pedal firmly. This will complete the adjustment.

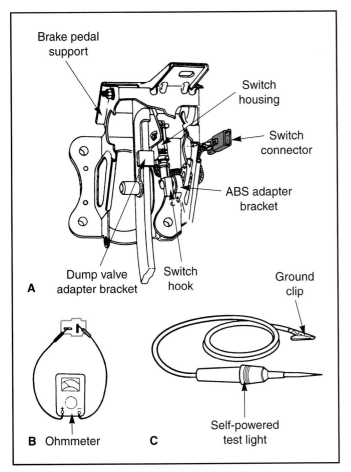

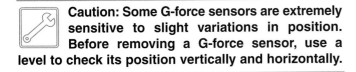

Figure 22-17. *Checking a brake pedal travel switch. A—Switch wiring harness connector. B—Checking switch continuity with an ohmmeter. C—A test light may also be used.*

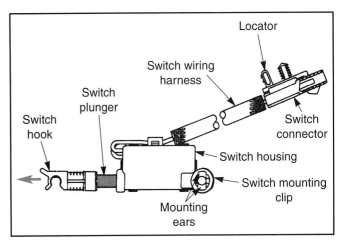

Figure 22-18. *A brake pedal travel switch with the switch plunger pulled out to the end of its travel. The switch assembly is now ready to reinstall.*

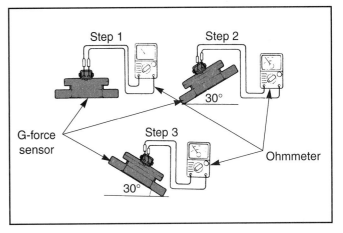

Figure 22-19. *Checking one particular G-force sensor for proper operation with an ohmmeter. Test the sensor lying flat and tilted. The sensor should be tilted to 30° or more. If there is no continuity change in the proper test positions, replace the sensor. Follow the vehicle manufacturer's testing procedures.*

Checking G-Force Sensors

Caution: Some G-force sensors are extremely sensitive to slight variations in position. Before removing a G-force sensor, use a level to check its position vertically and horizontally.

To check the G-force sensor, remove it from the vehicle. Place it on a level surface and check continuity between the terminals. Then tilt the sensor as shown in **Figure 22-19** and recheck continuity. Continuity should change as the sensor is tilted. Note that some G-force sensors have continuity when level, while others have continuity when tilted. What is important is that continuity changes from zero to infinity or from infinity to zero when the sensor is tilted.

Checking Control Module and Related Circuits

If previous testing or trouble codes indicate that the control module is faulty, first check for voltage at all module input circuits. Remember, there may be several electrical inputs to the control module, and all should have sufficient voltage. If any electrical input problems are found, check and repair all related fuses, relays, and wiring before proceeding. Remove the control module wiring harness and check circuits at the harness, **Figure 22-20.** Follow the manufacturer's instructions to perform these tests. Do not attempt any electrical checks to the control module unless specifically called for by the manufacturer. The electronic circuitry inside the module can be destroyed by improper electrical tests.

On many vehicles, it is possible to check the control module using a scan tool. In most cases, a trouble code indicating a defective control module or electronic relay means the part must be replaced. Scan tools can also operate the control module by inputting test signals through the data link connector. The control module should produce the proper outputs based on the inputs of the scan tool. If it does not and there are no wiring problems, the module is probably defective.

Caution: Before installing a new control module, check all grounds and related electrical devices, such as output relays and the hydraulic actuator. If one of these parts has shorted internally and is drawing high current, it can destroy the new module.

Checking Integral Hydraulic Actuators

If a faulty hydraulic actuator is suspected, it should be tested according to the manufacturer's recommendations. Before any other steps are taken, take a sample of brake fluid and inspect it for metal particles and other debris. Place the fluid in a clear container, as in **Figure 22-21**.

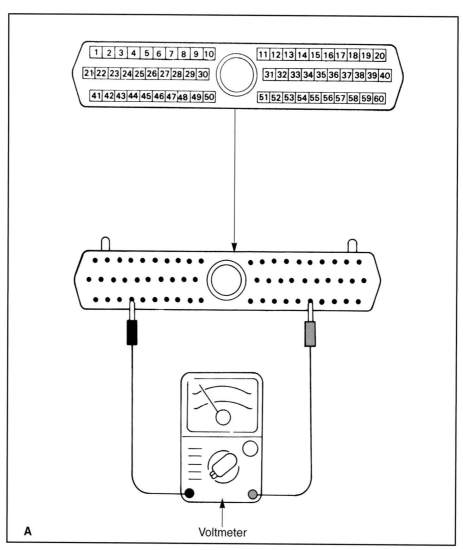

A

Voltmeter

Connector Pin Arrangement

Cavity	Function	Cavity	Function	Cavity	Function	Cavity	Function
1	-RR SENSOR	11	DATA IN	41	SOL +BATT	51	—
2	+RR SENSOR	12	DATA OUT	42	LF BUILD/DECAY	52	—
3	-LR SENSOR	13	STOP LAMP	43	RF BUILD/DECAY	53	—
4	+LR SENSOR	14	—	44	—	54	—
5	GROUND	15	YELLOW LIGHT	45	LF/RR ISO	55	—
6	-RF SENSOR	16	MP RELAY	46	LR Build/DECAY	56	—
7	+RF SENSOR	17	—	47	SOL +BATT	57	SYSTEM RELAY
8	-LF SENSOR	18	—	48	RR BUILD/DECAY	58	—
9	+LF SENSOR	19	—	49	RR/LR ISO	59	—
10	—	20	MP MONITOR	50	—	60	IGNITION +BATT

B

Figure 22-20. *Checking the control module wiring harness. Follow the service manual directions; do not use an ohmmeter to check the control module. (Hyundai)*

Warning: Be sure to depressurize the system before accessing any portion of the hydraulic actuator.

Allow the sample to sit for at least 10 minutes and then inspect it for contaminants and settling. Metal particles are evidence of component damage (usually the hydraulic pump). Non-metallic debris indicates brake fluid overheating or contamination. If there appear to be layers of dissimilar fluids in the container, the system is contaminated with the wrong type of brake fluid, a petroleum based liquid, or water. All components must be rebuilt or replaced and the entire system completely flushed.

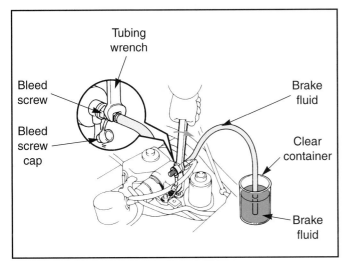

Figure 22-21. *Checking the brake fluid for metal, water, dirt, by bleeding fluid into a clear container. Flush and replace with new fluid if any contamination is found. (Lexus)*

Making Pressure Tests

Some actuators and accumulators can be pressure tested. To begin the pressure tests, depressurize the system by pumping the brake pedal at least 40 times with the ignition switch off. Then attach a high pressure gauge to the system according to service manual instructions, as shown in **Figure 22-22.**

Warning: Integral ABS brake systems can produce pressures of 2500 psi (17 238 kPa) or more. Therefore, to perform any kind of brake pressure test, a gauge capable of registering pressures of at least 4000 psi (27 580 kPa) is needed. Do not attempt to use an engine or transmission pressure gauge for this test. Equipment damage and/or personal injury will result.

Since the system was depressurized to connect the gauge, turn the ignition switch to the *on* position and wait at least 60 seconds for the hydraulic pump to repressurize the system. Note how long it takes before the hydraulic

pump shuts off, then begin the pressure test procedure. In most cases, the pump should take no more than 30-40 seconds to pressurize the system.

Closely follow the manufacturer's instructions and observe hydraulic pressures in various system operating modes. Record all pressures and compare them against specifications. Also note whether pressure develops smoothly, and if there are any sharp spikes or drops in pressure as the system passes through various operating modes. Then refer to the manufacturer's diagnostic charts to determine which components are defective.

The most obvious use of the pressure gauge is to determine whether the pump is producing sufficient pressure. However, other diagnostic steps can be performed using the pressure gauge. On some vehicles, the pressure gauge can be used to check solenoid operation. A typical procedure is to energize the solenoid while observing the pressure gauge. If pressure does not drop or increase as necessary, the solenoid is defective.

Pressure Testing the Accumulator

Test the accumulator by first depressurizing the system, attaching a pressure gauge, and allowing the pump to repressurize the hydraulic system. In some cases, such as the Powermaster III system, it may be necessary to remove the accumulator and install it in a fitting on the pressure gauge. Then turn the ignition switch to *off* and monitor the pressure gauge. If the pressure remains constant, the accumulator is working. Depress the pedal once. If the pump engages as soon as the pedal is pressed, the accumulator is defective. If the pressure begins to drop as soon as the ignition switch is turned off, the accumulator is defective.

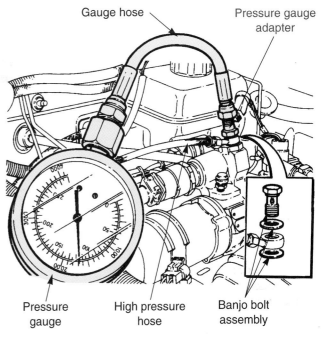

Figure 22-22. *Performing a pressure test. Use the recommended type of pressure gauge, and follow the vehicle manufacturer's testing procedures. Wear safety glasses.*

Testing the Pressure Switch

Some systems have one or more pressure switches. These switches prevent system overpressurization, or may signal the hydraulic pump motor to turn on when system pressure becomes too low. Some switches may also illuminate the warning light when a low pressure condition is detected. Follow manufacturer's directions to check pressure switch operation against actual system pressures.

Making Electrical Tests

Some system and vehicle manufacturers specify the use of ohmmeter tests to determine the condition of the solenoid valves and motor in the actuator. Other ABS system pumps and solenoids can be energized by a scan tool, or by using jumper wires from the battery. These tests should be done only when the manufacturer specifically recommends them. Do not test the actuator using battery voltage or powered test lights unless specifically instructed by the manufacturer. If jumper wires are used, make sure they are fused. Using a jumper wire without a fuse or circuit breaker can damage components or cause burns from overheated wires.

Disassembling the Actuator to Make Tests

Some hydraulic actuators can be partially disassembled to check for internal problems and to replace certain parts. Any disassembly should be performed according to manufacturer's procedures. **Figure 22-23** shows a cutaway

of an actuator that can be partially disassembled. It is often possible to check solenoid operation without disassembling the actuator. If there is any doubt as to whether an actuator can be repaired successfully, replace it.

Solenoid Tests

If the solenoid can be removed from the actuator, it can be checked with an ohmmeter or by energizing it with jumper wires, **Figure 22-24**. The ohmmeter readings across the windings should be within factory specifications. If readings are out of range, or read infinity or zero ohms, the solenoid is defective. If the solenoid cannot be energized with jumper wires, either the winding is defective or the solenoid plunger is stuck. In either case, the solenoid should be replaced.

Pump and Motor Tests

On some vehicles, the pump can be energized with jumper wires to check its operation. To prevent damage to other system components, the wiring harness connector must be removed from the pump before making this test. If the pump motor will not run, or if the pump will not produce pressure, it is defective.

If the pump runs continuously, check for a defective relay, hydraulic leak, or short to battery power. Also perform a pressure test to ensure the pump is not running due to the system not holding pressure.

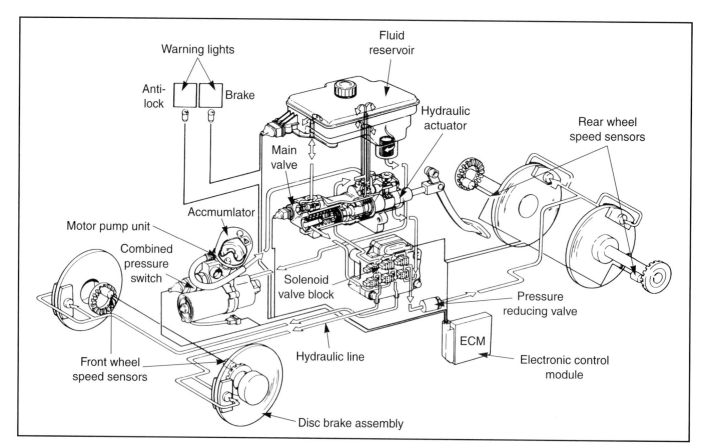

Figure 22-23. *A cutaway schematic view of one type of four-wheel anti-lock braking system. Study the various parts and the overall layout carefully.*

Checking Non-Integral Hydraulic Actuators

Many of the tests used for integral actuators can be used to check non-integral systems. Keep the following in mind when you are testing non-integral actuators:

❏ Non-integral actuators do not develop as much pressure as integral systems.
❏ Most non-integral systems cannot be disassembled for component testing.

More scan tool tests are available for non-integral actuators, since almost all of them have self-diagnostics. Use these along with pressure tests, electrical tests using a multimeter or scopemeter, and visual inspection. If the problem is in the actuator, it must be replaced as a unit.

Testing Output Devices Using a Scan Tool

Some scan tools allow the technician to diagnose output devices by bypassing the control module. Inputting commands through the tool will cause the device to operate, and the tool will monitor operating voltage and determine whether the device is operating properly, **Figure 22-25.** Some scan tools can simulate severe braking or loss of traction conditions to determine the operation and interaction of all system output devices. You should consult the scan tool and manufacturer's service manuals for detailed information on these diagnostic routines.

Actuator Tests
1. LF Solenoid
2. RF Solenoid
3. Rear Solenoid
4. Pump Motor
5. Warning Lamp
6. ABS Simulation
7. TCS Simulation

Figure 22-25. *Various output device tests are available by using a scan tool.*

Traction Control System Diagnosis

The brake portion of the traction control system can be diagnosed in the same manner as the ABS system. However, you must keep in mind that traction control systems also affect torque output through the engine ECM. Engine torque management is controlled by more than one method. Be sure to use the service manual and other resources to determine all possible causes of the problem.

Checking TCS Switch and Indicator Lamp

Begin by checking the TCS indicator lamp and control switch, if equipped. The lamp should illuminate for a

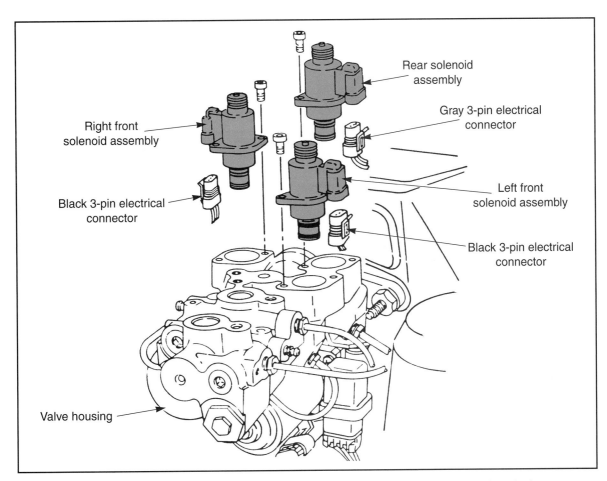

Figure 22-24. *Some ABS solenoids can be removed from the housing for testing and replacement. (Pontiac)*

few seconds when the vehicle is started. If the bulb does not illuminate, the problem is most likely an open bulb or in the instrument cluster.

If the bulb checks ok, proceed to test the switch. Begin by turning the TCS switch to off. The TCS light should illuminate when the switch is in the *off* position. If the light does not illuminate and the TCS system is disabled, the bulb is defective. If the light does not illuminate and the TCS system is active, the problem is either in the switch, wiring, or the control module. If the light illuminates, but the system remains active, the problem is in the control module.

Diagnosing Engine Torque Management Problems

If engine output is not reduced during periods of wheel slippage, the problem may be in the TCS system or the data bus to the engine ECM. If engine output is reduced at all times, the problem is either in the TCS system or the powertrain.

 Note: Most engine or drive train performance problems are unrelated to the traction control system.

If engine output is reduced, begin by checking the engine ECM for codes and other problems. Once the engine and drive train have been checked, you can proceed to check the traction control system.

Throttle Stepper Motor Problems

If the vehicle is equipped with a throttle stepper motor, check it by visual inspection first. Make sure the cables are not binding and the motor electrical connectors are tight. Ensure the throttle plate is not binding.

To check the stepper motor's operation, create an artificial wheel slip condition. Carefully raise the drive wheels off the floor. Have an assistant start the engine, put the transmission in Drive, and accelerate the engine. If the stepper motor is working properly, the motor will close the throttle plate in response to the wheel slip condition. If the motor does not move, recreate the wheel slip condition, but use a scan tool or meter to see if the stepper motor is receiving a signal. If a signal is being received, the motor is defective.

Replacing ABS and TCS System Parts

Many of the major components of the ABS or TCS system, such as wheel speed sensors, relays, control modules, and system modules, cannot be repaired. Some individual hydraulic actuator components can be serviced.

 Caution: Make sure that the ignition switch is in the off position, and is not turned on until all repairs are complete.

Replacing Speed Sensors

To replace a speed sensor mounted on a wheel assembly, check for clearance with the wheel and tire installed. It may be necessary to remove the wheel and tire assembly for access to the sensor. Remove the fasteners holding the sensor to the vehicle, **Figure 22-26.** Remove the sensor and its electrical connector. With the sensor removed, check the condition of the tone wheel, clean or replace as needed. Push the new sensor into place and attach the fasteners and electrical connector. Make sure that the sensor wiring cannot contact any rotating parts. If possible, check and adjust the sensor-to-tone wheel gap. Reinstall the tire and wheel assembly and recheck system operation.

 Note: Defective wheel speed sensors that are part of a sealed wheel bearing assembly usually require replacement of the hub and bearing assembly.

To replace a speed sensor mounted on the differential assembly or transmission, raise the vehicle and remove the electrical connector. Remove the fasteners holding the sensor to the transmission or differential, and pull the sensor out of the housing. On some vehicles, the clearance must be determined by measuring the distance between the housing and tone wheel with the sensor removed. Before installing the new sensor, check the mounting surface. Clean the surface and remove burrs if necessary. To install the new sensor, push it into place and install the fasteners. Then reconnect the electrical connector, making sure that all wiring is clipped in place properly and away from moving parts. Adjust the gap if possible, then check fluid level and ABS operation.

Replacing Tone Wheels

Tone wheels seldom require replacement, but can be damaged by road debris. On most vehicles, tone wheels are not removable, but must be serviced by replacing the brake rotor, CV joint, bearing, hub, or axle assembly. In a few cases, the tone wheel may be pressed or bolted to the rotor, axle assembly, CV joint, or axle shaft.

Remove the wheel and tire assembly, brake rotor, and axle, if needed, to gain access to the tone wheel. Tone wheels installed in the transmission or differential housing can only be removed after major unit disassembly, **Figure 22-27.** For these procedures, refer to the appropriate drivetrain sections of the service manual. Once all necessary components have been removed, the tone wheel can be replaced. Compare the old and new tone wheels before installation.

To replace a tone wheel that is bolted in place, remove the fasteners and remove the tone wheel. Check the mounting surface for dirt and burrs, and correct as necessary. If the tone wheel is pressed on, it can be hammered or heated to remove it. However, the replacement tone wheel should be carefully pressed into position using a suitable hydraulic

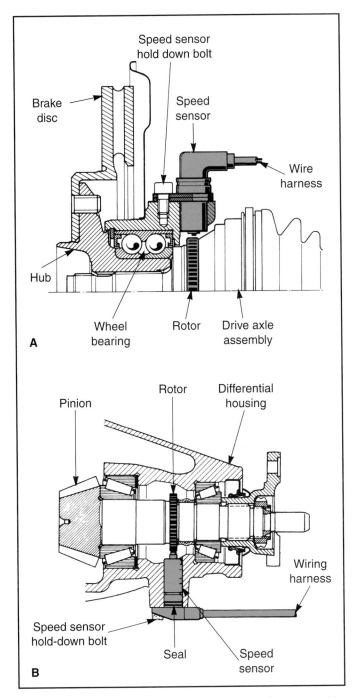

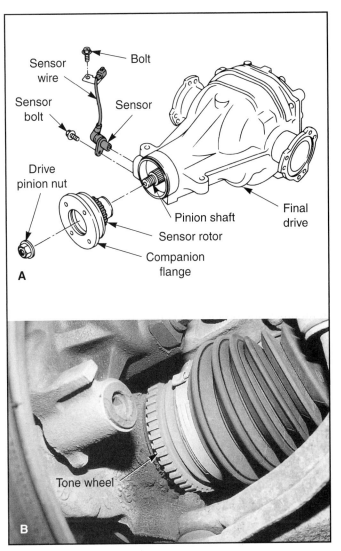

Figure 22-26. A—This wheel speed sensor can be removed by simply loosening and removing the hold-down bolt. B—A differential mounted speed sensor. Its removal requires the hold-down bolt be removed. Some gear lube may run out after the sensor is pulled free of its hole.

Figure 22-27. A—Placement of one type of rear differential mounted speed sensor and the sensor tone wheel. Note the sensor tone wheel is connected to the pinion shaft companion flange. B—This tone wheel is part of the CV joint.

press. Be sure to tighten any fasteners in a criss-cross pattern to reduce the chances of warping the tone wheel. Check wheel sensor gap and adjust as necessary.

 Caution: Replacement tone wheels should be treated carefully. Dropping, heating, or hammering may warp or damage a new tone wheel. Either condition can cause an erratic sensor signal.

Replacing G-Force Sensors

The G-force sensor is usually mounted in the passenger compartment, **Figure 22-28.** Begin by locating the sensor. It may be necessary to remove the seat or center console to gain access to the sensor. Then remove the wiring harness connector.

 Note: Before removing the sensor, carefully check the service manual to determine whether the angle of the sensor must be checked and recorded. This procedure was discussed under G-force sensor checking and adjusting.

To replace the sensor, loosen the bracket fasteners and remove the sensor. To install the new sensor, reverse the removal process. If necessary, adjust the sensor so it is at the same angle as the old sensor.

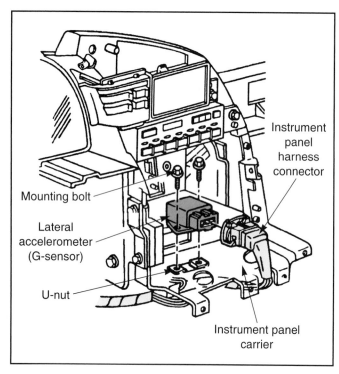

Figure 22-28. *Removing the G-force sensor for replacement. Handle these sensors with care. Just dropping on the shop floor may damage them.*

Replacing Brake Light and Pedal Travel Switch

Replacing the brake light and/or pedal travel switch uses almost the same procedure. A typical pedal travel switch installation is shown in **Figure 22-29**. Remove the clips holding the plunger-to-pedal assembly. Then loosen and remove the fasteners holding the switch to the vehicle frame and remove the switch. Reverse the installation procedure to install the new switch. Adjust the switch as outlined in the checking and adjusting the pedal travel switch section earlier in this chapter.

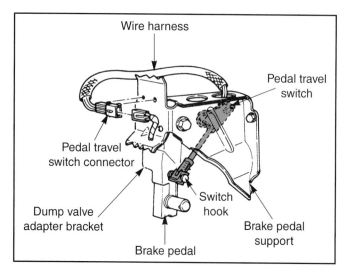

Figure 22-29. *A typical brake pedal travel switch and its mounting location.*

Most brake light switches simply pull out of the retainer and then the harness is removed. Installation is the reverse of removal. Adjust the brake light switch by pulling up on the brake pedal. In some cases where the brake light switch and pedal travel switch are part of the same unit, a special adjustment procedure may be needed.

Replacing the Control Module

The internal components of the control module cannot be serviced, and the control module is replaced as a unit. On many newer vehicles, the control module is part of the electronic control module (ECM) or powertrain control module (PCM). The module must be replaced as a unit. In a few cases, the ABS control module is part of the hydraulic actuator.

Before disconnecting any wiring, make sure the ignition switch is off, or disconnect the battery. Remove any components as needed to access the control module. Then remove the electrical connectors and fasteners and remove the module. Compare the old and new modules to ensure the new module is correct. Then install the new module and tighten the fasteners. Reattach the electrical connectors and install any components removed to access the control module. Depending on the model year, it may be necessary to enter certain vehicle parameters, such as tire size. Check the service manual for specific information. If all other repairs are complete, recheck ABS operation.

 Caution: To prevent static discharge damage to the new module, follow manufacturer's instructions concerning static discharge and unit grounding.

Reprogramming the Control Module

Many of the latest scan tools can be used to reprogram the ABS control module. Reprogramming is done to replace the original system program with an updated program. The updated program is designed to correct any problems that have been reported from the field. Often, problems that were not anticipated when the original programming was developed become evident after vehicles have been sold and operated for a few months or years. To avoid the expense of installing a new control module, methods of reprogramming the original module have been developed.

To reprogram the control module, the scan tool must have the new programming. In most cases, the scan tool is programmed with the information from a shop computer. The new programming will be sent to the shop on a CD-ROM disc and loaded into the scan tool using the shop computer. The scan tool can then load the information into the vehicle control module. On a few systems, the scan tool is used as a connector between the vehicle and another computer.

To reprogram the control module, load the new information into the scan tool and connect it to the vehicle data link connector. Then follow the menu directions to erase the old information and install new programming.

Servicing Hydraulic Actuator Assemblies

Hydraulic actuators may be installed in a single assembly, or as a separate unit connected by high pressure hoses and lines, as shown in **Figure 22-30.** Be sure to identify the proper components before beginning the removal process. Integral hydraulic system parts that can be replaced include the actuator assembly, accumulator, hydraulic pump and motor, pressure switches, and hydraulic lines.

Warning: Before attempting to replace any hydraulic system components, depressurize the hydraulic system. The hydraulic pump may attempt to repressurize the system if the ignition switch is turned on. This will spray fluid from any disconnected fittings. For maximum safety, turn off and remove the key from the ignition switch and disconnect the battery cable before beginning repairs. Once you are certain that battery power is disconnected, depressurize the brake system by pumping the brake pedal at least 40 times or until it is difficult to depress or becomes "hard." This will discharge the accumulator and remove pressure. This is especially important when working on integral ABS systems.

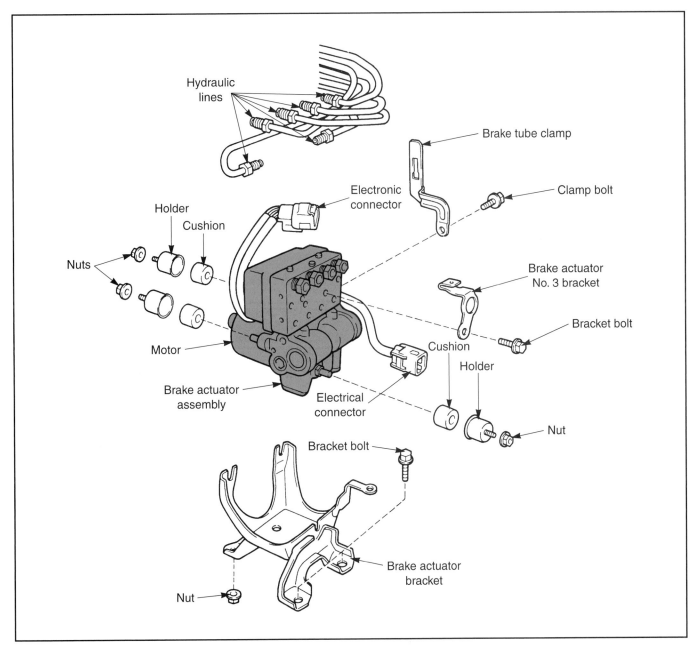

Figure 22-30. *ABS actuator components which may be installed as a single unit or separate from the master cylinder. (Toyota)*

424   Auto Brakes Technology

Integral Hydraulic System Service

Most integral hydraulic actuators can be partially disassembled without removal from the vehicle. Due to the great variety of actuators and service procedures, always consult the service manual for exact service information. If the component must be removed from the master cylinder, carefully remove all fasteners, according to manufacturer's directions. If any hydraulic actuator component shows signs of internal corrosion, contamination, or if there is any doubt as to its condition, the unit should be replaced as an assembly. When overhauling or servicing any integral hydraulic unit, be sure to use all new parts supplied, especially O-rings and gaskets.

Accumulator Replacement

Most integral ABS accumulators are threaded into the master cylinder or valve actuator body. To replace the accumulator, depressurize the system (if it has not been done). Then, using a wrench or socket bit, loosen the accumulator. Once loose, the accumulator should turn by hand. Carefully remove the accumulator from the ABS unit, **Figure 22-31.**

Once the old accumulator is removed, check at the base of the screw threads for the old O-ring. If it is still in the master cylinder or valve block, carefully remove using a machinist's pick. Lubricate the new O-ring with clean brake fluid and slip over the threads of the new accumulator. Install the new accumulator and tighten with a wrench or socket bit, whichever is called for. Some ABS accumulators are part of the valve body assembly. This setup requires replacement of the entire valve body.

Pressure Switch Replacement

Most pressure switches are threaded into the hydraulic control body. In some cases, they resemble oil pressure switches. To replace a pressure switch, disable battery power and depressurize the system. Locate the switch on the hydraulic system. Remove the electrical connector and unscrew the pressure switch, **Figure 22-31.**

 Note: In some cases, a special socket may be needed to loosen the switch. Consult the service manual if the switch will not loosen using a wrench.

After the old switch is removed, make sure any O-rings are present. If not, look inside the valve body and remove the O-ring. Lubricate and place a new O-ring on the replacement switch. Install the new switch, tightening to the proper torque. Reinstall the battery cable and start the engine, allowing it to pressurize the system. Recheck system operation and if everything is ok, clear any stored codes.

Pump and Motor Replacement

On some ABS systems, the pump and motor are serviced as a separate unit from the master cylinder, hydraulic valve, and other parts, **Figure 22-32.** Begin by making sure the unit is depressurized. Disconnect the electrical connections to the pump. Using a flare-nut wrench, loosen and remove the lines from the pump or hydraulic valve assembly. Make sure you retrieve any O-rings or copper seals used with the lines. Loosen and remove the bolts and pump assembly from the vehicle.

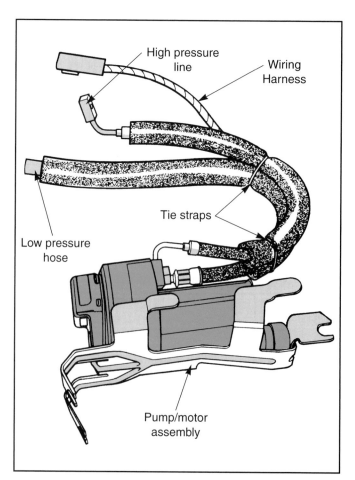

Figure 22-32. *Hydraulic pumps are usually connected by hoses or directly to the master cylinder. (Honda)*

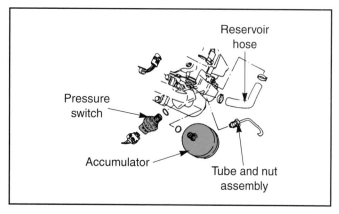

Figure 22-31. *The pressure switch and accumulator are screwed into the ABS unit and sealed with O-rings. Replace the O-rings when servicing these components. (Pontiac)*

In almost all cases, installation is the reverse of removal. Be sure to transfer over any switches, lines, hoses, etc., from the old unit to the new one. You may need to pour some brake fluid into the intake port for the pump to ensure it has a good prime charge. To bleed the new pump, depressurize and repressurize the system until the fluid in the reservoir comes out free of air.

Hydraulic Valve Body/Solenoid Replacement

To change a hydraulic control body and its related solenoids, begin by disconnecting the battery. Then, depressurize the system as outlined earlier. Remove any electrical connectors and use a flare-nut wrench to remove the lines. Remove the fasteners holding the hydraulic valve body and remove it from the vehicle, **Figure 22-33.** Make sure you account for all O-rings as you remove the valve body.

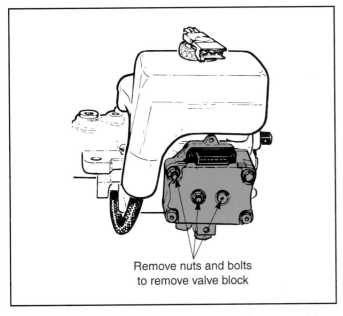

Remove nuts and bolts
to remove valve block

Figure 22-33. *The hydraulic valve body can be removed from the master cylinder. This makes servicing each individual part much easier. (General Motors)*

Compare the old valve body to the new assembly. If needed, transfer all related parts (solenoids, switches, reservoirs) to the new valve body. Use new O-rings as needed, lubricated with brake fluid. Place the new valve body in position and install the fasteners, but do not torque. Reinstall all hydraulic lines and electrical connectors, then torque the valve body fasteners. Reinstall the battery cable and start the engine. Bleed the brake system as needed and make sure there are no trouble codes.

Replacing Integral Hydraulic Actuator Assemblies

In some cases, it is necessary to remove and replace the entire assembly. After the system is depressurized, begin the replacement procedures by removing any electrical

connections. Then use the proper flare-nut wrench to loosen and remove all tubing connections. After all lines are removed, remove any fasteners holding the component to the vehicle and remove the actuator. In some cases, you may need to transfer some parts from the old unit.

Begin reinstallation by replacing all affected seals and gaskets. Lubricate all new seals with clean brake fluid before installation. If possible, bench test the unit before reinstallation. Then install the replacement actuator, using new gaskets, O-rings, and seals as needed. Tighten all fasteners to the correct torque, and use new fasteners where called for by the manufacturer. Carefully reinstall the hydraulic lines. Reattach the electrical connectors, and bleed the brakes according to the procedures given in the service manual.

 Note: It is not necessary to bleed some integral ABS systems after certain hydraulic actuator repairs. Check the service manual for details.

Non-Integral Hydraulic Assemblies

Most non-integral hydraulic actuators are not serviceable in the field and require replacement if defective. Depending on the design, there may be special service procedures needed for actuator replacement.

Replacing Non-Integral Hydraulic Actuator Assemblies

Start by checking the service manual for any special procedures. You may need to depressurize the system or disengage the actuator's motor(s) before beginning work. Place a pan under the actuator, then disconnect all electrical connections. Using a flare-nut wrench, loosen and remove the brake lines from the actuator. Unbolt the actuator and remove from the vehicle, **Figure 22-34.** You may need to remove other components, such as the master cylinder, to gain access to the actuator.

 Note: Do not remove the master cylinder lid at any time while the actuator is removed. This will minimize brake fluid loss.

In some cases, you may need to transfer parts, such as retainer brackets, from the old actuator to the new unit. Installation is the reverse of removal. Check the service manual for any specialized procedures as needed before installing the new actuator.

After all repairs and system bleeding are complete, start the vehicle and check ABS and traction control operation. The ABS warning light may remain on until system pressure builds up. After checking system operation, recheck the hydraulic system for leaks and add brake fluid to the master cylinder reservoir if necessary.

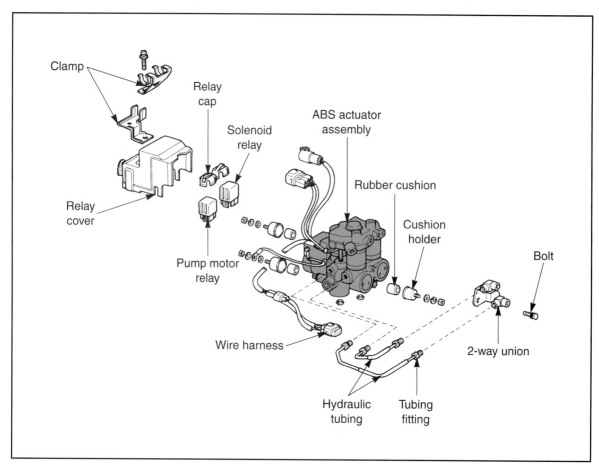

Figure 22-34. *A partially disassembled view of a non-integral hydraulic actuator. Service with care. Incorrect parts or service procedures may affect the brake system performance, leading to a driving hazard. (Toyota)*

Bleeding the System

Methods of bleeding an ABS or TCS system can vary greatly from bleeding procedures used on a standard brake system. Proper bleeding removes all air from the hydraulic actuator assembly, including the pump, solenoids, accumulator, and lines. Air must also be removed from the foundation brake parts, including the master cylinder, calipers, wheel cylinders, proportioning/metering, and pressure differential valves. If the entire system is not bled properly, air may be trapped in the system. This can cause a spongy brake pedal, and possible erratic system operation.

There are two general bleeding methods used on ABS/TCS systems. On some vehicles, either method can be used. On some systems, the system manufacturer recommends only one method. Exact bleeding procedures will also vary by system manufacturer. Check the master cylinder fluid level frequently, do not allow the reservoir to run dry.

The first method is manual bleeding, as used on standard brake systems, **Figure 22-35.** This type of bleeding was covered in earlier chapters. Note that it may be necessary to open lines at the hydraulic actuator or accumulator to remove additional trapped air. After the ABS or TCS is purged, bleed the wheel brakes to ensure that all air has been removed from the system.

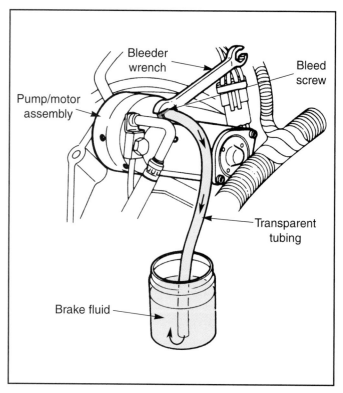

Figure 22-35. *Bleeding the pump/motor and hydraulic accumulator. (General Motors)*

⚠ **Warning: Follow the bleed procedure outlined in the service manual or other literature for each system. Failure to follow the proper bleed procedure may result in air becoming trapped in the hydraulic actuator. This trapped air may not become apparent until during or after the first ABS/TCS episode.**

ABS Bleeding Using a Scan Tool

Some ABS systems, such as the Kelsey-Hayes 4WAL system, must be bled using a scan tool to manually operate the hydraulic actuator. If this is not performed on these systems, a low brake pedal or improper ABS operation may result. Scan tool actuation must be used along with conventional bleeding techniques.

Begin by manually bleeding the master cylinder and lines to the actuator. Follow manufacturer's recommendations for manually bleeding the actuator, Then, install the scan tool, and follow manufacturer's directions to energize the actuator, **Figure 22-36.** As the actuator operates, it will purge itself of air by venting it to the brake calipers, wheel cylinders, or to the fluid reservoir. Repeat as needed to purge all air from the system.

Once the ABS and traction control system circuits are bled in this manner, the foundation brake hydraulic system can be pressure or manually bled by following the procedures covered in Chapter 6.

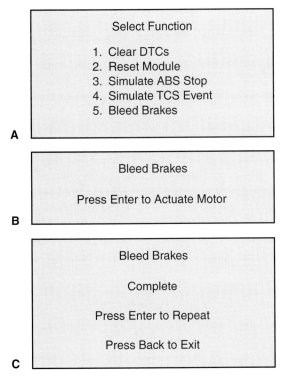

A

B

C

Figure 22-36. *Scan tools are sometimes needed to bleed certain ABS/TCS systems. A—Select the bleed brakes function from the scan tool menu. B—Follow instructions to operate the hydraulic modulator. Do not drive the vehicle during this process. C—Repeat as needed along with manual bleeding.*

Flushing the System

If the system is being flushed to replace old fluid, it can be satisfactorily handled by opening the bleeder valves while the system is manually pressurized. This was discussed in Chapter 6. However, if the system contains metal particles or other contaminants, it should be more thoroughly flushed. The most satisfactory way to completely flush the hydraulic system is to manually remove as much debris as possible, followed by flushing with a non-petroleum solvent, such as brake cleaner, followed by brake fluid to flush all solvent from the lines.

Begin the flushing procedure by disconnecting all accessible hydraulic lines and blowing them out with compressed air. In many cases, major components such as the master cylinder should be removed and washed out with approved brake cleaner. Make sure that all cleaning solution is removed before reinstalling the component. In cases of severe contamination, the master cylinder, calipers, wheel cylinders, and other sealed hydraulic system parts may require rebuilding or replacement. Also check for any system filters and screens and clean them thoroughly.

 Caution: Do not attempt to flush an ABS or TCS hydraulic actuator. If the hydraulic system has been exposed to a contaminant, replace the actuator.

Once the hydraulic system has been reassembled, finish flushing the system using brake fluid. This procedure is similar to the bleeding procedures discussed earlier. Once the fluid appears to be clean, finish removing any trapped air and recheck system operation.

Follow-up after ABS/TCS Repairs

After all repairs are complete, road test the vehicle to ensure the system is operating properly. Check for proper system operation and observe that the warning lights come on briefly when the engine is started and remain off thereafter. Perform at least 2-3 ABS assisted stops or TCS assisted accelerations from a stop. The brake pedal should remain firm after each test. Obey all traffic regulations. Be sure to accurately document the cause and what you did to correct the problem.

Summary

Hydraulic components on anti-lock brake and traction control equipped vehicles are not serviced in the same manner as conventional systems. Many common brake service procedures are not affected by the presence of ABS or TCS systems. However, you should carefully select replacement parts to ensure that they can withstand the higher pressures.

Troubleshooting ABS and TCS systems may require a scan tool or other special testers. However, many preliminary checks can be made by talking to the driver, and by using standard tools and test equipment. Always check tire sizes, the foundation brakes, system fuses and relays, and the vehicle electrical system before deciding that the anti-lock brake or traction control system is defective. Once you are certain that the system is at fault, retrieve trouble codes before proceeding. If possible, use a scan tool to retrieve codes and perform other diagnostic routines. On some systems, another type of special tool can be used to perform a series of diagnostic steps that will isolate problem areas or components.

After retrieving the trouble codes, proceed to check the indicated problem areas. Areas to be checked include the wheel speed sensors, pedal travel, G-force, and pressure switches, control module, relays, fuses, and hydraulic actuator components. Many electrical components can be checked with an ohmmeter or other testers. On many systems, the hydraulic components can be checked using a pressure gauge.

Most anti-lock brake service can be performed without special tools. Speed sensors and control modules can be replaced without depressurizing the hydraulic system. Handle speed sensor tone wheels carefully to avoid damaging them. Always depressurize the hydraulic system before doing any work on the hydraulic components. Ensure that the ignition key remains off until the hydraulic system is completely reassembled and bled.

Bleeding procedures are similar to those used on standard systems. Some bleeding procedures can be done using the system hydraulic pump. Flushing should be done to remove any metal particles from the hydraulic system.

Review Questions—Chapter 22

Please do not write in this text. Write your answers on a separate sheet of paper.

1. What type of brake fluid should *not* be used in an anti-lock brake/traction control system?

2. Chassis grease or road tar could interfere with the operation of the _____ sensors.

3. What effect will a normally operating ABS have on normal braking?

4. If the brakes lock up when the vehicle is braked hard at 5 mph (8 kph), but do not lock up under hard braking at 20 mph (32 kph), what is this a sign of?

5. Is it a normal condition for the ABS warning light to be on when the vehicle is first started?

6. Is it a normal condition for the ABS warning light to flicker as the vehicle is braked normally?

7. Mismatched tires will affect the readings provided by the _____ sensors.

8. The red or amber lights may be put on by a _____.

9. If the vehicle charging system voltage is below _____ volts, something is wrong.

10. List the three ways to read ABS/TCS diagnostic trouble codes.

11. Before proceeding with electrical diagnosis of an ABS or traction control system, check the system _____ and relays.

12. On some vehicles, the sensor _____ can be checked using a brass feeler gauge.

13. If the brake fluid contains metal, what is wrong with the system?

14. Before making any ABS repairs, be sure that the _____ switch is off.

15. Some ABS pumps can be energized with _____ for testing.

ASE Certification-Type Questions

1. Technician A says that ABS operation during a hard stop is a sign of a defective sensor. Technician B says that pulsation in the brake pedal is a sign of a defective actuator. Who is right?
   (A) A only.
   (B) B only.
   (C) Both A & B.
   (D) Neither A nor B.

2. Which of the following is *not* an ABS/TCS system defect?
   (A) Defective power booster.
   (B) Leaking hydraulic actuator.
   (C) Stuck actuator solenoid.
   (D) Warped speed sensor tone wheel.

3. On older vehicles, the trouble codes can be obtained by observing dashboard light flashes after the diagnostic terminal is _____.
   (A) energized with 12 volts
   (B) energized by an external 9 volt battery
   (C) energized by the scan tool battery
   (D) grounded

4. A scan tool could be used to retrieve trouble codes from ____.
   (A) the ABS system
   (B) the TCS system
   (C) the engine ECM
   (D) All of the above.

5. Technician A says that the scan tool keypad is used to select menu operations from the display screen menu.

Technician B says that all scan tools have an internal battery. Who is right?

(A) A only.

(B) B only.

(C) Both A & B.

(D) Neither A nor B.

6. Integral ABS systems have all of the following features, EXCEPT:

(A) accumulators.

(B) hydraulic pumps.

(C) operate at low pressure.

(D) master cylinders.

7. Technician A says that all ABS systems have the same trouble code format. Technician B says that a trouble code can indicate a defective wheel speed sensor. Who is right?

(A) A only.

(B) B only.

(C) Both A & B.

(D) Neither A nor B.

8. Low charging system voltage could cause which of the following?

(A) Erratic control module operation.

(B) Blown ABS fuses.

(C) Poor grounds.

(D) Defective relays.

9. To check brake hydraulic pressures, a gauge capable of reading at least _____ should be used.

(A) 300 psi (2069 kPa)

(B) 2000 psi (13 790 kPa)

(C) 4000 psi (27 580 kPa)

(D) 10,000 psi (68 950 kPa)

10. Wheel speed sensors can often be checked by attaching a(n)_____ to the sensor leads.

(A) voltmeter

(B) ohmmeter

(C) test light

(D) ammeter

11. To discharge the hydraulic system, pump the brake pedal at least _____ times.

(A) 10

(B) 20

(C) 40

(D) 80

12. Technician A says that a G-force sensor should be installed so that it is at the same angle as the old sensor. Technician B says that a wheel speed sensor should be installed so that it just touches the tone wheel. Who is right?

(A) A only.

(B) B only.

(C) Both A & B.

(D) Neither A nor B.

13. When adjusting a brake pedal travel switch, which of the following should be done *last?*

(A) Remove the switch plunger from its mounting clip.

(B) Pull the brake pedal to bring the plunger to the end of its outward travel.

(C) Press the brake pedal firmly.

(D) Reinstall the plunger on its mounting clip.

14. Some ABS and traction control systems must be bled using a _____ to operate the hydraulic actuator.

(A) jumper wires

(B) scan tool

(C) test light

(D) starter relay

15. When the ABS system contains metal particles, what should be done?

(A) Flush the system.

(B) Place a magnet in the reservoir.

(C) Replace all brake hydraulic system parts.

(D) Replace all brake friction parts.

Good diagnostics takes skill, patience, and a logical plan. This chapter will outline the strategies good technicians use to diagnose problems. (Ford)

Chapter 23

Troubleshooting Brake Systems

After studying this chapter, you will be able to:

- ❑ Use the seven-step procedure to troubleshoot brake problems.
- ❑ Question drivers concerning suspected brake problems.
- ❑ Safely road test vehicles to determine actual brake system problems.
- ❑ Separate brake system problems from other vehicle problems.
- ❑ Use correct diagnosis charts and test procedures.
- ❑ Inspect brake components for wear and damage related to the complaint.
- ❑ Determine what repairs must be made to correct brake problems.
- ❑ Recheck brake operation after repairs.

Important Terms

Strategy-based diagnostics
Logic
Seven-step troubleshooting process
Intermittent problems
Special tools
Follow-up
Uneducated guessing

Documentation
Noise, vibration, and harshness (NVH) diagnostics
Vibrations
Noise
Harshness
Transmitter

Transfer path
Receiver
Balancing
Dampening
Order
First-order vibration
Second-order vibration

This chapter is a general overview of the techniques needed to diagnose brake system problems. It combines the testing and diagnostic techniques discussed in earlier chapters with logical approach methods to troubleshooting. This is done by combining what you have learned in previous chapters with a seven-step troubleshooting process that will enable you to locate and correct brake system problems. Studying this chapter will enable you to develop an overall plan for finding the solution to any brake system problem.

Strategy-Based Diagnostics

In the past, it was fairly easy to find and locate a problem, since most vehicle systems were simple and common to many, if not all vehicles. As vehicles became more and more complex, the methods used to diagnose them became obsolete and in some cases, inapplicable. Technicians who were used to using the older diagnostic routines, or no routine at all, began to simply replace parts, hoping to correct the problem, often with little or no success. Unfortunately, this process was very expensive, not only to the customer, but to the shop owner as well.

In response to this problem, a routine involving the use of logical processes to find the solution to a problem was devised for use by technicians. This routine is called **strategy-based diagnostics.** The strategy-based diagnostic routine involves the use of a logical step-by-step process, explained in the next sections. Variations of strategy-based diagnostics is used in many fields outside of automotive repair. A flowchart of this process as recommended by one vehicle manufacturer is shown in **Figure 23-1.**

When troubleshooting a brake problem, always proceed logically. *Logic* is a form of mental discipline in which you weight all factors without jumping to conclusions. To work logically, the first thing you must know is how the brake system works. This has been covered in earlier chapters. The knowledge that you have gained can be put to use when a problem occurs.

The second thing that you need is a logical approach. To diagnose a problem, think about the things that could cause the problem, and just as important, what things cannot cause the problem. You can then proceed from the simplest things to check to the most complex. Do not guess at possible solutions, and do not panic if the problem takes a little while to find. If you remember these points, you will be able to diagnose most brake problems with a minimum of trouble.

The Seven-Step Troubleshooting Procedure

Troubleshooting is a process of taking logical steps to reach the solution to a problem. It involves reasoning through a problem in a series of logical steps. The great advantage of the **seven-step troubleshooting process** is that it will work in the troubleshooting of *almost any vehicle system*. Refer to **Figure 23-2** as you read the following sections.

Step 1—Determine the Exact Problem

Often the vehicle driver will state the problem in inexact terms, such as "It doesn't stop right" or "Something is wrong with the brakes." The first step is to determine the exact problem. This means finding out the exact complaint, what its symptoms are, and how it affects vehicle operation. This process involves talking to the driver and road testing the vehicle.

Talking to the Driver

Obtaining information from the driver is the first and most important part of troubleshooting. Information from the driver will sometimes allow you to bypass some preliminary testing and go straight to the most likely problem. In one sense, the driver begins the diagnostic process by realizing the vehicle has a problem and deciding it is serious enough to require service.

Question the driver as to the exact nature of the complaint. Try to get an accurate description of the problem before beginning work on the vehicle. Since the driver is usually the vehicle owner, he or she can provide some idea of past service problems and any maintenance that has been performed or neglected. Try to determine what is going on by a series of basic questions:

❑ Do the brakes work at all?
❑ Is the brake pedal low or spongy?
❑ Is the braking effort high or does the pedal seem hard?
❑ Does the pedal vibrate or pulse when the brakes are applied?
❑ Does the vehicle pull to one side during braking?
❑ Do you hear any noises during braking?
❑ Is there a burned smell after hard braking?
❑ Does the problem occur only during wet weather?
❑ Does the problem occur only when cold, or after the brakes are warmed up?
❑ Does the problem occur only after continued or hard braking?
❑ Did the vehicle have any recent brake, suspension, steering, or wheel bearing service?
❑ Did the problem start suddenly, or gradually develop?

You may think of other questions depending on the answers that you get to the above questions. Write down the driver's comments on an inspection form, **Figure 23-3,** the repair order, or a sheet of paper.

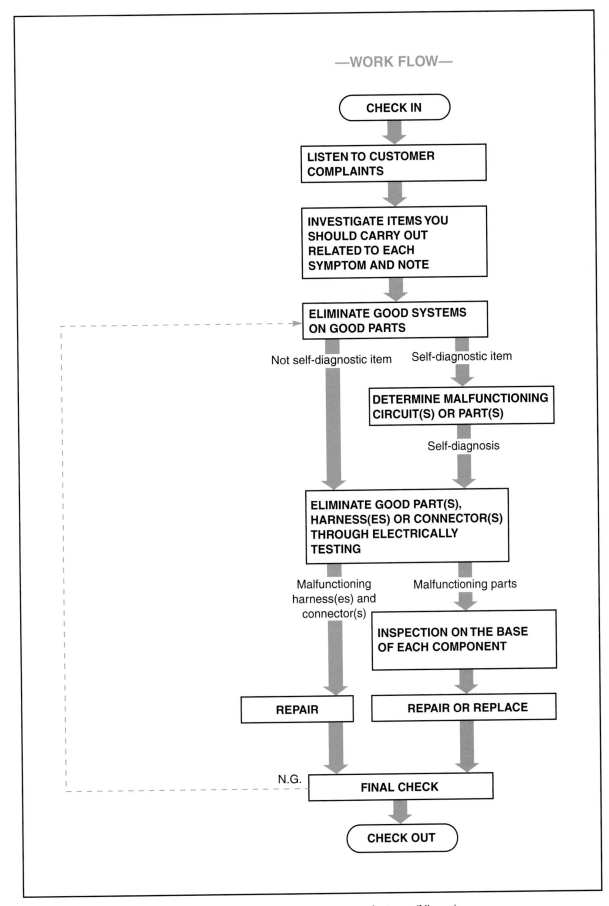

Figure 23-1. *Strategy-based diagnostic procedure as outlined by one manufacturer. (Nissan)*

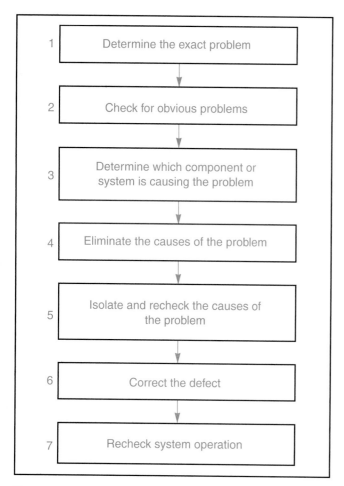

1 Determine the exact problem

2 Check for obvious problems

3 Determine which component or system is causing the problem

4 Eliminate the causes of the problem

5 Isolate and recheck the causes of the problem

6 Correct the defect

7 Recheck system operation

Figure 23-2. *As you are trying to find and diagnose a brake problem, it is essential that you follow some type of logical progression sequence. This will aid in locating the trouble in a faster manner, and will also prevent duplicating diagnostic procedures.*

Assessing Driver Input

While taking into account what the driver says, try to estimate his or her attitude and level of automotive knowledge. Because drivers are not usually familiar with the operation of automobiles, they often unintentionally mislead technicians when describing symptoms or may have reached their own conclusion about the problem. In describing vehicle problems, drivers have been known to use hand gestures, body language, and even simulate noises they have heard. While this can sometimes be fun to watch, keep in mind it is a part of the diagnostic process. Many times, important clues can be found simply by observing a driver's physical actions while describing a particular problem.

In many cases, the person bringing in the vehicle has already formed an opinion as to what is wrong. These opinions are a common occurrence, often based on poor or incomplete understanding of vehicle operation, advice from uninformed friends, or other failures to fully comprehend the problem. The best course is to listen closely to the driver's description of the symptoms. Some drivers will be sensitive to even slight changes, and may be overreacting to a normal condition. Never accept a driver's or another shop's diagnosis until you can verify it.

Often, the owner is concerned about the cost of repairs. Some will even downplay the symptoms, hoping for an inexpensive repair. Very few vehicle owners are unconcerned about the cost of vehicle repairs and maintenance. Do not give any type of uninformed estimate, even though you may have a good idea of the problem. Giving an estimate without diagnosis is a mistake made by many technicians. This practice invites one of two things to occur; either the recommended repair will not correct the problem or it will frighten the driver, who may decide to take his or her vehicle to another shop or not have the repair done at all. Explain that the charge for diagnosing the problem is actually more cost effective than paying for a service, which in many cases may not fix the problem. Before going on to the road test, be sure you have a good idea of the driver's complaint.

Road Testing

If possible, perform a road test to confirm the problem. A short test drive will usually confirm the driver's complaint. In some cases, the road test may reveal the problem is not in the brake system. Keep in mind that it is dangerous to road test any vehicle that may have a potential brake problem.

⚠ **Warning: Do not attempt to road test a vehicle if the brakes are in extremely poor condition or do not work at all. Before beginning the road test, apply the brake pedal to ensure that it does not go all the way to the floor. Before leaving the shop area make several slow speed brake applications to ensure the brakes are safe for driving.**

Before beginning a road test, make a few quick checks to ensure the vehicle can be safely road tested. Walk around the vehicle's exterior and make a note of any damage that is present. Check each tire to ensure they are inflated properly and in good condition. Also make sure that all safety-related equipment, such as the turn signals and horn are working properly.

Turn the steering wheel and make sure the steering system does not have excessive play. Also make sure the vehicle has enough fuel to conduct a road test. Wear your seat belt at all times during the road test.

Try to duplicate the exact conditions under which the driver says the problem occurs. Unfortunately, duplicating some conditions is not always possible. Always try to road test the vehicle with the owner. This will ensure you are both talking about the same problem, and will save valuable diagnostic time.

 Note: Do not adjust anything in the passenger compartment, such as mirror, seat, and tilt steering wheel position, than is absolutely necessary. If the radio is on, turn it off so that you can listen for unusual noises.

Customer Input Checklist

Name _____

Phone (H) _____ (O) _____ Date _____

In order to help us analyze your car's brake system problems and provide accurate cost estimates for repairs/replacements, please answer the following questions:

1. Does your car pull (steer) toward one side when the brakes are applied?

    Yes _____ No _____. If yes, which side? R _____ L _____.

2. Does brake pedal travel seem to be excessive? Yes _____ No _____.

3. Does the brake pedal pulsate (move up and down) when the brakes are applied?

    Yes _____ No _____.

4. Do any wheels lockup or skid easily when the brakes are applied? Yes _____ No _____.

5. Are there any unusual noises when you apply the brakes? Yes _____ No _____.

6. Does the pedal fade (gradually drop toward the floor) while the brakes are applied?

    Yes _____ No _____.

7. Additional information: _____

    _____

    _____

    _____

Cost estimate _____ Approved for service _____

Figure 23-3. *Be sure to always write down any customer comments on a sheet of paper, or use a customer input checklist. This will help to give you a basic idea of the possible problem before road testing the vehicle. If necessary, take the customer along, so he or she can help to describe the problem. (Bendix)*

Drive slowly as you leave the service area to ensure that no obvious mechanical problems exist that could further damage the vehicle or cause personal injury. Drive the vehicle carefully and do not do anything that could be remotely constituted as abuse. Tire squealing takeoffs, speed shifts, fast cornering, and speeding can all be interpreted as misuse of the vehicle. Remember: any vehicle with a suspected brake problem should be treated as an unsafe vehicle.

Find a level road with light traffic to make the brake test. It is best to let traffic pass you before beginning the test. Do not make any hard stops when vehicles are behind you. Make several light and medium stops at various speeds, bringing the vehicle to a complete stop each time.

When pushing on the brake pedal, observe how much effort is required each time you stop. Check for low pedal, poor braking effect, hard pedal, grabbing, pulling, and pedal pulsation. Check for noise by opening the driver's window and turning off the radio and blower motor. If you hear any squealing or rubbing when the brakes are applied, try to determine the affected wheel. Manually apply and release the parking brake to check for binding or failure to hold. If the brakes are pulsating, gently apply the parking brake while the vehicle is moving to determine whether the front or rear wheels are causing the pulsation.

 Caution: Do not make panic stops unless you are testing the ABS system. Take the vehicle to an isolated area for ABS road tests.

While road testing, obey all traffic rules, and do not exceed the speed limit. It is especially important to keep in mind that you are under no obligation to break any laws while test driving a customer's vehicle. Also be alert while driving. It is easy to become so involved in diagnosing the problem, that you forget to pay attention to the road or the traffic around you. If it is necessary to monitor a scan tool's readout or look for a problem while the vehicle is driven, get someone (not the vehicle's owner) to drive for you. Once you have determined the exact nature of the problem, proceed to Step 2.

Step 2—Check for Obvious Problems

Most of your time in Step 2 will be spent checking for obvious problems, or problems that can be easily tested. Visual checks and simple tests take only a little time, and might save more time later.

Performing Visual Inspections

The actual visual inspection begins during the road test. In many cases, seemingly complicated problems are caused by a defect that can be easily seen, smelled, or heard. Be sure to stop the engine before investigating any component near hot or moving engine parts.

During this step, you should check the master cylinder level, look for brake fluid leaks, and observe whether or not any brake system related dashboard lights are on. You can also look for loose vacuum hoses and get a general idea of overall engine condition and frequency of maintenance.

Depending on the brake complaint, you might want to make a pedal effort test using a force meter, **Figure 23-4.** If the problem is in the anti-lock brake or traction control system, you should check the charging system voltage, system electrical connections, and grounds. Look for prior work done on the vehicle if no service history is available. Finally, take a close look for signs of abuse or tampering. Any components that have been modified, removed, or replaced by non-stock parts may affect vehicle operation.

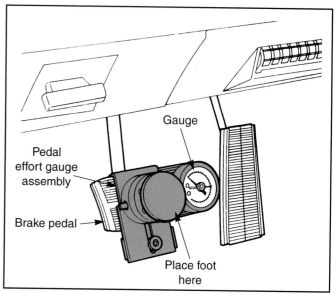

Figure 23-4. *A brake pedal effort force gauge (meter) installed on the brake pedal. Follow the vehicle and/or tool manufacturer's operating instructions. (Pontiac)*

Look for aftermarket add-on equipment that can affect brake system operation. This includes alarm systems, CB and police radios, fog lights, spotlights, fluorescent lights, car stereo amplifiers, compact disc players, trailer wiring harnesses, and other external lighting and wiring. Any of these add-on components, if installed improperly, can affect vehicle operation. In many cases, these simple checks will uncover the problem, or give you a likely place to start in Step 3.

Step 3—Determine Which System Is Causing the Problem

The third step is to set about determining which vehicle components or systems could cause the problem. The first reaction to what appears to be a brake problem is to decide if the brake system is actually defective. However, the brake system is composed of mechanical, hydraulic, and electrical subsystems. To determine which of these is the source of the problem, you must combine the information you obtained in Step 1 with the knowledge of what brake system parts could cause the symptom(s).

Instead of looking for something obviously wrong, as you did in Step 2, you are looking for something that could cause the specific problem. This will help you to eliminate many areas that could not cause the problems, so that in Step 4, you can concentrate on the areas that could cause the problem. This would be a good time to refer to any available diagnostic and troubleshooting charts. A typical troubleshooting chart is shown in **Figure 23-5.**

In most cases, you will have to place the vehicle on a lift to check the brake system and related components. Make a thorough inspection of the brake lines, hoses, and backing plates to ensure that no leaks are present. Check electrical connectors and mechanical linkages. Based on the results of the first two steps, you may want to proceed to Step 4 and remove a wheel to check the brake components.

WARNING LAMP SEQUENCE	SYMPTOM DESCRIPTION	SEE FIGURE

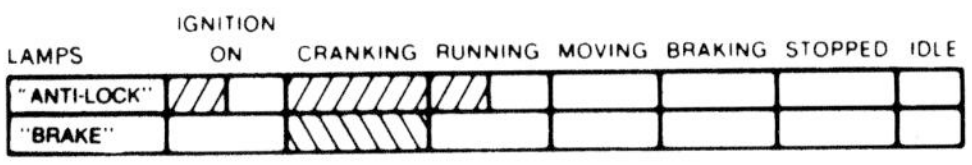	NORMAL WARNING LAMP SEQUENCE WITH
	— EXCESSIVE PEDAL TRAVEL OR SPONGY PEDAL
	— ANTI-LOCK BRAKING OPERATION OR
	    VALVE CYCLING DURING NORMAL STOPS
	    ON DRY PAVEMENT
	— POOR VEHICLE TRACKING DURING
	    ANTI-LOCK BRAKING
		24
		49
		
		63

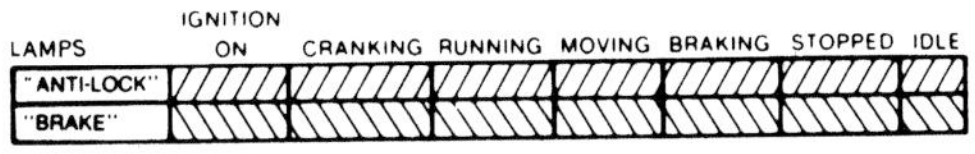	CONTINUOUS "ANTI-LOCK" WARNING LAMP CONTINUOUS "BRAKE" WARNING LAMP	
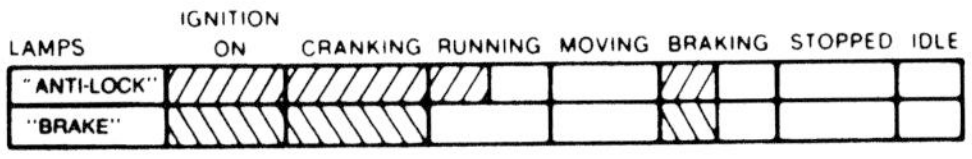	"ANTI-LOCK" AND "BRAKE" WARNING LAMPS COME ON WHILE BRAKING	25
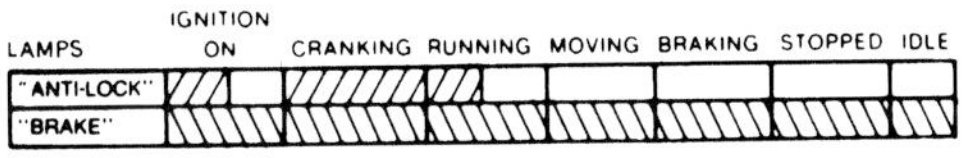	NORMAL "ANTI-LOCK" WARNING LAMP CONTINUOUS "BRAKE" WARNING LAMP	25
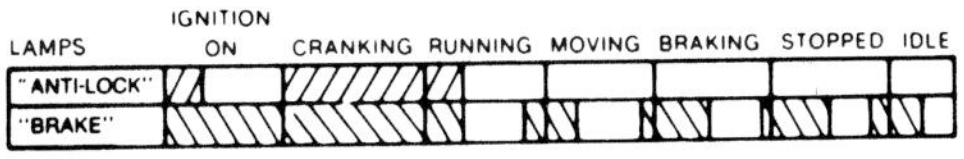	NORMAL OR CONTINUOUS "ANTI-LOCK" WARNING LAMP FLASHING "BRAKE" WARNING LAMP	25
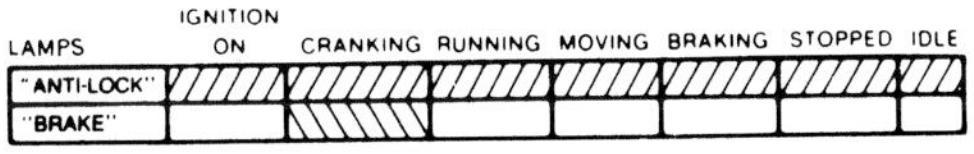	CONTINUOUS "ANTI-LOCK" WARNING LAMP NORMAL "BRAKE" WARNING LAMP	29
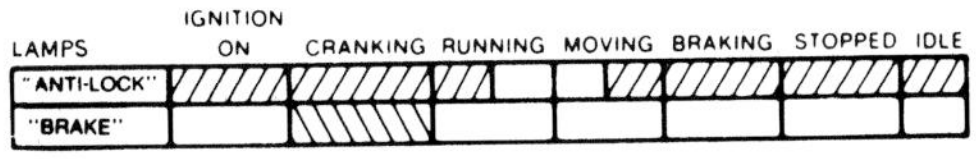	"ANTI-LOCK" WARNING LAMP COMES ON AFTER VEHICLE STARTS MOVING NORMAL "BRAKE" WARNING LAMP	49
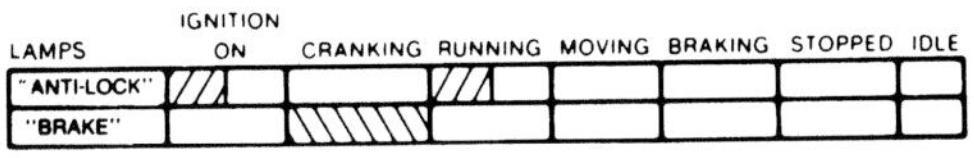	NO "ANTI-LOCK" WARNING LAMP WHILE CRANKING NORMAL "BRAKE" WARNING LAMP	68
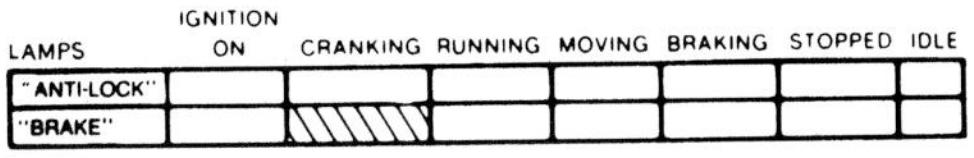	NO "ANTI-LOCK" WARNING LAMP NORMAL "BRAKE" WARNING LAMP	69
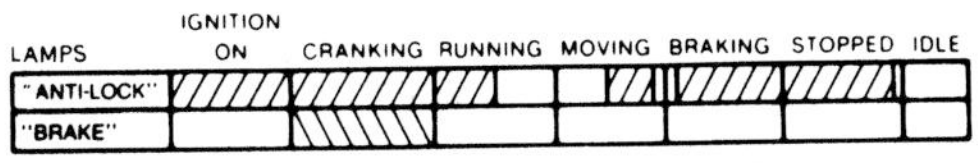	INTERMITTENT "ANTI-LOCK" WARNING LAMP WHILE DRIVING NORMAL "BRAKE" WARNING LAMP	70

LAMP STATUS   • SHADED AREAS: LAMP ON
              • BLANK AREAS (NO SHADING): LAMP OFF
              • PARTIALLY SHADED AREAS: LAMP ON FOR PART OF TEST PERIOD

Figure 23-5. *A trouble diagnosis chart as used by one vehicle manufacturer. These charts are like road maps to unfamiliar cities, without them you will quickly become lost, and travel many streets more than once. (General Motors)*

Diagnosing Intermittent Problems

If the problem does not occur during the road test, it is tempting to dismiss it as the owner's imagination or as normal vehicle operation, but the problem may well be real. **Intermittent problems** are the most difficult to diagnose, because they usually occur only when certain conditions are met. Intermittent malfunctions can be related to temperature, humidity, certain vehicle operations, or in response to certain tests performed by a vehicle computer. Intermittent problems in the brake system are very dangerous not only from the aspect of the driver, but in liability for the shop if it fails to find the problem.

When dealing with an intermittent malfunction, always try to recreate the *exact* conditions in which the problem occurred. Unfortunately, most drivers do not relate intermittent problems to external conditions. Intermittent problems cannot always be duplicated, even in an extensive road test. If a road test of reasonable duration does not duplicate the problem, it is time to try other types of testing. It is essential the principles of strategy-based diagnostics be followed closely when diagnosing intermittent malfunctions.

Step 4—Eliminate Other Causes of the Problem

In the fourth step, you begin eliminating possible causes of the brake problem, one by one. Always begin by checking the components or systems that are the most likely sources of the problem. In most cases, this step requires you to remove at least one tire and rim to gain access to the brakes, **Figure 23-6.** You should remove the wheel that seems to be the source of the problem. Then you can observe brake condition, measure linings and drums or rotors, and check for fluid leaks.

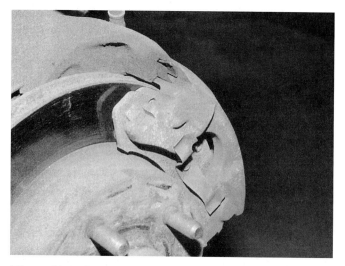

Figure 23-6. *By removing a tire and rim, your access to the brake parts will be much easier. You can check for leaks, damaged brake lines, broken speed sensor wires, worn pads/shoes, defective calipers, damaged rotors, etc.*

Be sure to make a thorough check, including measuring drum and rotor thickness. If necessary, remove the other wheels and check for problems. As a general rule, if you suspect a problem with the brakes at any wheel, you should check the brakes on all wheels.

If the wheel brakes seem to be okay, go on to the less likely causes, such as sticking valves or an ABS problem. This step takes the most time, and involves the various test procedures covered in previous chapters. You will eventually find the most likely cause, to be further investigated in Step five.

During this step, you should check for chassis related braking problems. These problems often show up as poor or erratic braking, but are actually caused by the suspension or steering system. As an example, a badly worn ball joint can cause an occasional pull to one side. If you do not inspect the suspension and realize the ball joint is worn, the result is time wasted looking at the brakes for a problem that lies elsewhere.

The following is a list of chassis parts that should be thoroughly inspected both to ensure braking effectiveness and to avoid potentially dangerous safety defects.

- ❑ Incorrect wheel bearing adjustment.
- ❑ Wheel lug nuts loose.
- ❑ Improper tire pressure.
- ❑ Excessively worn tires.
- ❑ Mismatched tire sizes.
- ❑ Incorrect wheel alignment.
- ❑ Defective shock absorbers or struts.
- ❑ Incorrect shock absorbers or struts.
- ❑ Damaged or sagging springs.
- ❑ Damaged spring mountings.
- ❑ Bent or missing front or rear stabilizer bar.
- ❑ Worn or defective tierods, idler arm, or Pitman arm.
- ❑ Loose steering gearbox.
- ❑ Worn or defective ball joints, control arm bushings, or strut rod bushings.
- ❑ Damaged frame.

More than one of the above parts can be defective. In addition to a visual inspection, a good way to check for worn front end parts is to shake the wheel, as in **Figure 23-7.** If the wheel can be moved excessively, or has a loose feel, make further checks to determine which suspension or steering component is at fault. If you find a part that is worn out or damaged, do not stop there. Continue to check the other suspension and wheel parts.

Step 5—Recheck the Cause of the Problem

In this step, the most likely cause of the problem is isolated and rechecked. This step requires reviewing the various test procedures that were performed in the last step, and determining whether the suspect component is likely to be the source of the problem. It is often helpful to take a short break to consider all possible causes of the problem, and determine if what you have found is the only thing that could be defective. Review in your mind how the particular system works, and how the defect could cause the system problem.

Before leaving this step, closely recheck the condition of the suspect part, and as far as possible, all other related parts. This will ensure you have not condemned the wrong part or overlooked another defect.

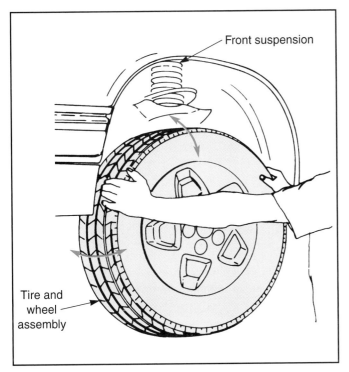

Figure 23-7. *Checking for worn or damaged front end parts by grasping and shaking the tire and wheel assembly, with the vehicle off the ground. Grasp the tire in several different locations, sides, top, bottom, etc., for best results. Be sure the lug nuts are tight. (Pontiac)*

Always Perform Additional Tests

Additional testing is especially important when the suspected part is a solid-state, or otherwise untestable device, such as an ABS hydraulic actuator. Most of these parts are too expensive to simply replace without knowing for sure whether they are good or bad. Making further checks to confirm the problem is always a good idea, if only to increase your confidence about finding the defective part. Not many technicians are sorry they made further checks, but a lot of them are sorry they did not.

Deciding on Needed Work

Deciding on needed work is a process of interpreting the results of all diagnostic tests. It is simply a matter of taking all test readings and deciding what they mean. As discussed earlier, the test results can be simple observations of visible defects, detailed readings from elaborate test equipment, or any procedure in between.

Before condemning any part based on test results, mentally review its interaction within the system and the various engine and vehicle systems. Then decide whether the part in question can cause the particular test reading or symptom. For instance, if the brake system is losing fluid and you have located a leaking wheel cylinder, do not assume that it is the only source of leaks. Check the other wheel cylinder as well as the calipers, hoses, lines, master cylinder, and valves.

Troubleshooting charts and other diagnostic information can be a great asset to this process. If researched and prepared correctly, the troubleshooting chart will list all the possible causes of the problem, allowing you to check everything in a logical sequence. Properly used, such information will speed up the checking and isolating process.

Deciding on the Proper Repair Steps to Take

The amount and type of corrective action must also be determined. In some cases, the repair is as simple as reattaching a vacuum hose, removing grease, dirt, or debris from a sensor, or tightening a belt. In other cases, major unserviceable parts, such as the proportioning valve, ABS control module, or hydraulic actuator must be replaced to correct the problem. To reduce the possibility of future problems, you should also service parts that interact with the defective part. Examples are rebuilding or replacing the wheel cylinders when the shoes are replaced, or flushing the hydraulic system when the master cylinder is replaced. In all cases, the technician must thoroughly determine the extent of the repairs before proceeding.

Factors that must be considered when deciding to adjust, rebuild, or replace a part are ease of adjustment, the need for special tools, cost of the replacement part, and the possibility the old part will fail again.

If a part is easily adjustable, you can try the adjustment procedure before rebuilding or replacing. Generally, most brake system parts cannot be adjusted. If adjusting the part does not restore its original performance, the part can still be rebuilt or replaced with little time lost. The important thing to remember is not to get too involved in trying to adjust away a stubborn problem. For example, do not spend time trying to adjust the parking brakes on a vehicle with worn rear pads or shoes. The extra time spent will be lost if the linings still requires replacement. If there is any doubt about whether an adjustment has corrected a problem, replace the part.

Rebuild or Replace?

In cases when a defective component can be rebuilt, the investment in materials and time must be weighed against the possibility that rebuilding the part may not fix the problem. At modern labor rates, it is often cheaper to install a new part than to spend time rebuilding the old one. Many repair shops, and even some new vehicle manufacturers, are going increasingly to a policy of replacing complete assemblies. You must determine if rebuilding is cost effective.

In many cases, the customer will come out ahead with a new or remanufactured assembly instead of paying to rebuild an old part. The price of the new or remanufactured part is often less than the charge to rebuild the old part. These parts often come with a limited warranty from the remanufacturer and the assurance the part was assembled in a clean, controlled environment. The technician will often come out ahead, since the labor time saved rebuilding the old part can be devoted to other work.

Therefore, when deciding what to do to correct a problem, make sure that all parts that could contribute to the problem have been tested. In one form or another, every possible component and system should be tested. Then you can decide with assurance what components are defective.

Special Tools

Special tools are often needed to adjust or disassemble a complex assembly, such as a vacuum power booster. Often, the cost of the tool may exceed the price of a complete replacement assembly. However, special tools can be used again for the same type of repairs in the future, and may be a good investment. You should also figure in the initial cost of the tool versus the number of jobs that will be possible using that tool. If you expect to do a lot of the same type of repairs in the future, and the special tools are reasonably priced, they should be purchased.

Contacting the Owner about Needed Work

After determining the parts and labor necessary to correct the problem and before proceeding to actually make repairs, contact the vehicle owner and get authorization to perform the repairs. *Never* assume the owner will want the work done. The owner may not have sufficient money for the repairs, may prefer to invest the money in another vehicle, or prefer to have someone else perform the repair work. The defective part or problem may be covered by the vehicle manufacturer's warranty or a guarantee given by another repair shop or chain of service centers. In these cases, the vehicle must be returned to an approved service facility for repairs. If your shop is not one of these approved facilities, you cannot expect to be reimbursed for any more than diagnosing the problem.

If the vehicle is leased, the leaseholder is the actual owner. Depending on the terms of the lease, the leaseholder may be the only one who can approve any expenses in connection with the vehicle. Be especially careful if the vehicle is covered by an extended warranty or service contract. Extended warranties and service contracts are a form of insurance, and like all types of insurance, it is necessary to file a claim for any expenses. In some cases, the owner can file a claim after repairs are completed; while in other cases, approval must be granted from the insurer before the repair work can begin. Sometimes, the insurer will send an adjuster to inspect the vehicle before approval is granted.

Before talking to the vehicle owner, leaseholder, or extended warranty company concerning authorization to perform needed repairs, you should make sure that you can answer three questions that will be asked. First, be prepared to tell exactly what work needs to be done, and why. Next, have available a careful breakdown of both part and labor costs. Third, be ready to give an approximate time when the vehicle will be ready. If you suspect a problem that requires further disassembly, be sure the customer understands that further diagnosis (and costs) may be needed before an exact price is reached.

Step 6—Correct the Defect

In Step 6, you correct the defect by servicing the brake system as necessary. This service can be as simple as tightening a bleeder screw or may require a complete overhaul of the brake system. For this step, refer to the repair procedures in the previous chapters.

Before beginning repairs, be sure the customer is informed of what defects you have found, and has approved the cost for the needed parts and service. Do not assume the customer will want the work done.

Be sure to completely fix the problem. Do not, for instance, adjust worn brake shoes if they are actually ready for replacement. In addition, tearing down the brake system to make repairs often uncovers other problems. Be sure to inform the customer about additional charges and get an ok before proceeding.

Step 7—Recheck System Operation

The final step is to recheck system operation to determine whether the problem has been corrected. The simplest way to do this is to conduct another road test, making stops from various speeds. Check for adequate stopping power, straight stops, and lack of noises and vibration.

 Caution: Do not make panic stops to test the brakes. Not only is this unnecessary, it can overheat and ruin the new pads and linings.

Do not skip this step, since it allows you to determine whether the previous steps accomplished the task of correcting the problem. If necessary, repeat Steps 1 through 6 until this step indicates the problem has been fixed.

Follow-Up

Once the seven-step checking process has isolated and cured the immediate problem, your first thought may be to park the vehicle and get on to the next job. However, it is worth your time to rethink the repair for a minute and decide whether the defect that you found is really the ultimate cause of the problem. This process is known as **follow-up.**

For example, a customer brings a vehicle into the shop complaining of a low brake pedal. The front pads were replaced and the rear brakes adjusted 10,000 miles (16 000 km) ago. If you find the front pads are worn again, replace them, but do not assume the vehicle is fixed until you ask yourself why the pads wore out so quickly. If the pads were not obviously defective, then you must find out why they wore down so quickly. For instance, a defective proportioning valve could keep pressure from reaching the rear brakes, causing the front brakes to do all the work. If the valve is not replaced, the new pads will soon wear out. It is also possible the rotors could have been turned too thin and as a result, cannot absorb enough heat. Be sure to check out all possibilities. If you do not locate an underlying problem, the vehicle will be back soon, along with a dissatisfied customer.

Hidden defects are common, and may cause a vehicle to return again and again with the same defective part. Do not let the vehicle leave until you are reasonably sure the observed defect is the real source of the problem. Some hidden problems can be tricky, such as a binding caliper that causes the pads on one wheel to wear out again and again, or a bad ground that ruins a series of ABS control modules. This is where good observation skills and customer feedback can be helpful.

Whenever you work on a brake system, always try to determine the real cause of a failure, even when the problem appears to be simple.

Additional Diagnostic Tips

When diagnosing any problem, *always check the simple things first.* It will save a lot of time and annoyance to check all of the visible, obvious possibilities before going on to the difficult ones that require a lot of test equipment and time. You must check every part in a system, without assuming anything about the condition of any part. Knowing that all the easy things are okay will allow you to zero in on the hard things. While diagnosing a problem, never assume a particular part or system is functioning properly until it has been checked.

Avoid making snap judgments about the cause of the problem. Sometimes you will hear of making an "educated guess." However, an educated guess is a reasonable decision, based on experience, testing, and the process of elimination. **Uneducated guessing,** or jumping at the first possible cause that comes to mind, is a dangerous way of diagnosing problems. Unfortunately, it can quickly become a habit, done over and over no matter how many times it leads to disaster.

Documentation of Repairs

Part of the follow-up process includes writing on the repair order what the problem was, and what was done to correct the problem. This is called **documentation** and it is a vital part of the diagnostic process, **Figure 23-8.** Every repair order line should have three things. These three things are:

- ❑ What the driver's complaint was.
- ❑ The cause of the complaint.
- ❑ What was done to correct the complaint.

This type of documentation not only allows the driver to clearly see what was done to correct the vehicle's problem, it also supplies a good history of what has been done. If the vehicle should come back with a similar problem, it gives you or the technician working on the vehicle a place to start looking, without having to repeat some of the steps you took to find the problem.

Remaining Calm

One of the hardest principles of diagnosis is to remain calm. Mastering your own emotions is often the hardest thing to do, especially if you meet with a series of dead ends while looking for a problem or are having to deal with an angry customer, but it is necessary. Nothing will be accomplished by losing your composure. If you lose your composure, you will waste valuable time and possibly upset the customer. If you have picked up a tendency to overreact to situations, you will have to unlearn this behavior and teach yourself to remain calm. Only a calm person can think logically.

Noise, Vibration, and Harshness

The passenger compartment is heavily insulated to deaden noises and some vibrations from normal vehicle operation. Unfortunately, unitized bodies easily transmit noises that can be heard and felt, even with the great amount of insulation used in today's vehicles.

Noise, vibration, and harshness (NVH) diagnostics is usually associated with drive train and suspension systems. Originally, NVH training was given only to engineers. Today, training in noise, vibration, and harshness diagnostics is becoming increasingly important for all technicians as vehicles become smaller. Most vehicle manufacturers now have dedicated NVH training classes for their technicians. Some post-secondary automotive programs are beginning to include NVH diagnostics training in their classes. Types of vibrations, noises, and harshness are listed in **Figure 23-9.**

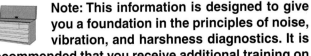

Note: This information is designed to give you a foundation in the principles of noise, vibration, and harshness diagnostics. It is recommended that you receive additional training on this subject.

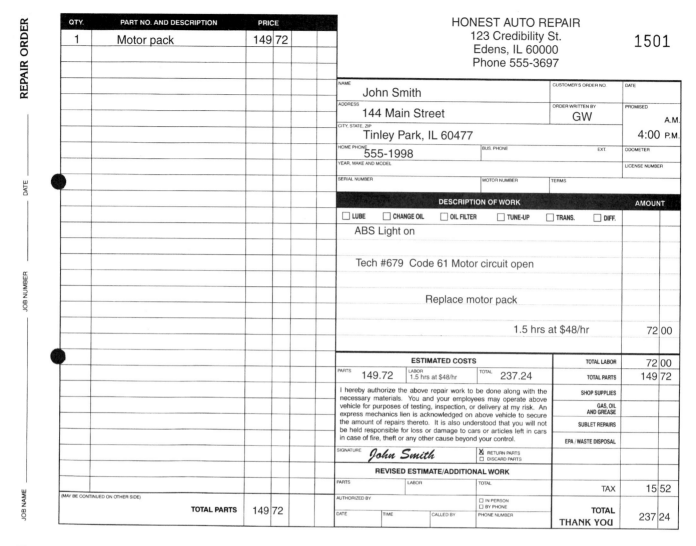

Figure 23-8. *Be sure to document all repairs properly. This will avoid confusion and also provide a starting point for you or others should the vehicle come back with a similar problem.*

Principles of Vibration

All vehicles produce some *vibrations,* with the majority coming from two sources. Most vibrations come from rotating components, such as the crankshaft, wheels, and drive axles. The other major source of vibration is the firing pulses produced by the engine during the combustion cycle. It is important to note that some vibrations are part of normal vehicle operation while others are caused by problems, such as defective or imbalanced components. Another factor in NVH diagnostics is that some vibrations, such as normal engine pulses, cannot be completely eliminated.

Order of Vibration

Order refers to how many times a vibration, noise, or harshness event occurs in one revolution of a rotating component. For example, a drum with a high spot would create a vibration once every revolution. This is called a *first-order vibration.* An oval shaped tire would create a vibration twice per revolution, or a *second-order vibration.* Three high spots would create a third-order vibration, and

so on. However, two first-order vibrations do not equal a second-order vibration or two second-order vibrations do not equal a fourth-order vibration.

Creating and Transmitting Noise

Vibrations also create audible *noise.* Vibrations are the source of almost all noise complaints since a vibration must exist for a noise to occur. Some vibrations begin with a part coming loose, becoming misaligned, or breaking. When a vibrating part touches another part, for example, an exhaust pipe vibrating and touching a heat shield, noise is created. Sometimes noise can be created without any noticeable vibration, however, a vibration is usually present. Noise can originate from almost any vehicle system.

Sources of Harshness

Harshness is when the operation of a component feels hard or more harsh than usual. While harshness is considered a form of vibration, it is more readily felt and heard

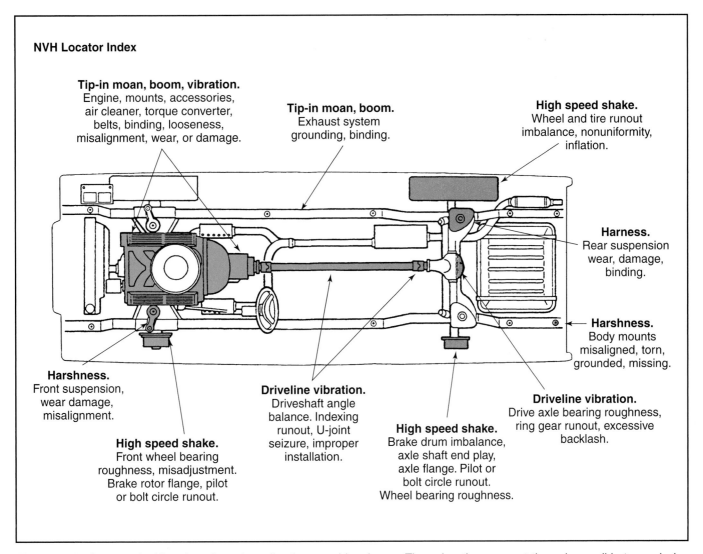

NVH Locator Index

Tip-in moan, boom, vibration.
Engine, mounts, accessories,
air cleaner, torque converter,
belts, binding, looseness,
misalignment, wear, or damage.

Tip-in moan, boom.
Exhaust system
grounding, binding.

High speed shake.
Wheel and tire runout
imbalance, nonuniformity,
inflation.

Harness.
Rear suspension
wear, damage,
binding.

Harshness.
Body mounts
misaligned, torn,
grounded, missing.

Harshness.
Front suspension,
wear damage,
misalignment.

High speed shake.
Front wheel bearing
roughness, misadjustment.
Brake rotor flange, pilot
or bolt circle runout.

Driveline vibration.
Driveshaft angle
balance. Indexing
runout, U-joint
seizure, improper
installation.

High speed shake.
Brake drum imbalance,
axle shaft end play,
axle flange. Pilot or
bolt circle runout.
Wheel bearing roughness.

Driveline vibration.
Drive axle bearing roughness,
ring gear runout, excessive
backlash.

Figure 23-9. *Some typical locations for noise, vibrations, and harshness. These locations are not the only possible transmission sources. (Ford)*

than a vibration. Harshness is often accompanied by noise. Sources of harshness include the transmission, tires, and rear differential. Harshness is caused by excess play between two components and unusual component wear.

Eliminating Noise, Vibration, and Harshness

To eliminate noise, vibration, or harshness, it is important to understand how they are transmitted. The origin of the vibration is usually called the **transmitter** or the *source*. The path the vibration or noise takes is called the **transfer path** or *conduit*. The part where the vibration, noise, or harshness is emitted is called the **receiver**, *emitter*, or *responder*. An example of this transfer is when a drum becomes out of round. The drum transmits vibration through the hydraulic system and unibody to the brake pedal and seats where the driver and passengers can feel it. In this example, the drum is the transmitter, the hydraulic system and the unibody is the transfer path and the brake pedal and seats are the receiver. Noise,

vibrations, and harshness can be measured in frequency or hertz. Specialized equipment is available to measure and convert vibrations to a frequency.

There are many specific methods of eliminating noises and vibrations, depending on the affected system. However, the two main groups are balancing and dampening. **Balancing** is performed when weight is added or removed opposite the source to compensate or cancel out the vibration. An example is when weight is added to a wheel to compensate for tire imbalance. **Dampening** is usually performed when a vibration cannot be eliminated, such as engine firing pulses. Engine mounts dampen the vibrations created by engine firing pulses. It is very important in NVH correction to find the root cause of the problem. Often, many technicians simply dampen an unusual vibration or noise, leaving the transmitting source uncorrected.

Checking Drums and Rotors for Vibration

In cases where a drum or rotor is suspected of causing a vibration other than brake pulsation, they can be checked using an off-vehicle wheel balancer, **Figure 23-10.** Many rotors cannot be tested, however, almost all drums can. Mount the drum in the same manner as you would mount a wheel.

Drums can only be checked for static imbalance. Ignore the balancer's dynamic readings. Keep in mind there is no set tolerance for drum imbalance. A good rule to follow is any drum calling for .75 ounces or more weight for balance may be the cause of a vibration. Any drum or rotor suspected of causing a vibration should be replaced.

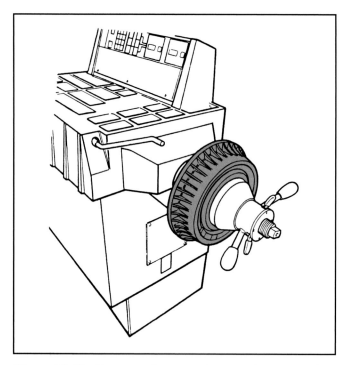

Figure 23-10. *Drums and some rotors can be mounted on wheel balancers to check them as a source of vibrations other than brake pulsation. (General Motors)*

Summary

Always use logical procedures when diagnosing a problem. Be sure that you know how the brake system works, and apply this knowledge to the brake problem at hand. Avoid jumping to conclusions or panicking.

The seven-step procedure will quickly locate most problems. Step one is to determine the exact problem. Step two is to check for obvious causes of the problem. Step three is to determine which system or component could cause the problem. Step four is to make checks of the most likely systems. In step five, the problem is isolated and rechecked to make sure that it really is the source of the trouble. In step six, repairs are made. In step seven, the vehicle is rechecked to make sure that the problem has been fixed.

Another important part of diagnosis is follow-up. Always find out if the problem that you discovered is the real problem. Make sure the part that failed was not dam-

aged by some other vehicle system or operator action. Ensuring that the real cause of the problem has been found and corrected will eliminate much wasted time and customer dissatisfaction.

Review Questions—Chapter 23

Please do not write in this text. Write your answers on a separate sheet of paper.

1. Try to talk to the vehicle driver to find out the _____ complaint.

2. If the brakes are almost completely inoperable, should you go on a road test?

3. A low fluid level in the master cylinder is an obvious cause of a low spongy brake pedal. What should you do after refilling the reservoir?

4. A troubleshooting _____ will often simplify diagnosis.

5. Wheel brake components cannot be satisfactorily checked unless the _____ is removed.

6. The thickness of rotors and drums should always be checked with a _____.

7. Shaking the wheel with the tire installed will check for worn _____ parts.

8. After repairing the vehicle, always make a _____ to recheck.

9. Panic stops can ruin new _____.

10. Even when the cause of a brake failure appears to be obvious, always take a minute to think about what could be the _____ cause of the problem.

ASE Certification-Type Questions

1. Technician A says that to successfully diagnose a brake problem, you should try to question the vehicle owner. Technician B says that a road test should be conducted very carefully if the brakes do not operate at all. Who is right?
   (A) A only.
   (B) B only.
   (C) Both A & B.
   (D) Neither A nor B.

2. To use logic in your diagnostic procedures, always do all of the following, EXCEPT:
   (A) panic when you do not figure out the problem.
   (B) carefully consider all possible problems.
   (C) test all parts carefully.
   (D) recheck your original findings.

3. Technician A says that visual checks can supply clues to braking problems. Technician B says that brake problems are sometimes obvious. Who is right?
    (A) A only.
    (B) B only.
    (C) Both A & B.
    (D) Neither A nor B.

4. Before road testing a vehicle for proper brake operation, always make certain that _____.
    (A) the driver can pay for repairs
    (B) the brakes are not completely inoperative
    (C) the master cylinder is full
    (D) there is a high pedal

5. In step 5 of the seven-step process, you will _____ a suspected part.
    (A) check
    (B) recheck
    (C) replace
    (D) adjust

6. In step 6 of the seven-step process, you _____ a suspected part.
    (A) check
    (B) recheck
    (C) replace
    (D) reinstall

7. Technician A says that the first thing that you come across is usually the real problem. Technician B says that a brake system may have more than one problem. Who is right?
    (A) A only.
    (B) B only.
    (C) Both A & B.
    (D) Neither A nor B.

8. After the brakes are repaired, a road test indicates that there is a new brake problem. Which of the following steps should the technician take?
    (A) Allow the customer to take the vehicle.
    (B) Recheck the parts that were replaced, and let the vehicle go.
    (C) Write a brief description of the new problem on the repair order.
    (D) Perform the seven-step diagnosis procedure.

9. Technician A says that hidden defects occur on most brake jobs. Technician B says that hidden defects can cause repeat brake failures. Who is right?
    (A) A only.
    (B) B only.
    (C) Both A & B.
    (D) Neither A nor B.

10. Failure to conduct a thorough follow-up often results in _____.
    (A) comebacks
    (B) satisfied customers
    (C) profits
    (D) goodwill

ASE certified technicians are in constant demand by employers. Obtaining and maintaining your certification will help to ensure your employability. (Hunter Engineering Company)

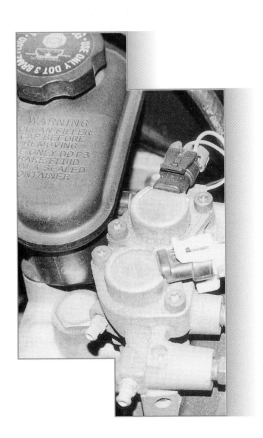

Chapter 24

ASE Certification

After studying this chapter, you will be able to:
- ❑ Explain why certification is beneficial to the technician, shop owners, and the driving public.
- ❑ Explain the process of registering for ASE tests.
- ❑ Explain how to take the ASE tests.
- ❑ Identify typical ASE test questions.
- ❑ Identify the format of the ASE test results.
- ❑ Explain how the ASE test results are used.

Important Terms

National Institute for Automotive Service Excellence (ASE)

Standardized tests

Certified

Master Technician

ACT

Registration booklet

Pass/fail letter

Test score report

Recertified

In this chapter, the purpose and organization of the National Institute for Automotive Service Excellence (ASE) will be explained. This chapter also covers ASE certification and the advantages of being ASE certified. In this chapter are directions for applying for and taking the ASE tests. When you have finished studying this chapter, you will know the purpose of ASE, the design of the ASE tests, how the test results are delivered, and the purpose of the test results.

Reasons for ASE Tests

The concept of setting standards of excellence for skilled jobs is not new. In ancient times, metalworkers, weavers, potters, and other artisans were expected to conform to set standards of product quality. In many cases, this need for standards resulted in the establishment of associations of skilled workers who set standards and enforced rules of conduct. Ancient civilizations had such associations, and many medieval industries were regulated by guilds. Many modern American labor unions are descended from early associations of skilled workers. Certification processes for aircraft and aerospace workers and electronics technicians have existed since the beginnings of these industries.

However, this has not been true of the automotive industry. Automobile manufacturing and repair began as a fragmented industry, made up of many small vehicle manufacturers, and thousands of small repair shops. Although the number of vehicle manufacturers decreased, the number of repair facilities continued to grow in number and in variety. Due to its fragmented, decentralized nature, standards for the automotive repair industry were difficult to establish. For over 50 years, there was no unified set of standards of automotive repair knowledge or experience. Anyone could claim to be an automotive technician, no matter how unqualified. This situation resulted in much unneeded or improperly done repair work. As a result, a large segment of the public came to regard technicians as unintelligent, dishonest, or both.

This situation changed in 1972, when the **National Institute for Automotive Service Excellence,** now called **ASE,** was established to provide a certification process for automobile technicians. ASE is a non-profit corporation formed to encourage and promote high standards of automotive service and repair. ASE does this by providing a series of written tests on various subjects in the automotive repair, auto body, engine machinist, and truck and school bus repair areas. These tests are called **standardized tests,** which means the same test in a particular subject is given to everyone throughout the United States. Any person passing one of these tests, and meeting certain experience requirements, is **certified** (officially recognized as meeting all standards) in the subject covered by that test. If a technician can pass all of the tests in the automotive or heavy truck areas, he or she is certified as a **Master Technician** in that area.

The purpose of the ASE certification test program is to identify and reward skilled and knowledgeable technicians. Periodic recertification provides an incentive for updating skills, and also provides guidelines for keeping up with current technology. The test program allows potential employers and the driving public to identify good technicians, and helps the technician advance his or her career. The program is not mandatory on a national level, but many repair shops now will only hire ASE certified technicians. Close to 500,000 persons are now ASE certified in one or more areas.

Other activities that ASE is involved in are encouraging the development of effective training programs, conducting research on the best methods of instruction, and publicizing the advantages of technician certification. ASE is managed by a board made up of persons from the automotive and truck service industries, motor vehicle manufacturers, state and federal government agencies, schools and other educational groups, and consumer associations.

The advantages ASE certification has brought to the automotive industry include increased respect and trust of automotive technicians, at least of those who are ASE certified. This has resulted in better pay and working conditions for technicians, and increased standing in the community. Thanks to ASE, automotive technicians are taking their place next to other skilled artisans.

Applying for the ASE Tests

You do not have to be employed in the automotive service industry to apply for and take an ASE test. However, to become certified, the applicant must have two years experience working as an automobile or truck technician. In some cases, training programs or courses, an apprenticeship program, or time spent performing similar work may be substituted for all or part of the required work experience.

ASE tests are given twice each year, in the Spring and Fall. Tests are usually held during a two-week period on nights during the work week. The actual test administration procedures are performed by **ACT,** a non-profit organization experienced in administering standardized tests.

The tests are given at designated test centers in over 300 places in the United States. If necessary, special test centers can be set up in remote locations. However, there must be a certain minimum of potential test takers before a special test center can be set up.

To apply for the ASE tests, begin by obtaining an application form like the one shown in **Figure 24-1.** To obtain the most current application form, contact ASE at the following address:

National Institute for Automotive Service Excellence
13505 Dulles Technology Drive, Suite 2
Herndon, VA 22071-3415

ASE will send the proper form, along with a **registration booklet** explaining how to complete the form. When you

Registration Form
ASE Tests

If extra copies are needed, this form may be reproduced.
Important: Read the Booklet first. Then print neatly when you fill out this form.

1. a. Social Security Number:

1. b. Telephone Number: (During the day)
Area Code   Number

1. c. Previous Tests: Have you ever registered for any ASE certification tests before? ❏ Yes ❏ No

2. Last Name: **First**   **Middle Initial**

3. Mailing Address:

Number, Street, and Apt. Number

City

State   ZIP Code

4. Date of Birth: Month   Day   Year   **5. Sex:** ○ Male ○ Female

6. Race or Ethnic Group: Blacken one circle. (For research purposes. It will not be reported to anyone and you need not answer if you prefer.)

1 ○ American Indian   2 ○ African American   3 ○ Caucasian/White   4 ○ Hispanic/Mexican
5 ○ Oriental/Asian   6 ○ Puerto Rican   7 ○ Other (Specify)

7. Education: Blacken only one circle for the highest grade or year you completed.

Grade School and High School   After High School: Trade or
(including Vocational)   Technical (Vocational) School   College

○7 ○8 ○9 ○10 ○11 ○12   ○1 ○2 ○3 ○4   ○1 ○2 ○3 ○4 ○More

8. Employer: Which of the below best describes your current employer? (Blacken one circle and fill in subcodes from pp. 19-20 as appropriate.)

1 ○ New Car/Truck Dealer/Distributor, Domestic—⎫ Enter code for
2 ○ New Car/Truck Dealer/Distributor, Import—⎭ make of car or truck . . . .

3 ○ Service Station—Enter code . . . .   13 ○ Volume Retailer—Enter code . . . . .

4 ○ Military—Enter Code . . . .   14 ○ Tire Dealer—Enter code . . . .

5 ○ Indp't Repair Shop— Enter code for ⎫
6 ○ Fleet Repair Shop— vehicle type ⎭   16 ○ Utility
                                              17 ○ Lift Truck Dealer/Repair Shop

7 ○ Specialty Shop—Enter code . . . .   18 ○ Machinist Facility—Enter code . . . . .

8 ○ Government/Civil Service
                                              19 ○ Leasing & Rental Shop—Enter code
9 ○ Student

10 ○ Educator/Instructor   15 ○ Other

11 ○ Manufacturer

12 ○ Independent Collision Shop

9. Test Center:
Center Number   City   State

10. Tests: Blacken the circles (regular) or squares (recertification) for the test(s) you plan to take. (Do not register for a recertification test unless you passed that regular test before.) Note: Do not register for more than four regular tests on any single test day.

Regular	Recertification	Regular	Recertification
○ A1 Auto: Engine Repair	❏ A	○ A2 Auto: Automatic Trans/Transaxle	❏ E
○ A8 Auto: Engine Performance	❏ B	○ A3 Auto: Manual Drive Train and Axles	❏ F
○ A4 Auto: Suspension & Steering	❏ C	○ A6 Auto: Electrical/Electronic Systems	❏ G
○ A5 Auto: Brakes	❏ D	○ A7 Auto: Heating & Air Conditioning	❏ H
○ F1 Alt. Fuels: Lt. Vehicle CNG		○ L1 Adv. Level: Auto Adv. Engine Perf. Spec.	
○ S4 School Bus: Brakes		○ T1 Med/Hvy Truck: Gasoline Engines	❏ I
○ S5 School Bus: Suspension/Steering		○ T2 Med/Hvy Truck: Diesel Engines	❏ J
		○ T6 M/Hvy Truck: Elec./Electronic Systems	❏ K
○ T3 Med/Hvy Truck: Drive Train	❏ L	○ S6 School Bus: Elec./Electronic Systems	
○ T4 Med/Hvy Truck: Brakes	❏ M	○ B2 Body: Painting & Refinish	❏ P
○ T5 Med/Hvy Truck: Suspension/Steering	❏ N	○ B3 Body: Nonstructural Analysis	
○ T8 Med/Hvy Truck: PMI		○ B4 Body: Structural Analysis	
○ M1 Machinist: Cylinder Head Specialist	❏ Q	○ B5 Body: Mechanical & Electrical Components	
○ M2 Machinist: Cylinder Block Specialist	❏ R		
○ M3 Machinist: Assembly Specialist	❏ S		

11. Fees: Number of **Regular** tests **(except L1)** marked above _____ x $18* = $_____

If you are taking the L1 Advanced Level Test, add $40 here   + $_____

If you marked **any** of the **Recertification** tests, add $20 here   + $_____

Registration Fee   + $ **20**

Important: School Bus Technicians, see page 8 for a money-saving offer! If you qualify, check here ❏ and adjust your fees (Item 11, first line) accordingly.

TOTAL FEE   = $_____

❏ MasterCard   ❏ VISA   Expiration Date

Credit Card #   Month   Year

Signature of Cardholder _____

12. Fee Paid By: 1 ○ Employer   2 ○ Technician

13. Experience: Blacken one circle. (To substitute training or other appropriate experience, see page 3.)

1 ○ I certify that I have two years or more of full-time experience (or equivalent) as an automobile, medium/heavy truck, school bus or collision repair/refinish technician or as an engine machinist. (Fill out #14 on opposite side)
2 ○ I don't yet have the required experience. (Skip 14.)

14. Job History: Required. Provide job history and job details on the opposite side of this form.

15. Authorization: By signing and sending in this form, I accept the terms set forth in the *Registration Booklet* about the tests and the reporting of results.

Signature of Registrant   Date

Do not send cash. Use credit card, or enclose a check or money order for the total fee, made payable to ASE/ACT and mail with this form to:
ASE/ACT, P.O. Box 4007, Iowa City, IA 52243

Figure 24-1. *A sample form for registering for ASE certification tests. Be sure you fill in all pertinent information required. Include the correct payment for all applicable test fees. (ASE)*

get the form, carefully fill it out, recording all needed information. You may apply to take as many tests as are being given, fewer tests, or only one test. Work experience, or any substitutes for work experience, should also be included, according to the instructions in the registration booklet. If there is any doubt about what should be placed in a particular space, consult the registration booklet. Be sure to determine the closest test center, and record its number in the appropriate space. Most test centers are located at local colleges and schools.

When you send in the application, you must include a check, money order, or credit card number to cover all necessary fees. A fee is charged to register for the test series, and a separate fee is charged for each test to be taken. See the latest registration booklet for the current fee structure. In some cases, your employer may pay the registration and test fees. Check with your employer before sending in your application. Recently, it has become possible to register over the Internet to take ASE tests. ASE's World Wide Web site also includes information on ASE,

the certification process, study materials, and guidelines for program certification, **Figure 24-2.**

To be accepted for either the Spring or Fall ASE tests, your application and payment must arrive at ASE headquarters at least one month before the test date(s). To ensure that you can take the test(s) at the location of your choice, send in your application as early as possible.

After sending the application and fees, you will receive an admission ticket to the test center. This should arrive by mail within two weeks of sending the application. See **Figure 24-3.** If your admission ticket has not arrived, and it is less than two weeks until the test date(s), contact ASE using the phone number given in the registration booklet. If the desired test center is filled when ASE receives your application, you will be directed to report to the nearest center that has an opening. If it is not possible to go to the alternate test center that was assigned, contact ACT immediately, using the phone number given in the latest ASE registration booklet.

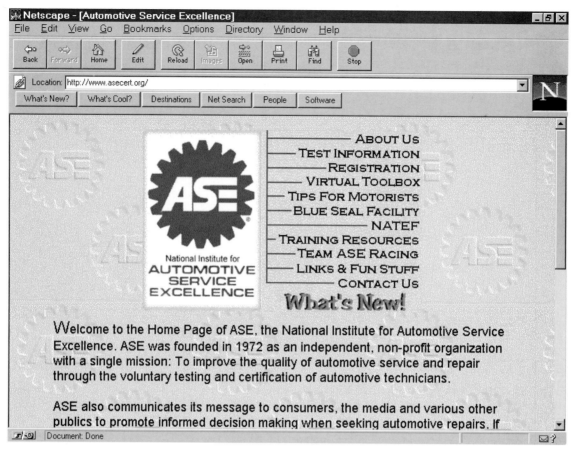

Figure 24-2. *You can register for ASE tests or find out more information about ASE over the information superhighway at ASE's official World Wide Web site. (ASE)*

Preparing for ASE Tests

ASE tests are designed to measure your knowledge of three things:

- ❑ Basic information on how automotive systems and components work.
- ❑ Diagnosis and testing of systems and components.
- ❑ Repairing automotive systems and components.

You should study the basic principles of how automotive systems work, and also study the latest information about diagnostics and repair techniques. Good sources of this material are your textbook, service manuals and factory training material, publications such as Motor or Chilton magazines, and service bulletins. Other technical training aids are available. These allow you to study at home at your own pace. This is an excellent way to secure the knowledge needed to pass ASE tests, **Figure 24-4.**

Remember that ASE tests are designed to test your knowledge of correct diagnosis and repair procedures. Do not assume the way you have always done something is the correct way.

Taking the ASE Tests

Be sure to bring your admission ticket with you when reporting to the test center. When you arrive at the test center, you will be asked to produce the admission ticket and a driver's license or other photographic identification. In addition to these items, bring some extra Number 2 pencils. Although pencils will be made available at the test center, extra pencils will save you time if the original pencil breaks.

After you enter the test center and are seated, listen to and follow all instructions given by the test administrators. During the actual test, carefully read all test questions before making a decision as to the proper answer.

Types of ASE Test Questions

Each ASE test will contain between 40 and 80 test questions, depending on the subject to be tested. All test questions are multiple-choice, with four possible answers. These types of multiple-choice questions are similar to the multiple-choice questions used in this textbook. The following section discusses the types of ASE questions that you are likely to encounter.

National Institute for Automotive Service Excellence

ACT, P.O. Box 4007, Iowa City, Iowa 52243, Phone: (319) 337-1433   017910042 T

Admission Ticket

Test Center to which you are assigned:

A

John Smith
123 Main Street
Edens, Il. 60000

REGULAR TESTS (Late arrivals may not be admitted.)		
DATE	REPORTING TIME	TEST(S)
11/14	7:00 PM	A1, A5, A8

RECERTIFICATION TESTS (Late arrivals may not be admitted.)

TEST CODE KEY

A1 Auto: Engine Repair
A2 Auto: Automatic Trans/Transaxle
A3 Auto: Manual Drive Train & Axles
A4 Auto: Suspension & Steering
A5 Auto: Brakes
A6 Auto: Electrical/Electronic Systems
A7 Auto: Heating & Air Conditioning
A8 Auto: Engine Performance

M1 Machinist: Cylinder Head Specialist
M2 Machinist: Cylinder Block Specialist
M3 Machinist: Assembly Specialist
T1 Med/Hvy Truck: Gasoline Engines
T2 Med/Hvy Truck: Diesel Engines
T3 Med/Hvy Truck: Drive Train
T4 Med/Hvy Truck: Brakes
T5 Med/Hvy Truck: Suspension & Steering

T6 Med/Hvy Truck: Elec./Electronic Systems
T8 Med/Hvy Truck: Preventive Main. Inspec.
B2 Coll.: Painting & Refinishing
B3 Coll.: Non-structural Analysis
B4 Coll.: Structural Analysis
B5 Coll.: Mechanical & Elec. Components
B6 Coll.: Damage Analysis & Estimating
P1 Parts: Med/Hvy Truck Parts Specialist

P2 Parts: Automobile Parts Specialist
F1 Alt. Fuels: Lt. Veh. Comprsd. Nat. Gas
L1 Adv. Level: Adv. Engine Perf. Spec.
S1 School Bus: Body Sys. & Spec. Equip.
S4 School Bus: Brakes
S5 School Bus: Suspension & Steering
S6 School Bus: Elec./Electronic Systems

See Notes and Ticketing Rules on reverse side. An asterisk (*) indicates your certification in these areas is expiring.

SPECIAL MESSAGES

```
-REVIEW ALL INFORMATION ON THIS TICKET.  CALL IMMEDIATELY TO REPORT AN
 ERROR OR IF YOU HAVE QUESTIONS.
-IF YOU MISS ANY EXAMS, FOLLOW THE REFUND INSTRUCTIONS ON THE BACK OF THIS SHEET.
 THE REFUND DEADLINE IS
-YOU HAVE BEEN ASSIGNED TO AN ALTERNATE TEST CENTER.  THE CENTER
 ORIGINALLY REQUESTED IS FULL.
8010-IL/LOCAL 150 IS LOCATED ON JOLIET AVE, THREE DOORS W. OF LAGRANGE RD ON
SOUTH SIDE OF THE ST.  ENTER THROUGH BACK DOOR.  NO ALCOHOL ON PREMISES.
```

MATCHING INFORMATION: The information printed in blocks B and C at the right was obtained from your registration form. It will be used to match your registration information and your test information. Therefore, the information at the right must be copied EXACTLY (even if it is in error) onto your answer booklet on the day of the test. If the information is not copied exactly as shown, it may cause a delay in reporting your test results to you.

IF THERE ARE ERRORS: If there are any errors or if any information is missing in block A above or in blocks B and C at the right, you must contact ACT immediately. DO NOT SEND THIS ADMISSION TICKET TO ACT TO MAKE SUCH CORRECTIONS.

Check your tests and test center to be sure they are what you requested. If either is incorrect, call 319/337-1433 immediately. Tests cannot be changed at the test center. **ON THE DAY OF THE TEST,** be sure to bring this admission ticket, positive identification, several sharpened No. 2 pencils, and a watch if you wish to pace yourself.

B FIRST FIVE LETTERS OF LAST NAME

S M I T H

C SOCIAL SECURITY NUMBER OR ACT IDENTIFICATION NUMBER

1 2 3 4 5 6 7 8 9

SIDE 1

Figure 24-3. *A copy of your Admission Ticket that you will receive in the mail. If your ticket does not arrive within two weeks after sending in your registration form, contact ASE immediately.*

To order any of these courses, or for more information, visit your NAPA AUTO PARTS store or call 1-800-292-NIAT (6428).

N.I.A.T. Self-Study Course

ASE Automobile Technician Test	Understanding Electronic Engine Management Systems	Hydraulics & Modern Braking Systems	Strategies of Exhaust and Emission Control	Understanding Modern Heating & Air Conditioning Systems	Understanding Antilock Brake Systems	Effective Diagnosis of No-Code Driveability Problems	Using the Labscope (DSO) to Master Driveability & Emissions Diagnosis	Using 4 & 5 Gas Analyzers for Emissions & Driveability Diagnosis	Troubleshooting Distributorless Ignition Systems (DIS)
Brakes (A5)		✓			✓				
Heating & Air Conditioning (A7)				✓					
Engine Performance (A8)	✓		✓			✓			
Advanced Engine Performance Specialist (L1)	✓		✓			✓	✓	✓	✓

Figure 24-4. *Other technical training is also available to help you secure extra knowledge. This will also aid in taking your ASE certification tests. When working on today's vehicles, it is extremely important to have the skill and knowledge to be a "professional technician." Take pride in yourself and your work. (Napa Institute of Automotive Technology)*

One-part Question

One-part questions require that you answer a single question:

1. Which of the following brake system parts creates hydraulic pressure?

    (A) Brake pedal.

    (B) Master cylinder.

    (C) Wheel cylinder.

    (D) Caliper piston.

Notice that the question calls for the best answer out of all the possibilities. The brake pedal (A) transfers foot pressure to the master cylinder. The wheel cylinder and caliper piston (C and D) are moved by hydraulic pressure, but do not create it. The master cylinder (B) is the only part that creates hydraulic pressure. Therefore, "B" is correct.

Two-part Question

Two-part questions used in ASE tests involve two imaginary technicians, Technician A and Technician B. Each technician makes a statement. You are asked to determine whether each statement is true:

1. Technician A says that the brake system changes movement into heat to stop the vehicle. Technician B

says that the coefficient of friction is the measurement of the brake linings' ability to absorb heat. Who is right?

(A) A only.

(B) B only.

(C) Both A and B.

(D) Neither A nor B.

Note that both statements can be true, both can be false, or only one of them can be false. In this case, the statement of Technician A is correct since the brakes do change motion into heat through the use of friction. The statement of Technician B is incorrect, since the coefficient of friction is a measure of the linings' ability to create friction, not to absorb heat. Therefore, the correct answer is "A."

A variation of the two-part question sets up a situation or states a subject under discussion. Again, Technician A and Technician B each makes a statement and you are asked to determine whether each of the statements is true:

1. A vacuum brake booster operates properly when the engine is running. When the engine is turned off, the booster immediately stops providing power assist. Technician A says that this is normal operation for some types of vacuum brake boosters. Technician B says that the vacuum check valve may be defective. Who is right?

(A) A only.

(B) B only.

(C) Both A and B.

(D) Neither A nor B.

In this case, you must read the statement carefully to determine the exact nature of the situation. You can then determine whether the statements made by Technicians A and B are correct. In this case, A's statement is incorrect, since all vacuum boosters provide one or two power assisted stops when vacuum is lost. Technician B's statement is correct, since a failed vacuum check valve can cause a loss of reserve vacuum. Therefore, the correct answer is "B."

Negative Questions

Some questions are called negative questions. These questions ask you to identify the incorrect statement from four possible choices. They will usually have the word "EXCEPT" in the question. The following is an example of a negative question.

1. The brake hydraulic system is composed of all of the following parts, EXCEPT:

(A) master cylinder.

(B) calipers.

(C) retracting springs.

(D) hoses and lines.

Since retracting springs are installed on the brake shoes, they are the only parts not included in the brake hydraulic system. Therefore, the correct answer is "C."

A variation of the negative question will use either the word "most" or "least," such as the following.

1. A late-model 4x4 pickup truck pulls to the right during braking. Which of the following defects is the least likely cause?

(A) Swelled brake hose on the front left brake.

(B) Worn out brake pads on the front left brake.

(C) Worn out brake pads on the front right side.

(D) Oil or grease on the brake pads on the front right side.

Swelled brake hoses or worn pads on the left side (A and B) would be very likely to cause a pull to the right. Oil or grease on the right side pads (D) would also be likely to cause the brakes to pull to the right. In this case, the least likely cause of pulling to the right would be worn out brake pads on the right side, which would be much more likely to cause a pull to the left. Therefore, the correct answer is "C."

Incomplete Sentence Questions

Some test questions are incomplete sentences, with one of four possible answers correctly completing the sentence. An example of an incomplete sentence question is given here.

1. A scan tool can be used to read _____.

(A) hydraulic pressures

(B) coefficient of friction

(C) trouble codes

(D) lining temperatures

Once again the question calls for the best answer. A scan tool is used to retrieve trouble codes from the ABS system, so "C" is correct.

The ASE Brakes Test

The ASE brakes test contains approximately 50-60 questions. The categories in this test include:
- Hydraulic System Diagnosis and Repair.
- Drum Brake Diagnosis and Repair.
- Disc Brake Diagnosis and Repair.
- Power Assist Units Diagnosis and Repair.
- Miscellaneous Diagnosis and Repair.
- Anti-lock Brake System Diagnosis and Repair.

The section on hydraulic system diagnosis and repair is subdivided into four sections covering master cylinders; fluids, lines, and hoses; valves and switches; and bleeding, flushing, and leak testing.

After completing all the questions on the test, recheck your answers one time to ensure you did not miss anything that would change an answer, or that you did not make a

National Institute for
AUTOMOTIVE SERVICE EXCELLENCE

December 20, XXXX   032527

John Smith
123 Main Street
Edens, Il 60000

Dear ASE Test Taker:

Listed below are the results of your November XXXX ASE Tests. You will soon be receiving a more detailed report.

If your test result is "Pass", and if you have fulfilled the two-year "hands-on" experience requirement, you will receive a certificate and credential cards for the tests you passed.

If your test result is "More Preparation Needed", you did not attain a passing score. Check your detailed score report when it arrives. This information may help you prepare for your next attempt.

If you do not receive your detailed report within the next three weeks, please call.

Thank you for participating in the ASE program.

A1	ENGINE REPAIR	PASS
A5	BRAKES	PASS
A8	ENGINE PERFORMANCE	PASS

123-45-6789

13505 Dulles Technology Drive · Herndon, Virginia 22071-3415 · (703) 713-3800

Figure 24-5. *A confidential pass/fail letter will be sent to you shortly after your ASE tests. Note that three tests were taken, A1, A5, and A8, and that all were passed. (ASE)*

careless error on the answer sheet. In most cases, rechecking your answers more than once is unnecessary and may lead you to change correct answers to incorrect ones. The time allowed for each test is usually about four hours. However, you may leave after completing your last test and handing in all test material.

Test Results

Test collection and grading takes from six to eight weeks. After this time, you will receive a *pass/fail letter* from ASE. A typical pass/fail letter is shown in **Figure 24-5**. This letter will tell you only whether or not you have passed each test. This letter will be followed in about two weeks by a *test score report.* The test score report is a confidential report of your performance on the test. The report will list the number of questions that must be answered correctly to pass the test, and the number of questions that you answered correctly. The test questions are also subdivided into the various categories to help you to determine which areas, if any, that require more study. A typical test diagnostic report is shown in **Figure 24-6.**

Also included with the report is a *certificate in evidence of competence.* This certificate lists all the areas in which you are certified. In addition, a pocket card and a wallet card are provided. Like the certificate, they list all the areas in which you are certified. Also included is an order form for shoulder patches, wall plates, and other ASE promotional material.

Note: If you did not indicate that you have two years of automotive experience on the test application, you will not receive a certificate in evidence of competence. After you have met the experience requirement, you should provide ASE with the necessary information to receive your certificate.

ASE takes the position that all test results are confidential information, and provides them only to the person who took the test. This is done to protect your privacy. The only test information that ASE will release is to confirm to an employer that you are certified in a particular area. Test results will be mailed to your home address, and will not be provided to anyone else. This is true even if your employer has paid the test fees. If you wish your employer to know exactly how you performed on the tests, you must provide him or her with a copy of your test results.

If you fail a certification test, you can retake it again as many times as you wish. However, you or your employer must pay the registration and test fees again. You should study all available information in the areas where you did poorly. A copy of the ASE Preparation Guide may be helpful to sharpen your skills in these areas. The ASE Preparation Guide is free, and can be obtained by filling out the coupon at the back of the registration booklet.

Recertification Tests

Once you have passed the certification test in any area, you must be *recertified* every five years. This assures that your certification remains current, and is proof that you have kept up with current technology. The process of applying to take recertification tests is similar to that for the original certification tests. Use the same form and enclose the proper recertification test fees. If you allow a certification to lapse, you must take the regular certification test(s) to regain that certification.

Summary

The automotive service industry was one of the few major industries that did not have testing and certification programs. This caused a lack of professionalism in the automobile industry, often leading to poor or unneeded repairs,

Your score is 47   Passed
The total score needed to pass A5 is *36 out of 55.*

Test A5 Brakes Content Area	Number of Questions Answered Correctly	Total Number of Questions
Hydraulic System Diagnosis & Repair	14	16
Drum Brake Diagnosis & Repair	3	6
Disc Brake Diagnosis & Repair	10	13
Power Assist Units Diagnosis & Repair	4	4
Miscellaneous Diagnosis & Repair	7	7
Anti-lock Brake System Diagnosis & Repair	9	9
Total Test	47	55

Figure 24-6. *A diagnostic report will follow the pass/fail letter. This allows you to see how you performed in each section and, if needed, which areas you need further study. (ASE)*

and decreased status and pay for automobile technicians. The National Institute for Automotive Service Excellence, or ASE, was started in 1972 to improve the status of the automotive service industry. ASE tests and certifies automotive technicians in major areas of automotive repair. This has increased the skill level of technicians, resulting in better service and increased benefits for technicians.

ASE tests are given two times each year, in the Spring and Fall. Anyone can register to take the tests by filling out the registration form and paying the test fees. The registrant must also select the test center that he or she would like to go to. To be considered for certification, the registrant must have two years of hands-on experience as an automotive technician. Proof of this should also be included with the registration form. About three weeks after applying for the test, the technician will receive a test entry ticket which he or she must bring to the test center.

The actual test questions will test your knowledge of general system operation, problem diagnosis, and repair techniques. All of the questions are multiple-choice questions with four possible answers. The questions must be read carefully. The entire test should be gone over one time only to catch careless mistakes.

Test results will arrive within six to eight weeks after the test session. A letter will be sent telling you whether you have passed the tests or need more preparation. This is followed by a detailed report showing the test scores for all areas of the tests taken. Results are confidential and will be sent only to the home address of the person who took the test. If a test was passed, and the experience requirement has been met, the technician will be certified for five years. Anyone who fails a test can take it again in the next session. Tests can be retaken as many times as necessary. Recertification tests are taken at the end of the five year certification period.

Review Questions—Chapter 24

Please do not write in this text. Write your answers on a separate sheet of paper.

1. An ASE Master Technician has passed all of the tests in the _____ or _____ areas.

2. During a given test session, everyone in the United States takes the _____ test in a particular automotive area.

3. _____ administers the ASE tests.

4. *True* or *False?* Test and registration fees are charged.

5. *True* or *False?* If you cannot get to the assigned test center, call ASE immediately.

6. *True* or *False?* The way that you have always done something should be the correct way for test purposes.

7. *True* or *False?* A one-part question features Technician A and B.

8. *True* or *False?* A negative question contains the word "always."

9. *True* or *False?* It may take up to 12 weeks to get your test results.

10. *True* or *False?* You must recertify every year.

ASE Certification-Type Questions

1. ASE was founded in _____.
   (A) the Middle Ages
   (B) 1975
   (C) 1925
   (D) 1972

2. Technician A says that ASE encourages high standards of automotive service and repair by providing a series of true-false tests. Technician B says that ASE encourages high standards of automotive service and repair by providing a series of standardized tests. Who is right?
   (A) A only.
   (B) B only.
   (C) Both A & B.
   (D) Neither A nor B.

3. A standardized test is given to everybody in _____.
   (A) a particular test room
   (B) a particular state
   (C) the United States
   (D) need of an automotive test

4. If a technician can pass all of the tests in the automotive or heavy truck areas, he or she is certified as a _____.
   (A) Technician
   (B) Master technician
   (C) General technician
   (D) Knowledgeable technician

5. The advantages that ASE certification has brought to automotive technicians includes all of the following, EXCEPT:
   (A) increased respect.
   (B) better working conditions.
   (C) lower pay scales.
   (D) increased standing in the community.

6. ASE tests are given _____ each year.
   (A) once
   (B) twice
   (C) four times
   (D) 12 times

7. You should bring all of the following to the ASE test center, EXCEPT:

(A) two #2 pencils.

(B) your admission ticket.

(C) any needed study materials.

(D) photographic identification.

8. ASE tests are designed to measure your knowledge of all of the following, EXCEPT:

(A) basic information on how automotive systems and components work.

(B) customer relations and dealing with difficult customers.

(C) diagnosis and testing of automotive systems and components.

(D) repairing automotive systems and components.

9. ASE provides test results to _____.

(A) the technician who took the test

(B) whoever paid for the test

(C) the technician's employer

(D) the Environmental Protection Agency

10. Technician A says that a technician can retake any certification test twice only. Technician B says that certified technicians must take a recertification test every ten years. Who is right?

(A) A only.

(B) B only.

(C) Both A & B.

(D) Neither A nor B.

To keep up with changes in technology, such as the rear disc brake system on this truck, technicians must maintain constant training. (Ford)

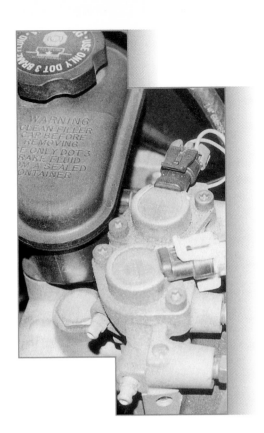

Chapter 25

Career Preparation

After studying this chapter, you will be able to:
- ❏ Identify three classifications of automotive technicians.
- ❏ Identify the major sources of employment in the automotive industry.
- ❏ Identify advancement possibilities for automotive technicians.
- ❏ Explain how to fill out a job application.
- ❏ Explain how to conduct yourself during a job interview.

Important Terms

Helper	Certified technicians	Used car superstore
Apprentice	Vehicle dealers	Entrepreneur

This chapter is an overview of career opportunities in the automotive service industry. It discusses types of automotive technicians, and what kind of work they perform. This chapter also includes information on the types of repair outlets, and the type of work, working conditions, and pay scale the beginning technician can expect to find in each place. Also included are some of the ways you can locate jobs and become employed in the auto repair industry. It includes information on the types of automotive related jobs that you can move into. Studying this chapter will help you to find and secure a job in the automotive service industry.

Automotive Servicing

The business of servicing and repairing cars and trucks has provided employment for millions of people over the last 100 years. It will continue to provide good employment opportunities for many years to come. Like any career, it has its drawbacks, but it also has its rewards.

Most persons in the auto service business work long hours, and the diagnosis and repair procedures can be mentally taxing, physically hard, and often hot and dirty. Automotive service has never been a prestige career, although this is changing as vehicles become more complex, and technicians become better trained.

The advantages of the auto service business are the opportunity to work with your hands, much less confinement than with many other professions, and the enjoyment of taking something that is not working properly, and making it work again. Auto repair salaries are usually competitive with those for similar jobs, and it is a secure profession where the good technician can always find work. To ensure that you stay employable, always seek to learn new things, and become ASE certified in as many areas as possible.

Levels of Automotive Service Positions

Although the public tends to classify all automotive technicians as "mechanics," there are many types of auto service professionals. The types of auto service professionals range from the helper who changes oil or performs other simple tasks; through apprentices who remove and install parts; to the certified technician, capable of diagnosing and repairing almost any automotive system. Although these levels are unofficial, they tend to hold true throughout the automotive repair industry. It would be possible to further break these levels down into more sublevels, but these three will adequately cover the general skill classifications.

Helpers

The *helper* performs the easier types of service and maintenance, such as installing and balancing tires, changing engine oil and filters, and installing batteries. The skills required of the helper are low, and the pay will be less than the other levels. However, the helper position is a good way for many people to start. In fact, many technicians started out in automotive service work as helpers when they were in their teens.

Apprentices

The *apprentice,* **Figure 25-1,** removes and installs parts on vehicles. These parts include suspension components such as shock absorbers or struts; mufflers and other exhaust system components; and master cylinders, batteries, and alternators. Apprentices seldom do any more complicated repair work, and do not generally diagnose vehicle problems. As the apprentice's skills increase, they may begin to do more complicated repair work and even diagnose vehicle problems under a certified technician's guidance. Apprentices are paid more than helpers, but less than certified technicians. Many apprentices take the opportunity to improve their knowledge and skills, and eventually become certified technicians.

Figure 25-1. Apprentices are generally service persons who attach the various parts, such as exhaust system components, brakes, suspension struts, trailing arms, control arms, stabilizer bars, etc.

Certified Technicians

Certified technicians, **Figure 25-2,** are at the top level, and have the skills to prove it. Most modern technicians are ASE certified in at least one automotive area, and many are certified in all car or truck areas. The certified technician is able to successfully diagnose and repair every area that he or she is certified in, and can perform many other service jobs. The certified technician is at the top of the pay scale.

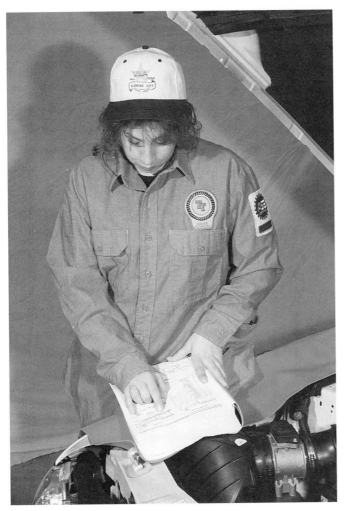

Figure 25-2. An ASE Certified Technician performing brake system diagnostic work by referencing a service manual. (Jack Klasey)

Types of Auto Service Facilities

There are many types of auto service facilities where the technician can work. The traditional place to get started in automotive repair, the corner service station, has been replaced by self-serve stations. However, many opportunities to repair vehicles still exist. Even the smallest community has many types of automotive repair facilities.

Vehicle Dealers

All new *vehicle dealers* must have well-equipped service departments to meet the warranty service requirements of the vehicle manufacturer. These service departments are equipped with all special testers, tools, and literature needed to service their vehicles. Dealership service departments are also equipped with lifts, parts cleaners, hydraulic presses, brake lathes, electronic test equipment, and other equipment for efficiently servicing vehicles. However, the technician has to provide his or her own hand and air tools. Dealers stock all of the most common parts, and are usually tied into a factory parts network, which allows them to quickly obtain any part. See **Figure 25-3.**

Pay scales at most dealerships are based on flat rate hours. The rate per hour is competitive between dealerships in the same area, and usually higher than local industry in general. The number of hours the technician is paid depends on what work comes in and how fast he or she can complete it. If you can work fast, and enough work comes in, the pay can be excellent. Most modern dealers offer some sort of benefits package.

Figure 25-3. New vehicle dealers offer excellent working condition and benefits. However, you do have to deal with all problems on the vehicle, and there is not much opportunity for specialization.

Dealership working conditions are relatively good, and most of the vehicles are new or well cared for older models. Since the dealer must fix any part of the vehicle, the technician can perform a large variety of work. Although many dealer service departments have technicians who work only in specific areas, the trend is toward training all technicians to handle any type of work. Most repairs will be on the same make of vehicle, although many large dealerships handle more than one make. Most modern dealerships are tied into manufacturer technical assistance lines. They are used to access factory diagnosis and repair information, which makes it easier to troubleshoot and correct problems.

The disadvantages of dealership employment are the lack of salary guarantees, low pay rates for warranty repairs, and fast paced, often hectic working conditions. If you welcome the challenge of being paid by the job, and do not mind working under deadlines, a dealership may be the ideal employer. Also check out the local large truck dealerships. Although the work is much heavier, the pay is usually somewhat higher, and working conditions are not as fast paced.

Used Car Superstore

A recent addition to the list of places for technicians to find work is the **used car superstore.** Used car lots are not new to the automotive industry, however, they were not known as good places to work. Used car superstores are national chains that service and sell all makes and models of vehicles. Typically, the work at these chains includes most of the repairs performed at new vehicle dealers. You may also be responsible for inspecting and, if needed, performing repairs on vehicles coming in for sale. Pay at used car superstores is very good and benefits are excellent, with plenty of opportunity for additional training and advancement.

Chain and Department Store Auto Service Centers

Many national chain and department stores have auto service centers where various types of automotive repairs are performed, **Figure 25-4.** These centers often hire technicians for entry-level jobs and may have several classifications above entry-level. Technicians in most of these situations are paid a salary, plus commission for work performed. Pay scales for the various classifications of work are competitive, and most companies offer generous benefit packages. One advantage of working for large companies such as these is the chance for advancement into other areas, such as sales or management.

One disadvantage of working at the average auto center is the lack of variety. Most auto centers concentrate on a few types of repairs, such as brake work and alignment, and turn down most other repairs. The work can become monotonous due to the lack of variety. Although the job pressure is usually less than at dealerships, customers still expect their vehicles to be repaired in a reasonable period

Figure 25-4. *Specialty shops allow you to concentrate on 1-2 areas. However, work at these shops can become monotonous.*

of time. However, if you enjoy working on only one or two areas of automotive repair, this type of employer may be ideal for you.

Tire, Muffler, and Other Specialty Shops

Tire, muffler, and other specialty shops can offer good working conditions and good pay. Most of these shops concentrate on their major specialty, with a few other types of repair, such as brake service. These shops usually offer a base salary, plus commission for work performed. Pay scales are generally competitive. A disadvantage of specialty shops is the lack of variety. Since they concentrate on a few types of repairs, working at these shops can become monotonous.

Many of these shops are franchise operations, and the demands of the franchise can create problems. If the prime purpose of the shop is to sell tires, for example, the technician who was hired to do brake repairs may be forced to spend time installing tires. This can be annoying to some technicians who want to be doing the job for which they were hired. However, if this does not bother you, the specialty shop could be a good work situation.

Independent Shops

There are millions of independent auto repair shops. As places to work, they range from excellent to terrible. Many shops are run by competent and fair-minded managers, and have first-rate equipment and good working conditions. Other independent shops have almost no equipment, low pay rates, and extremely poor, even dangerous working conditions. The prospective employee should carefully check all aspects of the shop environment before agreeing to work there. Technicians at independent shops are usually paid on a salary plus commission basis.

There are two major classifications of independent repair shop—the general repair shop and the specialty shop. The general repair shop takes in most types of work, and offers a variety of jobs. General repair shops may avoid some repair jobs that require special equipment, such as automatic transmissions, front end alignment, or air conditioning service. However, they will usually take a variety of other repair work on different makes and types of vehicles. This can be a good place to work if you like to be involved in many different types of diagnosis and repair.

Specialty shops confine their repair work to one area of repair, such as transmissions, engine performance, or brake repair. They are fully equipped to handle all aspects of their particular specialty. These shops may occasionally take in other minor repair work when business in their specialty is slow. These shops can be ideal places to work if you want to concentrate on one area of repair.

Government Agencies

Many local, state, and federal government agencies maintain their own vehicles. Government operated repair shops can be good places to work. Pay is straight salary, usually set by law, with no commission. Although civil service pay scales are lower than private industry, the benefits are usually excellent. Pay raises, while relatively small, are regular. Most government shops work a 35-40 hour week, and have the same holidays as other government agencies.

The working conditions in most of these shops are good, without the stress of deadlines, or having to deal with customers. Hiring procedures are more involved than with other auto repair shops. Prospective employees must take civil service examinations, which often have little to do with automotive subjects. Some government agencies require a certain level of education, a thorough background check, lists of former employers or other references, and may require the employee be a registered voter.

If you think you would be interested in working for a government owned repair shop, contact your local state employment agency for the addresses of local, state, and federal employment offices in your area.

Self-Employment

Many persons dream of going into business for themselves. This can be a good and profitable option for the good technician. However, in addition to mechanical and diagnostic ability, the person with their own business must have a certain type of personality to be successful. This type of person must be able to shoulder responsibilities, handle problems, and look for practical ways to increase business and make a profit. This type of person must maintain a clear idea of what plans, both short- and long-term, need to be made. A person like this is often called an **entrepreneur,** in other words, a person who has the energy and skill to build something out of nothing.

When you have your own business, all of the responsibility for repairs, parts ordering, bookkeeping, debt collection, and a million other problems are yours. Starting your own shop requires a large investment in tools, equipment, and working space. If the money must be borrowed, you will be responsible for paying it back. However, many people enjoy the feeling of independence, of not having to answer to an employer. If you have the personality to deal with the problems, you may enjoy the feeling of being your own boss.

Another possible method of self-employment is to obtain a franchise from a national service chain. A franchise operation removes some of the headaches of being in business for yourself. Many muffler, tire, transmission, tune-up, and other nationally recognized businesses have local owners. They enjoy the advantages of the franchise affiliation, including national advertising, reliable and reasonable parts supplies, and employee benefit programs. Disadvantages include high franchise fees and startup costs, lack of local advertising, and some loss of control of shop operations to the national headquarters.

Other Opportunities in the Auto Service Industry

Many other opportunities are available to the automotive technician. These jobs still involve the servicing of vehicles, without some of the physical work. If you like cars and trucks, but are unsure whether you want to make a career of repairing them, one of these jobs may be for you.

Shop Foreman or Service Manager

The most likely automotive promotion that you will be offered is shop manager. Many repair facilities are large enough to require one or more foremen or service managers. If you move into management from the shop floor, your salary will increase, and you will be in a cleaner, less physically demanding position. Many technicians enjoy the management position because it lets them in on the fun part of service—troubleshooting, without the drudgery of actually making the repairs.

The disadvantage of a move to management is that you will no longer be dealing with the logical principles of troubleshooting and repair. Instead, you will deal with the personalities of people. Both customers and technicians will have problems and attitudes that you will have to deal with. Unlike a vehicle problem, these problems require considerable personality and tact. Sometimes the manager has to compromise, which can be hard for the person who is used to being right.

The paperwork load is large for any manager, and may not be something that a former technician can adjust to doing. Record keeping requires a good bit of desk, and usually computer time. Automotive record keeping is like balancing a checkbook and writing a term paper every few days. If you do not care to deal with people or keep records, a career in management may not be for you.

464

Salesman or Service Advisor

Many people enjoy the challenge of selling. The salesperson or service advisor performs a vital service, since repairs will not be performed unless the owner is sold on the necessity of having them done. Salespersons are not necessary in many small independent shops, but are often an important part of dealership service departments, department store service centers, and specialty repair shops.

The salesperson may enjoy a large income, and be directly responsible for a large amount of business in the shop. However, selling is a people oriented job, and takes a lot of persuasive ability and diplomacy. If you are not interested in dealing with the public, you would probably not be happy in a sales job.

Parts Person

One often overlooked area of the automotive service business is the process of supplying parts. It is as vital as any other area. There are many types of parts outlets, including dealership parts departments, independent parts stores, parts departments in retail stores, and combination parts and service outlets. All of these parts outlets meet the needs of technicians and shops, as well as the do-it-yourself needs of the vehicle owner. The many different makes of vehicles, as well as the complexity of the modern vehicle, means a large number of parts must be kept in stock and located quickly when needed. See **Figure 25-5.**

Parts persons are trained in the methods of keeping the supply of parts flowing through the system until they reach the ultimate endpoint, the vehicle. Parts must be carefully checked into the parts department, and stored so they can be found again. When a specific part is needed, it must be located and brought to the person requesting it. If more of the same parts are needed, they must be ordered. If the part is not in stock, it must be special ordered. This can be a challenging job.

The job of the parts person appeals to many people. The job of the parts person does not pay as much as some other areas of automotive service, but rates are comparable with other jobs with the same skill level. If this type of work appeals to you, it may be a good job choice.

Getting a Job

There are many automotive jobs available at all times. The problem is connecting with the job when it is available. There are essentially two hurdles to getting a job: finding a job opening, and successfully applying for the job. Below are some hints for overcoming these hurdles.

Finding Job Openings

Before applying for a job, you must know about a suitable opening. A good place to start is with your instructors. They often have contacts in local industry and may be able to recommend you to an employer. Another good place to begin your search is the classified section of your local newspaper. Automotive jobs are often advertised in newspapers, especially those for automotive dealerships, specialty and franchise shops, and independent repair shops.

Also visit your local state employment agency or job service, **Figure 25-6.** Most of these agencies keep records of job openings, usually throughout the state, and may be able to connect to a data bank of nationwide job openings. Some private employment agencies specialize in automotive placement. If there is an agency of this type in your area, arrange for an interview with one of their recruiters.

Visit local repair shops where you are interested in working. Sometimes these shops have an opening that they have not advertised. If you are interested in working for a chain or department store, most of these stores have a personnel department where you can fill out an application.

Figure 25-5. A parts department is a good place to find work if you do not care for the pace of the auto shop.

Figure 25-6. State employment agencies have many job listings not normally advertised.

Even if no jobs are available, your application will be placed on file in the event a job becomes available in the future.

Applying for the Job

No matter how good your qualifications, you will not get a job if you make a poor first impression on the job interviewer. If you do not impress your potential employer as competent and dedicated, you will *not* get the job.

Start creating a good impression when you fill out the employment application. Type or neatly print when filling out the application. Complete all blanks, and explain any significant breaks in your education or employment history. List all of your educational qualifications, including those which may not apply directly to the automotive industry. **Figure 25-7** shows a typical employment application.

When you are called for an interview, try to arrange for a morning interview, since your potential employer is

---

ABT Brakes   **Employment Application**

600 S Street Edens, IL 60000

Date: _____ Social Security Number: _____

Name: _____

Address: _____

Phone: _____ United States citizen? _____ Can your furnish proof? _____

Employment Desired

Position: _____ Date you can start: _____ Expected salary: _____

Are you currently employed? _____ May we inquire of your present employer? _____

Education

Circle the number for the highest level completed:

High School   Trade/Technical school   Community/Junior College   University

1 2 3 4   1 2   1   2   1 2 3 4

Other: _____

Specialized Training or Certifications: _____

_____

Employment Record

Current/Last employer: _____ From: _____ To: _____

Address: _____ Phone: _____

Salary: _____ Job description: _____

Reason for leaving: _____

Previous employer: _____ From: _____ To: _____

Address: _____ Phone: _____

Salary: _____ Job description: _____

Reason for leaving: _____

Previous employer: _____ From: _____ To: _____

Address: _____ Phone: _____

Salary: _____ Job description: _____

Reason for leaving: _____

Figure 25-7. Typical job application. Most job applications are longer than this one.

most likely to be in a positive mood in the morning and less likely to be overwhelmed by last minute problems in the shop. Dress neatly, and arrive on time or a little early. When introduced to the interviewer, make an effort to repeat and remember their name. Speak clearly when answering the interviewer's questions. Do not smoke or chew gum during the interview. State your qualifications for the job without bragging or belittling your accomplishments. At the conclusion of the interview, thank the interviewer for his or her time. If you do not hear from the interviewer in a few days, it is permissible to make a brief and polite follow up call.

Summary

The automotive service industry provides employment for many people, and will continue to do so. Automotive service has some disadvantages, such as long hours; hard work, both mentally and physically; lack of status; and difficulties in dealing with the public. Advantages include interesting work, the security of a guaranteed career, and the enjoyment of diagnosing and correcting problems. Always stay employable by learning new things and taking the ASE tests.

The three general classes of technicians are the helper, the apprentice, and the certified technician. The helper does the simplest tasks, such as tire changing and lubrication. Many helpers move up into the other classes after a short time. The apprentice installs new parts, such as shock absorbers and strut assemblies, and sometimes moves into brake and front end repair. The certified technician performs the most complex diagnosis and repair jobs on vehicles, and makes the most money.

There are many places to work as an automotive technician. Among the most popular are new car and truck dealers, auto centers affiliated with department or chain stores, specialty shops, independent repair shops, and government agencies. Some people prefer to have their own businesses, either as independent owners, or as part of a franchise system.

Other opportunities in the automotive service field include moving into management as a foreman or service manager, or into sales as a service advisor or specialty salesperson. Another often overlooked employment possibility is in the automotive parts business.

To obtain a job in the automotive business, first locate possible job openings. Try the local newspaper and state job service, and visit repair shops in your area. Most department or other large stores with attached auto service centers have personnel departments where you can fill out a job application. To get a job, you must make a good impression, no matter how qualified you are. Fill out all job applications carefully and neatly, listing your qualifications honestly. When invited to a job interview, dress neatly, arrive on time, and be courteous. Answer all questions without over or understating your abilities and experience. Follow up the interview with a brief phone call within a few days.

Review Questions—Chapter 25

Please do not write in this text. Write your answers on a separate sheet of paper.

1. For how many years has it been possible to make a living in the automotive service business?

2. Match the type of automotive service technician with the job that he or she will perform.

   _____ Change oil and filter.   (A) Helper
   _____ Replace alternator.   (B) Apprentice
   _____ Diagnose ABS problem.   (C) Certified
   _____ Replace shock absorber.   technician
   _____ Troubleshoot brake pulling.

3. Battery installation is a job most likely to be done by a _____.

4. To work on newer models of the same make, consider seeking a job at a _____.

5. One disadvantage of working for a specialty shop is _____.

6. Which of the following places is the automotive technician most likely to move up to managing?

7. Needed parts can be special ordered if they are _____.

8. A shop that works only on mufflers or brakes is a _____ shop.

9. When filling out a job application, be sure to _____ all blanks, and strive to write as _____ as possible

10. The time to start making a good impression on a possible employer is _____.

ASE Certification-Type Questions

1. All of the following are advantages of working in the automotive service field, EXCEPT:
   (A) the opportunity to work with your hands.
   (B) lack of prestige.
   (C) competitive salaries
   (D) the pleasure of making something broken work again.

2. Technician A says that vehicle owners can be difficult to deal with. Technician B says that repair procedures can be hot and dirty. Who is right?
   (A) A only.
   (B) B only.
   (C) Both A & B.
   (D) Neither A nor B.

3. All of the following are businesses where the automotive technician is very likely to find his or her first job, EXCEPT:

(A) chain stores.

(B) corner service stations.

(C) specialty stores.

(D) new vehicle dealers.

4. Technician A says that a technician being paid by the flat rate method is being paid for the number of hours spent in the shop. Technician B says that commission is paid on the amount of repair work that the technician does. Who is right?

(A) A only.

(B) B only.

(C) Both A & B.

(D) Neither A nor B.

5. All of the following statements about working in large new vehicle dealerships are true, EXCEPT:

(A) dealers may sell more than one make of vehicle.

(B) the service department repairs all parts of the new vehicles sold.

(C) the technician may be paid by one of several methods.

(D) many repairs are done under warranty.

6. All of the following are advantages of working for a local, state, or federal government operated repair shop, EXCEPT:

(A) steady pay.

(B) good benefits.

(C) demanding deadlines.

(D) Both A & B.

7. Parts stores meet the needs of which of the following groups?

(A) Technicians.

(B) Repair shops.

(C) Vehicle owners.

(D) All of the above.

8. Technician A says that a big hurdle to getting a job is finding a job opening. Technician B says that a big hurdle to getting a job is making a good impression during the interview. Who is right?

(A) A only.

(B) B only.

(C) Both A & B.

(D) Neither A nor B.

9. Automotive service jobs are advertised in all of the following, EXCEPT:

(A) local newspapers.

(B) flyers and handbills.

(C) state employment agency data banks.

(D) some private employment agencies.

10. All of the following help to make a good impression during the job interview, EXCEPT:

(A) dress neatly.

(B) arrive a little late.

(C) do not smoke.

(D) make an effort to remember the interviewer's name.

Service manuals contain diagnostic charts and information on a vehicle's brake system. (Jack Klasey)

Diagnostic Charts

This chart lists the most common brake problems and symptoms, their causes, and recommended corrections. Also listed is the chapter in which information relative to the problem is located. This chart does not list every possible problem and should not be used as a substitute for the troubleshooting charts in the vehicle's service manual.

Brake Hydraulic and Friction Systems

System Loses Brake Fluid

Symptoms: Fluid level low in master cylinder. May be accompanied by low or spongy pedal.

The master cylinder fluid level will drop slightly as the brake pads and linings wear, causing the caliper and wheel cylinder pistons to move outward in their bores. This wear increases caliper and wheel cylinder brake fluid volume, reducing the amount of fluid in the reservoir. This is a normal condition.

Cause	Correction	Chapter
Fluid leaks in wheel cylinder.	Rebuild or replace wheel cylinder.	6
Fluid leaks in caliper.	Rebuild or replace caliper.	6
Bleeder screw loose.	Tighten bleeder.	6
Manifold vacuum sucking brake fluid through rear of master cylinder and power brake unit.	Replace master cylinder.	6
Fluid leaks in hoses or lines.	Replace or repair lines and hoses.	10
Leaks at hydraulic valves.	Replace hydraulic valves.	10

Brake Pedal Spongy or Low

Symptoms: Excessive pedal travel; pedal goes almost to floor before brakes apply. Pedal feels soft or springy when pressed. Pedal falls slowly when foot pressure is held.

Pedal travel will gradually increase as the linings and pads wear out from use. This is also a normal condition.

Cause	Correction	Chapter
Fluid low in master cylinder or reservoir.	See *System Loses Brake Fluid.*	
Air or gas in lines, wheel cylinders, or caliper.	Repair leak, bleed brakes.	6, 10
Master cylinder loose on mounting.	Tighten mounting bolts.	6
Master cylinder leaking primary and/or secondary cup.	Replace master cylinder.	6
Master cylinder ports closed or restricted.	Flush or replace master cylinder.	6
Excessive play in brake pedal push rod.	Flush or replace master cylinder.	6
Vapor lock (brake fluid boiling).	Flush brakes, check linings, advise driver.	6, 13, 15
Plugged master cylinder filler cap.	Replace cap.	6
Improper brake fluid.	Flush system, replace hydraulic components.	6, 10
Excessive pedal free play (lash).	Adjust or replace pedal assembly.	6
Partial brake system failure (dual master cylinder).	Replace master cylinder.	6
Damaged caliper piston seal.	Rebuild or replace caliper.	6
Weak (ballooning) brake hose.	Replace brake hose.	10
Excessive disc (rotor) runout.	Resurface or replace rotor.	13
Loose caliper attachment.	Tighten caliper bolts.	13
Thin drum.	Replace drum.	15
Excessive lining-to-drum clearance.	Adjust rear brakes.	15
Defective self-adjusting mechanism.	Lubricate or replace adjusting mechanism.	15
Bent brake shoe and lining.	Replace brake shoes.	15
Loose wheel bearing adjustment.	Adjust wheel bearings.	17

Brake Pedal Goes to Floor Intermittently or No Braking Action

Symptoms: Brake pedal occasionally goes to floor when pressed, no braking action; brakes work normally other times. Brake pedal pressed, but no braking action, pedal does not go to floor; brakes work normally other times.

Cause	Correction	Chapter
Debris in master cylinder between seals and cylinder bore.	Flush or replace master cylinder.	6
Debris in anti-lock brake hydraulic actuator.	Replace actuator.	22
Power booster mounting loose.	Tighten booster mounting nuts.	8
Damaged or broken pedal linkage.	Replace pedal assembly.	8

Brake Pedal Hard

Symptoms: Excessive pedal pressure required to stop vehicle. Power brakes do not have a minimum of 1-2 power assisted stops once the engine stops.

Cause	*Correction*	*Chapter*
Frozen caliper pistons and/or wheel cylinders.	Rebuild or replace calipers or wheel cylinders.	6
Plugged master cylinder hose or tubes.	Flush or replace hose, tube, or master cylinder.	6
Frozen piston(s) in master cylinder.	Replace master cylinder.	6
Partial brake system failure (dual master cylinder).	Repair hydraulic leak.	6, 10
Improper brake fluid.	Flush system, replace components as needed.	6, 10
Brake pedal binding, bent, or pivots damaged.	Replace pedal assembly.	6
Calipers binding or dragging.	Check caliper clearance; rebuild or replace calipers.	6
Master cylinder compensator ports blocked.	Flush or replace master cylinder.	6
Oil in brake fluid.	Flush system, rebuild or replace hydraulic components.	6
Poor manifold vacuum supply to power brakes.	Check hose and check valve; engine condition.	8
Defective power brake vacuum check valve.	Replace check valve.	8
Power brake vacuum hose leaking or loose.	Repair or replace vacuum hose.	8
Defective vacuum pump.	Replace vacuum pump.	8
Brake tubing, fittings, hoses, clogged or restricted.	Replace component.	9
Contaminated brake linings.	Replace pads or shoes.	13, 15
Glazed shoes and/or pads.	Replace pads or shoes.	13, 15
Worn shoes and/or pads.	Replace pads or shoes.	13, 15
Heat-checked rotor or drum.	Replace rotor or drum.	13, 15
Wrong type of lining.	Replace pads or shoes.	13, 15
Damaged brake pads or shoes.	Replace pads or shoes.	13, 15
Tapered (bellmouthed) drum.	Resurface or replace drum.	15
Inadequate lining-to-drum clearance.	Adjust rear brakes.	15
Cast iron dust from newly machined drum.	Clean drum and brake assembly.	15
Brakes adjusted too tightly.	Adjust rear brakes.	15
Low engine vacuum.	Check valve; engine condition.	8
Internal vacuum leaks in power booster unit.	Replace power booster.	8
Rusted or dry vacuum piston.	Replace power booster.	8
Broken counter reaction spring in power booster.	Replace power booster.	8
Defective vacuum valve or linkage.	Replace power booster.	8
Power steering pump problem (hydraulic power brakes).	Check power steering system, service as needed.	8
Control valve stuck.	Rebuild or replace Hydro-boost.	8
Defective gas charged accumulator.	Replace accumulator or Hydro-boost.	8
Defective accumulator valves (gas or spring).	Replace Hydro-boost.	8
Contaminated power steering fluid.	Flush power steering and Hydroboost system.	8
Internal leakage in accumulator system.	Replace accumulator or Hydro-boost.	8
Hydraulic leak in Hydro-boost system.	Repair leak.	8

Brakes Pulsate

Symptoms: Pedal moves up and down when braking.

Severe front brake pulsation may cause the steering wheel to vibrate. If rear brakes are pulsating, vibration will be felt in the seat of the pants. Brake pedal pulsation during ABS-assisted stops is considered normal.

Cause	Correction	Chapter
Excessive brake rotor lateral runout.	Resurface or replace rotor.	13
Excessive rotor out-of-parallelism.	Resurface or replace rotor.	13
Loose caliper mounting bolts.	Tighten caliper mounting bolts.	13
Bent drive axle.	Replace axle, check bearings.	13,15
Eccentric (egg-shaped) drum.	Resurface or replace drum.	15
Loose drum on hub.	Adjust bearings, replace drum or hub.	15
Backing plate, anchors or wheel cylinders loose.	Tighten or replace.	15
Worn or defective wheel bearings.	Replace wheel bearings.	17
Bent hub.	Replace hub, check wheel bearings.	17
Worn or defective front end parts.	Consult service manual.	
Bent wheel rim.	Consult service manual.	

Grabbing Brakes

Symptoms: Severe reaction to normal pedal pressure.

Cause	Correction	Chapter
Pistons sticking or seized in master cylinder.	Replace master cylinder.	6
Master cylinder compensator ports plugged.	Flush or replace master cylinder.	6
Intermittent binding of power brake booster.	Replace power brake booster.	8
Defective proportioning valve.	Replace proportioning valve.	10
Caliper anchor bolts loose.	Tighten bolts.	13
Contaminated brake pads or shoes.	Replace pads or shoes.	13, 15
Wrong type of pads or shoes.	Replace pads or shoes.	13, 15
Drum brakes not properly adjusted.	Adjust brakes.	15
Loose backing plate(s).	Tighten or replace backing plate.	15
Lining loose on shoe.	Replace brake shoes.	15
Dirt, dust, or metal shavings in drums.	Clean brakes.	15
Damaged drum braking surface.	Resurface or replace drum.	15
Brake shoes binding on backing plate pads.	Lubricate backing plate.	15
Damaged brake shoes.	Replace brake shoes.	15
Weak retracting springs.	Replace springs.	15
Primary and secondary shoes installed in wrong position.	Reinstall shoes.	15
Loose wheel bearings.	Adjust wheel bearings.	17
Binding parking brake linkage.	Lubricate or replace linkage.	19
Parking brake cables not properly adjusted, or seized.	Adjust or replace cables.	19
Faulty suspension parts.	Consult service manual.	
Tires underinflated, defective, or mismatched.	Consult service manual.	
Faulty hydraulic unit control valves.	Replace Hydro-boost.	8

Brakes Pull to One Side

Symptoms: Vehicle moves to one side during braking. In severe cases, wheel lock up will occur. May occur only when linings are cold, or after several severe stops.

Cause	*Correction*	*Chapter*
Sticking or seized pistons in wheel cylinder, caliper.	Rebuild or replace wheel cylinder or caliper.	6
Restricted brake tubing or hoses.	Flush or replace tubing or hoses.	10
Disc brake pads improperly positioned in caliper.	Reposition pads.	13
Excessive rotor runout.	Resurface or replace rotor.	13
Loose brake caliper bolts.	Tighten caliper bolts.	13
Worn linings.	Replace pads or shoes.	13, 15
Contaminated linings and/or pads.	Replace pads or shoes.	13, 15
Broken or missing retaining springs or clips.	Replace hardware.	13, 15
Wrong size and/or type of lining on one wheel.	Replace pads or shoes.	13, 15
Brake lining not seated to drums and/or rotors.	Bed-in brake linings.	13, 15
Drum brakes not adjusted correctly.	Adjust drum brakes.	15
Cracked or bent brake shoes.	Replace brake shoes.	15
Broken or weak retracting springs.	Replace springs.	15
Primary and secondary shoes reversed.	Reinstall shoes.	15
Backing plate bent or adjuster plug missing (water entry).	Replace backing plate; install plug.	15
Defective self adjuster.	Replace self adjuster.	15
Backing plate loose.	Tighten backing plate retainers.	15
One drum turned oversized.	Replace drum.	15
Damaged or worn wheel bearings.	Replace bearings.	17
Defective ABS valve body or actuator.	Replace valve body or actuator.	22
Tires underinflated, defective, or mismatched.	Consult service manual.	
Improper front end alignment.	Consult service manual.	
Front end not properly aligned, defective parts.	Consult service manual.	

Brakes Will Not Release

Symptoms: Vehicle will not move, problem appears to be in brake system.

Always check that the engine strains when the vehicle is placed in gear. If the engine is not loaded down when placed in gear (races easily), the problem is in the vehicle drive train, not the brakes.

Cause	*Correction*	*Chapter*
Sticking or seized wheel cylinder and/or caliper pistons.	Rebuild or replace calipers or wheel cylinders.	6
Master cylinder defect.	Replace master cylinder.	6
Swollen caliper piston seal(s).	Rebuild or replace caliper, check for contamination.	6
Power booster has poor release.	Replace power booster unit.	8
Defective hose or tubing preventing fluid return.	Replace hose or tubing.	10
Brake shoe adjustment too tight.	Adjust rear brakes.	15
Broken or weak return springs.	Replace springs.	15
Sticking or seized parking brake cables.	Replace parking brake cables.	19

Brake Fade

Symptoms: Increasing pedal pressure is needed to apply brakes. May reach the point where no amount of pressure will stop the vehicle. Often accompanied by smell of overheated brake linings.

Cause	Correction	Chapter
Sticking or seized wheel cylinder and/or caliper pistons.	Rebuild or replace calipers or wheel cylinders.	6
Oil in brake fluid.	Rebuild or replace hydraulic system components.	6
Poor power brake release.	Replace power brake booster.	8
Incorrect or defective lining and/or pads.	Replace pads or linings.	13, 15
Drums and/or rotors worn or machined too thin.	Replace drums or rotors.	13, 15
Dragging brakes.	See Brakes Dragging.	
Drums out-of-round.	Resurface or replace drum.	15
Poor lining contact.	Adjust brakes.	15
Broken or weak retracting springs.	Replace springs.	15
Wheels plugged with debris, stopping ventilation.	Clean wheels, advise driver.	15
Shoes hanging up on backing plate.	Lubricate backing plate.	15
Brake lining radius larger than drum arc.	Replace brake shoes.	15
Car operated with parking brake on.	Advise driver.	
Driver resting foot on brake pedal.	Advise driver.	
Excessive braking or loads.	Advise driver.	

Brake Dragging

Symptoms: One or more wheel brakes remain completely or partially applied when the brake pedal is released. Wheel may be locked.

Cause	Correction	Chapter
Oil in brake fluid.	Rebuild or replace hydraulic components, flush brake system.	6
Swelling wheel cylinder or caliper piston seals.	Rebuild or replace calipers or wheel cylinders, check for contamination.	6
Wheel cylinder push rod out of position.	Reposition pushrod.	15
Brake pedal binding.	Free or replace brake pedal assembly.	6
Compensating ports in master cylinder plugged.	Flush or replace master cylinder.	6
Master cylinder piston blocking compensating port.	Replace master cylinder.	6
No brake pedal free play.	Adjust pushrod or replace booster.	6
Inferior rubber cups and/or seals.	Replace hydraulic components.	6
Internal binding of power brake booster.	Replace power brake booster.	8
Defective tubing or hose preventing fluid return.	Replace hose or tubing.	10
Incorrectly installed metering valve.	Reinstall metering valve.	10
Sticking or seized caliper piston(s).	Rebuild or replace caliper.	13
Grease or oil on pads or shoes.	Clean or replace pads or shoes.	13, 15
Incorrect pads or shoes on one or more wheels.	Replace pads or shoes.	13, 15
Bent or damaged brake shoe(s).	Replace brake shoes.	15
Brake drum out-of-round.	Resurface or replace brake drum.	15
Broken or reversed automatic adjuster.	Reinstall or replace adjuster.	15
Sticking or seized wheel cylinder piston(s).	Rebuild or replace wheel cylinder.	15

Brake Dragging *(continued)*

Cause	Correction	Chapter
Broken or weak return spring(s).	Replace return spring.	15
Improper drum brake adjustment.	Adjust rear brakes.	15
Dry, rusted, or grooved, backing plate.	Clean and lubricate backing plate.	15
Loose wheel bearings.	Adjust wheel bearings.	17
Parking brake cable sticking or seized.	Replace parking brake cable.	19

Brakes Lockup

Symptoms: Wheels lock up under light pedal pressure. May be accompanied by front or rear tire squeal, severe front end dive or rear end sway.

Cause	Correction	Chapter
Defective metering or combination valve.	Replace valve.	10
Loose caliper or caliper support.	Tighten caliper or support.	13
Oil or grease on brake pads or shoes.	Clean or replace pads or shoes.	13, 15
Wrong type of pads used on front.	Replace pads.	13
Pads loose.	Replace anti-rattle spring or pads.	13
Loose rear wheel brake parts.	Tighten rear brake components.	15
Improper lining-to-drum clearance.	Adjust or replace shoes.	15
Excessive debris in drums.	Clean brakes.	15
Defective drum(s).	Replace drum.	15
Rear tires excessively worn.	Consult service manual.	

Excessive Lining and/or Pad Wear

Symptoms: Pad or shoe linings on one or more wheels (front or rear), wear out before their expected mileage. May be accompanied by signs of overheating.

Cause	Correction	Chapter
Sticking master cylinder pistons.	Replace master cylinder.	6
Sticking power brake booster.	Replace power brake booster.	8
Incorrect or defective shoes or pads.	Replace pads or shoes.	13, 15
Drums and/or rotors worn or machined too thin.	Replace drums or rotors.	13, 15
Brakes adjusted too tight.	Adjust brakes.	15
Poor wheel ventilation.	Clean wheels, advise driver.	13, 15
Driver resting foot on brake pedal.	Advise driver.	
Excessive braking or loads.	Advise driver.	
Sticking wheel cylinder and/or caliper piston.	Rebuild or replace caliper or wheel cylinder.	6
Misaligned backing plate.	Realign or replace backing plate.	15
Reversed primary and secondary shoes.	Reinstall shoes.	15
Parking brake stuck in partially applied position.	Replace or free linkage.	19

Uneven Lining and/or Pad Wear

Symptoms: Pad or shoe linings wear unevenly. May be accompanied by grabbing or pulling brakes.

Cause	Correction	Chapter
Misaligned caliper.	Reinstall caliper.	13
Bent, incorrect or defective shoes and/or pads.	Replace shoes or pads.	13, 15
Fasteners broken or loose.	Tighten or replace fasteners.	13, 15
Hold down springs weak or disconnected.	Replace springs.	15
Reversed primary and secondary shoes.	Reinstall shoes.	15
Tapered drum.	Resurface or replace drum.	15

Drum Brake Self Adjusters Do Not Work

Symptoms: Drum brake clearance becomes excessive, self adjusters do not compensate for wear.

Cause	Correction	Chapter
Adjusters frozen.	Lubricate or replace adjusters.	15
Adjusters swapped between right and left sides.	Reinstall adjusters.	15
Self-adjuster lever or linkage broken or disconnected.	Reinstall or replace linkage.	15

Brake Noise

Symptoms: Brake noises include squealing, chatter, grinding, clunking, or scraping noises when the brakes are applied, and squealing or scraping noises when brakes are released. The following sections address each type of sound.

Squealing

Cause	Correction	Chapter
Wear indicators on pads contacting rotor.	Replace pads.	13
Pad insulators missing.	Install insulators or add anti-squeal compound.	13
Brake shoes and/or pads glazed or incorrect.	Replace pads or shoes.	13, 15
Excessive brake dust build up.	Clean brakes, add anti-squeal compound.	13, 15
Scored, overheated, or glazed drums and/or rotors.	Resurface or replace drums or rotors.	13, 15
Loose mounting parts.	Tighten loose parts.	13, 15
Weak or worn anti-rattle clips or springs.	Replace springs or clips.	13, 15
Primary and secondary shoes reversed.	Reinstall shoes.	15
Loose wheel bearings.	Adjust or replace bearings.	17

Chatter

Cause	Correction	Chapter
Air in hydraulic system.	Bleed brakes, check for leaks.	6
Damaged or lost disc brake pad insulators.	Install insulators or anti-squeal compound.	13
Caliper binding.	Check caliper clearance, rebuild or replace caliper.	13
Excessive rotor runout.	Resurface or replace rotor.	13
Rotor out-of-parallel.	Resurface or replace rotor.	13

Chatter (continued)

Cause	Correction	Chapter
Brake shoes and/or pads worn out, glazed, or incorrect.	Replace pads or shoes.	13, 15
Contaminated brake shoes and/or pads.	Replace shoes or pads.	13, 15
Mixed, lining friction material grades.	Replace pads or shoes.	13, 15
Loose lining and/or pads on shoes.	Replace pads or shoes.	13, 15
Weak or broken shoe retracting springs.	Replace springs.	15
Primary and secondary shoes reversed.	Reinstall shoes.	15
Broken or bent brake shoes.	Replace brake shoes.	15
Loose caliper anchor plate or support plate.	Tighten or replace support plate.	15
Loose or damaged wheel bearings.	Adjust or replace wheel bearings.	17
Worn or loose suspension/steering parts.	Consult service manual.	

Grinding or Scraping

Cause	Correction	Chapter
Caliper mounting bolts too long.	Replace bolts.	13
Rotor making contact with caliper housing.	Check bearings, caliper clearance, replace components as needed.	13
Pad shoes bent.	Replace pads.	13
Pads worn to metal or wear indicator touching rotor.	Replace pads.	13
Debris imbedded in pads.	Replace pads.	13
Warped or cracked rotor.	Resurface or replace rotor.	13
Brake drum or rotor has rough brake surface.	Resurface drum or rotor.	13, 15
Brake shoes bent, cracked, or incorrect.	Replace shoes.	15
Lining rivets in contact with brake drum or rotor.	Replace pads or shoes, resurface drums or rotors.	15
Lining charred from overheating.	Replace pads or shoes and drums or rotors.	15
Warped or cracked brake drum.	Resurface or replace drum.	15
Loose drum brake mounting parts.	Tighten mounting bolts.	15
Weak shoe return or hold down springs.	Replace springs.	15
Oil or grease on brake lining.	Clean or replace brake pads or shoes.	15
Loose wheel bearings.	Adjust or replace bearings.	17

Thumping or Klunking

Cause	Correction	Chapter
Loose disc brake mounting hardware.	Tighten bolts, clips, or springs.	13
Lathe tool marks (defective lathe bit).	Replace tool bit, resurface drum or rotor.	13, 15
Cracked brake drum.	Replace drum.	15
Excessive clearance between shoes and anchors.	Replace shoes.	15
Return springs weak, unequal, or defective.	Replace springs.	15
Excessive clearance between lining and drum.	Adjust brakes, replace shoes.	15
Loose wheel bearings.	Adjust or replace bearings.	17
Worn or loose suspension/steering or driveline parts.	Consult service manual.	

Clicking or Snapping

Cause	Correction	Chapter
Incorrect disc brake pads.	Replace pads.	13
Grooves in backing plate support pads.	Replace backing plate.	15
Loose backing plate.	Tighten backing plate.	15
Threaded drum braking surface (improper turning).	Replace tool bit, resurface drum.	15
Weak shoe springs or clips.	Replace springs or clips.	15
Shoes bent, twisted, or broken.	Replace shoes.	15
Cracked brake drum(s).	Replace drum.	15
Worn or loose suspension/steering or driveline parts.	Consult service manual.	

Noise with Brakes Released

Cause	Correction	Chapter
Wear indicators on pads contacting rotor.	Replace pads.	13
Disc brake anti-rattle clip improperly installed.	Reinstall clip.	13
Loose caliper mounting.	Tighten caliper bolts.	13
Warped or cracked rotor.	Resurface or replace rotor.	13
Bent backing plate.	Replace backing plate.	13, 15
Weak or broken brake shoe retracting springs.	Replace springs.	15
Lining loose on shoes.	Replace shoes.	15
Hold down springs broken or loose.	Replace springs.	15
Shoes binding on grooved backing plate.	Replace backing plate.	15
Loose or damaged wheel bearings.	Adjust or replace wheel bearings.	17
Loose lug nuts or wheel cover.	Consult service manual.	

Hissing Noise as Brake Pedal Is Held

Cause	Correction	Chapter
Defect in valve assembly in vacuum brake booster.	Replace vacuum brake booster.	8

Vibration/Chatter

Cause	Correction	Chapter
Power steering pump belt slipping.	Tighten or replace belt.	8
Power steering pump reservoir fluid level low.	Fill reservoir, check for leaks.	8
Spool operation faulty.	Replace Hydro-boost.	8
Contaminated fluid.	Flush system, replace fluid.	8

Parking Brakes

Since parking brakes are seldom used, most defects show up as stuck linkage. Either the parking brake will not apply, will not hold, or will not release. Always check the service brakes before assuming that the parking brake is defective.

Parking Brakes Will Not Hold

Symptoms: Vehicle can be moved after parking brake is applied.

Cause	Correction	Chapter
Service brake linings worn out.	Replace pads or shoes.	13, 15
Loose, bent, or missing linkage or cables.	Replace linkage or cables.	19
Linkage improperly adjusted.	Adjust linkage.	19
Driver operated lever defective.	Replace lever.	19
Wheel parking brake components loose or broken.	Replace parking brake components.	19

Parking Brakes Will Not Release

Symptoms: Parking brake remains applied after lever is released.

Cause	Correction	Chapter
Service brakes sticking.	Lubricate or replace binding components.	13, 15
Brake cable adjusted too tightly.	Adjust cable.	19
Cables sticking.	Replace cables.	19
Parking brake lever assembly binding.	Replace lever assembly.	19
Wheel parking brake components binding or frozen.	Replace parking brake components.	19

Foot or Hand Lever Will Not Hold

Symptoms: Parking brake lever returns to the unapplied position when the driver removes foot or hand.

Cause	Correction	Chapter
Ratchet teeth worn or broken off.	Replace lever assembly.	19
Release button binding in housing.	Replace lever assembly.	19
Broken or missing release button return spring.	Replace lever assembly.	19
Brake release handle and/or cable stuck in release position.	Replace handle or cable.	19
Automatic vacuum release stuck in release position.	Replace vacuum release.	19
Ratchet pawl broken off or missing.	Replace lever assembly.	19

Automatic Release Will Not Release

Symptoms: Parking brake remains applied when vehicle is shifted into drive or reverse.

Cause	Correction	Chapter
Disconnected or plugged vacuum supply line.	Reconnect or unclog supply line.	19
Binding or frozen release linkage.	Replace linkage.	19
Vacuum diaphragm pull rod missing or broken.	Replace pull rod.	19
Defective or loose steering column vacuum switch.	Replace switch.	19

Brake Electrical

Most brake system electrical diagnosis involves the brake lights.

No Brake Lights

Symptoms: One or more brake lights do not illuminate when the brakes are applied.

Cause	Correction	Chapter
Defective or misadjusted stoplight switch.	Adjust or replace switch.	20
Wire leads to stoplight switch loose or disconnected.	Reinstall wiring.	20
Brake lightbulb(s), burned out.	Replace bulb.	20
Light bulb sockets corroded.	Replace socket.	20
Blown fuse.	Replace fuse, locate short.	20
Malfunctioning turn signal switch.	Replace turn signal switch.	20

Brake Lights Always On

Symptoms: Brake lights do not go off when the brake pedal is released.

Cause	Correction	Chapter
Stoplight switch misadjusted or defective.	Adjust or replace switch.	20
Brake pedal binding in partially applied position.	Free or replace pedal assembly.	20
Disconnected, broken, or missing brake pedal return spring.	Reinstall or replace spring.	20

Front Parking Lights Flash When Brakes Are Applied

Symptoms: Front parking lights come on whenever the brake pedal is pressed.

Cause	Correction	Chapter
Short between brake and parking light wiring.	Locate and repair short.	20
Bulb filaments touching (dual filament bulb).	Replace bulb.	20
Defect in turn signal switch.	Replace turn signal switch.	20

Red Brake Warning Light Does Not Come On

Symptoms: Brake warning light on dashboard does not illuminate when the parking brake is applied.

Cause	Correction	Chapter
Bulb burned out.	Replace bulb.	20
Bulb missing or out of socket.	Replace bulb.	20
Blown fuse.	Replace fuse, locate short.	20
Wires broken or disconnected.	Repair wiring.	20
Defective or misadjusted parking brake switch.	Adjust or replace switch.	20

Red Brake Warning Light Does Not Go Off

Symptoms: Brake warning light on dashboard on at all times.
Make sure there are no problems in the brake hydraulic system before proceeding.

Cause	Correction	Chapter
Pressure differential valve out of position.	Check hydraulic system, bleed brakes, replace valve.	10
Pad wear sensor grounded (when used).	Replace pads.	13
Misadjusted parking brake linkage.	Adjust parking brakes and linkage.	19
Defective or misadjusted parking brake switch.	Adjust or replace switch.	20
Short circuit in wiring.	Repair wiring.	20

Anti-Lock Brake and Traction Control System

Symptoms will vary greatly between the many types of systems in use. Make sure the base brake hydraulic and friction systems are in good condition before proceeding with ABS/TCS diagnosis. See Chapter 22 for more information.

Complaints of brake pedal or accelerator pulsation, tire chirp, reduced engine torque, and clicking or humming noises during ABS/TCS operation are all considered normal.

ABS or TCS System Light Does Not Come On When Vehicle Is Started

Symptoms: The warning light on the dashboard does not come on when the vehicle is first started.

Cause	Correction	Chapter
Defective bulb or bulb out of socket.	Replace bulb.	20
Fuse blown.	Replace fuse, locate short.	20
Electrical leads open to lamp circuit.	Repair open lead.	20
Defective control module.	Replace control module.	22

ABS or TCS System Light On at All Times

Symptoms: Warning light on dashboard illuminated at all times. The red Brake warning light may or may not be on.

Cause	Correction	Chapter
Defect in base brake system.	Inspect base brake system.	See appropriate chapter
Defect in vehicle electrical system.	Repair electrical problem.	20
System fuses blown.	Replace fuse, check electrical system.	20
Speed sensor(s) defective or misadjusted.	Replace wheel speed sensor.	22
Defective control module.	Replace control module.	22
Defective hydraulic actuator solenoids.	Replace solenoids.	22
Defective hydraulic actuator pump.	Replace pump, check pump relay.	22
Defective accumulator.	Replace accumulator.	22

No ABS or TCS System Operation

Symptoms: System does not operate when called for. Warning light may or may not be on.

Cause	Correction	Chapter
Defect in base brake system.	Inspect base brake system.	See appropriate chapter
System fuses blown.	Replace fuse, check electrical system.	20
Speed sensor(s) defective or misadjusted.	Replace wheel speed sensor.	22
Defective control module.	Replace control module.	22
Defective hydraulic actuator solenoids.	Replace solenoids.	22
Defective hydraulic actuator pump.	Replace pump, check relay.	22
Defective accumulator.	Replace accumulator.	22
Wiring disconnected or defective.	Repair wiring.	22

Excessive ABS or TCS Engagement

Symptoms: ABS or TCS system operates in situations where it is not needed. System operates under light braking or acceleration.

Cause	Correction	Chapter
Defect in base brake system.	Inspect base brake system.	See appropriate chapter
Misadjusted brake light switch.	Adjust or replace brake light switch.	20
Defective wheel speed sensor.	Replace sensor.	22
Defective control module.	Replace module.	22
Tires worn.	Replace tires.	

No Engine Torque Management during TCS Operation

Symptoms: TCS system immediately applies the brakes on the driving wheels, does not reduce engine torque. Not all vehicles equipped with TCS use engine torque management.

Cause	Correction	Chapter
Defective control module.	Replace control module.	22
Defect in vehicle data bus.	Repair damaged wiring.	20
Stepper motor inoperative.	Replace stepper motor.	22

Useful Tables

CONVERSION CHART

METRIC/U.S. CUSTOMARY UNIT EQUIVALENTS

Multiply:	by:	to get:	Multiply:	by:	to get:
ACCELERATION					
feet/sec^2	x 0.3048	= meters/sec^2 (m/s^2)	x 3.281	= feet/sec^2
inches/sec^2	x 0.0254	= meters/sec^2 (m/s^2)	x 39.37	= inches/sec^2
ENERGY OR WORK (watt–second = joule = newton–meter)					
foot–pounds	x 1.3558	= joules (J)	x 0.7376	= foot–pounds
calories	x 4.187	= joules (J)	x 0.2388	= calories
Btu	x 1055	= joules (J)	x 0.000948	= Btu
watt–hours	x 3600	= joules (J)	x 0.0002778	= watt–hours
kilowatt–hrs.	x 3.600	= megajoules (MJ)	x 0.2778	= kilowatt–hrs
FUEL ECONOMY AND FUEL CONSUMPTION					
miles/gal	x 0.42514	= kilometers/liter (km/L)	x 2.3522	= miles/gal

Note:
235.2/(mi/gal) = liters/100km
235.2/(liters/100 km) = mi/gal

LIGHT					
footcandles	x 10.76	= lumens/meter^2 (lm/m^2)	x 0.0929	= footcandles
PRESSURE OR STRESS (newton/sq meter = pascal)					
inches Hg(60 °F)	x 3.377	= kilopascals (kPa)	x 0.2961	= inches Hg
pounds/sq in	x 6.895	= kilopascals (kPa)	x 0.145	= pounds/sq in
inches H$_2$O(60 °F)	x 0.2488	= kilopascals (kPa)	x 4.0193	= inches H$_2$O
bars	x 100	= kilopascals (kPa)	x 0.01	= bars
pounds/sq ft	x 47.88	= pascals (Pa)	x 0.02088	= pounds/sq ft
POWER					
horsepower	x 0.746	= kilowatts (kW)	x 1.34	= horsepower
ft–lbf/min	x 0.0226	= watts (W)	x 44.25	= ft–lbf/min
TORQUE					
pounds–inches	x 0.11298	= newton–meters (N-m)	x 8.851	= pound–inches
pound–feet	x 1.3558	= newton–meters (N-m)	x 0.7376	= pound–feet
VELOCITY					
miles/hour	x 1.6093	= kilometers/hour (km/h)	x 0.6214	= miles/hour
feet/sec	x 0.3048	= meters/sec (m/s)	x 3.281	= feet/sec
kilometers/hr	x 0.27778	= meters/sec (m/s)	x 3.600	= kilometers/hr
miles/hour	x 0.4470	= meters/sec (m/s)	x 2.237	= miles/hour

COMMON METRIC PREFIXES

mega	(M)	= 1 000 000	or 10^6	centi	(c)	= 0.01	or 10^{-2}
kilo	(k)	= 1 000	or 10^3	milli	(m)	= 0.001	or 10^{-3}
hecto	(h)	= 100	or 10^2	micro	(μ)	= 0.000 001	or 10^{-6}

METRIC/U.S. CUSTOMARY UNIT EQUIVALENTS

Multiply:	by:	to get:	Multiply:	by:	to get:
LINEAR					
inches	x 25.4	= millimeters (mm)	x 0.03937	= inches
feet	x 0.3048	= meters (m)	x 3.281	= feet
yards	x 0.9144	= meters (m)	x 1.0936	= yards
miles	x 1.6093	= kilometers (km)	x 0.6214	= miles
inches	x 2.54	= centimeters (cm)	x 0.3937	= inches
microinches	x 0.0254	= micrometers (μm)	x 39.37	= microinches
AREA					
inches^2	x 645.16	= millimeters^2(mm^2)	x 0.00155	= inches^2
inches^2	x 6.452	= centimeters^2(cm^2)	x 0.155	= inches^2
feet^2	x 0.0929	= meters^2(m^2)	x 10.764	= feet^2
yards^2	x 0.8361	= meters^2(m^2)	x 1.196	= yards^2
acres^2	x 0.4047	= hectares (10^{4}m^2)		
		ha	x 2.471	= acres
miles^2	x 2.590	= kilometers^2 (km^2)	x 0.3861	= miles^2
VOLUME					
inches^3	x 16387	= millimeters^3 (mm^3)	x 0.000061	= inches^3
inches^3	x 16.387	= centimeters^3 (cm^3)	x 0.06102	= inches^3
inches^3	x 0.01639	= liters (L)	x 61.024	= inches^3
quarts	x 0.94635	= liters (L)	x 1.0567	= quarts
gallons	x 3.7854	= liters (L)	x 0.2642	= gallons
feet^3	x 28.317	= liters (L)	x 0.03531	= feet^3
feet^3	x 0.02832	= meters^3 (m^3)	x 35.315	= feet^3
fluid oz	x 29.57	= milliliters (mL)	x 0.03381	= fluid oz
yards^3	x 0.7646	= meters^3 (m^3)	x 1.3080	= yards^3
teaspoons	x 4.929	= milliliters (mL)	x 0.2029	= teaspoons
cups	x 0.2366	= liters (L)	x 4.227	= cups
MASS					
ounces (av)	x 28.35	= grams (g)	x 0.03527	= ounces (av)
pounds (av)	x 0.4536	= kilograms (kg)	x 2.2046	= pounds (av)
tons (2000 lb)	x 907.18	= kilograms (kg)	x 0.001102	= tons (2000 lb)
tons (2000 lb)	x 0.90718	= metric tons (t)	x 1.1023	= tons (2000 lb)
FORCE					
ounces—f (av)	x 0.278	= newtons (N)	x 3.597	= ounces—f (av)
pounds—f (av)	x 4.448	= newtons (N)	x 0.2248	= pounds—f (av)
kilograms—f	x 9.807	= newtons (N)	x 0.10197	= kilograms—f
TEMPERATURE					

°F -40 ... 32 ... 0 ... 40 ... 80 ... 98.6 ... 120 ... 160 ... 200 ... 212 ... 240 ... 280 ... 320 °F

°C -40 ... -20 ... 0 ... 20 ... 40 ... 60 ... 80 ... 100 ... 120 ... 140 ... 160 °C

°Celsius = 0.556 (°F – 32)   °F = (1.8 °C) + 32

TAP/DRILL CHART

COARSE STANDARD THREAD (N.C.) Formerly U.S. Standard Thread					FINE STANDARD THREAD (N.F.) Formerly S.A.E. Thread				
Sizes	Threads Per Inch	Outside Diameter at Screw	Tap Drill Sizes	Decimal Equivalent of Drill	Sizes	Threads Per Inch	Outside Diameter at Screw	Tap Drill Sizes	Decimal Equivalent of Drill
1	64	.073	53	0.0595	0	80	.060	$^3/_{64}$	0.0469
2	56	.086	50	0.0700	1	72	.073	53	0.0595
3	48	.099	47	0.0785	2	64	.086	50	0.0700
4	40	.112	43	0.0890	3	56	.099	45	0.0820
5	40	.125	38	0.1015	4	48	.112	42	0.0935
6	32	.138	36	0.1065	5	44	.125	37	0.1040
8	32	.164	29	0.1360	6	40	.138	33	0.1130
10	24	.190	25	0.1495	8	36	.164	29	0.1360
12	24	.216	16	0.1770	10	32	.190	21	0.1590
$^1/_4$	20	.250	7	0.2010	12	28	.216	14	0.1820
$^5/_{16}$	18	.3125	F	0.2570	$^1/_4$	28	.250	3	0.2130
$^3/_8$	16	.375	$^5/_{16}$	0.3125	$^5/_{16}$	24	.3125	I	0.2720
$^7/_{16}$	14	.4375	U	0.3680	$^3/_8$	24	.375	O	0.3320
$^1/_2$	13	.500	$^{27}/_{64}$	0.4219	$^7/_{16}$	20	.4375	$^{25}/_{64}$	0.3906
$^9/_{16}$	12	.5625	$^{31}/_{64}$	0.4843	$^1/_2$	20	.500	$^{29}/_{64}$	0.4531
$^5/_8$	11	.625	$^{17}/_{32}$	0.5312	$^9/_{16}$	18	.5625	0.5062	0.5062
$^3/_4$	10	.750	$^{21}/_{32}$	0.6562	$^5/_8$	18	.625	0.5687	0.5687
$^7/_8$	9	.875	$^{49}/_{64}$	0.7656	$^3/_4$	16	.750	$^{11}/_{16}$	0.6875
1	8	1.000	$^7/_8$	0.875	$^7/_8$	14	.875	0.8020	0.8020
$1^1/_8$	7	1.125	$^{63}/_{64}$	0.9843	1	14	1.000	0.9274	0.9274
$1^1/_4$	7	1.250	$^{17}/_{64}$	1.1093	$1^1/_8$	12	1.125	$1^3/_{64}$	1.0468
					$1^1/_4$	12	1.250	$1^{11}/_{64}$	1.1718

BOLT TORQUING CHART

METRIC STANDARD						SAE STANDARD/FOOT POUNDS						
Grade of Bolt	5D	.8G	10K	12K		Grade of Bolt	SAE 1 & 2	SAE 5	SAE 6	SAE 8		
Min. Tensile Strength	71,160 P.S.I.	113,800 P.S.I.	142,200 P.S.I.	170,679 P.S.I.		Min. Tensile Strength	64,000 P.S.I.	105,000 P.S.I.	133,000 P.S.I.	150,000 P.S.I.		
Grade Markings on Head	5D	8G	10K	12K	Size of Socket or Wrench Opening	Markings on Head	◆	◆	◆	◆	Size of Socket or Wrench Opening	
Metric						Metric	U.S. Standard				U.S. Regular	
Bolt Dia.	U.S. Dec Equiv.	Foot Pounds				Bolt Head	Bolt Dia.	Foot Pounds			Bolt Head	Nut
6mm	.2362	5	6	8	10	10mm	1/4	5	7	10	10.5	3/8	7/16
8mm	.3150	10	16	22	27	14mm	5/16	9	14	19	22	1/2	9/16
10mm	.3937	19	31	40	49	17mm	3/8	15	25	34	37	9/16	5/8
12mm	.4720	34	54	70	86	19mm	7/16	24	40	55	60	5/8	3/4
14mm	.5512	55	89	117	137	22mm	1/2	37	60	85	92	3/4	13/16
16mm	.6299	83	132	175	208	24mm	9/16	53	88	120	132	7/8	7/8
18mm	.709	111	182	236	283	27mm	5/8	74	120	167	180	15/16	1.
22mm	.8661	182	284	394	464	32mm	3/4	120	200	280	296	1-1/8	1-1/8

DECIMAL CONVERSION CHART

FRACTION	INCHES	M/M	FRACTION	INCHES	M/M
1/64	.01563	.397	33/64	.51563	13.097
1/32	.03125	.794	17/32	.53125	13.494
3/64	.04688	1.191	35/64	.54688	13.891
1/16	.6250	1.588	9/16	.56250	14.288
5/64	.07813	1.984	37/64	.57813	14.684
3/32	.09375	2.381	19/32	.59375	15.081
7/64	.10938	2.778	39/64	.60938	15.478
1/8	.12500	3.175	5/8	.62500	15.875
9/64	.14063	3.572	41/64	.64063	16.272
5/32	.15625	3.969	21/32	.65625	16.669
11/64	.17188	4.366	43/64	.67188	17.066
3/16	.18750	4.763	11/16	.68750	17.463
13/64	.20313	5.159	45/64	.70313	17.859
7/32	.21875	5.556	23/32	.71875	18.256
15/64	.23438	5.953	47/64	.73438	18.653
1/4	.25000	6.350	3/4	.75000	19.050
17/64	.26563	6.747	49/64	.76563	19.447
9/32	.28125	7.144	25/32	.78125	19.844
19/64	.29688	7.541	51/64	.79688	20.241
5/16	.31250	7.938	13/16	.81250	20.638
21/64	.32813	8.334	53/64	.82813	21.034
11/32	.34375	8.731	27/32	.84375	21.431
23/64	.35938	9.128	55/64	.85938	21.828
3/8	.37500	9.525	7/8	.87500	22.225
25/64	.39063	9.922	57/64	.89063	22.622
13/32	.40625	10.319	29/32	.90625	23.019
27/64	.42188	10.716	59/64	.92188	23.416
7/16	.43750	11.113	15/16	.93750	23.813
29/64	.45313	11.509	61/64	.95313	24.209
15/32	.46875	11.906	31/32	.96875	24.606
31/64	.48438	12.303	63/64	.98438	25.003
1/2	.50000	12.700	1	1.00000	25.400

SOME COMMON ABBREVIATIONS

U.S CUSTOMARY		METRIC	
UNIT	ABBREVIATION	UNIT	ABBREVIATION
inch	in.	kilometer	km
feet	ft.	hectometer	hm
yard	yd.	dekameter	dam
mile	mi.	meter	m
grain	gr.	decimeter	dm
ounce	oz.	centimeter	cm
pound	lb.	millimeter	mm
teaspoon	tsp.	cubic centimeter	cm^3
tablespoon	tbsp.	kilogram	kg
fluid ounce	fl. oz.	hectogram	hg
cup	c.	dekagram	dag
pint	pt.	gram	g
quart	qt.	decigram	dg
gallon	gal.	centigram	cg
cubic inch	in^3	milligram	mg
cubic foot	ft^3	kiloliter	kl
cubic yard	yd^3	hectoliter	hl
square inch	in^2	dekaliter	dl
square foot	ft^2	liter	L
square yard	yd^2	centiliter	cl
square mile	mi^2	milliliter	ml
Fahrenheit	F°	square kilometer	km^2
barrel	bbl.	hectare	ha
fluid dram	fl. dr.	are	a
board foot	bd. ft.	centare	ca
rod	rd.	tonne	t
dram	dr.	Celsius	C°
bushel	bu.		

Acknowledgments

The production of a textbook of this type is not possible without assistance from the automotive industry. In procuring and compiling the materials for this text, the industry has been extremely helpful. The authors would like to thank the following firms and individuals for their assistance in the preparation of **Auto Brakes Technology.**

ABS Education Alliance; AC-Delco; Acura Automobile Division; Alfa-Romero, Inc.; Allied-Signal Inc.; Ammco Tools, Inc.; ATE; Audi of America; Autolite; Automotive Diagnostics; Balco Inc.; Bean Tools; Barret Company; Bear Automotive Service Equipment; Beauron Industries, Inc.; Bee Line Company; Bendix; Black and Decker; BMW of North America; Robert Bosch Corporation; Bower/BCA; Bridgestone U.S.A.; British-Leyland; Buick Motor Division; Cadillac Motor Division; Central Tools; Champion; Channellock, Inc.; Chevrolet Motor Division; Chief Automotive Systems; Chief Industries; Clayton Manufacturing; Continental Teves; Cooper Industries; Cooper Tire and Rubber; CR Industries; Daihatsu America Inc.; DaimlerChrysler; Dana Publication Services; Dayton; Deere and Company; Delphi Corporation; Dorman Products, Inc.; Douglas Components Corporation; Dover Corporation; EMI-Tech; Fafnir Bearing Division of Textron Inc.; Federal-Mogul Corporation; Ferrari North America; John Fluke Manufacturing Company; FMC Corporation; Ford Motor Company; Gates Rubber Company; Gencorp; General Motors Corporation, Service Technology Group; Girling Brake; GMC Trucks; Goodyear Tire and Rubber Company; Graymills Corporation; H and M Dreyer, Inc.; Honda Motor Company; Hunter Engineering; Huot Manufacturing; Hyundai Motor Corporation; Imperial-Eastman; Infiniti North America; Jaguar Cars; John Bean Corporation; Kelly-Springfield Tire Company; Kent-Moore Tool Group; Kia Motors; Land Rover of North America; Lexus Automobile Division; M&M Forbes, Inc., Mac Tools; Lisle Corporation; Loctite Corporation; Mac Tools; Maremont Corporation; Mazda Motors of America; Mercedes-Benz of North America; Mitsubishi Motors of America; Mohawk Resources; Moog Automotive; NAPA; National Institute for Automotive Service Excellence; New Departure, Div. of General Motors; Niehoff Company; Nissan North America; Oldsmobile Motor Division; OTC Division of SPX Corp.; Paprl Automotive Industries; Parker Hannifin Corporation, EIS Division; PBS Corporation; Pontiac Motor Division; Porsche Cars of North America; Prestolite; Raybestos; Rubber Manufacturers Association; Saab Cars USA; Saturn Corporation; SFK Industries; SK Hand Tools; Snap-On Tools; Star Machine and Tool Company; Subaru of America; Sun Electric Corporation; Timken Company; Torrington Bearing Company; Toyota Motor Sales; Truecraft Tools; TRW Inc.; Vaco Tool Company; Vetronix Corporation; Volkswagen of America; Volvo Corporation of North America; Wagner Brake; Walker Company; Warner Company; Wright Studio and Camera; Wynn's Company; Yokohama Tire Corporation; ZF Industries.

The authors would also like to thank the following persons and organizations who provided vehicles, parts, and test equipment for the photographs as well as other items used throughout this text.

Paul Anderson; Brian Cook, Saturn of Greenville, SC; James Dalby, Goodyear Auto Service Center, Greenville, SC; JoAnn Ferguson, Sears Auto Center, Greenville, SC; Corey Glassman, John Fluke Corporation; Mash Hayward; Jimmy Kellett, Kellett's Auto Salvage, Hickory Tavern, SC; Hervert and Dorothy MacMillian; Palmetto Chapter, Pontiac-Oakland Club International; Mike Payne, C.W. Weeks.

"Portions of materials contained herein have been reprinted with permission of General Motors Corporation, Service Technology Group."

Glossary

A

Abbreviations: Letters or letter combinations that stand for words. Used extensively in the automotive industry.

ABS: Abbreviation for *Anti-lock Braking System.*

ABS warning light: Amber indicator lamp mounted in the instrument cluster. Illuminates when there is a problem with the anti-lock brake system.

ABS/TCS controller: See *Electronic Brake and Traction Control Module (EBTCM).*

Accumulator: Gas-charged cylinder or chamber used to store brake fluid. Used in anti-lock and Hydro-boost brake systems.

Actuator: Any output device controlled by a computer.

ALDL: Abbreviation for *Assembly Line Diagnostic Link.* Also referred to as a *data link connector.*

Allen wrench: A hexagonal wrench, which is usually "L" shaped, designed to fit into a hexagonal hole.

Alternating current (ac): An electrical current that moves in one direction and then the other.

Ammeter: Instrument used to measure the flow of electric current in a circuit in amperes. Normally connected in series in a circuit.

Ampere: The unit of measurement for the flow of electric current.

Analog: A signal that continually changes in strength.

Anchors: Solid attaching points for the brake shoes and springs installed on the backing plate. Often called *anchor pins* or *anchor plates.*

Anodized: Electrically deposited coating used to reduce wear on aluminum parts.

Anti-lock brake system (ABS): A computer-controlled system that is part of the base brake system. The system "cycles" the brakes on and off to prevent wheel lockup and skidding.

Anti-rattle clip: One or more metal components designed to keep brake pads from vibrating and rattling.

Anti-rattle spring: Spring used in drum brake assemblies to reduce rattle and clicking noises.

Anti-squeal compound: Chemical insulation that can be applied to the metal side of brake pads.

Antifriction bearing: A bearing that uses balls or rollers between a journal and bearing surface to decrease friction.

Arc grinder: Machine used to change the radius or arc of a brake shoe to match the drum. Also known as a *brake shoe grinder.*

Armored line: Standard steel tubes around which steel wire is coiled.

Asbestos: A mineral material that has great heat resistant characteristics. Once widely used in brake and clutch linings. Asbestos is a known cancer causing agent.

ASE: Abbreviation for *National Institute for Automotive Service Excellence.*

Atmospheric pressure: The pressure exerted by the earth's atmosphere on all objects. Measured with reference to the pressure at sea level, which is around 14.7 psi (101 kPa).

AWG: Abbreviation for *American Wire Gage,* which is a standard measure of wire size. The smaller the wire, the larger the number.

Axial load: Sideways load on a bearing.

Axial ribs: Metal ribs which run parallel around the drum.

Axle: Shaft or shafts used to transmit driving force to the wheels.

Axle housing: A metal enclosure for the drive axles and differential. It is partially filled with differential fluid and other additives as needed.

Axle ratio: The relationship or ratio between the number of times the propeller shaft must revolve to turn the axle drive shafts once.

B

Backing plate: A metal plate that holds the wheel cylinder, shoes, and other parts inside a drum brake.

Backlash: The clearance or "play" between two parts, such as the teeth of two gears.

Ball bearing: A bearing consisting of an inner and outer hardened steel race separated by a series of hardened steel balls.

Ball-and-ramp disc parking brake: Caliper parking brake. Used with single piston, sliding brake calipers. It consists of three steel balls between an operating shaft and a thrust screw.

Banjo fitting: A fitting consisting of two washers, a drilled bolt, and a compression fitting. It is installed by placing one washer on each side of the banjo fitting, then installing and tightening the bolt.

Base brakes: Also referred to as *foundation brakes*. The conventional hydraulic and friction brake parts minus the ABS unit.

Bearing collar: A retainer plate on the brake backing plate which holds the rear axle bearings in some axle housings.

Bedding-in: The process of wearing in brake pads or shoes that have been replaced or rotors or drums that have been reconditioned.

Bench bleeding: Bleeding technique used to fill the master cylinder's internal passages, reducing the amount of time needed to bleed the brake system once the master cylinder is reinstalled.

Bleed: The process of removing air or fluid from a closed system, such as the brakes.

Bleeder screws: Air removal devices installed at the highest point in a hydraulic device. Used to bleed the brake system of air.

Bleeder wrenches: Tool used to open the bleeder screws found at various points throughout the brake system.

Bluing: Dark bluish spots on a rotor or drum braking surface caused by overheating. May be accompanied by heat-checking.

Body control module (BCM): On-board computer responsible for controlling such functions as interior climate, radio, instrument cluster readings, and on some vehicles, cellular telephones. May also interact with the engine's electronic control unit.

Bonding: Process that glues the lining to the pad or shoe.

Brake bleeding: Technique used to remove air from the brake hydraulic system.

Brake cylinder hones: Tool used to remove glaze and deposits from wheel cylinders, master cylinders, and disc brake caliper bores.

Brake drum: A cast iron or aluminum housing upon which the brake shoes contact to provide braking action.

Brake dust: Dust which contains friction material, mixed with road dirt, that has been worn off pads or shoes by the normal braking process.

Brake fade: Term given to gradual brake failure caused by brake overheating. Condition occurs when brake linings become so hot they can no longer create friction.

Brake feel: The discernible relationship between the amount of brake pressure and the actual force being exerted by the driver.

Brake fluid: A special fluid compound used in hydraulic brake systems. It must meet exacting specifications, such as resistance to heating, freezing, and thickening.

Brake lathe: Device used to remove the glazed and scored outer layer from rotors and drums.

Brake lathe adapters: Cones, hubs, and other tools used to mount drums and rotors to a brake lathe.

Brake lathe arbor: Metal shaft on a brake lathe on which drums and rotors are mounted.

Brake light switch: Switch used to operate the brake lights, installed on the brake pedal linkage.

Brake lining: The friction material which contacts the rotor or drum.

Brake multiplier: Also called an *intermediate lever*. Used in some parking brake systems to increase cable pull leverage.

Brake pedal linkage: Increases foot pressure by simple mechanical leverage.

Brake rotor burnisher: Machine designed to burnish, or polish, the rotor braking surface after turning.

Brake shoe: Friction lining assemblies used with drum brake.

Brake shoe adjuster: Often called a *brake spoon*. Used to adjust brake shoes with the drum installed.

Brake warning light: Red warning light used on almost all modern vehicles to inform the driver of a brake problem or when the parking brake is applied.

Braking ratio: A comparison of the front wheel to rear wheel braking effort. On most vehicles, the ratio is 55%-60% for the front brakes and 45%-50% for the rear brakes.

Breakout box: A box containing numbered test terminals and a harness designed to mate with a system's main wiring harness connector.

Brinelling: A series of unwanted lines on the races of roller bearings.

Bus: Pathway for data inside a computer. Can also be used to refer to a circuit used to connect two on-board computers.

C

Cable adjusters: A small braided cable and associated parts used in some drum brakes to maintain proper lining-to-drum clearance.

Cable pulley: Pulley wheels used to carry brake cables around sharp bends.

Cable sheath: Conduits made of steel wire wrapped tightly together inside a sheath which guides and protects the cable.

Cam disc parking brake: Used on some Asian-built vehicles. It consists of a lever operated cam, which pushes on a pin or actuator rod attached to the caliper piston.

Carcinogen: A substance that can cause cancer.

Center of gravity: The point on an object on which it could be balanced.

Central processing unit (CPU): Microprocessor inside a computer that receives sensor information, compares the input with information stored in memory, performs calculations, and makes output decisions.

Centrifugal force: A force which tends to keep a rotating object away from the center of rotation.

Chamfer: To bevel an edge on an object at the edge of a hole.

Channel: A single system of ABS sensor input and hydraulic output. There are a maximum of four channels, one for each wheel. One-, two-, and three-channel systems are also used.

Check valve: A valve that permits flow in only one direction.

Chromium: A substance added to steel to help it resist oxidation and corrosion.

Circuit breaker: Circuit protection device consisting of a contact point set attached to a bimetallic strip. The bimetallic strip will heat and bend as current flows. Unlike a fuse, it does not blow out.

Class 2 serial communications: Electronic data transfer medium used on newer vehicles. Operates by toggling the line voltage from 0-7 volts, with 0 being the rest voltage, and by varying the pulse width.

Coefficient of friction: The amount of friction produced as two materials are moved against each other. Coefficient of friction is calculated by dividing the force needed to push a load across a given surface.

Combination valve: A single assembly containing two or more hydraulic valves, including metering valve, proportioning valve, residual pressure valve, and pressure differential valve.

Compensating port: Sometimes called the *bypass* or *vent port*. Allows pressurized fluid in the hydraulic system to return to the reservoir when the brakes are released.

Composite master cylinder: Master cylinders which use a plastic reservoir installed directly on the master cylinder.

Concentric ground lining: Also referred to as a *full contact lining*. Used on large trucks or other medium- and heavy-duty vehicles. Brake shoe ground so that it contacts the drum evenly.

Conductor: Any material or substance that provides a path for electricity to flow. Examples of good conductors are copper and aluminum.

Constant velocity axle: An axle that utilizes two constant velocity joints to affect torque transfer to the driving wheels.

Contour ground lining: Also referred to as *eccentric ground*. Brake lining ground thicker at the center than at the ends.

Control module: A computer that controls the operation of a system, based on sensor inputs. Usually less complex than the computer that controls the engine and other vehicle systems.

Cotter pin: A soft metal pin that is used to secure components in place.

Cross ribs: Metal ribs that run across the drum at right angles to the braking surface.

Cup: Rubber seal installed on the master cylinder. Sometimes called *lip seals.*

Current: The number of electrons flowing past any point in the circuit.

D

Dampen: A unit or device used to reduce or eliminate vibration, oscillation, of a moving part, fluid, etc.

Dead axle: An axle that does not rotate, but merely forms a base on which to attach hubs and wheels.

Density: The relative mass of an object in a given volume.

Diagonal split system: Brake system in which one of the hydraulic systems operates the left front and right rear brake system, while the other piston operates the right front and left rear system. Used in all front-wheel drive and some rear-wheel drive vehicles.

Diaphragm: A flexible rubber sheet that divides two sides of a chamber.

Diode: Semiconductor device that allows current to flow in one direction.

Disc brake: A brake assembly that uses a hydraulic caliper to actuate brake pads against a metal rotor. Used for both front and rear brakes.

Disc brake caliper: Cast iron or aluminum cylinder and piston assembly used to receive, contain, and convert hydraulic pressure from the master cylinder to mechanical force against brake pads.

Disc brake pads: Disc brake friction lining. Designed to stop the vehicle when forced into contact with the rotor.

Disc/drum parking brake: Brake assembly used on some cars and light trucks. Incorporates disc brakes used for normal stopping and drum brake used for parking.

Documentation: Process of writing on the repair order a vehicle's problem, what caused the problem, and what was done to correct the problem.

Double-lap flares: A flare which, when made, utilizes two wall thicknesses. Used on the end of some steel brake lines.

Drive-fit: A fit between two parts that is so tight that they must be driven together.

Drum brake: Sometimes called *internal expanding brakes*. A brake system that uses a wheel cylinder to force two brake shoes against a rotating drum. Used primarily as rear brakes.

Drum brake springs: Are holddown springs and return springs.

Drum grinder: Device used to obtain a final finish on brake drums.

Dual master cylinder: Also referred to as a *tandem master cylinder*. Master cylinder used with a split brake system.

Dust boot: Rubber seal which keeps dust and water out of calipers and wheel cylinders.

E

Eccentric: A circle within a circle that has a different shape and center.

Electro-hydraulic control unit (EHCU): Type of ABS system that incorporates the hydraulic actuator and control module into one unit.

Electromagnet: A magnet that is produced by placing a coil of wire around a steel or iron bar. When current flows through the wire, the bar becomes magnetized.

Electromagnetic interference (EMI): Electronic noise caused by voltage spikes and stray magnetic fields created by defects in such circuits as the ignition and charging systems.

Electron: A negatively charged particle that makes up part of the atom.

Electronic Brake Control Module (EBCM): A computer that controls the operation of the ABS system.

Electronic Brake and Traction Control Module (EBTCM): Computer that controls the operation of the ABS and TCS system.

Electronically erasable programmable read only memory (EEPROM): A type of microprocessor whose programming can be changed by special electronic equipment that "burns in" the new programming.

Engine torque management: Traction control protocol used on some vehicles to reduce wheel spin before the TCS control module energizes the hydraulic actuator.

Equalizer: Bracket or pulley installed in the parking brake system to provide a balanced, or equal pull on the brake levers.

Erasable programmable read only memory (EPROM): A type of microprocessor whose programming can be altered only by erasing it with special equipment and reprogramming.

Extreme pressure lubricant (EP): A lubricant compounded to withstand heavy loads such as that imposed between gear teeth.

F

Fiber optic: A path for electricity or data transmission in which light acts as the carrier.

Finish cut: A cut made to remove a small amount of metal at slow feed speed.

Fixed caliper: Brake caliper that is rigidly bolted to the spindle and does not move during braking.

Flare-nut wrench: Sometimes called a *tubing wrench* or *line wrench*. Tool used to remove tube fittings used on all brake line connections.

Floating calipers: A caliper that can move or "float" back and forth in relation to the rotor.

Flushing: Process similar to brake bleeding and can be carried out by using the same methods. The only difference is that instead of removing air, the flushing procedure removes old fluid.

Four-piston fixed caliper: Caliper which uses four pistons; two located on the inboard side of the rotor and two on the outboard side. The two sides of the caliper are connected by internal fluid passages, or on a few vehicles, by external tubing.

Four-wheel anti-lock (4WAL): Also referred to as *four-wheel systems*. ABS system which controls brake actuation on all four wheels.

Frame: The steel structure that supports or is part of the vehicle's body.

Friction: Any resistance to movement between two objects placed in contact with each other.

Friction bearing: A bearing made of babbitt or bronze with a smooth surface.

Friction materials: Heat resistant linings and surfaces that perform the actual act of stopping the vehicle.

Front/rear split hydraulic system: Brake system in which the rear master cylinder piston operates the rear brakes, while the front master cylinder piston operates the front brakes. Used on rear-wheel drive vehicles.

Fuse: Circuit protection device made of a soft metal that melts when excess current flows through them, before the current can damage other components or circuit wiring.

Fusible link: A special calibrated wire installed in a circuit. Will allow an overload condition for short periods. A constant overload will melt the wire and break the circuit.

G

G-force sensor: Mercury switch which measures the rate of deceleration by comparing vehicle tilt during braking to the normal ride position.

Galvanize: To coat a metal with a molten alloy mixture of lead and tin. Used to prevent corrosion.

Gasket: A material used to prevent leaks between two stationary parts.

Glaze: A highly smooth, glassy finish on a friction lining. Produced by brake overheating.

Graphite: A soft form of the element carbon which adds flexibility, corrosion resistance, and helps retard scoring caused by grit.

Gravity bleeding: Procedure involving clamping off a brake hose leading to a caliper or wheel cylinders to be removed and then removing the clamp and allow the brake system to bleed itself.

Groove: Slot intentionally cut or formed in the brake lining. Used to relieve stress on the shoe web and prevent uneven contact pressure.

Ground: The part of a circuit connected to a body terminal or the battery.

H

Handling control system: Type of traction control system which uses G-force and lateral acceleration sensors to ensure the vehicle does not spin out of control during hard braking or acceleration.

Hard code: Trouble code set by an on-going problem that still exists in the vehicle.

Hazardous waste: Any chemical or material that has one or more characteristics that makes it hazardous to health, life, and/or the environment.

Heat dissipation: Heat removed from a friction surface by direct transfer to the surrounding air.

Heat-checking: Tiny surface cracks on the braking surface of a rotor or drum caused by excessive heat. Often seen with bluing.

Heat-shrink tubing: Plastic tube used to insulate electrical solder joints.

Height-sensing proportioning valve: A brake valve that can change the braking ratio in relation to vehicle load.

Height sensor: An electronic switch used to measure changes in vehicle ride height. Normally used with electronic ride control systems.

High efficiency particulate air (HEPA) vacuum: Type of vacuum cleaner that can trap extremely small particles, including asbestos particles.

High lead screw: Used on screw type rear disc brake calipers. Connected to a cable actuated lever. When the cable moves the lever, the screw rotates and moves the piston outward, tightening the pads against the disc.

Holddown springs: Drum brake spring assembled used to keep the nibs against the support pads, without holding them so tightly that they cannot move or wear excessively.

Holding brake: A simple mechanical brake which keeps the vehicle from moving when it is already stationary.

Honing: Process of removing deposits from the master cylinder bore by spinning a set of abrasive stones inside of the bore.

Hoses: Hydraulic conduits made of braided rubber that connect the brake system parts which move in relation to each other.

Hot spots: Sometimes called *hard spots*. Rotor sections that have been overheated by severe brake operation and become much harder than the surrounding metal. Can be accompanied by bluing.

Hub: Mounting plate for the wheels on the end of an axle, spindle, or bearing.

Hydraulic actuator: Anti-lock brake system consisting of solenoid valves, a hydraulic pump, an accumulator, and various tubing connections and electrical connectors.

Hydraulic booster: Brake booster that uses the pressure of hydraulic fluid to provide braking assist.

Hydraulic modulator: Anti-lock brake system which does not have a motor driven hydraulic pump. It uses movable pistons operated by small electric motors through reduction gears to vary the pressure.

Hydraulic system: System which uses fluid to transmit force and movement by sealing the fluid in a closed set of tubes and actuators.

Hydraulics: The study of liquids and how they work.

Hydro-boost: Also referred to as Hydro-boost II, Hydra boost, or Bendix hydraulic booster. Hydraulic booster unit powered by the vehicle's power steering pump.

Hygroscopic: Able to absorb water.

I

Induction: The imparting of electricity by magnetic fields when a wire moves through a magnetic field.

Inertia: Force which tends to keep stationary objects from moving and keeps moving objects in motion.

Information center: Display that shows vehicle condition to the driver.

Inlet port: Sometimes called the breather or intake port. Used to allow the entry of brake fluid into the rear section of the piston cylinder.

Inner race: Part of an antifriction bearing. The smooth inner metal cup on which the rolling elements move.

Input sensors: Any sensor that provides information to a computer.

Insulators: Any material or substance which resists the flow of electricity. Glass and plastic are examples of good insulators.

Integral hydraulic actuator: Large, complex anti-lock brake systems that take the place of the conventional master cylinder and power booster.

Interference angle: Flare angle which causes the seat and flare to wedge together when the nut is tightened into the fitting.

Intermittent code: Computer diagnostic code that does not return immediately after it has been cleared.

Intermittent problem: System malfunction that occurs infrequently, usually when certain conditions are met, such as temperature, humidity, vehicle operation, or in response to tests performed by a computer.

ISO flare: Type of flare used on the end of some steel brake lines. Used on modern vehicles.

J

Jumper wire: A wire used to make a temporary electrical connection.

K

Kinetic friction: Sometimes called *sliding* or *dynamic friction*. Friction that slows a moving object by converting momentum to heat.

Knockout plug: A metal or plastic slug in the backing plate which allows access to the adjustment mechanism.

Knuckle: A machined pivot point on which a hub is installed or bearing races rest on a spindle.

L

Lamp driver module: Electronic module used to operate the ABS trouble lamp.

Lateral acceleration sensor: Sensor used to measure cornering speed. These units contain two mercury switches, one for each side direction.

Lever adjuster: Drum brake adjustment assembly that uses a combination lever and link to contact the star wheel. Applying the brakes in reverse causes the lever to pivot on its attaching point, moving the star wheel.

Liner: Cast iron braking surface used with aluminum drum housings.

Lines: Hydraulic conduits made of steel that connect the stationary parts of the brake hydraulic system.

Link adjuster: Rods mounted near the center of the brake shoe, used to maintain drum-to-lining clearance as the linings wear.

Lithium grease: Often called *extreme pressure (EP) lithium*, or *lithium soap*. Lithium-based grease used on the wheel bearings of some modern vehicles.

Load-sensing proportioning valve: Also referred to as a *height-sensing proportioning valve*. This valve varies the maximum pressure to the rear brakes according to the load placed on the vehicle.

Loaded caliper: New caliper with the pads already installed.

Low brake fluid switch: Switch usually installed in the reservoir cap or in the side of the reservoir. The contacts close when the brake fluid or a float falls below a certain point.

Low brake pedal: Condition where the brake pedal approaches too close to the floorboard before actuating the brakes.

Lug nuts: Fasteners installed on the wheel studs and tightened to hold the wheel in place.

M

Machining: Also referred to as *cutting* or *turning*. The removal of a layer of metal from a rotor or drum to restore the surface finish.

Malfunction indicator light (MIL): Amber-colored light in the instrument cluster used to indicate that a problem exists in a vehicle's computer control system. Generalized term used for any instrument cluster light used to indicate a problem in a system.

Manual bleeding: Bleeding procedure which requires no special equipment, but does require the presence of a helper.

Master cylinder: Brake system device which stores fluid and provides the pressure to operate the other hydraulic components.

Material Safety Data Sheet (MSDS): Information on a chemical or material that must be provided by the material's manufacturer. Lists potential health risks and proper handling procedures.

Maximum diameter: Also referred to as the *discard diameter*. Number marked on a brake drum indicating the maximum amount of metal that can be removed.

Metallic brake linings: Brake pads used on high performance and competition vehicles. They are made from sintered metal (powdered metal that is formed into linings).

Metering valve: Used to keep the front brakes from applying before the rear brakes.

Momentum: The combination of the vehicle's weight and speed. Often referred to as *kinetic energy*.

Multimeter: An electrical test meter that can be used to test for voltage, current, or resistance.

Multiplexing: A method of using one communications path to carry two or more signals simultaneously.

N

Neutral safety switch: A switch that prevents starter engagement if the transmission is in gear.

Nibs: Raised spots on brake shoes which contact raised spots on the backing plate when the shoe is installed.

Non-directional finish: Rotor finish that eliminates the microscopic tool marks made by the cutting bits, replacing them with a series of extremely fine random scratches.

Non-integral hydraulic system: Anti-lock brake and traction control systems used with a conventional brake master cylinder and power booster.

Non-servo brakes: Used on smaller vehicles, often with front-wheel drive. On these vehicles, the rear brakes carry only a small portion of the braking load.

O

O-ring: A rubber or neoprene ring which serves the same function as a gasket in a smaller space.

OBD II: On-Board Diagnostics-Generation Two. Protocol adapted by vehicle manufacturers for standardization of diagnostic trouble codes and automotive terminology.

Ohm: Unit of measurement for resistance to the flow of electric current in a given unit or circuit.

Ohm's law: Formula for computing unknown voltage, resistance, or current in a circuit by using two known factors to find the unknown value.

Ohmmeter: An electrical instrument used to measure the amount of resistance in a given unit or circuit.

Oil seal: A device used to prevent oil leakage past certain areas, such as around a rotating shaft.

Open circuit: A circuit that is broken or disconnected.

Order: The number of times a vibration, noise, or harshness event occurs in one revolution of a rotating component.

Organic brake linings: Brake pads or shoes whose linings are made from high-temperature organic materials, such as asbestos.

Out-of-round: Condition where a cylinder or other round object, such as a drum, has greater wear at one point in its diameter.

Outer race: The outer cage in which the rolling elements are retained and held to the inner race.

Overhaul kit: A set of the most commonly needed seals, gaskets, and parts. Used to rebuild an assembly such as a brake caliper or wheel cylinder.

P

Pad: Disc brake friction lining.

Parking brake: A hand- or foot-operated brake which prevents vehicle movement while parked by actuating the rear brakes.

Parking brake cable: Stranded steel wire cable used to apply the parking brakes. Cable thickness is usually around 3/16" (4.76 mm).

Pascal's law: Law of hydraulics which states the pressure in a closed hydraulic system is the same everywhere in the system.

Pedal feedback: An upward pressure or pulsation of the brake pedal against the driver's foot that occurs during ABS braking.

Pedal travel: Sometimes called *free play*. The amount of brake pedal movement before the hydraulic system is affected.

Pedal travel switch: Switch installed on the brake pedal assembly to measure pedal travel and determine how much pulsation (up and down movement) is occurring per second.

Permanent magnet: A magnet capable of retaining its magnetic properties over a very long period of time. Used in wheel speed sensors.

Piston: Used in the master cylinder and disc brake calipers to create movement. They are constructed of aluminum or high impact plastic.

Pitting: Small holes in the rollers and races of a bearing.

Power brakes: A brake system that has a vacuum or hydraulic powered booster as part of the overall system.

Power steering: A steering system that utilizes hydraulic pressure to increase the driver's turning effort.

Power steering pump: Rotor or vane pumps, belt driven from the engine. Used to provide hydraulic fluid that operates the power steering and to provide power assist to a hydraulic brake booster.

Powermaster: Sometimes called an *electrohydraulic assist*. Hydraulic power brake unit installed on some mid-1980s General Motors vehicles. Uses pressure supplied by an electrically driven hydraulic pump.

Preload: The amount of torsional pressure placed on the bearing before it is put into service.

Press-fit: Condition of fit between two parts that requires pressure to force the two parts together.

Pressure bleeding: Bleeding technique similar to manual bleeding, with a pressure bleeder unit eliminating the need for a helper.

Pressure differential valve: A double-sided valve used as a warning device in all split brake systems.

Pressure seal: Seal used to hold fluid under pressure.

Primary cup: Rubber master cylinder seal which prevents brake fluid from leaking past the piston which could prevent the development of hydraulic pressure.

Primary shoe: Also referred to as the *leading shoe*. Brake shoe which faces the front of the vehicle.

Primary wiring: Small insulated wires which serve the low voltage needs of the ignition and vehicle systems.

Programmable read only memory (PROM): A semiconductor chip that contains instructions that are permanently encoded into the chip. Instructions contain base operating information for how a system's components should operate under various conditions.

Proportioning valve: Hydraulic valve used to equalize system pressure between the front and rear brakes to prevent wheel lockup. Installed in the rear brake line, a calibrated spring holds the valve away from the opening to the rear brakes.

Pulsation: A type of vibration usually felt as a side-to-side motion in the steering wheel, or an up-and-down motion in the brake pedal, or both, when the brakes are applied.

Pulse width: The length of time an output device is operated by a computer.

Pumping: Alternately applying and releasing the brake pedal.

Pushrod length: The length of pushrod that sticks out from the vacuum booster.

Q

Quick take-up master cylinder: Sometimes called the fast-fill master cylinder. Master cylinder installed on vehicles with no- or low-drag calipers. Has a large third bore located at the rear of the cylinder to create a low pressure, high flow chamber.

R

Radial load: Loading that occurs at a right angle to the bearing and shaft.

Radio frequency interference (RFI): Electronic noise caused by stray signals from electronic transmitters such as police and CB radios.

Ratchet adjusters: Type of brake adjuster used on non-servo drum brakes of some smaller cars.

Rear-wheel anti-lock (RWAL): Also referred to as *two-wheel systems*. Provides ABS braking for the rear wheels only. Primarily used on light trucks.

Receiver: Also referred to as an *emitter* or *responder*. Point where a vibration is heard or felt.

Reference voltage: A known voltage (can vary from 0.5-5v) that is sent to a sensor by a computer. The changes in sensor resistance will change the voltage, which is read by the computer as a change in temperature, airflow, etc.

Relay: Switching device which uses a magnetic field that closes one or more sets of electrical contacts, causing electrical flow in a circuit.

Residual pressure valve: Used on vehicles with drum brakes to maintain a small amount of pressure to keep the wheel cylinder lip seals from collapsing.

Resistance: The opposition of atoms in a conductor to the flow of electrons.

Retaining prong: A set of metal fingers used to hold the brake cables, pressed through the backing plate.

Return springs: Coil springs connected between the brake shoe and a stationary support or from one shoe to another. Used to return the brake shoes to the unapplied position when hydraulic pressure is removed from the wheel cylinder.

Riding the brakes: Condition where the driver rests his or her foot on the brake pedal while the vehicle is moving.

Riveting: Process using brass rivets to attach the lining to the pad or shoe.

Rolling element: Ball or roller, part of an antifriction bearing.

Rotor: A flat metal disc that serves as the friction surface for the front brake assemblies.

Rough cut: A cut made on a rotor or drum to remove a great deal of metal.

S

SAE: Abbreviation for Society of Automotive Engineers.

Scoring: Also referred to as *grooving*. Deep cuts in the rotor surface as a result of worn linings, foreign object damage or sand.

Screw disc parking brake: A caliper parking brake actuator. The piston contains a nut and cone assembly. The nut is threaded into a lever operated actuator screw.

Secondary cup: Seal installed on the master cylinder pistons to keep hydraulic pressure from leaking between the two chambers and out the back of the master cylinder.

Secondary shoe: Also referred to as the *trailing shoe*. Brake shoe which faces the rear of the vehicle.

Self-energizing brakes: A drum brake assembly that, when applied, develops a wedging action that actually assists or boosts the braking force developed by the wheel cylinders.

Self test: Cycling of hydraulic motor(s) and relays in an ABS system. Occurs for 1-2 seconds when the vehicle reaches approximately 3-9 mph (4-12 kph).

Semi-metallic linings: Brake pads or shoes whose linings are made from a combination of non-metallic materials and iron, mixed and molded into the proper shape.

Serial data: Sensor and actuator information that is shared by the various vehicle on-board computers.

Servo brakes: Sometimes called *duo-servo brakes*. Drum brake assembly in which the primary and secondary shoes contribute to the normal braking process.

Shielded wiring: Type of wiring used in circuits that are particularly sensitive to magnetic interference, such as anti-lock brake systems. Made by surrounding the signal carrying wire with Mylar tape and a grounding wire.

Shock loads: Rapid, severe increases in load, usually caused when the vehicle drives over bumps or potholes.

Shoe and drum gauge: Measuring device used to adjust brake shoes before the drum is reinstalled.

Silencer band: Elastic or leather band which reduces noise and vibration during rotor and drum machining.

Single diaphragm vacuum booster: A vacuum power booster unit that uses a single diaphragm attached to a diaphragm plate (or support plate).

Single master cylinder: Older design which has only one piston. The hydraulic system of both the front and rear brakes are operated by pressure developed by this piston.

Single-piston floating caliper: A one piece casting with a single piston which can apply both brake pads.

Smearing: A metal transfer between the rolling elements and races.

Snap ring: A split ring snapped in a groove to hold a bearing, thrust washer, gear, etc., in place.

Soft code: Also referred to as an *intermittent code*. Trouble code that is set by a problem that occur, occasionally, or only once.

Solenoid: Switching device which uses a magnetic field to perform a mechanical task, such as opening or closing anti-lock brake actuator valves.

Solid baffle reservoir: Master cylinder reservoir which contains a solid divider between the two sides of the hydraulic system. One compartment is used to supply brake fluid to the front brakes and the other to the rear.

Solid rotor: A brake rotor which has no openings between the friction surfaces. The rotor is cooled by air passing over the outside surfaces of the rotor.

Solvent: A liquid used to dissolve or thin another material.

Spalling: Severe pitting or cratering on a bearing's surfaces.

Special tools: Custom tools designed to perform a specific task, such as adjusting or disassembling a system.

Specific gravity: A relative weight of a given volume of a specific material as compared to an equal volume of water.

Splash shield: A sheet metal stamping or molded plastic assembly installed over the inner surface of a rotor.

Split baffle reservoir: Also known as a *semi-baffle reservoir*. This type has a semi-separate compartment that supplies brake fluid to both the front and rear brakes.

Spring removal tool: Tool used to remove retracting springs on drum brakes.

Sprung weight: The weight of all the parts of the vehicle that is supported by the springs and suspension system.

Staked nut: Type of nut whose edges can be bent downward into a slot to secure it on a shaft.

Staking: Also referred to as *clinching*. Used to secure the outer pad by pinching a portion of the outer pad shoe against the caliper.

Star wheel adjuster: Threaded assembly placed at the bottom or middle of a drum brake assembly. Used to adjust the lining-to-drum clearance.

Static friction: Friction produced by a holding action which keeps a stationary object in place.

Stopping brake: Brake system designed to smoothly slow and eventually stop a moving vehicle.

Straight roller bearing: Used on the rear axles of rear-wheel drive vehicles.

Support pads: The flat spots on the backing plate.

Swirl grinding: Process is used to make a final non-directional finish.

T

Tandem diaphragm vacuum booster: Sometimes called a *dual diaphragm booster*. Vacuum booster that consists of two diaphragms in a single housing.

Tapered roller bearing: A bea ring that utilizes a series of tapered steel rollers that operate between an outer and inner race.

TCS: Abbreviation for *Traction Control System*.

TCS warning light: Blue or green indicator lamp mounted in the instrument cluster or driver's information center. Illuminates when the traction control system is deactivated or when there is a problem.

Technical service bulletins (TSB): Information published by vehicle manufacturers in response to vehicle conditions, problems, etc., that may not be diagnosed by normal methods.

Tire chirp: The sound tires make during an ABS assisted stop.

Tire slip: Any tire skidding or spinning. Tire slip is measured between 0-100% and referred to in terms of negative and positive slip.

Tone wheel: A toothed rotor installed on the wheel hub, transfer case, CV joint housing, brake rotor, with the sensor inside a sealed wheel bearing, or inside the rear axle or transmission extension housing.

Traction control system (TCS): A computer-controlled system that reduces engine speed and selectively applies the brakes to reduce excessive wheel spin.

Transfer path: Also referred to as a *conduit*. The path a vibration or noise takes to the receiver.

Transmitter: Also referred to as the *source*. The point of origin of a vibration.

Trouble code: Numeric or alpha-numeric designator identifying the general location of a problem or defective component.

Two-piston floating caliper: A one piece casting with two pistons. Uses either integral cylinders (part of the caliper) or removable cylinders.

U

Universal asynchronous receive and transmit (UART): Electronic communications medium used between the ECM, off-board diagnostic equipment, and other control modules. UART is a data line that varies voltage between 0-5 volts at a fixed pulse width rate.

Unsprung weight: The weight of all of the vehicle's parts not supported by the springs, such as wheels and tires.

V

Vacuum: Negative pressure; pressure in an enclosed area that is lower than atmospheric.

Vacuum booster: A relatively large metal chamber assembly installed between the firewall (bulkhead) and the master cylinder. Uses vacuum to create additional force to the master cylinder, making the stopping process easier on the driver.

Vacuum pump: A motorized pump used to provide additional vacuum to the brake booster and other vacuum powered accessories.

Vacuum release: Diaphragm canister attached to the brake release lever used on some vehicles with automatic transmissions. Releases the parking brake when the transmission is shifted into Drive.

Vacuum runout point: The point in a vacuum brake booster when it is providing the maximum boost pressure possible.

Vehicle control module (VCM): On-board computer which controls ABS/TCS functions as well as engine and powertrain operation.

Vehicle identification number (VIN): Individual series of letters and numbers assigned to a vehicle by the manufacturer at the factory.

Vehicle speed sensor (VSS): Sensor placed in the transmission/transaxle or the rear axle assembly. Used by the engine's ECM to monitor vehicle speed.

Ventilated rotor: A rotor which has internal fins and/or holes drilled between the two friction surfaces.

Volt: Unit of measurement of electrical pressure or force that will move a current of one ampere through a resistance of one ohm.

Voltage: Electrical pressure, created by the difference in the number of electrons between two terminals.

Voltage drop: A lowering of circuit voltage due to excessive lengths of wire, undersize wire, or through a resistance.

Voltmeter: Instrument used to measure voltage in a given circuit.

W

Wear indicator: A small piece of flat spring steel attached to one of the brake pads. As the pad linings wear down, the end of the spring contacts the spinning rotor, notifying the driver of the need for brake service.

Wet cleaning station: Closed system brake cleaner which uses water mixed with a solvent or detergent to clean drum brakes.

Wheel bearing: Ball or roller bearing assemblies that reduce friction and support the wheels and axles as they rotate.

Wheel bearing grease: High temperature grease used to pack (fill) wheel bearings.

Wheel cylinder: Hydraulic device used in drum brakes to change hydraulic pressure from the master cylinder into mechanical force that applies the brake shoes against the rotating drum.

Wheel hub: Assembly containing the hub and bearings and forms a mounting surface for the wheel and drum or rotor.

Wheel speed sensor: Permanent magnet sensor used to detect wheel speed by monitoring the movement of the tone wheel.

Wheel studs: Threaded bolts or studs pressed into the hub or flange.

Index

B